botany

A Functional Approach

Walter H. Muller

University of California, Santa Barbara

Selected Illustrations by
ISABELLE GREENE

FOURTH EDITION

botany
A Functional Approach

Macmillan Publishing Co., Inc.
NEW YORK

Collier Macmillan Publishers
LONDON

To the instructors
who stimulated me,
the colleagues
who encouraged me,
and the students
who challenged me.

Macmillan Publishing Co., Inc.
866 Third Avenue, New York, New York 10022
Collier Macmillan Canada, Ltd.

Library of Congress Cataloging in Publication Data

Muller, Walter H
    Botany : a functional approach.

    Includes bibliographies and index.
    1. Botany. I. Title.
QK47.M8   1979   581   77-20083
ISBN 0-02-384700-X

Printing: 1 2 3 4 5 6 7 8   Year: 9 0 1 2 3 4 5

Preface to
the Fourth Edition

In any of the sciences one is involved with a continuing study: new data, or more data, are obtained; interpretations fluctuate; conclusions are revised; new discoveries are made; hypotheses are tested; and changes of various sorts appear with almost dazzling rapidity. It is obvious that a textbook must be revised periodically to keep pace with such changing aspects of its field. In addition colleagues and students make frequent suggestions with regard to improving the presentation of topics, new areas to discuss, and even material to delete. As all these details and considerations accumulate, an author eventually realizes that a "new edition" is advisable, warranted, desirable, and finally required. Thus, I arrived at a fourth edition.

The general purpose and the specific goals of the text are not changed. The book is written for both science and nonscience students. For the former there is a sound basis for further studies in biology, especially in botany. They are exposed to basic biological concepts, with the hope that they will come to understand plant processes and grasp the importance of plants to all life. The interrelationships between structure and function are indicated to facilitate an understanding of what plants do, how they do it, how this influences other organisms, how humans can utilize plants, and, possibly, how humans can manipulate plants. Photosynthesis, respiration, transpiration, growth regulators, and genetics are covered in considerable detail because of their bearing on plant functions, distribution, variations, and utilization. As in previous editions, the functional and ecological aspects of plants are emphasized although structural detail is covered as necessary to clarify the manner in which cells and tissues participate in the various activities of the plant.

Flowering plants are the focus of most discussions because they are not only the dominant plants of most environments but also the plants most utilized by humans. The main groups of the plant kingdom are covered, and selected types are discussed in more detail to emphasize relationships and evolutionary advances. No attempt has been made to survey the plant kingdom completely. There are several fairly recent texts devoted specifically to that feature of botany, and they should be consulted if other groups or more details are desired.

The student who is not a science major finds that plants are responsible for his existence and for most of his pleasures. He becomes aware of the importance of decision making as it influences plants because other organisms must then also be affected. Highway routes, building constructions, dam sites, forest fires, etc., entail decisions that take on much more meaning. And one hopes that students using this book will take a more sympathetic attitude toward plants in the environment and to the environment itself.

Over the past few years, people have become much more cognizant of their surroundings. The terms "environment" and "ecology" are in common usage, as are "environmentalist" and "ecologist." It is true that the words are frequently misused and there are many misconceptions as to what is entailed by "ecology" or "an ecosystem." But this new awareness presents an opportunity for botanists, especially, because people are really interested. And a course in botany lends itself very well to the teaching of biological concepts of general educational value as well as such items as conservation, erosion, food chains and food supplies, and population problems.

Every portion of the text has been considered carefully; revisions have been made in most areas in an attempt to clarify concepts; and many of the discussions have been expanded by including more recent findings. This is especially true in the chapter on photosynthesis where C_3, C_4, and CAM plants are discussed, photorespiration is handled, and comparisons are made with regard to the advantages and disadvantages of these respective pathways. I resisted the temptation to write a chapter entitled "ecology" because this aspect of botanical studies appears in most of the chapters. Throughout the text there is an effort to incorporate environmental effects into the discussion. What is the effect of temperature and light on photosynthesis and respiration? How do soil structure and texture influence plant growth? What about soil types? Rainfall? These are all interwoven—as they are for plants growing in the field. Because of many questions raised by students, there now is a chapter that presents an introduction to the myriad ways that humans utilize plants and plant parts and that serves primarily as a quick reference point.

Although I am responsible for what finally went into the book, there are many individuals who made useful suggestions, contributed illustrations, considered various portions of the text, and sent advice and recommendations. I am especially indebted to Isabelle Greene, who created most of the original line drawings; Peter Jankay for various photographs; and Robert Gill for the more recent photomicrographs and electron micrographs. The following individuals read, discussed, criticized, and suggested changes in various portions of the manuscript: R. G. S. Bidwell, D. W. Bierhorst, Jean Cummings, George W. Burns, William C. Dickison, N. W. Gillham, Joe R. Goodin, M. E. McCully, Katherine Muller, David K. Northington, William Purves, James Sandoval, Jon Sanger. Special thanks are due C. H. Muller, who interested me in new perspectives in botany; M. F. Moseley, after whose morphology courses I patterned the plant kingdom portion of this text; and the late D. G. Clark who encouraged me to enter the teaching profession. But most of all I am indebted to my wife, Veronica, who has not only typed all four versions of my book, proofread all of the editions, put up with me all these years, but also raised our fine family. Both sons (James and John) are now raising their own families and so Jim is no longer proofreading, but our two daughters

(Geri Lee and Debra Joan) have stepped in to help with the proofreading chores. Thus the book is still a family project. My wife's mother, Lillian Champagne, has passed away but was of inestimable help by handling the household during two of the revisions. I also want to thank the editorial and production staffs of the Macmillan Publishing Company, especially Woodrow Chapman and Hurd Hutchins, for their help and patience during the final stages of preparing this book.

Without such individuals, and the students, of course, the book never would have been completed—or probably never even started.

W. H. M.

Contents

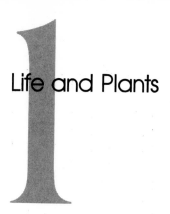

Life and Plants

1.1 Why Study Plants?

Those parts of the earth on which are found the greatest numbers and kinds of animals, including humans, are also those areas on which plants are most easily cultivated. In fact, the presence of plants makes possible the very existence of animals, primarily because plants provide food for animals, and the earth should really be thought of as an area suitable for plant life and therefore for animal life. Since animals are surrounded by and dependent upon plants, the factors that influence plant processes, plant structure, and plant distribution affect the animal world as well. Even though a plant scientist may be concerned with one of these facets of plant growth for itself alone, all humans are concerned directly or indirectly with similar problems. The beauty of a flower garden, the value of a vegetable patch, the simplicity of a green lawn—all are possible only as a result of investigations by many individuals and by their interest in plant growth and development.

Man's entire existence is closely correlated with plant growth and his increasing understanding of such growth. The development of modern agricultural practices, storage facilities, and rapid means of transportation has allowed man to build and live in towns and cities. But this is a recent change. Early man lived only where he could hunt and gather plants. Farming, as such, did not exist, nor did domestic animals. During this early period of civilization the practical aspects of food, shelter, clothing, and heat were of primary importance. Man migrated to those areas where such facilities could be obtained readily. Even when grazing animals were used, movement from one location to another was necessary as the grass was consumed. In fact, such movement is still practiced in some undeveloped and primitive areas of the world. The growing of plants specifically as a food supply became only gradually a part of man's life and has been practiced for possibly 8000 or 10,000 years. This development, and the use of tools, led to a more stable existence and the beginnings of village life. With the beginning of agriculture man had the advantage of a more permanent and dependable food supply. The actual planting and care of crops removed man from the nomadic class and made possible the development of higher civilizations. Individuals were now able to devote time and thought to pursuits other than obtaining food. There is a very close corre-lation between the areas in which cultivated plants originated and the areas in which early civilizations arose. Knowledge at this early stage was a hit-or-miss affair, with few investigations to solve specific problems. Certain plants had food value and could grow in certain areas, but why? The tremendous advances of civi-lization resulted when the question of *why* became important and people at-tempted to find the answer.

It is not known how or where **agriculture,** the tilling of land and the deliberate planting of crops, originated. It is quite likely that primitive man may have ob-served that plants grew larger and produced a greater yield if they happened to germinate near his habitation, an area which would probably be rich in organic matter and nitrogen from rubbish and excretions. Removal of unwanted plants in the area would constitute a weeding process. The sowing of seeds, cultivation of the soil, and general care of desirable plants in specific locations was a gradual development and it may have occurred independently in several regions of the world: southeast Asia, southwest Asia, and the American continents are likely areas of origin. Skilled archeologists have unearthed ancient tools for grinding grain, digging sticks, and various other types of evidence that a primitive type of agriculture was practiced in these regions probably by 8000 B.C.

A very simple type of agriculture could have utilized roots or underground stems of plants even during a nomadic existence. Whatever was not eaten would likely increase in amount and could be utilized later. It is also possible that man soon discovered that new plants would grow if pieces of these underground por-tions were planted, as we now know occurs with the ''Irish'' or white potato tuber. The use of cereals also occurred very early in the development of agricul-ture, but this necessitated a more sedentary existence. Most of these plants (e.g., wheat, barley, corn, rye) must be sown and harvested at definite times of each year, and they are not long lived. Their edible portions or grains, which are really one-seeded fruits, contain proteins, carbohydrates, fats, minerals, and vitamins. They are exceedingly nutritious; much more so than the underground edible por-tions of various plants. In addition, grains are dry and small. They store well and can be carried about with ease. It is no wonder that cereals became the basic food in most areas, although the dominant grain was not necessarily the same in each location. For example, the origin of wheat appears to have been in central or southwest Asia from where it was introduced into the Mediterranean region and then into Europe. Wheat grains have been found in archeological remains that ap-pear to be about 9000 years old. Rye and oats were apparently carried along as weeds, but were eventually selected as the cereal crops in northern regions be-cause of their greater ability to grow at lower temperatures than wheat. In the New World the cereal of importance was maize or corn,[1] which most likely had its origin in Mexico, and which was a well-established crop by 2000 B.C.

Primitive cultures were able to produce improved types of crop plants by adopt-ing selection techniques based upon trial and error in which the more desirable

---

[1] The term ''corn'' originally referred to the seedlike fruit of the cereals, including wheat, oat, bar-ley, and rye. To avoid confusion, the term ''maize'' (from *Zea mays*) is preferable.

plants were retained. As a result, many of them are so altered that their wild ancestors cannot be traced with certainty. It is, however, interesting that many crop plants with rather high yields were developed by peoples who had no knowledge of hereditary characteristics and very little knowledge of the influence of environmental factors upon plant growth.

Thousands of years passed before man came to realize that all life on this earth is dependent upon green plants. The food we eat, the shelter, the clothing, and the very oxygen of the air we breathe all come to us through these organisms. We eat cattle that eat grass; fish that eat seaweed; or we eat plants directly (e.g., rice, corn, wheat, beans). The basic food for all organisms is produced by green plants. In this process of food manufacture, oxygen gas is liberated. This oxygen, which we obtain from air as we breathe, is essential to life. The only sources of food and oxygen are plants; no animal alone can supply these.

Shelter, in the form of wood for houses, and clothing, in the form of cotton fibers, are obvious examples of uses of plant materials. But we must not forget coal, furniture, paper products, medicinals (including the famous antibiotics), chemicals and dyes of various sorts, and so forth. Many synthetic fibers and other materials that man produces utilize cellulose as the basic material of the manufacturing process, and cellulose is the main component of wood, which we obtain from various trees. The materials obtainable from plants and their products are almost infinite in number. Probably nothing that we use has not involved plants either directly or indirectly. (Note: Even the manufacture of steel and plastics requires heat.)

It was probably during the early Greek civilization that the study of plants received its first great impetus. Although most of Aristotle's (384–323 B.C.) botanical writings have been lost, he was very interested in plants and founded the first botanic garden. One of his students, Theophrastus (371–286 B.C.), is usually considered the founder of botanical science. He was the first to systematically classify plants, creating a system which lasted about 2000 years until the time of Linnaeus (1707–1778). There were various publications dealing with plants during these years and, after the Dark Ages, they culminated in the herbals and encyclopedias of the thirteenth to sixteenth centuries. These were basically vast compendiums of true and false information regarding plants, stressing their folklore and medicinal value. This interest in drug plants actually extended for thousands of years, even as early as 4000 B.C. in China, but was largely bound up with the beliefs of primitive peoples. During the period of the herbalists, there arose the curious "doctrine of signatures" according to which plants were supposed to possess some sign indicating the use for which they were intended. Thus a plant with heart-shaped leaves should be used for heart ailments, and a plant with three-lobed leaves was used for liver trouble. From these crude beginnings, however, many plant products of medicinal value have been discovered; many of these had been used in impure form by various primitive peoples. Curare is a case in point. Indians of the tropical American jungles have long prepared brews of plant material to make poisons for arrows and darts. The active material, from leaves of *Chondodendron tomentosum,* interferes with the transmission of nerve impulses. It is now used to treat lockjaw (tetanus) and to relax muscles during surgery. Another example is the use of *Rauvolfia* (or *Rauwolfia*) *serpentina* for the treatment of various afflictions in India. Reserpine, isolated from the roots of this shrub, is now used to relieve hypertension and high blood pressure.

These comments about the early period of agriculture are necessarily brief because this text is designed primarily to present information about today's world and discussions of the importance of plants. However, the reader should realize that many of our present methods and techniques for improving the growth of plants were established many years ago. We have usually been able to improve upon procedures because of our advanced technology and because of a clearer understanding of basic *reasons* for carrying out such procedures. But, after all, the ancient cultures of Mesopotamia developed profuse irrigation systems, using asphalt as a sealant, at least 4000 years ago. The Egyptians were draining some areas and irrigating others at least that long ago. They also developed hoes, plows, and cutting tools for harvesting. Many plants were long cultivated for purposes other than food: fibers, oil, spices, cosmetics, and ornamentals.

Although the rise of Greek culture did not result in contributions to practical agriculture, it did bring about a tremendous impetus to the formal study of plants. The Romans, on the other hand, were extremely interested in the practical aspects of agriculture. By 200 A.D. or so, crop rotation utilizing legumes, the use of manure, grafting and budding, and the use of different varieties of plants were well established in Rome. In fact the agricultural technology developed by the Romans was not really improved upon for the next 1000 years or so in Europe.

Plants are also important because of their usefulness in biological research. They can be grown in large numbers, treated in many ways, and their great diversity enables the investigator to select an organism best suited as experimental material for a particular problem. Studies utilizing plants have contributed significantly to our general knowledge concerning cell division and structure, cell function, inheritance, hormones, enzymes, essential minerals, viruses, and vitamins. In many of these areas fundamental discoveries were first made by investigators working with plants. For example, the inheritance of characteristics in a predictable manner was first shown by Gregor Mendel working with garden pea plants. His investigations eventually opened the door to orderly plant and animal breeding programs with their attendant advantages of improved quality, quantity, and disease resistance.

To these tangible beneficial aspects of the plant world must also be added the importance of beauty and the relaxation derived therefrom. Gardening is probably the fascinating hobby of more people than all other hobbies combined; almost everyone has tried to grow something at one time or another. The enjoyment and relaxation obtained from watching a plant develop from a seed or a cutting is one way of putting aside the worries and cares that appear to be so burdensome at times. Gardening is also a big business and an important part of our economy. Artists, poets, and engineers alike draw inspiration from the beauty and detail of plants individually and as part of the landscape enclosing our villages and cities.

Living things surround us on all sides, yet ''life'' is the most amazing and interesting and unknown factor in our very existence. The study of any living organism is certain to increase our understanding of the world around us and to make our days more interesting and enjoyable.

The present emphasis on environmental problems and population growth is an imperative reason for everyone to have an intelligible understanding of the plant sciences. Many of our present difficulties arise from an ignorance of plant growth and distribution, and lack of an appreciation of the influence of plants on the environment as well as the influence of the environment on plants. The word *ecology* is frequently heard in conversations now, but few people really understand the implications and ramifications of this term. Most people realize that ecology refers to plant-environment interactions. But this is not enough. To make valid decisions that affect our environment (e.g., urban development, grazing rights in National Parks, location of freeways, location of dams), the voting public as well as individuals in authority should have some basic understanding of the consequences of such decisions. This almost invariably means knowledge concerning plants. The environment is really plant oriented. The widespread dissemination of information about plants and how they influence our existence is essential to solving many of our problems, including pollution and other misuses of the environment. Without such information the problems cannot really be appreciated, much less solved. Much of this text has been written (or rewritten) with these thoughts in mind.

1.2 What Is Botany?

Botany is the study, or investigation, of plant structure, function, and evolution, just as **zoology** is a similar study of animals. Within this study are many specialties that have developed as knowledge concerning plants has increased. The following are some of the botanical disciplines, but it must be borne in mind that there may be considerable overlapping and subdividing of these areas. They are presented here to give the reader a general idea of the kinds of studies with which various botanists concern themselves. As with most categories established by man, there are few sharp boundaries and clear separations. Studies carried on by individual botanists frequently overlap with regard to these categories.

Morphology

Morphology is the study of plant form, structure, and development. If this concerns the external appearance and life histories, the investigator is termed a **morphologist.** If the emphasis is upon internal structural development, this study is called **anatomy,** whereas **cytology** concerns itself with the smallest units of structure, the cells. The concept of evolution is dependent primarily upon the evidence obtained through such morphological investigations.

Physiology

Physiology refers to the study of processes that take place within a plant. A **physiologist** is interested in the nutrition of plants, the influence of the environment upon plant processes, the products of plant activities, and the sequence of events that result in the growth and development of plants. A **cell physiologist,** of course, concerns himself primarily with the activities of cells or their components.

Systematics

Systematics refers to the identification, classification, and evolutionary relationships of plants. The **taxonomist** uses reproductive parts (e.g., flowers and cones) as far as possible in his work, since these are more stable and less influenced by environmental conditions than are the vegetative (or nonreproductive) structures. Each plant is designated by a scientific name that is used throughout the world in accordance with a series of internationally accepted rules. In classifying plants every attempt is made to indicate relationships among the various groups, and this involves the study of the evolution of plants. In recent years a whole new area of "biochemical systematics" has developed in which plant products and functions are utilized to indicate relationships.

All botanists are dependent upon taxonomic information in their work. They must know the identity of the plants with which they are working because of the variations found among species (or types). These identities are accepted throughout the world and make it possible for other investigators to corroborate and extend the findings. This is possible, of course, only if each investigator is careful to use the accepted scientific name in his published articles, rather than a common (trivial or vernacular) name for the plant. (See Section 19.1 and 19.2 for a more thorough discussion.)

Genetics

Genetics is the study of inheritance and variation. The **geneticist** is interested in how characteristics are transmitted from parents to **progeny** (offspring) and the mechanisms that control such a process. Plant breeding is dependent upon the information obtained from such studies. We look to the plant breeders for new varieties of plants that are more suitable for human purposes than the wild-types (the original types that humans first find usable in various ways). It may be possible by inbreeding and cross-breeding to obtain plants with higher yields (for food, for fibers, for wood, for beverages, etc.), shorter growing seasons, greater resistance to disease, greater resistance to adverse environmental conditions, or other characteristics which humans find desirable. There are many individuals involved in plant breeding investigations, and their success really depends upon an understanding of genetics.

Various other subdivisions of botany can be listed that include aspects of more than one of the above or are even more specialized than has been indicated. **Phycology** is the study of **algae** (singular, **alga**), extremely simple forms of green plants. **Mycology** is the study of simple nongreen plants, the **fungi** (singular, **fungus**). **Bacteriology** deals with the bacteria. **Plant pathology** is concerned with plant diseases and their control. **Plant ecology** involves a study of the influences of the environment upon plant communities or upon individual plants. These are just a few of the many specialized disciplines in botany.

As more and more knowledge accumulates, a greater tendency toward specialization is inevitable. However, the introduction to any field of study begins on the general level. Only after a general understanding of plant structure and function is obtained is it possible to delve more deeply into one of the specialized aspects of botany.

1.3 Science, Life, and Plants

Botany can be considered the science of plant life. To understand exactly what this statement means, we must know the meanings of the three words used in the definition, i.e., science, plant, and life. (The discovery of unknown words in the definition of an unknown term can be one of the most frustrating experiences.)

Science

The methods whereby data are accumulated and the organizing of these data in an attempt to discover general relationships that may exist are both incorporated in the term **science**. The **scientific method** is merely a method of common sense that usually entails several general features.

1. Observations. A scientist carefully examines a situation or problem and accumulates factual material, or **data** (singular, **datum**). This must be done in an accurate and objective manner with care taken that the investigator himself does not inadvertently bias or influence the results. Such objectivity must be maintained or we are not in the realm of science. One must be honest and fair in collecting data or there would be no way of comparing and discussing the results obtained by two investigators working on the same problem.

2. Hypothesis. He then attempts an explanation of the interrelations among the data. Such an explanation, while in the tentative stage, is a hypothesis.

3. Experimentation. The scientist next plans experiments that will test the validity of his hypothesis by comparing actual results under controlled conditions with predictions based upon his explanation. Experiments are designed so that there is but one variable; all other factors must remain constant. (See Section 12.11 for a discussion of limiting and retarding factors.) A **control** group, set up at the same time as the treated group, is one in which all conditions are identical with the treated group except for the one condition that is being tested. Factors are kept constant in the control group, while the treated group is exposed to similar factors with *one* being varied. This is an extremely important part of any experiment because it will indicate whether or not the variable factor is bringing about a response in the treated group. In other words, if the treated group responds differently from the control group, then we can say that the one variable factor in the former group elicited the response. Groups of similar plants must be used in the investigation because of the individual variations found in living organisms. It is essential that every effort be made to ensure that any response that is elicited is due to the variable factor and not to some inherent variations within the organisms. The groups must be similar not only in the type (kind) of plants used, but also in terms of age, size, and growth conditions during their development. If only two organisms were being compared, a difference in response to a stimulus may occur, even though the stimulus was the same for both, simply because of basic differences in the organisms. However, if a large number of similar organisms are randomly separated into two groups, the two groups may be compared. Each group would be composed of similar organisms with their various individual variations. After all, humans are basically similar but there are variations in susceptibility to colds, for example. To determine the effectiveness of a cold remedy, it would be worthless to compare only two humans. One might be resistant to that cold whether or not he was treated with the "remedy." Therefore, we would use two groups; all members of one group would receive the remedy and all members of the second group would not. The results could then be analyzed statistically to determine whether any difference in response was the result of the treatment.

4. Revision and further experimentation. The results obtained in step 3 may confirm the original hypothesis or may result in a modification (revision) or discarding of the hypothesis. If a revision is made, further experiments are designed as before in order to test the modified hypothesis.

Hypotheses may be revised many times as more and more information is obtained. If the results of a series of experiments by various individuals over a period of years substantiate the ideas put forth in the hypothesis, this explanation is on a firmer foundation and may be termed a **theory** or a **conceptual scheme.** If no exceptions are found, the theory may eventually become a scientific law or principle.

A simplified example might serve to clarify the way in which scientific problems are approached. A person noted that potted plants in his home always seemed to grow so that they were bent toward the windows. These observations led him to suspect that light is a stimulus that causes these plants to bend. He decided to test the validity of his hypothesis. A uniform batch of seeds (e.g., all from one seed package) were planted at random in three large pots of similar soil so that each pot had 30 seeds. The amount of water added to each pot was the same. After the seeds germinated under similar conditions, one pot was placed in a dark chamber. The second pot was placed in a chamber with overhead lights that gave fairly uniform lighting. The third pot of seedlings was placed in a chamber that was illuminated from one side only. After several days of growth at similar temperatures, all three groups of plants were examined. Plants from chambers one and two were found to grow in an upward direction, while the plants in the third pot curved toward the source of light. Remember that the seeds and their treatments were alike except for the light conditions. The investigator's conclusions were that darkness or uniform light (his controls) did not affect plant curvature, but that light shining from one direction brought about the curvature of his plants toward that source of light. He had verified his idea. Further and more elegant experiments could involve different kinds of plants, different colors of light, or other factors. If all three lots of seeds had grown in a similar manner, the hypothesis would have been discarded—temperature might then be a factor to investigate.

The results of investigations and the hypotheses resulting from them are published in various scientific journals (e.g., *The American Journal of Botany, Plant Physiology, Journal of Ecology*) in order to disseminate this information to all who are interested. Others can then repeat the experiments and verify or refute the hypothesis. Probably the most important standard of science is **repeatability.** If the hypothesis is sound, the experimental results will not vary significantly on repetition, and it can be verified by competent critics.

The scientist approaches problems critically and objectively and records observations accurately. He must be willing to revise his ideas if acceptable data disagree with his preconceptions. But the data must be examined carefully and critically before they are accepted. The scientist continually asks: Why, where, how? The conclusions he forms after examining sets of data may differ from those of a colleague. This necessitates further experimentation and the collection of additional data. Variations in techniques and procedures may clarify the situation although this might take years. It may not be possible to solve a particular problem until certain technological difficulties are overcome. For example, much of the very fine structure of the cell cannot be seen with the light microscope. Until the electron microscope was developed, it simply was not possible to examine, or even discover, some of the structures that we now know to be present.

Life

Certain characteristics are associated with all living organisms except viruses:

1. Cellular structure. With one exception (the viruses), all living creatures are composed of one or more discrete, tiny masses of living material (**protoplast**) enclosed by a membrane. Each of these tiny structures is a **cell.**

2. Assimilation. The production of the protoplast from nonliving material (i.e., food and minerals) is termed assimilation.

3. Growth. The increase in size that results from assimilation is growth.

4. Reproduction. The production of like individuals, including the possibility of genetic changes, is entailed in reproduction. (See Chapter 18 for a discussion of genetics and Section 28.11 for a discussion of mutations.) All cells come from preexisting cells. (See, however, Chapter 33 concerning the origin of life.)

5. Respiration. This is a complex series of chemical reactions that makes energy available for the work being done by each cell as it assimilates, grows, and reproduces. The term **metabolism** includes assimilation and respiration.

6. Responsiveness. The capacity to respond to a stimulus is also a characteristic of life.

"Life" is actually rather difficult to define. The characteristics that have been mentioned apply to the great majority of living things, but one famous (or infamous) example of an organism exhibits but few of these. A **virus** is considered to be alive mainly because it reproduces and undergoes genetic changes. Assimilation undoubtedly occurs, since virus particles increase in number, but the source of energy may be their own respiration or a parasitizing of the host-cell respiratory mechanism. The virus is invisible when an ordinary microscope is used, but examination with an electron microscope at a magnification of approximately 100,000 diameters shows a definite structure, even though it is not cellular. Various types of evidence demonstrate that viruses are nucleoproteins (nucleic acid surrounded by protein), and certain of them have been purified as a crystalline material. Possibly this is the stage between what we are sure is definitely living and what we are sure is certainly nonliving. (See Chapter 33 with regard to the origin of life.)

Increased knowledge about living things has shown that these complex organisms are governed by basic principles of chemistry and physics. Such a **mechanistic** viewpoint allows for greater ease in investigations than does the **vitalistic** approach, which assumes that "vital forces" are necessary for the activities we associate with life. It is rather difficult to measure or investigate forces that cannot be perceived by the senses. **Biologists** (individuals, both botanists and zoologists, who study living things) attempt to explain the "why" of living activities by physical and chemical principles rather than on the basis of a purpose. Activities occur because certain mechanisms are present and not "in order to" achieve a result. For example, the cactus plant grows readily in the desert because it happens to have certain structural features that result in low water losses and thus a low water requirement; it does not (or did not) evolve such structures in order to grow in the desert. These features resulted from random variations over extremely long periods of time, and the environment determined survival. Some form of plant life has existed on earth for hundreds of millions of years. During this exceedingly long period of time, those random variations that made it possible for a plant to gain some advantage over its neighbors in a specific environment would result in that plant leaving more offspring (or survivors). This is the way that the environment is selective. If certain characteristics happened to develop and if these happened to be advantageous, this plant would benefit, and so would its offspring. The cactus plant cannot determine its own future. Explanations based upon a goal or purpose are **teleological** and not scientific and should be rejected.

Plant

Since botany refers to a study of plants, we should indicate what differentiates a plant from an animal. Certain general characteristics are to be found in the former and not in the latter. The higher and more complex plants and animals can be distinguished from one another without much difficulty, but the very simple types are not so readily distinguishable. The following characteristics refer primarily to higher forms.

1. Food manufacture. Most plants can synthesize complex food from simple substances (such as carbon dioxide, water, and minerals) if **chlorophyll,** a green pigment, is present and the plant is exposed to light. Animals require ready-made food in the form of plants or other animals that in turn have eaten plants. The fungi, however, are simple nongreen plants that cannot make their own food but obtain it from external sources.

2. Cell walls. Most plants have rigid cell walls of **cellulose,** a carbohydrate material. This results in the stiff and sturdy framework and lack of motility of plants. Animal cells in general are flexible, since they lack cell walls. Here again we find exceptions, in that many of the simplest types of plants are motile and some do not have cellulose walls. Many animals have rigid internal or external skeletons but the material involved is not cellulose.

3. Unlimited growth. Most plants have unlimited (or indeterminate) growth because the **meristematic** tissue (i.e., tissue consisting of actively dividing cells) remains active so long as the plant lives, whereas growth in animals is limited (or determinate). After the mature individual attains a certain maximum size and characteristic form, there is no further growth in the animal but a plant continues to grow until it dies. This is readily observed in a tree. As long as the tree remains alive, it produces more roots and branches and the diameter increases. This results from the production of more and more cells by the meristematic tissues and the enlargement and differentiation of these cells; growth is indefinite and continuous so long as the environment is suitable. Some plant organs, however, upon reaching their mature sizes, cease growth. This is true of leaves and petals, which exhibit determinate growth. The situation is very different in the case of animals, such as humans. Once a human reaches a mature size, there is no further growth really; except for rather minor changes caused by overeating, for example. Certainly humans do not continue growing throughout their lives; there is a rather definite limit to the size of humans (within a general range), and the adult does not change much.

No single criterion separates all plants from all animals. This should not be surprising, since they are but the two main branches of a single family tree and have common ancestors in those dim ages when life first originated and evolved. The more complex the organism, the more readily it is recognized as a plant or an animal because the relationship is so distant. The simple plants and animals, however, are usually more closely related and thus more difficult to distinguish one from another. For example, *Euglena* is a one-celled organism that is flexible, motile by means of a hairlike appendage, and some are capable of ingesting solid food particles by means of a small gullet. These characteristics are obviously animal-like. However, *Euglena* has chlorophyll and can manufacture its own food, a definite plantlike characteristic. Then what is this organism? It is studied as an animal by zoologists and as a plant by botanists. The important thing is not what we call it, but that *Euglena* exists and that it is studied. Remember that classifications are the result of manmade rules and regulations and that man cannot insist upon an organism *being* a plant or an animal—he can merely insist upon *calling* it a plant or an animal. Plant classification will be discussed more thoroughly in Chapter 19.

1.4 Kinds of Plants

Now that the terms *botany* and *plant* have been discussed, mention should be made of the great variety of organisms that botanists study. There are well over 300,000 distinct kinds of plants, which range in size from the microscopic bacteria to the gigantic seaweeds and California redwoods. The former may be small spheres with a diameter of 0.5 micron (about $\frac{1}{50,000}$ of an inch), while the latter range up to at least 110 meters (about 330 ft) in length or height and hundreds of tons in weight.

Some familiar categories of plants are trees, shrubs, flowering annuals, crop plants, ferns, and grasses. Some of the less well-known plants include bacteria, toadstools, mushrooms, mosses, algae, rusts (on roses, for example), and mildews. Not only do the sizes vary tremendously, but so do the rates of growth and types of nutrition. Some bacteria can reproduce within 20 or 30 minutes, while other plants may grow for seven to ten or more years before producing seeds. Oxygen is required for most, but others (yeast, certain bacteria) grow readily in the absence of oxygen; in fact, some bacteria will not grow in the presence of oxygen. The nutritional aspects vary from those plants that produce their own food to those that require an external supply of food.

These immense differences between plants constitute one of the most fascinating aspects of botany. This is almost as fascinating as the fact that all life depends upon the diverse plants found on and in this earth. There is such a great variety of plants and so many adaptations have evolved that plants are found in virtually any environment that will support life at all. They occur in the hottest and driest of deserts, in the heat of the tropics, the cold of the arctic, and some lichens can grow on bare rock surfaces with almost no water. With such a great diversity of plants and with them being able to exist in such a variety of environments, it is not strange that humans became familiar with plants, began to utilize them, and then attempted to manipulate the environments or the plants in order to arrange for more and better productivity. The next chapter will discuss some of the more common and most used plants. This will include plants used for food, fiber, lumber, beverages, and medicinals; but only a small percentage of usable plants and their products can be discussed because of space limitations.

Summary

1. Plants are studied because they provide food and oxygen for animals, many useful products for humans, and because of the interest people have in living things.

2. Botany is the study of plant structure, function, and relationships. Many specialized areas exist within this broad framework of study.

3. The scientific method is basically one of common sense and consists of observations, hypotheses, experimentation using control and treated groups, and revisions of hypotheses. The scientist is concerned with factual material and a mechanistic approach.

4. The characteristics of most living organisms are cellular structure, assimilation, growth, reproduction, respiration, and responsiveness.

5. Plants are generally differentiated from animals by their ability to manufacture food, the presence of cell walls, and their unlimited type of growth.

Review Topics and Questions

1. List ten items that you use or consume during the day and explain how each is dependent upon plants. Be explicit as to whether plants are involved directly or indirectly.
2. List three specialties in the study of plants and indicate how they differ.
3. Describe an experiment that would enable you to determine the effect of temperature on the rate at which seeds germinate or sprout.
4. Discuss the difficulties involved in defining life.
5. Discuss the difficulties involved in distinguishing plants from animals.
6. List ten careers in which a knowledge of plants would be essential.
7. Of what importance are scientific journals or periodicals?

Suggested Readings

ANDERSON, E., *Plants, Man and Life* (London, Andrew Melrose, 1954).

BAKER, H. G., *Plants and Civilization*, 2nd ed. (Belmont, Calif., Wadsworth, 1970).

EDITORS OF *Scientific American, Plant Life* (New York, Simon & Schuster, 1957).

Use of Plants by Humans

As mentioned in the previous chapter, it is impossible to discuss more than a very few examples of plants or plant parts that are valuable to humans. However, an attempt will be made at least to indicate the diversity of uses and the great variety of plants that can be considered to be practicable in some ways to humans. The amount of a particular plant material that is used varies considerably and our discussion will include only those materials used in relatively large amounts. In some cases almost the entire plant may be utilized whereas only certain parts may be used in other cases. For ease of discussion, a rather arbitrary grouping will be employed.

2.1 Food Plants

The importance of green plants in providing the food for all organisms has already been emphasized. Green plants are the producers and other organisms are the consumers. The latter may utilize plants directly, or indirectly by eating organisms that eat plants. Humans do both; we will discuss some plant foods first and then discuss some plants that provide food indirectly.

Grains

Three members of the grass family probably provide the greater proportion of the food supply for humans. The **grain** that is consumed is the one-seeded fruit of these cereal plants: rice, wheat, and maize (corn). Each accounts for about one fourth of the world's annual supply of approximately one billion metric tons. The rest of the cereal supply is derived from oats, rye, sorghum, barley, and millets.

Rice (*Oryza sativa*) is the principal food for tropical populations, and more than half of the world's population depends heavily upon rice for sustenance. It is typically grown in paddy land, which can be alternately irrigated and drained. The rice seedlings are started in nurseries and then transplanted to flooded paddies; about four months later, the paddies are drained and the rice harvested. Yields average about 4 tons per hectare (about 3600 lbs per acre) but may be at least twice that much with intensive fertilization as in Japan, and total production is about 320 million tons annually. There are upland varieties that can be grown as any small grain, but yields are less and most rice is grown in standing water.

In the United States, where only about 1 per cent of the world's rice is grown, the grain is polished to remove the outer brown layers after the hulls are removed. These outer layers are rich in protein and the B vitamins. New rice varieties are much more responsive to fertilizers because they are short and stiff stemmed. Most rice elongates greatly when fertilized, and the stems frequently collapse under the weight of the developing grain. This *lodging* results in loss of the grain, which does not occur with the newer varieties.

Wheat (*Triticum aestivum*) can be grown in areas that are too dry—15 to 35 inches of rainfall—for rice and maize or too cold. The area corresponds roughly to the original large grasslands of the temperate regions of the world with the U.S.S.R. being the leading producer and the United States second. It is the world's most widely cultivated crop, with about 360 million tons produced each year, and it is a good source of protein, carbohydrates, and B vitamins.

Wheats can be divided into spring and winter types. *Spring wheats* are relatively fast growing, requiring about 90 to 100 days to mature, are planted in the spring, and harvested in the fall. These are the types grown in the northern wheat belt of the United States. The slower growing *winter wheats* are sown in the autumn, are sufficiently hardy to withstand the milder winters of the southern wheat belt, and mature in early summer. These varieties give higher yields than the spring wheats. As a result of the many strains of wheat that are available, this crop is cultivated in all continents except Antarctica. Yields vary, of course, but 50 bushels per acre (and more) are produced if soils are adequately fertilized in humid regions. Much wheat grain is now used for livestock feeds along with corn. As with rice, the outer layers of the wheat grains are removed when they are processed for white bread. As one would expect, whole wheat breads are much more nutritious.

Maize (corn; *Zea mays*) is cultivated most intensively in the United States where a bit more than half of the annual 300 million tons of corn grain is grown. It is a widely distributed crop and ranks second only to wheat in acreage planted throughout the world. It is the most important crop of the United States, being grown in every state, and can yield up to 300 bushels of grain per acre under favorable circumstances. Over 85 per cent of the corn crop is fed to livestock; approximately half of this amount to hogs, the remainder to cattle, poultry, horses, and mules.

Most grains are not a good sole protein source for humans because they are low in lysine and tryptophan, two amino acids which are essential in our diet because they cannot be synthesized by humans. Some of the newer varieties of grain have more adequate proportions of the essential amino acids, but it is still necessary to supplement the diet with other protein sources. Dairy products and meats are the best protein sources for humans, but supplementing grains with legumes will also provide adequate amounts and proportions of essential amino acids.

Legumes

Plants belonging to the family Leguminosae produce a fruit, termed a **legume,** which is a capsular pod that opens along two sides when ripe. These plants, particularly the fruits and seeds, are rich sources of proteins especially as a result of the nitrogen-fixing bacteria that live in nodules on the roots. Such bacteria convert atmospheric nitrogen gas into amino acids (see Section 20.6), which are the building blocks of proteins, and make the legume plants the least dependent upon nitrogenous fertilizers of all crop plants. Legume seeds have protein contents of 25–35 per cent as compared to 5–10 per cent for cereal grains; and the former usually have adequate amounts of lysine and tryptophan.

Soybean (*Glycine soja*) is the world's most abundantly grown seed legume with the United States (60 per cent) and Mainland China accounting for more than 90 per cent of the world's production, which is about 60 million metric tons annually. It is well adapted to warm temperate climates as a summer annual and is grown throughout the world. Soybeans are about 20 per cent oil and much of this oil is used in the manufacture of margarine and other food products, soaps, lubricants, and a variety of other materials. The residue, soybean meal or flour, is about 50 per cent protein (by dry weight) and is becoming increasingly important in human foods as an extender for more expensive forms of protein. It is also a valuable livestock feed. Soybean protein is an excellent dietary supplement to balance some of the nutritional deficiencies of such grains as maize and wheat.

World production of the peanut (*Arachis hypogaea*) is about 20 million tons, most of it in the Far East (India about 35 per cent) and Africa, and much of the crop becomes stock feed. Peanuts contain about 50 per cent oil and 30 per cent protein; the oil is used much as is soybean oil. The protein is a relatively poor source of lysine and thus not as suitable for a dietary supplement as is soybean protein. It is, however, a rather inexpensive source of protein.

"Peas and beans" include many different members of the legume family that provide seeds that are directly consumed for food; probably 40 million tons annually. They are primarily warm-season annuals that do well in hot weather as long as there is ample moisture. Beans (*Phaseolus* species) may be used as a fresh vegetable when the immature pod is eaten entire or as a dry bean (cooked) when matured and shelled. They are probably the main source of protein for less affluent people, and are much grown in Latin America and the Far East. The familiar green pea (*Pisum sativum*) is usually shelled and usually it is only the seed that is eaten. Beans, especially lima beans, are somewhat better sources of nutritious protein than are peas.

Root Crops

In the case of grains and legumes, fruits and/or seeds are eaten whereas it is the fleshy storage root that is consumed in the plants considered to be root crops. The latter have abundant starch or sugar, but they tend to be low in protein and fats; thus, they are useful energy foods but they do not provide a balanced diet. This is unfortunate because the cassava, for example, is an especially important food staple in certain tropical areas. Numerous roots are consumed in addition to the ones we will discuss: radish, carrot, yam, turnip, parsnip, and so forth. The common potato is not a root and will be discussed later with other "stem crops."

Cassava (*Manihot esculenta*) grows well in tropical lowlands with but little cultivation, and it is one of the world's most important foods in poor and less-developed areas. About 100 million tons are produced annually, but the roots contain very little protein or fat although they are almost 30 per cent starch. It is necessary to add some protein source to obtain a nutritionally balanced diet. In most areas where cassava is an important portion of the diet (for example, parts of Brazil, Indonesia, Nigeria, and the Congo) money is scarce and additional food materials are impossible to obtain.

The sweet potato (*Ipomoea batatas*) is similar to the cassava root in general appearance and nutritional content. About 130 million tons are produced annually, but most of this is used as a nourishing livestock feed in many parts of the world. It grows well in tropical lowlands in Asia, Africa, and Latin America although Japan is the leading producer.

There are several kinds of beets (*Beta vulgaris*); some are eaten directly and some are used as livestock feed, but most of the 150 million tons harvested consist of sugar beets. About one third of the sugar produced in the world comes from this source with the U.S.S.R. being the largest producer, followed by the United States and then Germany. Sugar beets contain about 20 per cent sugar. After shredding and extraction, the pulp is pressed, dried, and used primarily for livestock feed although it is poor in protein.

Stem Crops

The common potato ("Irish" potato, white potato, *Solanum tuberosum*) is as important in the temperate climates as the cassava is in tropical climates. Although the potato tuber is dug from the ground, it is not a root; it is an enlarged underground stem consisting primarily of storage tissue and containing about 25 per cent starch and 3 per cent protein. It is primarily an "energy" food. The plants grow best in a cool environment, with a poor yield when temperatures average above 21°C (70°F), and Europe is the primary area in which they are grown. Annual production is about 300 million tons and potatoes represent the world's fourth largest food crop, after maize, wheat, and rice, but about 80 per cent of the tuber is water. Also, in Europe much of the potato crop is fed to livestock. Potatoes are grown in all states of the United States, but almost three fourths of the crop is grown in the northern half of the nation. With adequate fertilizer treatment the yields average about 10 tons per acre.

Sugar cane (*Saccharum officinarum*) is a plant of the humid tropical lowlands and large numbers of plants are grown in Hawaii, Cuba, Brazil, the Philippines, and India. The sugar-rich juice is pressed out of the cut canes, evaporated, and the sugar refined. The yield is about 20 tons per hectare, and world production totals about 50 million tons annually.

Fruits

Except for the banana and coconut in parts of the tropics, fruits in general are relatively unimportant as food staples. They add variety and flavor to the diet and are frequently excellent sources of vitamin C, but they tend to have low concentrations of nutrients. This does not mean that fruits are not consumed by humans. Tomatoes, apples, and citrus, for example, are produced at the rate of about 30 million tons annually *each*. More than 60 million tons of grapes are produced per year, leading all fruits, but these are largely used in making wine. Various members of the gourd family (*Cucurbitaceae*) comprise the diet of some primitive tropical societies. These include squashes, pumpkins, and melons. Technically, or botanically speaking, grains are fruits. They form such a very different type of food supply, however, that they have been discussed separately from other fruits.

The banana (*Musa* species) is a plant of the humid tropics that can yield as high as 10 tons per acre annually without much attention. Total world production is about 35 million tons with about three fourths of this being grown in Latin America. When ripe, the banana provides a fairly well-balanced source of nutrition with a high content of carbohydrate (as much as 22 per cent), some oil, minerals, a little protein, and a good source of several vitamins. As a result, it is eaten by many tropical peoples.

Although the coconut (*Cocos nucifera*) is used primarily for the oil that is obtained from the dried meat, some of the 30 million tons produced annually forms the main dietary component of some tropical societies especially those along sea coasts. These trees usually start bearing after 5–6 years and continue until they are about 50 years old. The husk of the coconut yields a fiber from which ropes, baskets, mats, and so forth, may be manufactured. The dried and compressed "cake," after the coconut oil (the world's ranking vegetable oil) is extracted, usually serves as livestock feed.

Leaves

There are not many plants whose leaves form any significant portion of the diet of humans. The cabbage (*Brassica oleracea*) varieties are the most important, about 15 million tons annually, and they do best in mild to cool climates. Long cultivation has developed several specialized forms from the original wild cabbage: head cabbage, Brussel sprouts, kale, and others in which the flower parts or stems are eaten. The edible leaves are low in calories but serve as sources of bulk, vitamins, and minerals. Considerable quantities are consumed in Europe and in the United States.

Forages

Even in a very brief discussion of plants that form food supplies for humans, a comment must be made about **forages,** that is, any plants consumed by livestock. Obviously, they are used indirectly by humans. However, livestock are exceedingly important as protein sources and as means whereby nonedible (by humans) plant materials are converted to nutritious foods. There are many forage plants, but only the grasses and legumes (e.g., alfalfa, clovers) contribute to any great extent. The carbohydrates, fats, and proteins of forages are converted to meat, eggs, and dairy products, which may then be consumed by humans. Frequently, forages can be grown on land that is not suitable for cultivated crops and thus make such lands productive in terms of human utilization. The proteins of eggs, milk, and most meats have excellent proportions of all the essential amino acids. The importance of forages is obvious when one considers that in the United States about 170 lbs of meat are consumed yearly per person. Even though most cattle are fattened on grain in feed lots before being sent to market, probably 75 per cent or so of livestock food is derived from such forage plants as alfalfa, clovers, vetches, and various grasses.

The chief forage plant in the United States is alfalfa (*Medicago sativa*), which grows best on well-drained soils where there are warm summers and cool winters. The main root attains great depths, frequently over six feet within the first five months of growth, and this accounts for the unusual ability to withstand drought. It also has a remarkable capacity for rapid and abundant regeneration of dense growths of new stems and leaves following cutting. With a long growing season and with irrigation, as in the southwestern United States, up to ten crops of hay and a yield of ten tons of dry hay per acre may be obtained. Alfalfa is a legume, rich in protein; green leafy alfalfa hay contains about 16 per cent proteins, 8 per cent minerals, and considerable amounts of various vitamins.

2.2 Lumber Plants

In this group will be considered those forest trees which are used in significant amounts for a variety of purposes; only a few uses can be mentioned of course. The greatest single use is as fuel. One third of the human race still relies on firewood as its principal source of fuel, and this accounts for half the timber cut in the world. A vast number of trees have been cut, and it has been claimed that people have already reduced the world's original forested area by at least one third and perhaps by as much as one half. The rising price of petroleum and ever-increasing numbers of people have accelerated this assault on the forests. Denuding the forests increases erosion, which also damages cropland. Whenever any vegetative cover is removed, the soil is no longer held in place and the pounding rain erodes the soil. In many areas, remote villages are no longer surrounded by forests; journeying to gather firewood may take a whole day. Other uses for wood have also contributed to the disappearance of forests, of course.

Lumber for various building purposes requires the cutting of enormous numbers of trees to provide furniture, houses, barrels, boxes, cabinets, and so forth. The production of lumber in the United States totals over 45 billion board feet annually, of which about 85 per cent is from softwoods and 15 per cent from hardwoods. (A board foot is equal to one foot square by one inch thick.) The softwoods are conifers, or cone-bearing trees, and the leading lumber species are Douglas fir (*Pseudotsuga menziesii*), Southern yellow pine (*Pinus palustris*), ponderosa pine (*Pinus ponderosa*), white pine (*Pinus strobus*), and redwood (*Sequoia sempervirens*). By far the most important hardwoods are the oaks (*Quercus* species); others used are yellow poplar (*Liriodendron tulipifera*), maple (*Acer saccharum*), beech (*Fagus grandifolia*), and ash (*Fraxinus* species).

In addition to its use as fuel and lumber, tremendous quantities of wood are converted to pulp for making paper. In the United States alone nearly 50 million tons of paper and paperboard are made each year, and two thirds of this is derived from softwoods. The wood is cut or ground into chips, cooked with steam in chemical solutions, and usually bleached for white paper. The purpose of the various treatments is to dissolve the lignin, separate and disperse the fibers, and then mat them into a thin layer that is filled, coated, and compressed. There are several processes for making wood pulp and paper, different treatments and coatings for the paper, and the result is a variety of papers ranging from inexpensive newsprint to the best quality and specialized papers, as for banknotes.

2.3 Fiber Plants

A **fiber** is a slender, very elongated, tapered cell with thick walls although string-like masses or clusters of cells are also termed fibers in industrial parlance. There are quite a few plants from which fibers may be obtained, but cotton is by far the most important natural fiber, and each fiber is a single elongated cell that protrudes from the surface layer of the cotton seed. Each seed bears about 10,000 such fibers, which usually range in length from one cm to seven cm depending upon the type of cotton (*Gossypium* species). After the cotton boll or pod (i.e., the fruit of the plant) matures, the fiber cells die and lose water; each fiber collapses into a form that looks like a twisted ribbon and is about 90 per cent cellulose. Fibers are strong and durable and are easily spun (twisted) into thread or yarn after they are combed out to ensure parallel orientation. Most fiber is used for weaving cotton cloth, oil is extracted from the seed, and the cake left behind when the oil is squeezed out is used as fertilizer or animal feed because of its high protein content. Because of its many uses, world production of cotton is high; more than 45 million bales, totaling over 21 billion lbs. The United States grows about 30 per cent of this, and most of it is exported.

Jute (*Corchorus capsularis*) is second to cotton in world production, and most is produced in India and Pakistan. It is inexpensive, but it is yellowish and difficult to bleach, coarse, and not very strong. The fibers occur as bundles in the stems, are not very long (about 2–4 mm), and only 75 per cent cellulose. The stems are tied in bundles and submerged in ponds where bacteria and fungi decompose the tissues. The freed fibers are dried and spun into yarn, which is used primarily for industrial products such as burlap bags and backing for linoleum and carpets. The commercial fiber actually consists of strands of overlapping fiber cells held together by natural plant gums.

Man-made fibers have replaced natural fibers in many instances, but we will discuss only those whose production directly utilizes plant materials; thus, various synthetic polymers will not be discussed (e.g., nylon, orlon, dacron). The cellulosic fibers, as their name implies, utilize cellulose which is dissolved and then reorganized as filaments that are spun into thread. **Rayon** was the first of these fibers and surpasses in poundage the production of all others—over 8 billion lbs of worldwide production. The cellulose is obtained from softwoods and purified. Different chemical solvents are used to dissolve the cellulose, depending upon the process used, and the solution is forcefully extruded through a group of fine nozzles (the spinneret). The hardened group of strands that results as the solvent evaporates or as they pass through a hardening process is twisted into a filament or thread. Various pigments may be incorporated into the fiber substance before spinning, resulting in yarns that are colorfast. Everyone is familiar, of course, with the great many uses to which rayon is put.

2.4 Beverage Plants

Humans have probably sucked or squeezed the juice out of many kinds of fruits over the ages, but coffee and tea are undoubtedly the most popular beverages derived from plant materials. Pleasant, and even somewhat nutritious, beverages are produced in many societies by allowing the natural alcoholic fermentation of fruit juices and moist grains; wines and beers are among the oldest of man's beverages, but they are not consumed in such copious quantities as coffee and tea. The fermentations would actually be the activity of wild yeast blowing about and landing in the juices. Although there are many ways in which humans now use alcohol, we will confine our discussion to coffee and tea.

The coffee plant (*Coffea arabica*) is a tropical evergreen shrub or small tree that thrives best in rain forests at moderate elevations. Although it originated in Africa, it is now mostly grown in the New World; world production is about 4 million tons annually with almost half of that coming from Brazil. The United States consumes almost an equal amount. Coffee berries (the fruit) are harvested by hand, the hard seeds (the so-called "beans") are separated from the pulp and dried; after being polished, the "beans" are roasted and ground. The roasting creates the brown color and develops the aromatic qualities and flavor. In recent years "instant coffee" has taken over almost 50 per cent of the market. It is made by drying or dehydrating coffee and is prepared simply by adding hot water.

Tea, prepared from the young leaves of a tropical broadleafed evergreen shrub (*Thea sinensis*), is drunk by almost one half of the world's population. India and Ceylon account for more than half of the total world production of over a million tons annually. The plant grows best in mild, moist climates—for example, hill country in the tropics with at least 60 inches of well-distributed rainfall annually. The stem tip (terminal bud) and the young first two or three leaves are picked for the best-quality tea. The leaves are dried, rolled, and allowed to ferment in producing black tea, which is the type most preferred in the western world; green tea is made from unfermented leaves. There are varieties and blends of teas just as there are with coffees, and tastes vary considerably.

2.5 Spices

Spices were responsible for the rise and fall of empires, explorations throughout the world, intrigue and conflict between nations, and the making of vast fortunes; most of this activity occurring between the fifteenth and nineteenth centuries. There are a great many spices, of course. Their function is to add variety so the diet will not become too monotonous, and they are used as preservatives and as flavorings to disguise the unpleasant flavor of meat that is not too fresh. Most of the flavors and aromas are caused by volatile oils. Some of the spices that were important during these difficult four centuries were: black pepper, cinnamon, clove, mints, allspice, nutmeg, mace, and ginger; a few brief comments will be made about the first three.

The true pepper or black pepper (*Piper nigrum*) is a perennial climbing vine extensively cultivated throughout Southeast Asia that requires a long rainy season, fairly high temperatures, and partial shade to produce its best growth. The fruits, small globose berrylike structures (called peppercorns), are hand picked and left in the sun to dry. Natural fermentation causes the fruits to turn black and they are ready for shipment; when ground, the peppercorns yield black pepper. White pepper, which has a milder flavor, is produced by removing the outer layers of dried peppercorns. Total world production approaches 100,000 tons annually and is primarily a home operation by small landowners in the Far East.

Cinnamon comes from the bark of a tree (*Cinnamomum zeylanicum*) native to Ceylon and India, where almost all of the world's annual production of 10,000 tons is grown. It was once the most profitable spice in the Dutch East India Company trade and was more valuable than gold. The bark is stripped from cut twigs, preferably second-year growth, and dried. These are the hollow tubelike "quills" of commerce; fragments are ground to yield powdered cinnamon.

Cloves are the dried unopened flower bud of a tropical tree (*Eugenia caryophyllata*) and were a major item in the early spice trade. They were much used as food preservatives and flavorings. Clove growing was confined almost entirely to Indonesia under the Dutch monopoly; early in the seventeenth century cloves were eradicated from all but one or two islands to sustain this control. The French eventually succeeded in smuggling cloves from the East Indies and the monopoly was broken by the end of the eighteenth century; greatest production is now in Zanzibar and Madagascar.

2.6 Medicinal Plants

Various parts of different plants, or extracts of them, have been used since before recorded history for the treatment of human ailments, real and imaginary. Under careful scrutiny, many of these so-called treatments have been found to be useless or of questionable value. However, enough useful medicinal materials have been obtained from plants to stimulate a continual search for others. There are also attempts to improve the production or quality of those already in use.

One of the best known medicinal plants is the opium poppy (*Papaver somni-*

ferum), whose use is as ancient as recorded history. When the petals of this annual plant have fallen, a milky juice is collected by making slight incisions in the mature fruit (capsule). This latex turns brown as it coagulates and dries; morphine is the chief active alkaloid of the 30 or so that are found in opium. In its power to relieve pain morphine is without rival, but it has undesirable side effects and is very addictive; it should be reserved for the more severe degrees of suffering. Heroin, a derivative of morphine having similar action, is even more dangerous with regard to addiction. Its manufacture and importation are forbidden in the United States. However, it has been estimated that most drug addicts are users of heroin. Thus far it has been impossible to regulate the cultivation of the opium poppy so that the alkaloids are used for medicinal purposes only, and drug addiction has become an extremely serious problem in many parts of the world.

Marijuana is not recognized as a medicine in Britain or the United States, but it has been used since ancient times as a stimulant especially in India, the Near East, and North Africa; it is much used in the United States at the present time. Use of the drug may result in hallucinations, euphoria, hilarity, nausea, dizziness, delusions, and other effects. There is a difference of opinion as to the seriousness of marijuana addiction, and withdrawal does not cause the extreme physical anguish of opiate withdrawal. It is prohibited in the United States, as well as in many other countries, and penalties for possessing marijuana may be quite severe. The active principles of marijuana, a mixture of complex alcohols, are obtained from a resinous exudate of the top leaves and flowers of the hemp plant (*Cannabis sativa*). Smoking of cigarettes made from such leaves results in the drug being absorbed into the system.

Quinine is the most important alkaloid of cinchona bark (*Cinchona* species) and is a specific and effective remedy for malaria. Although in use by South American Indians before colonization, it was first mentioned in European medical literature about 1643. Since that time, it has probably benefitted more people than any drug used for the treatment of infectious disease. The trees are native to the warm moist Amazonian slopes of the Andes at elevations of 1.5 to 2.5 kilometers but were introduced into Java about 1865. This country accounted for about 95 per cent of the world's commercial supply until capture by the Japanese during World War II intensified the search for substitutes. Several antimalarials have been synthesized and atabrine is the one used most commonly, but quinine is still the main treatment for the disease.

The dried flowers of several species of *Chrysanthemum* are the chief sources of the insecticide pyrethrum, which is a contact poison for insects and cold-blooded vertebrates but is nontoxic to plants and higher animals. Therefore, it is much used in household and livestock sprays as well as in dusts for edible plants. Most of these highly aromatic plants are cultivated in Africa, Japan, and India as a source of insecticide. Although pyrethrum is not a medicinal in terms of treating human ailments, it was mentioned in this section because of its effectiveness in controlling insects, many of which are so exceedingly important as carriers of disease.

Antibiotics are organic substances produced by living organisms that, in low concentrations, inhibit the growth of or kill other organisms. Probably everyone is familiar with at least some antibiotics that are used in the treatment of human diseases—penicillin is, of course, the best known. Most of the antibiotics are produced by microorganisms and are effective primarily against bacterial diseases. There are, however, a great variety of antibiotics, including those produced by flowering plants that inhibit growth of other flowering plants. There is a tremendous amount of effort exerted in the search for and development of new and more effective antibiotic agents. It is beyond the scope of this book to go into any greater detail of this vast field of endeavor.

2.7 A Few Other Useful Plants

Many articles and books have been written about plants that are utilized in one way or another by various humans. In this chapter I have tried to give a bit of an insight as to the diversity of plants that are useful and have just barely skimmed the surface of this facet of botany. It would be remiss to conclude the discussion without mentioning two other plants, important to human affairs, which do not quite fit into the categories used in this chapter.

The United States produces almost one fourth of the 4.5 million tons of tobacco, primarily *Nicotiana tabacum,* grown annually in the world. Most of the tobacco is used in the manufacture of cigarettes with the United States alone consuming almost 600,000,000 of them each year despite adverse publicity concerning the injurious effects of smoking. Smaller quantities are utilized as cigars, pipe tobacco, chewing tobacco, and snuff. The tobacco industry is enormous, employing thousands of individuals, and generating considerable revenues for various governments. The plants are frequently grown in shade; the leaves are usually harvested at intervals as they mature and "cured" by hanging in heated sheds to dry. Fermentations occur during this drying. After auctioning, tobacco is generally "aged," which is a longer fermentation process carried out in casks for periods that may extend to years. Tobacco, as it is consumed, usually consists of blends of different varieties with or without additives.

We will close our discussion with some comments about rubber, the treated latex of *Hevea brasiliensis,* a tree native to moist tropical South America. Latex is a whitish, somewhat viscous, fluid that exudes from cuts made in the bark. It is coagulated, usually by the addition of acids, heated with sulfur under pressure (vulcanized), and molded into various articles. Synthetic rubber has been prepared from petroleum sources and now accounts for about 80 per cent of the world's production of rubber. Of the two million tons or so of natural rubber that is produced annually, about 90 per cent comes from Malaya and Indonesia. Rubber trees were introduced to southeast Asia about 1877 by the British, who greatly increased yield by their breeding and selection programs and by their improved methods of cultivation. On well-tended plantations as much as a ton of latex per year can be obtained per tree; tapping usually is started when the tree is about 5 years old, yield is greatest at about 12–14 years, and the tree is abandoned when it reaches 25 years of age. Most of the rubber is used in the automobile industry, primarily as tires.

Summary

1. There are many ways in which plants, or plant parts, are used by humans and one of the most important is as food, of course. Three grains (rice, wheat, and maize) are consumed in greatest quantities, followed by the common potato and cassava.

2. Good sources of protein are found in the legumes (such as soybean, peanut, beans, and peas) as a result of the activity of nitrogen-fixing bacteria that are found in the roots.

3. Root crops (e.g., cassava, sweet potato, and beets) and stem crops (e.g., common potato, sugar cane) are high in energy-rich carbohydrates but low in protein content.

4. Except for the banana and coconut and grains, fruits are not particularly good sources of food. They add variety, flavor, and some vitamins to the diet.

5. Forages (e.g., alfalfa, grasses) are important primarily because they are consumed by livestock, which can then be eaten by humans. In effect, food that is nonedible or nonnutritious to humans is converted to an excellent source of proteins as well as other materials.

6. Lumber plants (i.e., forest trees) are used for building purposes, to manufacture paper, and as a raw material in the production of rayon.

7. Cotton fibers are still the basis of a multibillion dollar industry in spite of the many man-made fibers that are now in use.

8. Next to water, coffee and tea are the liquids that most people drink.

9. Various spices (e.g., true pepper, cinnamon, clove) and medicinals (e.g., morphine, quinine, antibiotics) are obtained from plants, as is tobacco.

10. Although synthetic rubber has replaced much of the natural rubber in the world market, there is still a considerable amount of the latter produced.

Review Topics and Questions

1. Select any plant growing in the vicinity (your campus or home neighborhood) and discuss how it may be useful to humans.
2. Assume you are an explorer who has found a plant never before reported; it is a "new" plant as far as anyone knows. The roots are large and bulky, and might possibly be a source of food for humans. Discuss a logical procedure that you should use to determine if this is a safe and nutritious food supply.
3. List four different plant parts and indicate how each of them may be useful to humans. These are not necessarily from the same plant.
4. Discuss the following statement: "If a plant has not yielded a product useful to man, it should be replaced in that area by plants that are useful."
5. Prepare a list of plants or plant parts that are useful to man but have *not* been mentioned in this chapter.

Suggested Readings

BAKER, H. G., *Plants and Civilization,* 2nd ed. (Belmont, California, Wadsworth, 1970).

DOVRING, F., "Soybeans," *Scientific American* **230,** 14–20 (February 1974).

FISHER, H. L., "Rubber," *Scientific American* **195,** 75–88 (November 1965),

HALL, F. K., "Wood Pulp," *Scientific American* **230,** 52–62 (April 1974).

HARLAN, J. R., "The Plants and Animals That Nourish Man," *Scientific American* **235,** 88–97 (September 1976).

SCHERY, R. W., *Plants for Man,* 2nd ed. (Englewood Cliffs, New Jersey, Prentice-Hall, 1972).

3

A Few Basic Physical and Chemical Principles

Explanations of the activities of living organisms have been found by the application of various concepts common to the fields of physics and chemistry. Obviously, a knowledge of these concepts is essential to an understanding of such explanations. Although this is an introductory text, the following discussions of a few basic principles are included to make it easier for the beginning student to understand more fully the activities with which plants are involved. It should be understood that these discussions have been simplified and include only such concepts and information as are considered to be useful in the context of an introductory course involving the study of plants. Much more information and more detailed coverage of these topics can be found in any good general chemistry textbook or in such texts as listed at the end of this chapter.

3.1 Molecular Structure

All matter (that which has mass and occupies space) is composed of submicroscopic particles, called **molecules,** which are in continuous motion. Such molecules are the smallest subdivision of a substance that still possesses all of the specific properties or characteristics of that substance. A **substance** is a material all samples of which have the same set of properties; it consists of only one kind of molecule. Because these molecules are not visible, their movement has been determined by various indirect methods. However, a visible indication that water molecules are moving can be obtained by observing **Brownian movement.** If small but visible particles (such as in India ink) are suspended in a drop of water and observed through a microscope, these particles will be seen to "jiggle" about. Their nondirectional trembling movement is a result not of their own molecular composition but of the motion of water molecules striking first from one direction and then from another. The unequal collisions push the ink particles first to one side and then to another. Molecules are much too small to be seen at this magnification, and the particles that are moving should not be confused with molecules.

The "particles" are masses or aggregates of thousands and thousands of molecules; visible as a "particle" because there are so many. The water molecules, which are not visible, cause the "jiggling."

The molecules of any material are attracted to each other in varying degrees, and this greatly influences the amount of activity of individual molecules. In solids molecules are strongly attracted, resulting in little movement. In liquids the attraction is less, molecular motion is greater, and the material is fluid. The molecules of a gas have very little attraction for each other, the individual molecules have great motility, and these molecules will move farther and farther apart until the gas completely fills the space available to it. This movement of molecules from one area to another is termed **diffusion** and is a result of the inherent continuous movement of all molecules.

Molecules in turn are composed of **atoms,** the building blocks. Atoms are not hard, homogeneous spheres but are complex systems made up of a number of smaller particles. Recent evidence indicates that atomic structure is more complicated than had been suspected, but for our purpose we may consider an atom to be made up of a swarm of **electrons** (negatively charged particles) surrounding a positively charged nucleus.[1] The positive charge is caused by **protons** (discrete units of positive charge), but the nucleus also contains particles having no charge at all, the **neutrons** (Figure 3-1). This central portion contains practically the entire mass of the atom but occupies an extremely tiny portion of the volume; practically all the volume is occupied by swarms of orbiting electrons. In a neutral atom, the negative charge of the electrons is exactly balanced by the positive charge of the nucleus; the number of electrons is equal to the number of protons. When an electron is removed from a neutral atom, the particle that remains behind no longer has all of its positive charges (or protons) balanced by electrons; it is positively charged, or a **positive ion:**

$$Na \rightarrow Na^+ + e^-$$

The electron is shown as e^-. When a neutral atom picks up an electron, it gains an additional negative charge and forms a **negative ion:**

$$Cl + e^- \rightarrow Cl^-$$

The symbols Na and Cl refer to sodium and chlorine, respectively. Such symbols are the chemists' method of abbreviating the name of an **element,** a substance that cannot be decomposed into simpler substances by ordinary chemical action. Table 3-1 indicates some of the elements that are important in cellular structure and function.

In addition to using letters to indicate atoms, the chemist uses other methods of identifying them. The **atomic number** is the number of protons in the nucleus of an atom. This also indicates a great deal about the chemical properties of the atom because, in a neutral atom, the number of protons equals the number of electrons, and it is the electrons that are basically involved in chemical reactions. The **mass number** is the sum of an atom's protons and neutrons. The mass of the electron is so small that it is ignored in calculating mass numbers. As shown in Figure 3-1, for

---

[1] The term *nucleus* as used here should not be confused with a cellular nucleus.

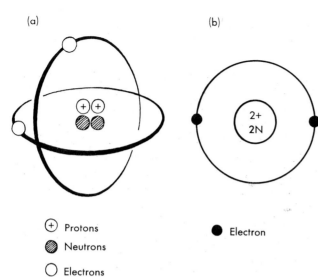

(a)　　　　(b)

Figure 3-1. **(a)** Diagrammatic sketch of the helium atom, which has two protons and two neutrons in the nucleus and two electrons orbiting around this nucleus. This represents a working model and should not be considered as a "picture" of the atom. **(b)** A more simplified diagram of the same atom. These diagrams utilize the planetary models of atoms as first proposed by Niels Böhr.

⊕ Protons

▨ Neutrons

○ Electrons

● Electron

example, the helium atom has two protons and thus an atomic number of two (which also indicates that this atom has two electrons); it has two neutrons in addition to the protons and thus a mass number of four.

The **atomic weight** is another feature used in referring to atoms. These weights are relative values using carbon with a mass number of 12 as the standard for comparison. All other elements are indicated as being lighter or heavier than carbon. The absolute weight values are not used simply as a matter of convenience; these cumbersome numbers would have to include on the order of 25 decimal places. Since the mass of the electron is so exceedingly small, it would appear that the atomic weight and the atomic mass number are the same. This is not so. The latter are always whole numbers, while the former are given to several decimal places.

The reason for the difference between mass number and atomic weight is that almost all elements exist in one or more isotopic forms. **Isotopes** are atoms of the

Table 3-1 Elements important to cell structure and function

Element	Symbol	Element	Symbol
Boron	B	Magnesium	Mg
Calcium	Ca	Manganese	Mn
Carbon	C	Molybdenum	Mo
Chlorine	Cl	Nitrogen	N
Cobalt	Co	Oxygen	O
Copper	Cu	Phosphorus	P
Fluorine[a]	F	Potassium	K
Hydrogen	H	Sodium	Na
Iodine[a]	I	Sulfur	S
Iron	Fe	Zinc	Zn

[a] These elements are essential for animals, but they have not been proved to be essential for plants. Quite possibly more elements will be added to this list as investigations produce further data (see Section 16.1).

same element that have the same atomic number and identical chemical properties, but differ in masses. The atoms have the same number of protons, but vary in the number of neutrons. The atomic weight is an *average* of the mass number values for all the naturally occurring isotopes of that element, and this depends upon the respective amounts of the different isotopes. For example, chlorine is found mainly as two isotopes: one with 17 protons and 18 neutrons, the other with 17 protons and 20 neutrons. Their chemical properties are the same because they each have 17 electrons orbiting around the nucleus. One has an atomic mass number of 35 (17 + 18), and that of the other is 37. All the samples of such elements contain the same proportions of the different isotopes. The atomic weight of chlorine is 35.453, which shows that the lighter isotope is present in greater abundance in this case. Table 3-2 lists some of the elements and their isotopes that are of importance in plant metabolism.

Many isotopes are unstable and decompose by radioactive decay to form a more stable atom. These radioactive elements emit one or more of three types of penetrating rays: alpha rays, which are composed of positively charged helium atoms (the helium nucleus stripped of its two planetary electrons); beta rays, composed of negatively charged electrons; and gamma rays, the most penetrating, composed of shortwave x-rays. Radioactive isotopes can be prepared in nuclear reactors or in cyclotrons, usually by bombardment with high-speed alpha particles. Since the emanations from such radioactive elements are relatively easy to detect, even when the element is in very low concentrations, these atoms are frequently used to "tag" molecules. The molecule can then be followed through various processes and movements in the cells or body of an organism.

Radioactive tracers have been used to determine which tissues in plants are involved in transporting various materials and where these materials are finally located. For example, minerals containing radioactive atoms may be applied in a water solution to the soil in which a plant is growing (plants in pots are usually used). These minerals are absorbed by the roots and are distributed through the plant. If the plant is small, the entire plant is removed from the soil and placed in contact with unexposed x-ray film. After a period of time, depending upon the amount of radioactivity, the film is developed. Darkened areas indicate where em-

Table 3-2 Atomic weights and stable isotopes of a few elements

Element	Symbol	Atomic number	Atomic mass numbers	Atomic weight
Calcium	Ca	20	40, 42, 43, 44	40.08
Carbon	C	6	12, 13	12.01
Chlorine	Cl	17	35, 37	35.45
Hydrogen	H	1	1, 2	1.01
Magnesium	Mg	12	24, 25, 26	24.31
Nitrogen	N	7	14, 15	14.01
Oxygen	O	8	16, 17, 18	16.00
Phosphorus	P	15	31	30.97
Potassium	K	19	39, 40, 41	39.10
Sulfur	S	16	32, 33, 34, 36	32.06

anations from the radioactive (tracer) atoms have exposed the film. The portion of the plant that had been placed at that location on the x-ray film is the site to which the minerals had been transported. In more delicate and sensitive techniques, various tissues or portions of the plant may be excised and placed upon unexposed film. This is also done if the plant is too large to fit the film. Evidence may thus be obtained as to which groups of cells are involved in the movements of materials. Additional use of radioactive tracers will be discussed in Section 12.9

3.2 Electron Orbitals and Energy

The electrons of an atom occupy certain fixed orbits or shells of different energies. Figure 3-2 shows the shapes of some of these **orbitals,** which represent a measure of the probability of finding an electron in a given region of space. The planar orbits of the planetary atom (see Figure 3-1) are abandoned in favor of certain volumes in space in which the electrons move. The innermost orbital is represented as a sphere, and all the electron shells beyond the first are further subdivided into subshells containing electrons that differ slightly in energy. One of these orbitals is shaped like a dumbbell; the illustration shows three of them oriented perpendicularly to one another. For representational purposes, a radius is usually chosen so there is a 90 per cent probability of finding the electron within the volume determined by that radius. The amount of energy an electron possesses depends upon the orbital in which it is found; the greater the distance from the nucleus, the more

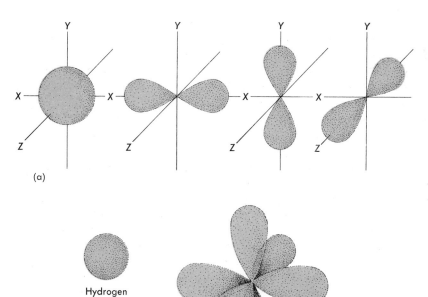

(a)

(b)

Hydrogen

Oxygen

Figure 3-2. Representations of orbitals. **(a)** The innermost orbital as a sphere; the next orbitals are dumbbell-shaped and oriented perpendicular to one another. **(b)** The sphere represents the space in which the single hydrogen electron is usually found; the three dumbbell-shaped orbitals of the oxygen atom are the areas in which the electrons of the outer shell are usually found.

potential energy the electron possesses. An attractive force exists between the negatively charged electrons and the positively charged nucleus of the atom. This force decreases with the distance from the nucleus, and the outer electrons are more easily removed from an atom. It is the outermost shell of electrons that determines the chemical behavior of the element.

An atom may have several shells of electrons, and each of these has a certain maximum number of electrons it can hold; the outermost level never has more than eight electrons and the innermost orbital has a maximum of two electrons. For any atom this is a stable configuration. Reactions between atoms result from rearrangements of electrons in the outermost orbitals, wherein one atom gives up one or more electrons to the other or each atom shares one or more electrons with the other. This interaction between outer electrons results in each atom approaching a more stable configuration of the outermost level.

When energy is supplied to an atom, by radiant energy (light), for example, electrons at certain energy levels (orbitals) jump to a higher level. The atom is now in an **excited** or **activated** state, and it has captured a certain amount of energy. When the electrons fall from the higher energy level, the same amount of energy is released. This movement may release energy as visible light or other wavelengths of radiation, or the energy may be released in such a way as to allow electrons in other atoms to make transitions. When we discuss photosynthesis by green plants, we will see how a molecule of chlorophyll (a green pigment) absorbs light, becomes excited, and makes possible the production of sugar by these plant cells.

3.3 Compounds and Formulae

Some substances are **compounds;** they consist of more than one kind of atom and they are capable of being decomposed into simpler substances. Water can be broken down into its constituents, hydrogen and oxygen, but the latter two materials are elements, which cannot be decomposed further. In order to indicate the composition of a compound, a **formula** is written (e.g., H_2O). The formula, utilizing symbols and subscripts, provides the following information: (1) the elements in the compound; (2) the relative number of each atom (subscripts indicate the number of atoms, the number 1 being omitted); (3) the combining weights of the elements, since the symbol refers to an atom and the atomic weights are known; and (4) the molecular weight of the compound, if it is known (by determining the sum of the individual atomic weights).

A **gram molecule,** or **mole,** is the amount of a substance in grams that is numerically equal to the molecular weight. These weights are relative values because they are determined by the sum of the individual atomic weights, which in turn are assigned values with relation to the carbon atom. Therefore, one mole of one substance contains exactly as many molecules as one mole of any other substance. This number, equal approximately to 6.02×10^{23}, or 602,000,000,000,000,000,000,-000, is known as Avogadro's number. One gram molecule of water (H_2O) weighs about 18 grams, one gram molecule of oxygen gas (O_2) weighs about 32 g, one gram molecule of carbon dioxide gas (CO_2) weighs about 44 g, and one gram molecule of sodium chloride (NaCl) weighs about 58 g. All of these amounts contain the same number of molecules—only the weights differ.

As mentioned in the previous section, the outermost electrons react when compounds are formed. Figure 3-3 represents the reaction between an atom of sodium (Na) and one of chlorine (Cl). The sodium atom has 11 electrons in the three shells, the outer shell being incomplete since only one electron is present. If this electron were lost, the outermost shell would be the second, which has a complete set of eight electrons—a stable configuration. In a similar manner, the addition of one electron to the chlorine outer shell would result in a change from an unstable condition of seven electrons to the stable configuration of eight electrons. The reaction produces a molecule of the compound sodium chloride (NaCl, or table salt). The sodium atom, having lost an electron, is a positively charged ion, while the chlorine atom is negatively charged as a result of gaining that electron. Thus, these atoms (in the NaCl molecule) are held together by a considerable electrostatic force between the ions of opposite charge. This attractive force between the ions is referred to as an **ionic bond.**

Electrons are not always lost or gained in this fashion; they may be shared. In Figure 3-4 is represented the reaction between four hydrogen atoms and one carbon atom, which yields a molecule of methane (CH_4). The atoms in such molecules are said to be connected by **covalent bonds.** In a similar fashion the hydrogen atoms and oxygen atom of a water molecule share electrons. Two hydrogen atoms are required because the oxygen atom has six electrons in its outer sphere, and two electrons (one from each hydrogen) are necessary to complete a stable configuration of eight.

When three or more atoms combine to form a molecule, the bonds hold these

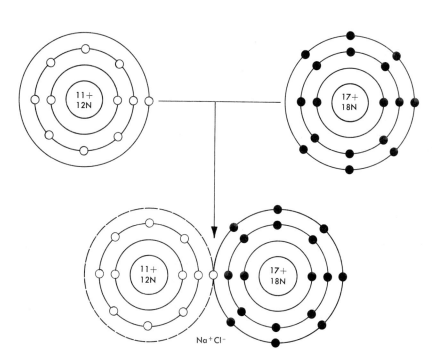

Figure 3-3. A simplified representation of the formation of a molecule of sodium chloride (NaCl) from an atom of sodium (Na) and an atom of chlorine (Cl). The oppositely charged atoms in NaCl are held together by electrostatic force.

Na^+Cl^-

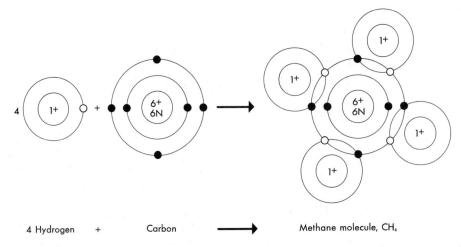

4 Hydrogen + Carbon ⟶ Methane molecule, CH₄

Figure 3-4. Diagram indicating the formation of a molecule of methane (CH₄) from a carbon atom (C) and four hydrogen atoms (H) by the sharing of electrons.

atoms in specific relationships with each other. As a consequence of the geometrical shapes of molecules, one portion may be positive or negative in relation to another portion of the same molecule; these are **polar molecules.** The water molecule is a good example of a molecule exhibiting polarity. An angle of 103° is formed between the two hydrogen atoms and one oxygen atom of water molecule (Figure 3-5). The nucleus of the oxygen atom attracts electrons more than the nuclei of the hydrogen atoms. This results in a positively charged region on one end of the molecule and a negatively charged region at the other end. Because of this polarity, water molecules are readily attracted to, and held to, negatively charged surfaces such as soil particles.

Polarity tends to bring molecules into specific orientations that enable chemical bonding to take place more readily. Large molecules may have numerous polar areas. In living organisms much of molecular (submicroscopic) structure results from polar attractions between molecules. One of the most common types of chemical bonds produced by polar attractions is the **hydrogen bond.** Such bonds are produced by the electromagnetic attraction between positively charged hydrogen atoms in one region of the molecule and negatively charged atoms of oxygen or nitrogen in another region of that molecule or neighboring molecules. A great deal of the structural aspect of the living portion of a cell depends upon very large protein molecules. In addition to other atoms, these molecules contain hydrogen, oxygen, and nitrogen. Hydrogen bonds are believed to be responsible for much of the basic configuration of protein molecules as well as the orientation of such mol-

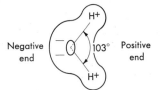

Negative end   103°   Positive end

Figure 3-5. Diagram indicating the polarity of a water molecule.

ecules into the submicroscopic structural pattern of the protoplast (the living cell material). Hydrogen bonds are rather weak, about one tenth as strong as the average covalent bond, but they are quite effective because of the large numbers of them that are usually involved.

Many compounds are found within plants, but only three of these will be mentioned now. **Carbohydrates** (e.g., sugars, starch, cellulose) are composed of carbon (C), hydrogen (H), and oxygen (O), with the last two occurring in a two-to-one ratio, as in water (H_2O). Molecules of the simple sugar glucose ($C_6H_{12}O_6$) are frequently combined by the cell to form more complex carbohydrates, such as starch, which are then stored in the cell. Later these complex molecules may be broken down into simple sugars and utilized. **Fats** and **oils** are also composed only of carbon, hydrogen, and oxygen, but relatively little oxygen is found in proportion to the other two atoms as, for example, in stearin ($C_{57}H_{120}O_6$). At ordinary room temperature, oils are liquids and fats are solids, but there is no general chemical distinction between them. The breakdown of fats and oils results in the production of fatty acids and glycerol (see Section 11.7). **Proteins** contain carbon, hydrogen, oxygen, nitrogen (N), frequently sulfur (S), and sometimes phosphorus (P), as in milk casein ($C_{708}H_{1130}N_{180}O_{224}S_4P_4$), and iron (Fe) as in heme proteins. These are the most complex molecules found in living cells and can be broken down to simpler substances known as amino acids. Just as complex compounds can be broken down to simpler ones, they can also be synthesized from the simple compounds by living cells. Such complex compounds are usually important constituents of the protoplast, or they may be storage products.

3.4 Ionization

Water will dissolve most ionic compounds. The polar bonds in water are attracted to the ions of the compound and reduce the force of attraction between such oppositely charged ions so that they separate (Figure 3-6). The molecules of the compound, thus, separate into two electrically charged particles, each surrounded by a number of attached water molecules. Positively charged ions are termed **cations,** and negatively charged ions are termed **anions.** The ions move about independently in the solution and are highly reactive. Common table salt is composed of

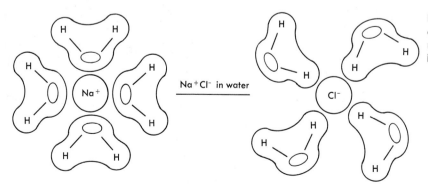

Na^+Cl^- in water

Figure 3-6. Sodium chloride dissolved in water; diagrammatic to show separation of ions.

sodium (Na) and chlorine (Cl) and **ionizes,** or **dissociates,** as follows (water molecules not shown):

$$NaCl \rightarrow Na^+ + Cl^-$$

The plus and minus charges balance. The charged particles are atoms, as above, or groups of atoms, as in the carbonate ion ($CO_3^=$) below:

$$Na_2CO_3 \text{ (sodium carbonate)} \rightarrow 2Na^+ + CO_3^=$$

All acids ionize to form hydrogen ions (H^+):

$$HCl \text{ (hydrochloric acid)} \rightarrow H^+ + C;^-$$
$$H_2SO_4 \text{ (sulfuric acid)} \rightarrow 2H^+ + SO_4^=$$

All bases dissociate to form hydroxyl ions (OH^-):

$$NaOH \text{ (sodium hydroxide)} \rightarrow Na^+ + OH^-$$
$$Ba(OH)_2 \text{ (barium hydroxide)} \rightarrow Ba^{++} + 2OH^-$$

When equivalent amounts of acids and bases are mixed, they are **neutralized,** in that the H^+ ions and the OH^- ions combine to form water (H_2O), which is only very slightly dissociated. During neutralization, **salts** are formed:

$$HCl + NaOH \rightarrow H_2O + NaCl \text{ (a salt)}$$

Salts do not form either H^+ or OH^- ions when they ionize.

The degree of acidity is usually expressed in terms of pH units, which are expressions of the number of hydrogen ions present in the solution. Specifically, **pH** is defined as the logarithm of the reciprocal of the number of moles of H^+ ions present in one liter of solution:

$$pH = \log \frac{1}{[H^+]}$$

For example, pure water dissociates to a very slight degree, and measurements show that there are 0.0000001 moles of H^+ ions present in a liter of water. There are just as many OH^- ions as there are H^+ ions; water is neither acidic nor basic, but it is chemically neutral.

$$H_2O \rightarrow H^+ + OH^-$$

$$pH \text{ (of water)} = \log \frac{1}{0.0000001} = \log \frac{1}{10^{-7}} = \log 10^7 = 7$$

When acids are neutralized by mixing them with an equivalent amount of a base, as shown previously, the pH is 7; the number of hydrogen ions is equal to the number of hydroxide (OH^-) ions.

The pH scale runs from 0 to 14; the lower numbers refer to acid solutions and the higher numbers to basic solutions. Any solution with a pH less than 7 has more hydrogen ions than hydroxide ions in solution; the lower the pH, the more acid the solution (Table 3-3). Note that the pH scale is a logarithmic progression. Thus, the number of H^+ ions in a liter of solution at pH 6 would be extremely small compared to that at pH 1. In the former there would be 0.000001 moles of H^+ per liter

Table 3-3 The pH scale

Concentration of H^+ Ions (Moles per Liter)		pH	Concentrations of OH^- Ions (Moles per Liter)	
	1.0	10^0	0	10^{-14}	
	0.1	10^{-1}	1	10^{-13}	
	0.01	10^{-2}	2	10^{-12}	
Acidic	0.001	10^{-3}	3	10^{-11}	
	0.0001	10^{-4}	4	10^{-10}	
	0.00001	10^{-5}	5	10^{-9}	
	0.000001	10^{-6}	6	10^{-8}	0.00000001
Neutral	0.0000001	10^{-7}	7	10^{-7}	0.0000001
	0.00000001	10^{-8}	8	10^{-6}	0.000001
		10^{-9}	9	10^{-5}	0.00001
		10^{-10}	10	10^{-4}	0.0001
Basic		10^{-11}	11	10^{-3}	0.001
		10^{-12}	12	10^{-2}	0.01
		10^{-13}	13	10^{-1}	0.1
		10^{-14}	14	10^0	1.0

while the latter solution would contain 0.1 moles per liter. The solution of pH 1 would contain 100,000 times more H^+ ions than a solution of pH 6.

Lemon juice and orange juice have a pH of about 2 and 3, respectively, and the former figure is also about the value for the stomach contents in humans. Although soils may range from pH 2 to 10 or so, most plants grow best in a pH range centering around 6.5 (soils at pH 6–7 usually produce the best crop yield). The improved growth at this pH range, as compared to higher or lower pH values, results primarily from the solubility of various soil minerals, which is influenced by pH. This will be discussed more fully in the chapter on soils.

3.5 Energy

Energy may be thought of as the ability to do work, or anything that can be converted into work (e.g., heat, electricity, or the energy residing in a coiled spring). One frequently speaks of **potential energy,** that which is stored or inactive. Water in a pond at the top of a hill has potential energy. If this water flows downhill, the potential energy becomes **kinetic energy** or active energy. This latter is the kind of energy that causes an effect upon matter. When the water reaches the bottom of the hill, all its potential energy has been converted into kinetic energy. It requires an expenditure of energy to pump the water back into the pond to reestablish its potential energy. Energy is frequently being changed from potential to kinetic and back again. For example, living organisms convert the potential energy of foods into the energy used for the synthesis of cellular components, and these latter materials contain potential energy that may be used by other organisms feeding upon the first.

There are various forms of energy: chemical, radiant, mechanical, electrical, and atomic. At any particular moment each of these forms may exist as either potential or kinetic energy, depending upon the situation at that time. One form of energy may be converted to another. For example, water flowing downhill (mechanical energy) may turn the armature of a generator that produces electrical energy. In the botanical field we are primarily concerned with chemical and radiant energy; the others will not be included in the following discussion.

Chemical energy is that which resides in chemical compounds. Green plants are capable of using sunlight (radiant energy) to synthesize sugar from carbon dioxide and water (see Chapter 12). As a result, potential chemical energy is stored in such sugar molecules. In later chemical reactions these compounds are utilized and their energy is released for use in the various processes that occur in living cells. Animals, of course, obtain food from plants and thus have a supply of chemical energy.

Compounds vary considerably in the amount of potential energy they contain. Fats, for example, contain more energy than carbohydrates. It is also possible to convert such energy into another form. Wood and coal, which are both plant products, may be burned to provide heat, a form of radiant energy.

Radiant energy is energy propagated through a medium as waves. Light, heat, sound, and x-rays are examples of this form of energy. As mentioned previously, the green plant converts radiant energy to chemical energy. Heat produces its effect by speeding up the motion of the molecules that compose matter. Just the reverse occurs as an object cools. This is the reason that evaporation cools the surface from which the water molecules are lost. The more rapidly moving water molecules are lost first. The ones that remain are moving more slowly, creating a cooler condition that our senses would detect if this evaporation were taking place from our body.

3.6 Light

Because light is such an exceedingly important form of radiant energy, it will be considered in more detail. Figure 3-7 represents the general spectrum of radiant energy of which visible light is only a small fraction and heat is indicated as infrared. (See Appendix I for units of measurement.)

In many of its properties light appears to travel in a manner analogous to waves on a pond. Light rays can be bent, as by a prism, with some rays bending more than others. The white light is thus broken up into the various colors of which it is composed: violet, blue, green, yellow, orange, and red. In the case of a rainbow the droplets of rain are acting as a prism. Such observations are best explained by the wave theory. However, if light beams are directed at the surfaces of certain substances, such as selenium, electrons are ejected from its atoms. The number of electrons ejected varies with the intensity or amount of light. The quantum theory takes such observations into account and holds that light is composed of tiny particles called **photons,** which have no mass. These two concepts each explain cer-

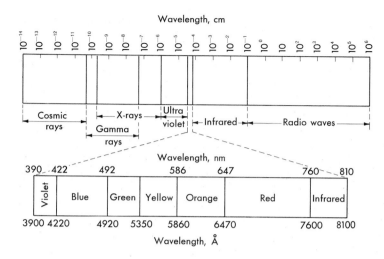

Figure 3-7. The electromagnetic spectrum. (From *Plant Physiology*, by R. M. Devlin, copyright 1966 by Reinhold Publishing Corporation, New York, reprinted by permission.)

tain characteristics of light, but not all of them, and it is probably best to think of light as consisting of waves of particles.

The distance from the crest of one wave to the crest of the next wave is the **wavelength.** The expression of each color of light depends upon the wavelength; of the visible rays, violet light has the shortest and red light has the longest wavelengths. As the wavelength decreases, the **frequency** (the number of crests per unit of time) increases; in equal distances, there will be more of the short waves than the long ones. The higher the frequency, the greater the energy. Photons can be thought of as packets of energy; the greater the frequency, the more such packets will impinge upon a surface. Violet light thus conveys more energy than red light, which has a longer wavelength and lower frequency.

It should be pointed out that the electromagnetic spectrum is continuous. There is no sharp distinction, for example, between blue and green or between x-rays and gamma rays. There is a gradual change and frequently an overlapping of our labels.

3.7 Oxidation and Reduction

When coal is burned, oxygen combines with the carbon, carbon dioxide (CO_2) is formed, and the chemically bound energy in the coal is liberated as heat and light. The combination of oxygen with other elements and the consequent release of energy was the original meaning of the term *oxidation*. However, many oxidations occur, especially in living cells, where free oxygen is not involved; sometimes there is a loss of electrons, and sometimes the electrons may be removed together with protons as hydrogen atoms. In any event, the basic characteristic of an **oxidation** is that there is a loss of electrons. As far as the living cell is concerned, the important result of oxidation is that bound energy (as in a food molecule) is liberated or made available.

Whenever one material is oxidized, another is reduced. **Reduction** refers to the gain of electrons. Sometimes this is accomplished in the form of the addition of hydrogen atoms. In other words, the electrons that are lost when a substance is oxidized are accepted by the substance that is reduced. Of course electrons are not really lost, they are transferred from one molecule to the other. In biological systems the removal of electrons or hydrogen atoms from one substance and their subsequent acceptance by others are the chief means of energy transfer and release. In cellular oxidations oxygen is frequently the acceptor, and its reduction results in the formation of water.

3.8 Hydrolysis and Condensation

The breakdown and the production of complex compounds that occur so frequently in living cells are actually examples of hydrolyses and condensations. The **digestion** of a complex starch molecule to sugar molecules actually utilizes water and thus is a type of **hydrolysis;** one molecule of water is added for every molecule of sugar that is produced:

$$\text{Starch} + n\ H_2O \rightarrow n\ \text{Sugar}$$

The symbol n is used because the size of the starch molecule is unknown, and so the number of water molecules used and the number of sugar molecules produced are unknown. The reverse process, whereby a large number of sugar molecules combine to form starch, with the attendant elimination of water molecules, is termed **condensation** or **synthesis.** Fats, oils, and proteins undergo similar syntheses and hydrolyses (see Section 11.7). The synthesis of such complex compounds from simpler materials entails the expenditure of considerable amounts of energy. Various oxidations occurring within the cell provide available energy for the formation of the many complex molecules that are essential to the normal functioning of that cell.

3.9 Adsorption, Capillarity, and Imbibition

As mentioned previously, many materials have charged areas, and molecules or ions are frequently concentrated, or held, on various surfaces, a phenomenon termed **adsorption.** For example, the clay particles of soils are negatively charged and hence attract and hold water and positively charged ions. This characteristic of soil particles is of tremendous importance to the growth of plants and will be emphasized in Section 17.4.

If the tip of a tube of small diameter is immersed in water, as in Figure 3-8, the level of water in the tube will rise higher than the surface into which the tube was placed. Such a rise of water in tubes, termed **capillary action** or **capillarity,** depends upon the cohesion of water molecules to each other and the adhesion of water molecules to the tube walls. The smaller the diameter of the tube, the higher

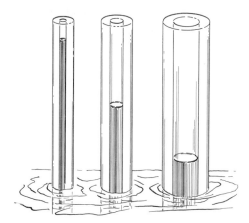

Figure 3-8. Diagram showing capillarity. The rise of water into tubes of small diameter depends upon the cohesion of water molecules and their adhesion to the walls of the tubes. The height to which water rises in such tubes depends upon the diameter of the tube, for one thing, as shown in the diagram.

will be the capillary rise. **Cohesion** refers to the attraction of similar molecules to each other (e.g., water molecule to water molecule). **Adhesion** refers to the attraction between dissimilar molecules (e.g., water molecule to molecule of tube wall).

The swelling of dry wood when it is placed in water results from the diffusion of water into the wood, the capillary action of water entering tiny cracks, crevices, and tubes of the wood, and the entrance of water between the particles of wood and its adhesion to these particles. **Imbibition** is this entrance of water into solids and the resultant swelling. The wood particles are forced farther and farther apart as the water enters, and the swelling forces may be considerable. The granite blocks used to build ancient pyramids were probably quarried by drilling holes in the stone, pounding dry wood tightly into these holes, and then pouring water on the wood. The resultant imbibitional forces split the granite. Leaky rowboats may become watertight when placed in water, as a result of the swelling of the dry wood. In a similar manner a door may stick as atmospheric humidity rises and the wooden door, or even the door jamb, swells. When a seed germinates, the imbibitional forces resulting from water absorption are sufficient to rupture the tough seed coat, allowing the enclosed embryo to grow. Proteins swell much more than starch, which swells more than cellulose. Therefore, the storage tissues of seeds (which contain much protein and starch) swell more than the seed coats (primarily cellulose), which results in rupturing the latter.

3.10 Surface-to-Volume Ratio

Adsorption and absorption are both dependent to a great extent upon exposed surfaces. In the former phenomenon, materials are held to surfaces, whereas in **absorption** materials pass through a surface. Therefore, the greater the amount of surface, the more adsorption and absorption will take place, other factors being equal. For example, a convoluted soil particle will adsorb more water molecules than a spherical soil particle of similar composition and volume because of the greater surface area in the former.

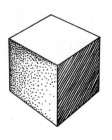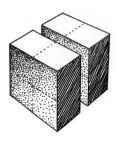

Figure 3-9. Diagram indicating the increase in exposed surface area when an object is subdivided into smaller pieces; the total volume is not changed. The surface-to-volume ratio increases as the size becomes smaller.

In general, the smaller or the more convoluted a structure, the greater is the surface-to-volume ratio. A simple way of visualizing this is to think of your textbook, which has a certain volume and six surfaces when closed. When the book is opened to expose each of the pages individually, the volume is unchanged, but the exposed surface area has increased tremendously, just as with any convoluted structure. If the pages are torn out the total volume does not change, but each page now has surfaces exposed. In effect, the extremely small soil particles, which have developed by disintegration of large particles, are similar to the torn book. Figure 3-9 indicates how the surface (S) of a cube-shaped object increases as the cube is cut, while the volume (V) remains the same; the S/V ratio has increased. See also Table 3-4 as an example of the great change in the surface area when specific volume is subdivided.

The importance of this principle of surface-to-volume (S/V) ratio will be emphasized several times throughout the text.

Summary

1. All matter is composed of molecules, which are in continuous motion. Spheres of electrons surrounding a central nucleus constitute the atoms of which molecules are composed. Reactions between atoms involve electrons of the outermost orbits. Atoms are identified by symbol (letters), atomic number, mass num-

Table 3-4 Relationship of surface area to particle size

Length of single edge of cube	Number of cubes in 1 cm³	Total surface area exposed per 1 cm³
1 cm	1	6 cm²
1 mm	10^3	60 cm²
0.1 mm	10^6	600 cm²
0.01 mm	10^9	6000 cm²
1.0 μm (micrometer)	10^{12}	6 m²
0.1 μm	10^{15}	60 m²
0.01 μm	10^{18}	600 m²
0.001 μm	10^{21}	6000 m²

ber, and atomic weight. Many atoms are found in isotopic form, some of which may be radioactive.

2. Molecules may separate into electrically charged ions. They may also exhibit polarity.

3. Energy is considered as the ability to do work and exists in several forms; chemical and radiant energy are of primary concern to biologists. Each form may exist as potential or kinetic energy. Light energy is considered to consist of waves of particles (photons). Oxidations provide the energy that is required for normal functioning of the cell.

4. Hydrolysis is the splitting of complex molecules into simpler ones with the utilization of water. Digestions are hydrolytic processes. Condensations, or syntheses, are the reverse of hydrolyses.

5. Adsorption is the concentration of molecules or ions on surfaces caused by the forces of attraction between them. Imbibition refers to the entrance of water into solids and the resultant swelling. Absorption is the passing of materials through a surface.

Review Topics and Questions

1. How could you determine whether or not the movement or particles in a drop of water was Brownian motion?
2. Describe a procedure that you could use to demonstrate diffusion.
3. The starch content of a cell increases while the amount of starch in a neighboring cell decreases. Since there is no other source of carbohydrate, the first cell presumably is obtaining materials from the second cell. Starch is not soluble and cannot penetrate cell membranes or surfaces. Explain how the observed phenomena are possible if the assumption is correct.
4. List six elements and explain how each is important to living cells.
5. Discuss the advantages of utilizing symbols and formulae, as the chemist does when discussing reactions that occur within a cell.
6. A cube with dimensions of 1 cm is divided into smaller cubes with 1-mm dimensions. What was the original volume? The final volume? What was the original surface area? The final surface area?
7. What are radioactive isotopes, and how may they be used in biological investigations?
8. As far as cells are concerned, what may be the significance of molecular polarity?
9. Select one form of energy and discuss the differences between its potential and kinetic states.
10. What should be your answer if you were asked to distinguish between blue and violet light?

Suggested Readings

BAKER, J. J. W., and G. E. ALLEN, *Matter, Energy, and Life* (Palo Alto, Calif., Addison-Wesley, 1965).

WHITE, E. H., *Chemical Background for the Biological Sciences,* 2nd ed. (Englewood Cliffs, N.J., Prentice-Hall, 1970).

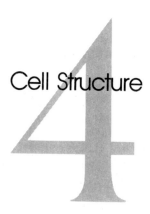

Cell Structure

One of the characteristics of living organisms is that they are composed of one or more cells and that all these cells come from preexisting cells (Rudolph Virchow, 1858). However, this concept did not arrive on the scientific scene because of a burst of brilliancy on the part of any one individual. Various observations and investigations over many years gradually resulted in a greater understanding of the structure and development of plants and animals.

4.1 The Cell Theory

Robert Hooke (1665) examined thin sections of cork and accurately described the cells of which it was composed. Though he observed only the walls of dead cells, he was aware of the three-dimensional aspect of these structures and the fact that they were distinct from one another. He also estimated that more than 1 billion cells were contained in a cubic inch of cork. Hooke first used the term *cell* and compared the appearance of cork to that of a honeycomb, but he did not stipulate the universal occurrence of cells, though he did observe living plant cells filled with liquid and observed that the pith and stalks of various plants appeared very much like cork with its many little boxes.

Subsequent investigations added to the developing concept of cellular structure, and R. J. H. Dutrochet (1824) presented one of the first clear statements of the idea that all living things are composed of cells. He further indicated that growth results from both the increased size of cells and the addition of new little cells. The cell was considered to be the primary unit, and the function of the organism was a result of a summation of the functions of individual cells. Dutrochet did not discover the existence of a nucleus within the cell, but this information was added a few years later by the observations of Robert Brown (1833), who first observed the nucleus in hairs and other cells of orchids. A few years later Matthias Schleiden (1838) stated that plants were composed of cells, and Theodor Schwann (1839) made a similar statement concerning animals. The latter also pointed out that cells must be capable of producing chemical changes in the materials they absorb because these materials are in other combinations outside the cell and because various parts of the cell are composed of materials that differ in chemical properties.

With additional investigations concerning the structure of organisms and the functioning of their component parts, plus the recognition of protozoa as single-celled animals (Karl von Siebold, 1845), many biologists were led to the modern concept of cellular structure: that multicellular organisms are subdivided into functional units called cells, which are variously modified and specialized, thus resulting in a division of labor, and that new cells are formed from preexisting cells by cell division. In the case of a unicellular organism, of course, the functions of the cell are the functions of the organism. This idea of structure emphasizes the coordination among cells and tissues in the explanation of the behavior of organisms. The mere summation of the activities of individual cells is not sufficient to explain the functioning of multicellular plants and animals.

Complex plants and animals contain many different kinds of cells. In order to understand the uses and functions of plants—how they grow, reproduce, synthesize materials, what man does with them—we must understand clearly the structures that are involved, because function and structure are so intimately tied together. For example, supporting or strengthening cells have thicker walls than conducting cells, while the latter are usually elongated. The possibility of a terrestrial existence depends upon the structure of land plants. The structure of the individual cells that comprise a tissue affects the efficient functioning of that tissue. Thus, we should start with a generalized cell, discuss its basic structure and function, and then undertake the more difficult task of studying the correlation of various cells functioning as tissues and organs.

4.2 The Generalized Cell

The size of individual cells varies greatly, from microscopic ones on the order of 0.5 μm in diameter (μm = micrometer; 1 μm = about $\frac{1}{25,000}$ in.) to macroscopic units up to 10 cm (cm = centimeter; 1 cm = about 0.4 in.). In the higher plants the general range in diameter is 10–100 μm. (See Appendix for units of measurement.) The number of cells in a plant is astronomical. Remember Hooke's estimate? A single leaf on a tree may have more than 40,000,000 cells. Count the number of leaves on a tree, multiply by 40 million, multiply by about 20 to include cells in the roots and stems, and you will have a general idea of the number of cells that are involved in the growth of a tree. Each cell consists of living and nonliving material, the latter being produced by the former as the cell is developing. In some cells the living material dies and disintegrates after the cells have matured. This occurs in cork cells and explains why Hooke thought of cells as little empty boxes.

Cell Wall

The outer boundary of plant cells consists of nonliving structure, secreted by the living part of the cell and termed the cell wall. It is a strong, porous, rather

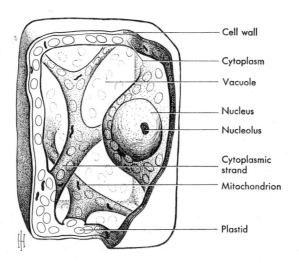

Figure 4-1. General features of a plant cell.

Cell wall

Cytoplasm

Vacuole

Nucleus

Nucleolus

Cytoplasmic strand

Mitochondrion

Plastid

rigid but somewhat elastic wall. (See Figure 4-1 for structures pertaining to the general plant cell.) The shape varies as the cell differentiates, but the young or immature cells are basically alike. In multicellular plants the walls form a rather continuous structure similar to the walls, floors, and ceilings in a large building. Much like layers of plaster on the walls of a room, cell walls are laminated, even though such layers may not be visible unless special staining techniques are used. The interconnected cell walls provide strength and support to the entire plant body.

Middle lamella. When the cell wall first develops during cell division, it is composed of granules that increase in size and number until they coalesce and form a thin layer called the **cell plate,** which eventually develops into the **middle lamella,** or **intercellular layer** (Figure 4-2). This layer is composed largely of **pectic substances,** mainly calcium pectate, which are viscous and gelatinous and function as the cementing material that holds the cells together. Commercial preparations of pectins are prepared from various fruits and are added to jellies to ensure the ''jelling'' of these materials.

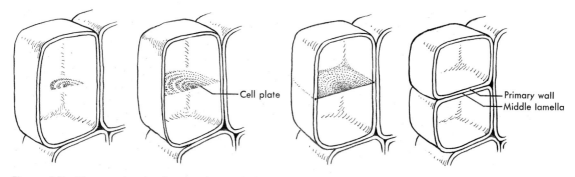

Cell plate

Primary wall
Middle lamella

Figure 4-2. Diagram showing the development of a new cell wall.

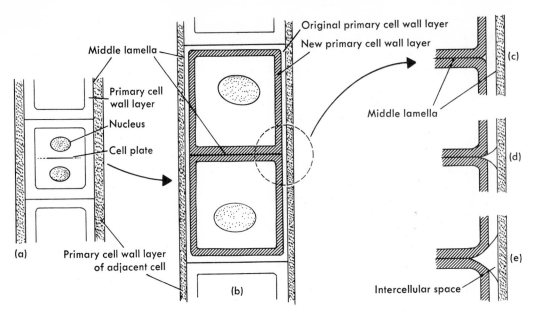

Figure 4-3. Diagram showing adjustments between old and new cell walls. **(a)** Cell plate forming. **(b)** Primary daughter-cell walls have been laid down on the inside of the primary mother-cell wall. Cells have expanded. **(c), (d), (e)** Establishment of continuity between old and new middle lamellae. (From *Plant Anatomy*, by Katherine Esau, John Wiley and Sons, 1965, New York, reprinted by permission.)

The middle lamella of the new cells comes in contact with the original primary cell wall layer. A small cavity arises at this point, and the original wall dissolves. The middle lamellae become continuous, and the cavity develops as an intercellular space. This sequence of events is shown in Figure 4-3.

Primary wall. The protoplast secretes an additional layer against the middle lamella. This second layer consists mainly of bundles of intertwined molecules of **cellulose,** a complex carbohydrate formed from simple glucose (sugar) molecules, and is rather plastic and capable of extension as the cell grows. In many cells no further layers are produced. The primary wall becomes more rigid as additional cellulose molecules are added after the cell has attained its mature size. The additional cell wall material may be added as layers or inserted as new particles between those already present.

The primary walls also contain hemicelluloses, some pectin, and small amounts of proteins as an amorphous matrix in which the cellulose is embedded. The hemicelluloses are made up of 40–200 repeating molecules of glucose attached end to end, while cellulose is similarly composed of 3,000–10,000 glucose units. The large cellulose molecules are packed and twined together along their long axes to form microfibrils, which, in turn, may coil around one another like strands on a cable. These fibrils form an irregular mesh or network that constitutes the basic structural framework of the cell. The cable-like arrangement results in a wall of considerable strength.

Cutin, a fatty material, is usually found as a layer (the **cuticle**) on the outer walls of the cells forming the exterior surfaces of land plants (see Figure 10-7). Waxes are also frequently found in this layer. The cuticle is relatively impermeable to water and protects leaves and stems against water loss.

Secondary wall. In those cells with relatively thick walls, a secondary wall, usually of three layers, is produced and deposited between the primary wall and the protoplast after cell enlargement ceases. These layers are basically cellulose, but additional materials may also be present, such as **lignin,** a complex material that is responsible for hardness and decay-resisting qualities of many woods, and **suberin,** a fatty material found in cork cells. The latter material renders the cells almost impermeable to water and gases. Thus, layers of cork cells protect underlying tissues from water loss. Most cells with secondary walls are devoid of living material at maturity. They are dead (empty) cells whose principal function is structural strength, support, and protection. In mature woody tissue, lignin may also be deposited in the primary wall and middle lamella.

Cellulose is more abundant in the secondary wall than in primary walls, and the former is more rigid. Within any one layer of the secondary wall, the cellulose

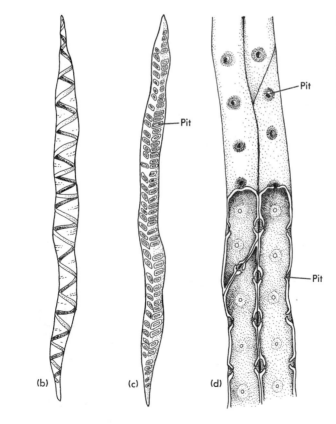

Figure 4-4. Tracheids. A variety of patterns of wall thickening: **(a)** annular, **(b)** spiral, **(c)** scalariform, **(d)** pitted. (From *Plants, An Introduction to Modern Botany,* by Greulach & Adams, 1975, Wiley, N.Y., reprinted by permission.)

(a)   (b)   (c) —Pit   (d) —Pit   —Pit

fibrils are essentially parallel but the orientation varies with the layer. This laminated structure greatly increases the strength of the wall. In some cell types the secondary wall is laid down over only a portion of the primary wall, resulting in rings, spirals, or other patterns of the wall (Figure 4-4).

Pits and plasmodesmata (Figure 4-5). The walls of cells vary greatly in thickness; supporting cells have relatively thick walls, and cells that are actively synthesizing materials have rather thin walls. Most cells also have thin areas in the primary walls (no secondary wall layers) that greatly facilitate the movement of dissolved materials from one cell to another. With special staining techniques and high magnification, many of these areas are seen to have exceedingly small pores through which delicate strands of cytoplasm (**plasmodesmata;** singular, **plasmodesma**) connect the protoplast (living material) of one cell with that of the adjacent cell. A **pit** is a region in the cell wall in which no secondary wall is deposited; the thinner, perforated region in the primary wall is termed the **primary pit field.** Cytoplasmic strands may also extend through cell walls in areas where there are no pits or pit fields. Plasmodesmata aid in the movement of materials and may possibly transmit stimuli.

Although the primary material is cellulose, cell walls vary considerably in composition and may contain, in addition to those materials already mentioned: min-

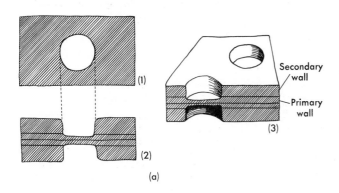

Figure 4-5. Pits and plasmodesmata. **(a)** Several views of a simple pit. **(b)** Several views of a bordered pit in which the secondary cell wall arches over the thin area. **(c)** Plasmodesmata extending through the pit area from the cytoplasm of one cell to the cytoplasm of a neighboring cell. Note that the pits are in opposing pairs, one belonging to each of two adjacent cells.

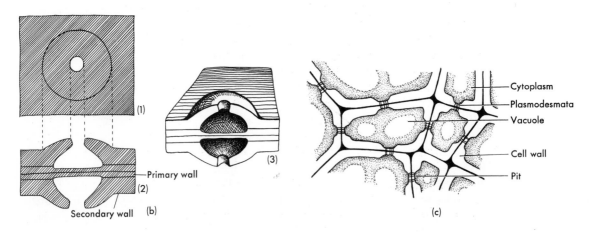

erals, tannins, resins, pigments, mucilages, and gums. It is beyond the scope of this book to consider such details, and one of the cytology texts listed at the end of this chapter may be consulted for additional information. The cell walls are quite porous, in effect, a rather open lattice-work of cellulose fibrils, and are freely penetrated by water and various dissolved materials except when cutin or suberin is present. These latter materials greatly retard penetration of substances through the walls.

The cellulose component of the wall is quite useful commercially. Since cellulose comprises approximately 80 per cent of the dry weight of wood, forests are the source of many valuable articles in addition to lumber for buildings and furniture. Wood pulp is used to manufacture paper; rayon is produced from cellulose and so are explosives, cellophane, buttons, and many other materials.

Plant fibers, which are twisted together to manufacture thread or yarn from which fabrics are woven (e.g., cotton, linen, burlap), are exceedingly long (up to at least 5 cm) cells of small diameter and with thick secondary walls. The fibers are obtained primarily from the cotton, flax, or hemp plants.

Protoplast

The living material that produces and is surrounded by the nonliving cell wall is commonly termed the **protoplast.** All the various complex functions of life are carried on in this small amount of viscous, colorless, transparent material. Its consistency varies from time to time from that of raw egg white to that of semisolid gelatin. It is not homogeneous, for various definite structures can be observed quite readily, even with low magnification, and particularly with the high magnification of an electron microscope. Those protoplasmic particles that are bounded by discrete membranes are frequently termed **organelles.** Though these structures are important (and will be discussed), the submicroscopic structure is equally important. Evidence indicates that large, complex protein molecules form the basic framework, much like intertwined twigs of a brush pile, and that other materials (e.g., water, fats, oils) are scattered among the proteins. Molecular polarity and hydrogen bonds are involved in preserving a structural framework (see Section 3.3). One should remember that even though a chemist can determine the composition of the protoplast, listing all the elements involved, this does not explain "life" or how a protoplast functions. The problem is similar to a jeweler describing a watch in terms of gold, silver, steel, and jewels—unless these elements are placed in the correct order or structure, no watch can be said to exist. The placement and interrelationships of the component parts are all important. The protoplast is, of course, more complicated than a watch, especially in view of the fact that its composition is not constant but variable.

Nucleus. In observing a cell with the use of a microscope, the most obvious living structure within the protoplast is the spherical or ovoid **nucleus** (plural, **nuclei**). Most organisms have one nucleus per cell, but some of the algae and fungi have cells with two to several nuclei. Well-organized and delimited nuclei are lacking in certain simple organisms, such as the blue-green algae and the bacteria,

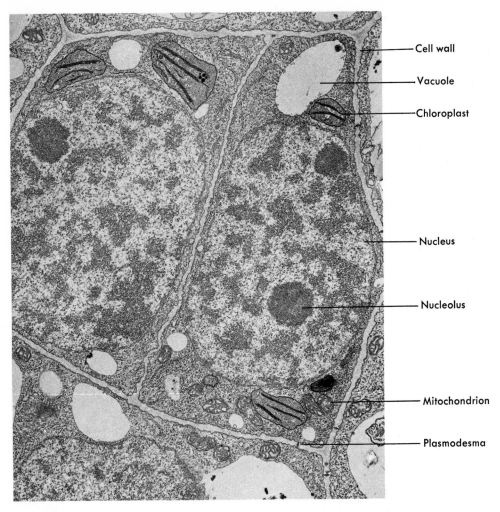

— Cell wall

— Vacuole

— Chloroplast

— Nucleus

— Nucleolus

— Mitochondrion

— Plasmodesma

Figure 4-6. Electron micrograph of parenchyma cells from tobacco (*Nicotiana tabacum*) leaf mesophyll (×12,000). (Courtesy of Katherine Esau and Robert Gill.)

but distinct genetic material is distributed through the cytoplasm although it is not limited by a membrane (see Section 4.3).

The nucleus is bounded by the **nuclear envelope,** and within this structure may be seen one or more spherical **nucleoli** (singular, **nucleolus**) and irregular strands of **chromatin.** The nucleoli appear to be reservoirs of ribonucleic acid (see later). The nuclear envelope has been shown, by electron microscopy (Figures 4-6 and 4-7), to be double and to contain tiny pores. The outer membrane of the nuclear envelope is connected to the extremely convoluted and flexible **endoplasmic reticulum,**[1] an interconnected system of membrane-bound cavities in the cytoplasm

---

[1] Because this structure cannot be seen with the ordinary light microscope, but only with the use of an electron microscope at exceedingly high magnifications, it is not indicated in Figure 4-1. For the same reason, the ribosomes have not been shown in this figure.

(Figures 4-14 and 4-15). The remaining part of the nucleus is a clear liquid, the **karyolymph.**

In order to see the various structures more clearly, the cells are treated with certain dyes. The chromatin material stains very deeply and is seen to be in the form of long, coiled, threadlike structures made up of deoxyribonucleic acid (DNA) combined with proteins. These threads are paired but are difficult to trace as individuals except when the cell is dividing. When cell division occurs, the paired chromatin threads become more tightly coiled and shorten, and a matrix forms around them. The condensed matrix, probably a proteinaceous material. stains very deeply and usually obscures the chromatin that it surrounds (Figure 4-8). This more distinct, rod-shaped structure is the **chromosome;** every type of organism that has a nucleus has a characteristic constant number of such chromosomes in each cell. The chromatin strands contain distinct submicroscopic particles, the **genes,** which control the characteristic growth and chemical processes of the organism. These genes are basically portions of the DNA of a chromosome. Chemically, then, the nucleus contains a relatively large amount of nucleoproteins (proteins in combination with nucleic acids) of which phosphorus is an important constituent.

The nucleus functions as the controlling center for the physiological activities of the cell and also governs the transmission of hereditary characteristcs. If the nucleus is removed from the cell, physiological abnormalities occur and death results. The red blood cells of man (erythrocytes) do not contain nuclei when they are mature, but such cells function for only a short period, being replaced by newly formed erythrocytes. The sieve tube elements (one of the conducting cells in vascular plants) also lack nuclei at maturity. All other cells of the higher plants contain at least one nucleus. Simple plants, like the bacteria and blue-green algae, have DNA, but it is not contained in an enclosing envelope or membrane.

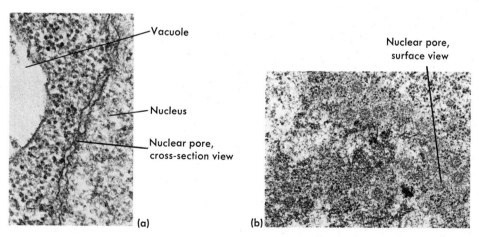

Figure 4-7. **(a)** Electron micrograph of parenchyma cell from bean (*Phaseolus vulgaris*) root tip (×60,000). **(b)** Electron micrograph of a nuclear surface in a tobacco (*Nicotiana tabacum*) root tip parenchyma cell (×30,000). (Both courtesy of Katherine Esau and Robert Gill.)

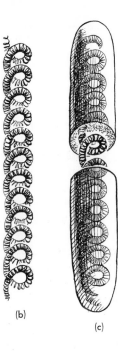

Figure 4-8. Diagrammatic representation of the relationship between chromatin and chromosome. For simplicity, only one of the paired chromatin strands is shown. **(a)** Chromatin strand expanded. **(b)** Chromatin strand coiling and becoming shorter. **(c)** Chromosome; matrix condensed about the chromatin strand. In most staining procedures the matrix stains so deeply that the chromatin strands are obscured.

(b)

(c)

(a)

Cytoplasm. That portion of the protoplast that surrounds the nucleus and in which other formed bodies are contained is the **cytoplasm.** Recent investigations have emphasized the importance of a submicroscopic structure, and the arrangement of molecules that compose the cytoplasm greatly influences the activities of the cell. Many cellular enzymes are located in the cytoplasm and are important in respiration and various syntheses. **Enzymes** are proteinaceous organic catalysts produced by living cells. A catalyst greatly speeds the rate of a chemical reaction without being used up in that reaction. Without enzymes the activities of living cells would cease. (See Section 11.4.) In young cells the cytoplasm occupies most of the volume of the cell, but in mature cells the cytoplasm usually occurs as a thin layer lining the walls and with thin strands penetrating throughout the cell.

The cytoplasm frequently can be seen streaming about in the cell, but not from one cell to another. Many of the distinct bodies found in the cytoplasm are carried about much like wood in a stream. The movement of one cytoplasmic strand may be opposite to that of another; lower layers of the same strand may flow in one direction while the upper layers flow in the opposite direction. Such movement indicates great activity and may be important in the transport of food and other materials.

The cytoplasm is bounded by a **membrane,** just as are the various organelles of the protoplast. All these membranes are **differentially permeable.** That is, when the membrane is alive, some materials penetrate the membrane quite readily, others penetrate more slowly, and still others do not penetrate at all. This property of membranes is variable, and under some conditions they are more permeable than under others. Upon dying, membranes become freely permeable. This characteristic holds true for all living membranes.

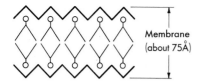

 Protein

Phospholipid

Membrane
(about 75Å)

Figure 4-9. Diagram indicating the probable arrangement of protein and phospholipid molecules of a membrane. Å = angstrom = 10^{-8} cm.

Membranes are basically very thin structures composed of proteins and phospholipids. The latter are fatty compounds containing phosphorus and usually a nitrogenous group. There is evidence indicating that phospholipids occupy the center of the membrane with proteins bound to such molecules and aligned on the exteriors as shown in Figure 4-9.

In electron micrographs the membranes appear as triple structures—a light (electron-transparent) area bounded by two dark bands (electron-dense), as shown in Figure 4-10. The electron-dense areas are considered to be composed largely of proteins; compare Figures 4-9 and 4-10. The differential permeability of membranes is related to this structure. Its lipid nature allows for the penetration of relatively nonpolar molecules, which are readily soluble in lipids, whereas polar, or ionized, molecules penetrate with difficulty. Exceedingly tiny pores in the membranes probably allow water molecules to penetrate. Such pores, however, have not been demonstrated. More recent evidence suggests that proteins are distributed at various intervals and may extend through one or both of the phospholipid layers (Figure 4-11). Some of the proteins may posses enzymic functions, while others may be solely structural in nature. The movement of materials through membranes will be discussed more thoroughly in Section 5.5.

Although there is still some controversy, it appears likely that the outer cell membrane (**plasma membrane** or **plasmalemma**), the nuclear membrane, and the endoplasmic reticulum are interconnected and form a system of membranes that

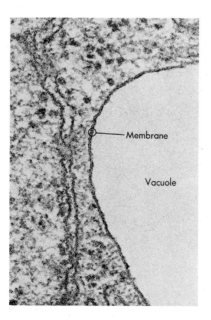

Membrane

Vacuole

Figure 4-10. Electron micrograph of membrane (vacuolar membrane or tonoplast), parenchyma cell in a leaf stalk (*Gossypium*) (×100,000). (Courtesy of Robert Gill.)

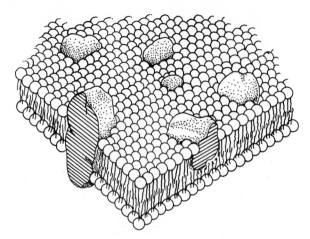

Figure 4-11. The phospholipid-globular protein mosaic model of cell membranes. The spheres represent polar portions and the wavy tails represent fatty acid portions of the phospholipids. The larger bodies represent globular protein molecules of the membranes. (From S. J. Singer and G. L. Nicholson, *Science* **175:** 723, 18 February 1972, with the permission of *Science* and the authors. Copyright 1972 by the American Association for the Advancement of Science.)

permeate the cytoplasm. The endoplasmic reticulum provides a large internal membrane surface along which enzymes are distributed in an orderly arrangement, allowing for the efficient functioning of a sequence of activities. The protoplast, thus, is compartmentalized with metabolites segregated and providing a pathway for the transport of materials from one part of the cell to another. Within the plasmodesmata the thin plasma membrane of a cell is continuous with that of an adjacent cell. The endoplasmic reticulum is also associated with plasmodesmata and may extend through to the adjacent cell. As a result of this structural arrangement, there are pathways for the exchange of materials between cells.

Mitochondria. Small rods or granules (about 7μm $\times$ 0.5μm), called **mitochondria** (singular, **mitochondrion**), are abundant in the cytoplasm but are visible only when the cells are stained with certain dyes or observed with an electron microscope. Respiratory enzymes are oriented in a specific manner in each mitochondrion and function efficiently as a group. Most of the energy made available to the cell is a result of activities occurring within these structures. As determined by a study of electron micrographs, the structure of a mitochondrion is basically that of a fluid-filled vessel with two membranes. The inner membrane of this double layer has many involutions or infoldings termed **cristae** (Figures 4-12, 4-13, and 4-14). The specific orientation of enzymes occurs in these inner membranes.

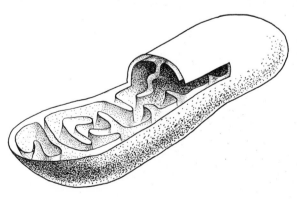

Figure 4-12. Mitochondrion, the inner membrane is very convoluted forming cristae. (From A. L. Lehninger, *Scientific American*, September 1961.)

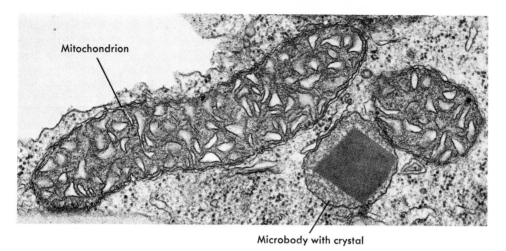

Mitochondrion

Microbody with crystal

Figure 4-13. Electron micrograph of mitochondria and a microbody in a tobacco (*Nicotiana tabacum*) leaf vascular parenchyma cell (×45,000). (Courtesy of Katherine Esau and Robert Gill.)

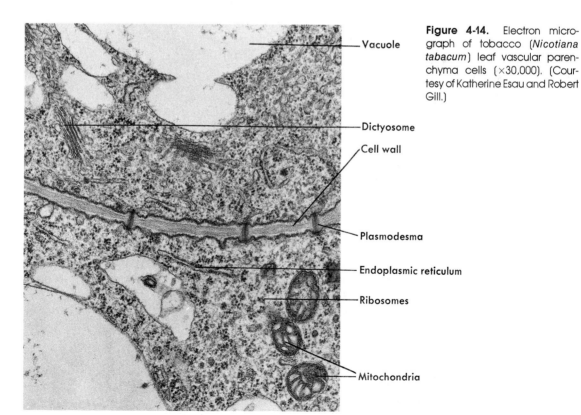

Vacuole

Dictyosome

Cell wall

Plasmodesma

Endoplasmic reticulum

Ribosomes

Mitochondria

Figure 4-14. Electron micrograph of tobacco (*Nicotiana tabacum*) leaf vascular parenchyma cells (×30,000). (Courtesy of Katherine Esau and Robert Gill.)

Most biologists believe that mitochondria arise by the growth and division of preexisting mitochondria. The discovery that these organelles contain DNA, which differs from the DNA in the nucleus, indicates that they contain genetic information and supports the idea of at least partially independent division. Evidence from electron microscopy seems to suggest that mitochondria may also be formed when cytoplasm fills small cavities of the endoplasmic reticulum, which then pinch off to form discrete bodies that develop into mitochondria. Additional investigations are needed to resolve this situation.

Ribosomes. Within the cytoplasm, located on the endoplasmic reticulum as well as free in the cytoplasm, are exceedingly small granules (15 to 25 nm in diameter) that are visible only with the use of the electron microscope (Figures 4-14 and 4-17). These **ribosomes** are exceedingly rich in ribonucleic acid (RNA) and are the main site for the synthesis of proteins, especially enzymes. One type of RNA, termed messenger RNA (mRNA), presumably acts as templates that determine the sequence in which amino acids are linked together to form proteins. This function will be discussed in more detail in Section 18.20. The ribosomes appear to be formed at the nucleoli and migrate out through pores in the nuclear envelope to their eventual location; at least there is some evidence that ribosomal RNA (rRNA) is derived from the chromosomal DNA in a special region related to the nucleolus. Ribosomes are also present in mitochondria and in plastids but are smaller than the ribosomes located in the cytoplasm.

Electron microscopy has revealed that ribosomes often occur in clusters or groups, called **polysomes** (or **polyribosomes**), held together by mRNA. It is believed that the several ribosomes are attached to a single strand of mRNA and that each synthesizes a separate protein. The information for such syntheses is carried by mRNA.

Plastids. Frequently present in the cytoplasm are one to many spherical, ovoid, or lens-shaped organelles (usually 4 to 10 μm in diameter and about 1 to 2 μm thick), which are clearly visible without staining. These **plastids** are sometimes collar-shaped, ribbonlike, or even very irregular in shape, but this is usually true only in some simple plant types. There are three kinds of plastids, all bounded by two membranes, and a cell may contain more than one kind: **chloroplast** (green), **chromoplast** (red, orange, or yellow), and **leucoplast** (colorless). The green color of plants, as in leaves, is caused by the presence of the green pigments, **chlorophyll a** and **chlorophyll b,** contained in chloroplasts. The yellow carotenoid pigments, **carotene** and **xanthophyll,** are also present in these plastids, but their color is usually masked by the chlorophylls. The chlorophyll absorbs light energy, which enables the plant to manufacture food.

With the aid of the electron microscope, the chloroplasts of mosses, ferns, and seed plants are found to be quite similar. There is an elaborate series of double membranes, **lamellae,** extending through the chloroplast. At intervals, additional membrane-bound flattened compartments or vesicles are formed and appear as stacked discs. These discs are interconnected by other lamellae; a stack of discs is termed a **granum** (plural, **grana**). The pigments of the plastid and various enzymes are located in the lamellae. The region between the grana and lamellae is the

stroma, and other enzymes are located in this area. (See Figure 4-15). The lamellar structure appears to arise by invagination of the inner chloroplast membrane. Ribosomes and DNA, distinct from cytoplasmic ribosomes and from the chromosomal DNA, have been found in chloroplasts and indicate considerable synthetic ability for these organelles.

Not all chloroplasts have the same internal structure. In the green and brown algae the plastids have outer membranes and lamellae but no grana. In the blue-green algae there are no plastids as such; the pigments are located in lamellae that extend through the cytoplasm. Lamellae are poorly developed in leucoplasts and chromoplasts, and no grana are formed.

The color of a chromoplast depends primarily upon the presence of carotene and xanthophyll. Such plastids are found in fruits, flower petals, and some other plant parts (e.g., carrot root). These pigments may participate in food manufacture when located in chloroplasts by absorbing light energy and transmitting this excitation to chlorophyll, but their significance in chromoplasts is not clear. Carotene is converted in animals to the essential vitamin A, but this vitamin is not necessary for normal functioning in plants. Leucoplasts contain no pigments and are located most commonly in storage cells where they act as centers for starch or fat (lipid) formation (food storage products). If the leucoplasts are specialized as starch-storing bodies, they are termed **amyloplasts;** if fats are formed, they are called **elaioplasts.** The cells of a potato tuber are gorged with starch grains, which, on careful observation, can be seen to be enclosed by the leucoplast membrane (Figure 4-16).

In some cells one type of plastid may be converted to another. For example, a developing tomato fruit contains leucoplasts that develop into chloroplasts and eventually into chromoplasts by the loss of chlorophyll and the accumulation of carotenoids. It has also been shown that the chromoplasts of carrot roots may develop into chloroplasts when exposed to light.

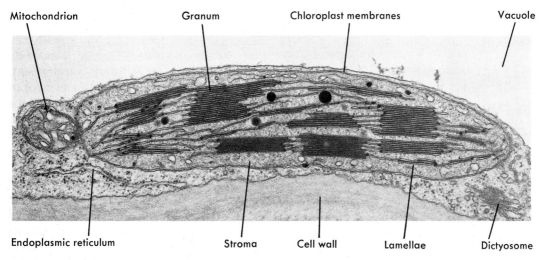

Figure 4-15. Electron micrograph of a chloroplast and mitochondrion in a tobacco (*Nicotiana tabacum*) leaf vascular parenchyma cell (×30,000). (Courtesy of Katherine Esau and Robert Gill.)

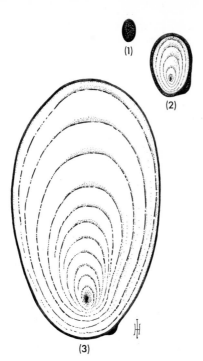

(1)

(2)

(3)

Figure 4-16. Starch grain forming within a leucoplast. The outermost margin represents the membrane of the plastid.

In some algae and mosses, chloroplasts multiply by division, in which the plastid constricts in the middle and separates into two parts, which then enlarge; this may be true of all plastids. In seed plants there is evidence that plastids arise from smaller, colorless bodies, termed **proplastids** (about 1—3 μm), which are capable of dividing. As the proplastids enlarge and develop into chloroplasts, the typical grana structure forms only in the presence of light. Chloroplast differentiation is also inhibited by deficiencies of oxygen, by the lack of certain essential mineral elements, or by genetic deficiencies (as in albino organisms or tissues).

Dictyosome. The electron microscope with its exceedingly high magnifications made it possible to demonstrate that **dictyosomes** (about 2 μm in diameter and about 0.5 μm in width) are universally present in plant cells (Figure 4-17). The

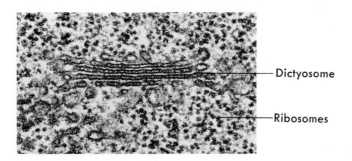

Dictyosome

Ribosomes

Figure 4-17. Electron micrograph of a dictyosome and ribosomes in a bean (*Phaseolus vulgaris*) root tip parenchyma cell ($\times$ 60,000). (Courtesy of Katherine Esau and Robert Gill.)

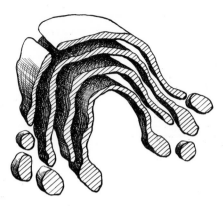

Figure 4-18. Diagrammatic representation of a dictyosome and vesicles.

central portion of this structure consists of a series of stacked discs, each composed of a flattened sack bounded by two membranes. Around the edges of these discs are small globular bodies, or vesicles, which appear to be pinched off from the margins of the discs (Figure 4-18). Dictyosomes are also called **Golgi bodies;** the term **Golgi apparatus** is used to refer collectively to all of the Golgi bodies, or dictyosomes, of a given cell.

The dictyosome is believed to be involved in the formation of the new wall when cells divide and in the general synthesis of wall material. The cell plate probably grows by the fusion of vesicles derived in part from the dictyosomes and in part from the endoplasmic reticulum. It also appears, from the results of some experiments, as though material collects in the enlarging vesicles of the dictyosome, which then move to the plasma membrane. Here the membrane of the vesicle fuses with the plasma membrane, and the contents are extruded and become part of the cell wall. In secretory cells this organelle may be involved in the secretion of macromolecules. It is hoped that the continuing investigations will eventually clarify the functioning of the dictyosomes.

Microtubules and microbodies. Present in the cytoplasm are small structures, the **microtubules,** of indefinite length and about 25 nm in diameter (Figures 4-19 and 4-20). They are found in the spindle fibers during mitosis (see Section 4.4), at the expanding margin of the cell plate, and in cilia and flagella. These latter are hairlike extensions responsible for the motility of certain cells. Although their specific function is not known, the microtubules may participate in the conduction of materials through cells. They are also believed to be involved in the orderly growth of the cell wall by controlling the alignment of the cellulose microfibrils in the various cell wall layers. Microtubules also appear to direct dictyosome vesicles toward the developing cell wall and may be responsible for the orientation of other cytoplasmic components (e.g., nucleus, plastids, mitochondria).

Various spherical organelles, about the size of mitochondria or smaller, are also located in the cytoplasm. These **microbodies** (0.5–1.5 μm) are bounded by a single membrane (Figure 4-13) and apparently contain oxidative or hydrolytic enzymes. There are probably several kinds of microbodies, but their function is not clear.

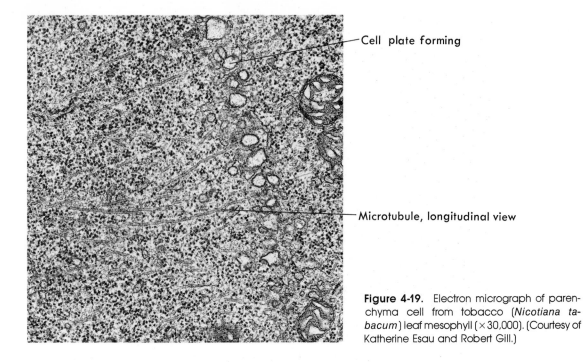

Cell plate forming

Microtubule, longitudinal view

Figure 4-19. Electron micrograph of parenchyma cell from tobacco (*Nicotiana tabacum*) leaf mesophyll (×30,000). (Courtesy of Katherine Esau and Robert Gill.)

Inclusions

Present within living cells are various nonliving structures, generally termed **inclusions.**

Vacuoles. In a mature cell the cytoplasm, nucleus, and plastids are frequently located adjacent to the cell walls, while the greater portion of the cell is occupied by one or more large cavities called **vacuoles.** Within the vacuoles is a fluid, the **cell**

Plasma membrane

Endoplasmic reticulum   Cell wall   Microtubule, cross-section view

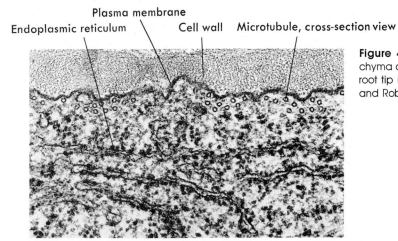

Figure 4-20. Electron micrograph of parenchyma cell from tobacco (*Nicotiana tabacum*) root tip (×50,000). (Courtesy of Katherine Esau and Robert Gill.)

sap, composed of water and a great variety of dissolved materials (i.e., sugars, inorganic salts, organic acids and their salts, pigments, alkaloids, and various other water-soluble materials). The pigment found most frequently is **anthocyanin,** which is responsible for many of the red, purple, and blue colors in flower petals, leaves, and even roots (e.g., beet); the color varies with the acidity of the cell.

Vacuoles are storage areas for food materials and depots for waste materials. In addition to this, they are important in the maintenance of cell turgidity, because of the presence of the differentially permeable **vacuolar membrane** (or **tonoplast**), which is really the inner boundary of the cytoplasm. The outer boundary of the cytoplasm, adjacent to the cell wall, is a similar membrane called the **plasma membrane.** The movement of materials through cell membranes will be discussed later (see Chapter 5).

Crystals. In the cells of various parts of the plant may be found **crystals,** differing considerably in size and shape (see Figure 4-21), which appear to be waste products, or excretory products, of the protoplast. They are located primarily in the vacuoles and usually are calcium oxalate crystals, though others are sometimes found.

Stored foods. Visible in many cells are **starch grains** (as in the cells of potato tubers and many green leaves), **oil globules** (as in the cells of avocado fruits and diatoms), and **aleurone grains** (stored protein found in the cells of many seeds; for example, wheat). These materials are formed when excess foods are produced by the plant, and may be broken down and utilized by the plant during those periods when food is not being produced in sufficient quantities (or when the seed germinates or the tuber sprouts to form a new plant).

Less frequently found in cells are tannins, gums, resins, and mucilages. These are probably all waste products resulting from the physiological activities of plant cells. Some of these materials are important in water retention (see Section 15.5).

Figure 4-22 is a schematic diagram of cell structure, based upon studies using the electron microscope. The electron micrographs, obtained during these investigations, indicate that the structure of the cell is exceedingly complex. The membranes, for example, extend throughout the cytoplasm as a fine network (the endoplasmic reticulum) from the outer cell membrane to the nuclear membrane. The basic components of the cell are, therefore, linked together by this framework. Also, the plasma membranes extend through the plasmodesmata from one cell to another. Although the cytoplasm appears to be a fluid mass when viewed with a light microscope, the greatly higher magnifications possible with an electron microscope reveal the orderly and integrated structural arrangements and compart-

Figure 4-21. Types of crystals frequently found in cells.

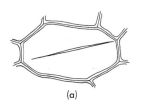
(a)

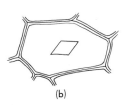
(b)

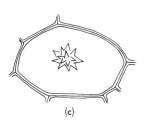
(c)

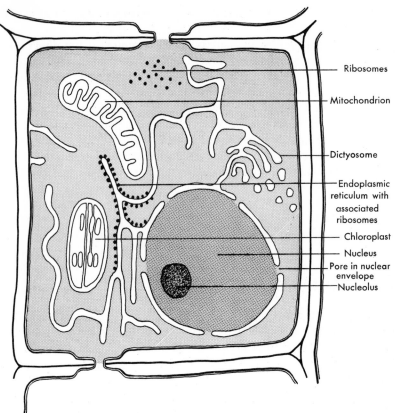

Figure 4-22. Schematic diagram of cell structure based upon studies using the electron microscope.

Ribosomes

Mitochondrion

Dictyosome

Endoplasmic reticulum with associated ribosomes

Chloroplast

Nucleus

Pore in nuclear envelope

Nucleolus

mentalizations that are actually present. The sequential reactions that are known to occur in cells are more readily understandable in the light of this knowledge of structural organization.

4.3 Prokaryotic Cells

The cell structure and organelles that have just been described are characteristic of **eukaryotic cells;** those which have membrane-bound nuclei. Most organisms are composed of such cells. However, the cells of some very simple organisms, the bacteria and blue-green algae, do not have such a nucleus; neither do these **prokaryotic cells** have plastids, dictyosomes, or mitochondria.

The DNA is present as thin strands and not associated with protein as in the eukaryotic cells. In bacteria there is a single, long, circular DNA molecule that contains the genetic information of the cell. Occasionally there may be two or four DNA molecules because the DNA replicates before the cell divides. Upon cell division, the DNA normally is shared equally by the daughter cells. The plasma membrane often contains many folds and convolutions that extend into the interior of the cell. In photosynthetic bacteria and the blue-green algae the pigments are in membranes that lack grana and extend through the cytoplasm, often parallel with the plasma membrane.

The ribosomes of prokaryotic cells are smaller than those of the eukaryotic cells, except for those in chloroplasts and mitochondria. As mentioned previously, the DNA in these latter two structures is quite distinct from chromosomal DNA. It is similar to prokaryotic DNA however. Because of the similarity between the ribosomes and DNA of chloroplasts and mitochondria with those of prokaryotic cells, as well as other types of evidence, it has been suggested that the invasion of larger cells by early bacteria and blue-green algae resulted in the origin of mitochondria and chloroplasts, respectively. The smaller cells were able to trap and convert energy from the surroundings, and this ability would then be available to the larger cells. The latter might protect their ''organelles'' from environmental extremes. These larger cells are considered to be the forerunners of the eukaryotes.

4.4 Cell Division

All plants are composed of cells, and, at least in the multicellular organisms, the number of cells increases greatly as the plant grows. Since new cells develop from preexisting ones, some means must exist for an individual cell to form more cells. This process is cell division. Cell division includes two distinct processes: **mitosis,** or nuclear division, and **cytokinesis,** which involves the division of the remainder of the cell and the formation of a new wall. Time-lapse photography[2] has enabled the biologist to speed up the apparent rate of cell division so that the tremendous activity can be observed more readily. It is fascinating to watch—much more so than one can possibly put in words on a printed page.

Mitosis

The division of a nucleus is a continuous process, and no pauses delimit the various phases that will be discussed. However, for ease of discussion, biologists subdivide mitosis, more or less arbitrarily, into several phases. This is similar to discussing the phases of the moon; one phase actually blends into the next. The entire process of mitosis may take from 30 minutes to 22 hours or more and is usually more rapid at higher temperatures.

Metabolic phase, or interphase. This phase refers to the condition of a nucleus when it is not dividing. Such a nucleus has at times been referred to as a ''resting nucleus,'' but this term is not acceptable since living activities do not stop merely

---

[2] In time-lapse photography motion-picture film is exposed one frame at a time with a lapse of one or more minutes between exposures. When the film is projected at normal speed, all activity is shown much more rapidly than it actually occurs. Such photography can greatly condense movements that take 30 to 60 minutes or even longer. The opening of buds is frequently photographed in this fashion to emphasize the fantastic movements undergone by unfolding leaves and petals.

because a cell is not dividing. In fact, it is during the metabolic phase that new nuclear material (specifically DNA) is being synthesized. The appearance of the nucleus was indicated earlier in this chapter and is shown in Figure 4-23 and in more diagrammatic form in Figure 4-24(a).

Prophase (Figures 4-23 and 4-24[c]).   The first indication that a nucleus is about to divide is the shortening and thickening of the chromosomes as the chromatin strands become more tightly coiled and a matrix forms about each (see Figure 4-8). The nuclear envelope and the nucleolus disappear, and **spindle fibers** become visible. These fibers extend from one end, or **pole,** of the cell to the other in the

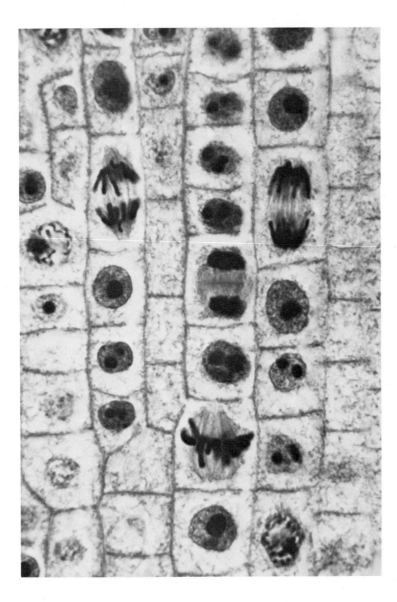

Figure 4-23. Photomicrograph or root tip cells in the process of mitosis. In the row at the left, just above the center, is a cell in prophase. In the next row, slightly higher, is a cell in early anaphase. A telophase with the new cell wall forming and a metaphase are visible in the central row of cells (×650). (Courtesy of General Biological Supply House, Chicago.)

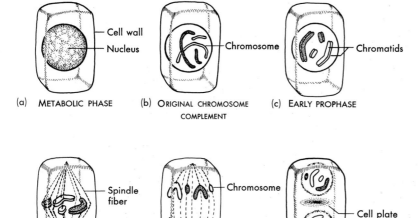

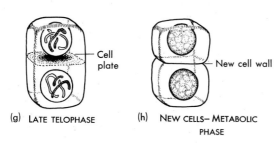

(a) METABOLIC PHASE   (b) ORIGINAL CHROMOSOME   (c) EARLY PROPHASE
                          COMPLEMENT

Cell wall
Nucleus
Chromosome
Chromatids

(d) METAPHASE   (e) ANAPHASE   (f) EARLY TELOPHASE

Spindle fiber
Chromosome
Cell plate forming

(g) LATE TELOPHASE   (h) NEW CELLS– METABOLIC PHASE

Cell plate
New cell wall

Figure 4-24. Cell division. In **(a)** and **(h)**, the chromosomes are in the greatly expanded and indistinct condition typical of nuclei that are not dividing. In the other diagrams the chromosomes and chromatids are shown as distinct entities. In **(b)** the original chromosome complement is shown for comparison with those in **(g)**. In **(c)** the doubled condition of the chromosomes is visible. The formation of the new cell wall is shown in **(f)**, **(g)**, and **(h)**.

shape of a spindle. This is a three-dimensional structure that is thickest in the equatorial plane and tapering to a point at each end. Electron microscopy has shown that each spindle fiber consists of a bundle of microtubules. The chromosomes now move toward the central, or equatorial, region of the cell, and it is apparent that each chromosome has doubled (reproduced itself). This duplication has actually taken place during the previous metabolic phase. Careful measurements of nuclei have shown that the deoxyribonucleic acid (DNA) content, one of the main constituents of chromosomes, has doubled shortly before prophase begins. Each longitudinal half of a doubled chromosome is termed a **chromatid,** and each chromatid is an exact duplicate of the original chromosome. This duplication possibly is brought about by the original chromosome functioning as a template or pattern, to which ''building blocks'' are added from the molecules abounding in the nucleus (see Section 4.5). The **kinetochore** (or **centromere**) is the region of the chromosome where the chromatids are joined until anaphase and where a spindle fiber is attached. Not all fibers are attached to chromosomes, some extend continuously from pole to pole. The kinetochore has a characteristic location on each chromosome and divides the chromosome into two arms of varying lengths.

Metaphase (Figures 4-23 and 4-24[d]). When the kinetochores of the duplicated chromosomes migrate to and reach the equatorial region, prophase is considered to be over, and nuclear division has reached the metaphase stage. The chromosomes vary considerably in length, and frequently, though the kinetochores are gathered in the central region, the "arms" of the chromatids may dangle in various directions.

Anaphase (Figure 4-23 and 4-24[e]). The kinetochore now duplicates and separates. As a result, the two parts (chromatids) of the duplicated chromosomes separate, forming two identical groups, each going toward a different pole of the cell. Once the chromatids separate, each is considered to be a distinct chromosome (a matter of terminology). Contraction of the fibers may aid in this movement of the chromosomal groups. In any event, the kinetochores "lead" during this movement, and the arms of the chromosomes appear to be pulled along. At the termination of anaphase, one group of chromosomes is located at each pole of the spindle.

Telophase (Figures 4-23 and 4-24[f]). As the spindle fibers disappear, the chromosomal groups at each pole of the cell become reorganized into new nuclei by the development of nuclear envelopes surrounding each group and by the reappearance of the nucleoli. The membranes appear to develop from the endoplasmic reticulum, which remains relatively intact during mitosis. The chromosome mass disperses to give rise to the chromatin material.

It should be emphasized that, because the original chromosomes duplicated, each of these two chromosomal groups at the poles has the same number and same kind of chromosomes as were present in the cell after the previous mitosis. When DNA measurements are made of nuclei in telophase, it is found that the amount in each nucleus is just half of that found in the original nucleus of the prophase, which marked the beginning of mitosis. This indicates that the duplicated chromosomes (chromatids) have separated, and that reproduction of chromosomes occurs during interphase.

Cytokinesis

As mitosis nears completion, vesicles, apparently produced by the dictyosomes and the endoplasmic reticulum, appear across the equatorial region. These appear as small granules when viewed with the relatively low power of the light microscope. As additional formation occurs, these vesicles become larger, more dense, and eventually coalesce to form a **cell plate,** which extends across the cell from wall to wall. The cytoplasm is thus divided into two portions, each containing one of the new nuclei, and the cell plate becomes the middle lamella of the new cell wall. The two protoplasts of the resultant daughter cells deposit additional cell wall layers on the middle lamella, and cytokinesis is completed (see Figures 4-2, 4-3, 4-23, and 4-24[f], [g], [h]). In animal cells and in some plant cells, cytokinesis is accomplished by a pinching in of the parent cell until two new cells are formed. Multinucleate cells result if cytokinesis does not follow mitosis, a condition found in various simple plants.

Significance of Mitosis

As a result of the duplication of chromosomes and the subsequent and separation of chromatids, each daughter cell contains the same **number** and the same **kind** of chromosomes as did the original (parent) cell. Since genes, which determine hereditary characteristics, are located within the chromosomes, each daughter cell has the same potentialities as the parent cell. The daughter cells enlarge to a mature size, at which time they also may undergo cell division. Enormous numbers of cell divisions occur during the development of the large multicellular organisms found on this earth, and each such division occurs basically as just discussed.

4.5 Chromosome Duplication

Although we are certain that chromosomes, and thus genes, are duplicated when a cell divides, the exact mechanism involved in such duplication is not known. Recent investigations, however, have suggested a possible answer based upon the molecular components of chromosomes.

The principal constituents of chromosomes are **proteins** and **nucleic acids,** combined as **nucleoproteins,** the nucleic acid fraction being the key in the transmission of hereditary characteristics from cell to succeeding cells. Nucleic acids are long sequences of **nucleotides,** just as proteins are long sequences of amino acids, and have two forms: **ribonucleic acid (RNA),** which occurs mainly in the cytoplasm, although it is synthesized in the nucleus, and **deoxyribonucleic acid (DNA),** found primarily in the chromosomes. The nucleotides of DNA consist of a phosphate group, **P,** linked to the sugar, deoxyribose, **S,** which is bound to any one of four different nitrogenous bases: adenine (**A**), guanine (**G**), cytosine (**C**), or thymine (**T**). Figure 4-25 represents a portion of the DNA molecule. This molecule is very long and has an enormously great number of possible sequences of nucleotides, even though they differ only in which of the four nitrogenous bases are involved. In other words the nucleotide sequence could be assembled in any order, and the order would be different for each separate DNA moleucle. One might be **TCCGAGTACG,** and so on, whereas another might be **AGCTTCGAGT,** and so on. Besides the great variety of arrangements that are possible, the situation is made even more complex by the fact that the DNA molecule actually consists of two parallel portions held together by hydrogen bonds. To oversimplify for a moment, we can think of the molecule as having the appearance of a ladder. The vertical and parallel sides of the "ladder" are made up of alternating sugar and phosphate groups with the phosphate of one nucleotide linked to the sugar of the next nucleotide. The "rungs" of the ladder are formed by the nitrogenous bases with two bases forming each rung. The paired bases are held together by weak hydrogen bonds between adenine and thymine or between guanine and cytosine; no bonds form between **A** and **G, A** and **C, T** and **G,** or **T** and **C.** The parallel portions actually twist around in a helical fashion, but for clarity they have been shown untwisted in this figure. The longitudinal pattern of **A-T** and **G-C** bonds apparently

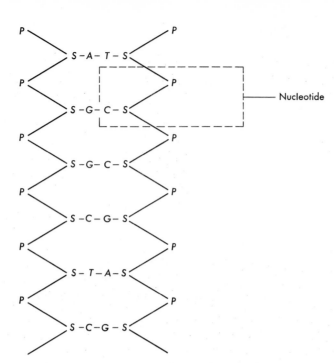

Figure 4-25. A representation of a portion of a deoxyribonucleic acid (DNA) molecule.

Nucleotide

determines the structure of proteins, including enzymes, that are important in determining the characteristics of an organism (see Section 18.20).

Adenine and guanine are **purines,** while cytosine and thymine are **pyrimidines** (Figures 4-26 and 4-27). Examples of nucleotides, utilizing adenine and thymine as the nitrogenous bases, are shown in Figure 4-28. The linkage of two nucleotides is shown in Figure 4-29; additional nucleotides would link at the positions indicated by R_1 and R_2 to form part of one strand of the large deoxyribonucleic acid molecule. Probably more than 1,000,000 nucleotides are linked in each DNA. Remember that the DNA molecule is two stranded and coiled like a circular staircase, forming a double helix, the "railing" on either side corresponding to the phosphate-sugar linkages and the "steps" consisting of the linked nitrogenous bases. In this depiction one half of each "step" is a purine base connected by hydrogen bonding to a pyrimidine base, which forms the other "half-step." The strands are

Figure 4-26. Purines found in nucleic acids.

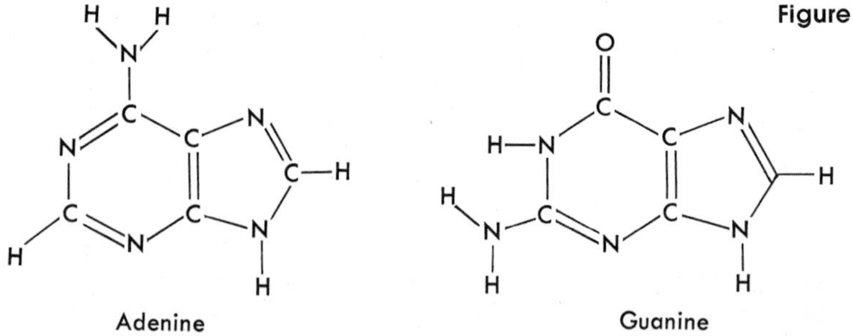

Adenine
$(C_5 H_5 N_5)$

Guanine
$(C_5 H_5 N_5 O)$

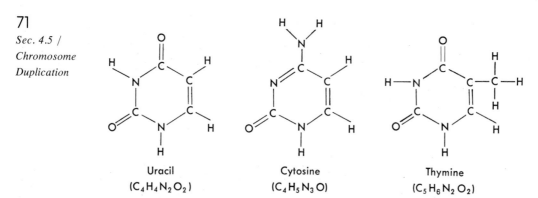

Figure 4-27. Pyrimidines found in nucleic acids.

Uracil
$(C_4H_4N_2O_2)$

Cytosine
$(C_4H_5N_3O)$

Thymine
$(C_5H_6N_2O_2)$

a uniform distance apart along the entire length of the molecule, a situation that could be accounted for only if a purine is paired with a pyrimidine as shown in Figure 4-30. Although hydrogen bonds are relatively weak, the large number of such bonds along the length of the DNA double helix gives a high degree of stability to the molecule. As a result, the structure is fairly stable so that heritable changes occur only rarely.

Many experiments with a variety of bacteria have shown that DNA extracted from one type can have a hereditary influence on bacteria of another type. For example, if DNA extracted from type *A* bacteria is mixed with a culture of type *B* bacteria, some type *A* bacteria can subsequently be found growing in the medium. The extracted DNA has transformed a few type *B* to type *A* bacteria. The modern

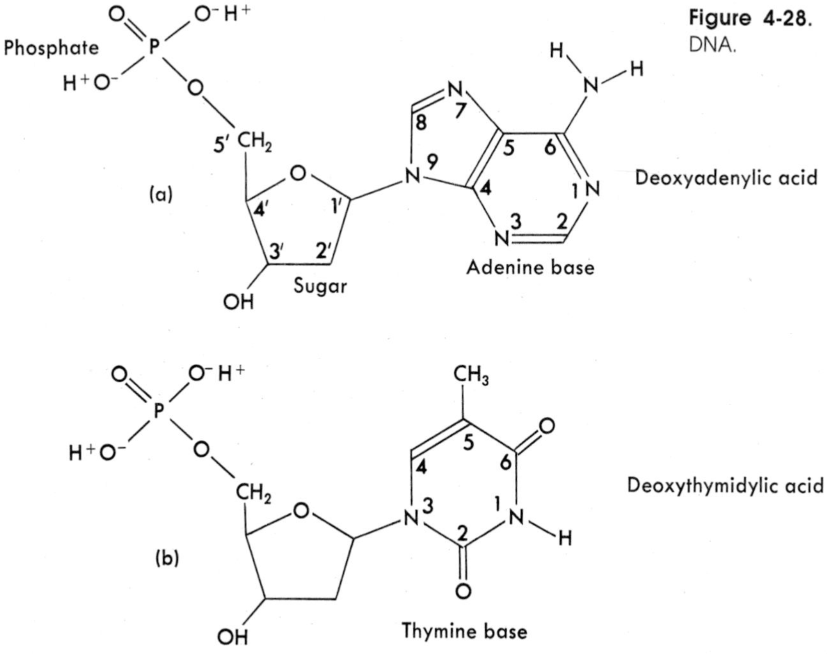

Figure 4-28. Examples of nucleotides in DNA.

Phosphate

(a)

Sugar

Deoxyadenylic acid

Adenine base

(b)

Deoxythymidylic acid

Thymine base

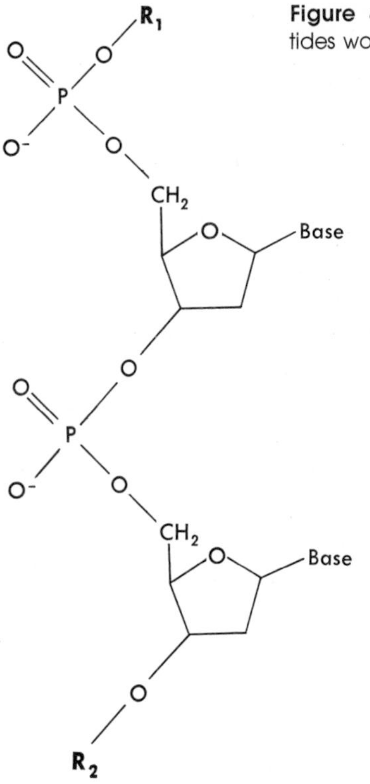

Figure 4-29. Linkage between nucleotides; additional nucleotides would link at the positions indicated by R_1 and R_2.

interpretation is that some of the DNA (type *A*) entered a few type *B* cells, was incorporated into the type *B* chromosome, and thus changed the genetic makeup of these cells. Labeling the transforming (extracted) DNA with radioactive phosphorus results in incorporation of this radioactivity into recipient bacteria. Such experiments are strong evidence that DNA is the genetic material. It is the master informational molecule of the cell, containing all the genetic information required for the production of each type of cell from generation to generation. This will be considered in more detail in Chapter 18, especially Section 18.20.

With this knowledge of DNA structure and its importance in governing the development of organisms, a very interesting suggestion has been made as to the possible mechanism of chromosome duplication. The bonds between the nitrogenous bases are weak and easily broken, as represented in Figure 4-31, giving rise to "half-chains" of DNA. Present within the nucleus, floating around in the nuclear karyolymph, are many nucleotide building blocks, which are synthesized by enzymes within the cell. Whenever nucleotide P-S-A collides with the P-S-T nucleotide portion of the half-chain, a bond forms; the same thing occurs when P-S-T collides with P-S-A, P-S-G with P-S-C, or P-S-C with P-S-G. The newly adjacent phosphate and sugar groups then link together. Each half-chain gradually accumulates an exact replica of the half-chain from which it separated, and, as a result, two DNA molecules—exactly alike and the same as the original, and each composed of half of the original—are formed and replace the original one. It appears that separation (unwinding really) and synthesis go on together so that only a

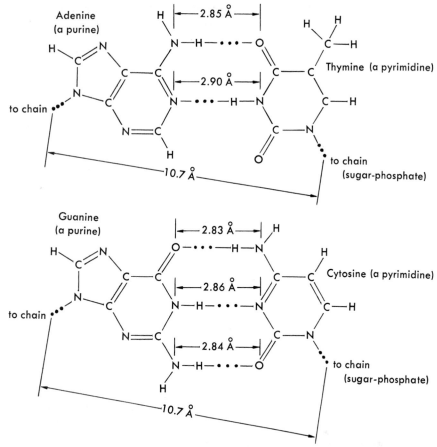

Figure 4-30. Hydrogen bonding between two nitrogenous bases. A purine is always paired with a pyrimidine, adenine with thymine, and guanine with cytosine.

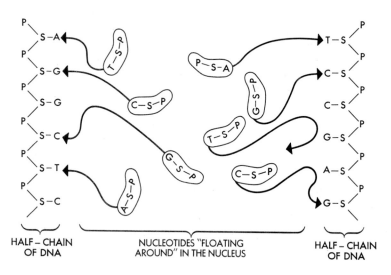

HALF – CHAIN
OF DNA

NUCLEOTIDES "FLOATING
AROUND" IN THE NUCLEUS

HALF – CHAIN
OF DNA

Figure 4-31. Possible mechanism of DNA duplication. The bonds between the nitrogenous bases (designated by letters A, T, G, C) have broken, giving rise to "half-chains" of DNA. The nucleotide building blocks are shown attaching to these half-chains. Exact duplicates accumulate, and two DNA molecules result.

small part of the chain is ever single stranded at any one time (Figure 4-32). No bonds form if two purines or if two pyrimidines collide; a purine is always paired with a pyrimidine.

A "mistake" in copying as DNA duplicates could cause the resultant cells to be quite different from the original cell because the DNA would be different, and this is the material that carries information. For example, if a P-S-T ends up next to a P-S-G, no bond will form, but the half-chains may be kept together by all of the *other* bonds between them. In this case the mistake is generally not sufficiently great so as to result in death. When the next duplication occurs, the "mistake" is perpetuated, and differences in subsequent cells may sometimes be apparent as variations in basic characteristics. (See Section 28.11 for a discussion of mutations.)

The information contained in chromosomal DNA is carried to the cytoplasm by ribonucleic acid molecules, called messenger RNA, as will be considered in more

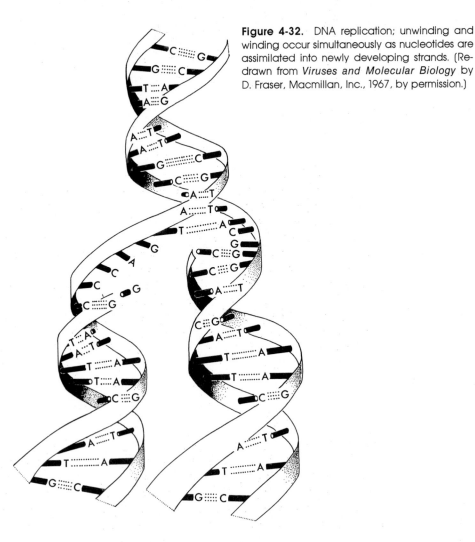

Figure 4-32. DNA replication; unwinding and winding occur simultaneously as nucleotides are assimilated into newly developing strands. (Redrawn from *Viruses and Molecular Biology* by D. Fraser, Macmillan, Inc., 1967, by permission.)

detail later (Section 18.20). RNA is very similar to DNA; the sugar is ribose, rather than deoxyribose (which has one less oxygen atom), and uracil substitutes for thymine as one of the nitrogenous bases. RNA also consists of a single strand rather than two parallel strands as found in DNA. Some of this RNA passes through the nuclear envelope to the cytoplasmic ribosomes where it is involved in protein synthesis. Since the formation of RNA is controlled by DNA, the latter effectively governs protein synthesis.

Summary

1. Organisms are composed of functional units called cells, which may be variously modified, and all cells come from preexisting cells.

2. The generalized plant cell consists of a nonliving wall enclosing the protoplast. The wall is either two or three layers thick and contains certain thin areas, or pits. The more fluid portion of the protoplast is the cytoplasm, within which various formed bodies are located. The latter are the nucleus, mitochondria, ribosomes, plastids, and dictyosomes. Various nonliving inclusions may also be present in the cytoplasm (e.g., vacuoles, stored foods, and waste products).

3. Nuclear division is termed mitosis and consists of several intergrading phases: prophase, metaphase, anaphase, and telophase. Cytokinesis refers to the division of the remainder of the cell and the formation of a new cell wall. Chromosomes are duplicated prior to nuclear division.

Review Topics and Questions

1. Diagram a generalized plant cell and label fully.
2. Discuss the functions of the various parts of a generalized plant cell.
3. Describe the changes that take place within a cell during prophase of mitosis, during anaphase, during telophase.
4. Compile three lists, indicating which structures of a green plant cell are visible (a) with an ordinary light microscope, (b) with an ordinary light microscope plus staining the cell, and (c) with an electron microscope.
5. A thin section is made through a plant tissue. When this is examined microscopically, some of the cells are observed to be without nuclei. Suggest two possible explanations.
6. What are plasmodesmata and what is their importance?
7. Cytoplasm is frequently seen to be in motion. What is the possible significance of this movement?
8. How can cells grow if the cell walls are rigid? Is there any practical advantage to the plant of having rigid cell walls?
9. Explain why the term *metabolic nucleus* is more appropriate than *resting nucleus* in describing the nucleus of a cell during interphase (i.e., between divisions).

10. Discuss the reasons for considering a multicellular plant as an organic whole rather than as a community of isolated cells.
11. Discuss the manner in which chromosomes are believed to duplicate. What is the evidence that duplication occurs during interphase? What is the importance of such duplication?

Suggested Readings

BRACHET, J., "The Living Cell," *Scientific American* **205,** 50–61 (September 1961).

BRYAN, J., "Microtubules," *BioScience* **24,** 701–711 (1974).

CAPALDI, R. A., "A Dynamic Model of Cell Membranes," *Scientific American* **230,** 26–33 (March 1974).

CRICK, F. H. C., "The Structure of the Hereditary Material," *Scientific American* **191,** 54–61 (October 1954).

Fox, C. F., "The Structure of Cell Membranes," *Scientific American* **226,** 30–38 (February 1972).

JENSEN, W. A., *The Plant Cell,* 2nd ed. (Belmont, Calif., Wadsworth, 1970).

LOEWY, A. G., and P. SIEKEVITZ, *Cell Structure and Function,* 2nd ed. (New York, Holt, Rinehart & Winston, 1969).

MAZIA, D., "The Cell Cycle," *Scientific American* **230,** 54–64 (September 1974).

MORRISON, J. H., *Functional Organelles* (New York, Reinhold, 1966).

RICH, A., "Polyribosomes," *Scientific American* **209,** 44–53 (December 1963).

SWANSON, C. P., and P. L. WEBSTER, *The Cell,* 4th ed. (Englewood Cliffs, N.J., Prentice-Hall, 1977).

TAYLOR, J. H., "The Duplication of Chromosomes," *Scientific American* **198,** 36–42 (June 1958).

5
Diffusion and the Entrance of Materials into Cells

Many materials pass into and out of the various cells of a plant when it is situated in a suitable environment. For example, in vascular plants water and mineral salts enter the roots from the soil and move through various cells as these materials become distributed within the plant. Carbon dioxide gas, on the other hand, enters the leaves from the atmosphere. The plant also loses various materials—water vapor and, at times, oxygen and carbon dioxide—to the atmosphere. Many such movements result from the diffusion of molecules or ions from one area to another. Gases move through the intercellular spaces, and water and dissolved materials move within the cells. Some of these movements within the plant are accomplished by more complicated mechanisms than diffusion, but these will be discussed in Chapter 8. For the present, we shall be concerned with diffusion, since few, if any, of the physiological processes occurring in plants do not involve diffusion at least in part.

5.1 Diffusion

In order to understand more clearly what is happening when a substance diffuses from one area to another, imagine a large auditorium that contains a great number of small rubber balls bouncing about in all directions. A ball's direction of movement will change only upon a collision with a wall or another ball. The result is a zigzag fluctuating movement, and the room becomes filled with flying rubber balls. Similarly, any chamber becomes filled with gas molecules (=rubber balls) because of their inherent motion. The average speed of the entire group will remain constant as long as the temperature or pressure does not change. An increase in temperature would result in an increased **kinetic energy** (or motion energy).

Now, if the wall between this auditorium and an adjacent one were removed, some of the balls would pass into the empty room because their motion would no longer be impeded by the wall that has been removed. The random movement finally results in an equal number of balls in each room. At this time, the number moving from the first room to the second will be equal to the number moving in the opposite direction, a condition termed **dynamic equilibrium.** Two things should be emphasized: (1) movement is still taking place from one area (room) to the other, and (2) the concentrations in the two areas are equal. The term **diffusion** is best used for the movement of molecules or ions from one location to another, while **net diffusion** is used for the direction of movement of the greatest number of molecules. In other words, net diffusion of the rubber balls was from the first room into the second, and net diffusion ceased (or is zero) at equilibrium, while diffusion continues even at equilibrium. That is, there is still a movement from one area to the other even though such movement is equal in the two directions.

That such diffusion of gases occurs throughout a room can be demonstrated quite readily by opening a bottle of perfume (or ether) at one end of a room. Persons at that end of the room will detect the odor before individuals at the other end, but eventually the entire room will be permeated with the odor. Such movement is due to random molecular motion (see Section 3.1). Diffusion is also exhibited by molecules of liquids and solutes (materials dissolved in liquids). An example of the diffusion of a solute can be obtained by dropping a cube of sugar in a cup of coffee. Careful sipping at the lip of the cup without tipping (to prevent stirring of the liquid) allows one to taste the increasing sweetness as sugar molecules diffuse, after the sugar dissolves, from the bottom of the coffee to the top layers. In all cases, the rates of diffusion are relatively slow because of the many collisions that occur and because of the vast spaces, in comparison to molecular dimensions, that humans use in measuring distances.

The direction of diffusion of one material is independent of the direction in which a second material is diffusing. The rate may slow down a bit because of collisions between the two different molecules, but the direction depends upon factors concerning each type of molecule separately. Quite possibly, and in fact usually, as certain molecules diffuse into a cell, others are diffusing out.

Diffusion and mass movements should be clearly distinguished from one another. If in our example of solute diffusion the coffee had been stirred with a spoon, sugar molecules would have been carried about by the mass movement of the coffee, much as twigs are carried by streams of water. Air currents would have the same effect on gas molecules.

Gases may exert considerable pressures upon the walls of containers. For example, if a rubber balloon is placed in a closed bottle of carbon dioxide gas, the balloon becomes distended because rubber is readily permeable to carbon dioxide molecules. If the temperature remains constant, the pressure exerted by any gas is directly proportional to the number of molecules per unit volume, in other words, its concentration. The net diffusion of gases results from a difference in gas pressures or concentrations in the two areas under discussion. When the pressures are equal, the condition is a dynamic equilibrium. Such pressures, in discussing diffusion, are termed **diffusion pressures.**

5.2 Factors Influencing Rate of Diffusion

Every component of a system possesses free energy, or molecular activity that is capable of doing work, and diffusion phenomena should really be considered in terms of differences in free energies. In other words the net diffusion of gases, or other substances like water, is from a region of its greater free energy (or total molecular activity) to a region of its lesser free energy. All those factors that influence diffusion are effective because they bring about changes in free energy.

Temperature

An increase in temperature results in an increase in the speed of diffusion. This is because of the increased activity of molecules as the temperature is raised; the average speed with which a molecule moves is greater at high temperatures than at low temperatures. Another way of putting it is that free energy increases directly with temperature, and there will be a net movement of molecules toward the regions of lower temperature (that is, lower free energy).

Concentration Gradients

Free energy also increases with an increase in concentration of the molecules, and differences in quantity of free energy in different parts of a system will occur if there are differences in concentrations. The greater the difference in free energy between two regions, the more rapid will be the net diffusion of molecules. However, this rate is also influenced by the distance between the two regions, so that the rate decreases as the distance increases. The term **concentration gradient** implies the application of concentration differences over a specific distance. To visualize what is happening, imagine a stone poised on top of a hill and then rolling to the bottom. The top of the hill represents the higher concentration (i.e., higher free energy) and the bottom of the hill represents the lower concentration of an adjacent area. In Figure 5-1(a) the stone rolls down a much steeper hill and reaches the bottom faster than the stone in Figure 5-1(b), even though both started at the same elevation (=concentration = free energy). In Figures 5-1(c) and 5-1(d), the horizontal distance is the same, but the stone in (c) moves faster because it started at a higher elevation (=concentration). The steepness of the gradient thus may be changed by a variation in either the concentrations or the distance.

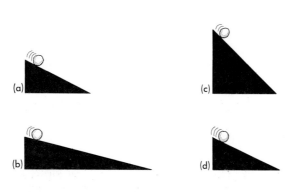

Figure 5-1. Representations of concentration gradients. The top of the hill represents the higher concentration, the bottom of the hill represents the lower concentration, and the stone represents the material that is moving from one area to the other. **(a)** and **(b)** The concentrations are the same but the distance is greater in **(b)**. Therefore, the material moves more rapidly in **(a)**. **(c)** and **(d)** The distances are the same but the concentration is greater in **(c)**. Therefore, the material moves more rapidly in **(c)**.

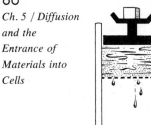

Figure 5-2. The effect of an external force on the diffusion of water. The greater force in **(b)** causes more water to penetrate the membrane than in **(a)**.

(a)   (b)

External Forces

This concept is particularly important in an understanding of water movements through plant cells. If pressure or force is imposed upon water, as by a piston in a closed system, the free energy of the water in that system is increased by the amount of force applied. In other words, the activity of the water is increased. As shown in Figure 5-2, the amount of water passing through the membrane increases as the weight on the piston is increased. The production of such forces in plant cells will be discussed in the next section.

Size of Molecules

In general, the larger the molecule, the slower is the rate of diffusion. Actually, Graham's Law of Diffusion states that the rates of diffusion of gases are inversely proportional to the square roots of their densities (density = mass per unit volume). For example, since the density of oxygen is 16 times that of hydrogen, the rate of diffusion of hydrogen gas is 4 times that of oxygen. The size of a molecule may not necessarily be directly related to its density, but for our purposes the statement at the beginning of this paragraph is sufficiently accurate.

Presence of Other Molecules

An increase in the number of "foreign" molecules causes the rate of diffusion to decrease because of the additional collisions that occur. In other words, a gas diffuses more rapidly through a vacuum than air. However, the *direction* of net diffusion is not influenced by the presence of other types of molecules.

5.3 Differentially Permeable Membranes and Diffusion

In describing a generalized cell (see Section 4.2), mention was made of various membranes, surface layers that have properties quite distinct from the specialized bits of protoplasm they enclose. Also stated was the fact that all such living membranes, and also some nonliving ones (e.g., cellophane, parchment, dialysis tubing), are characteristically **differentially permeable.** To reiterate, for any given membrane some molecules and ions penetrate quite readily, others more slowly, and some not at all. There are gradations in penetration rates.

The reasons for differential permeability are not fully known at the present time, but it is undoubtedly related to the structure of the membrane, especially its fat and oil (lipid) content; the size of the diffusing molecule; and the relative solubility of the diffusing substance in fatty materials and water. It is a fact that only dissolved materials will penetrate membranes. Even gases first dissolve in the water which impregnates cell walls before penetrating the cell membranes. However, materials more soluble in lipids tend to penetrate membranes more rapidly than do materials that are less soluble in lipids. This is probably related to the lipid structure of the membrane. In the case of nonliving membranes the size of the diffusing particle is the important factor, since such membranes are differentially permeable because the size of the pores (holes) is large enough for small molecules, such as water, but too small for large molecules, such as cane sugar (sucrose) or proteins, to pass through.

Cell membranes regulate the entrance or exit of materials. However, since these are living membranes, various external and internal factors bring about significant changes that influence the penetration of molecules from one moment to another. The general order of penetration remains fairly constant but individual rates may fluctuate. In other words, A may penetrate more rapidly than B, and under certain conditions both may diffuse through more rapidly, but A still faster than B.

The cell wall is quite porous with many microchannels existing between the interwoven cellulose fibers. Although these channels usually contain water, they present many pathways through which materials penetrate quite readily. As a result, the cell wall does not present a barrier to materials entering or leaving a cell.

Osmosis

One of the most important materials that enters cells is water. The term **osmosis** refers to the movement (net diffusion) of water through a differentially permeable membrane from a region of high free energy to a region of low free energy of water. In a strict sense, osmosis refers to the movement of a **solvent** through such a membrane. However, the only solvent involved in living cells is water; so the more restrictive definition will be used. In plants this movement of water may be from the environment into the cell, as from the soil into a root cell, or from one cell to another, or from the cell into the environment. There is nothing strange or unusual about osmosis; it is merely a special example of net diffusion.

Osmosis in nonliving systems can be demonstrated quite readily. Figure 5-3(a) indicates a capillary glass tube with an enlarged base submerged in a beaker of distilled (pure) water. Such a tube has a very small inner diameter. A small change in volume of the enclosed solution will be readily observed as a fluctuation of the level in the small tube. The enlarged portion of the tube is filled with a 50 per cent

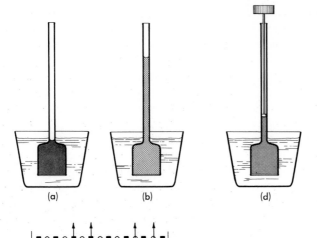

(a)   (b)   (d)

(c)

Figure 5-3. An osmometer. The sugar solution within the osmometer is separated from distilled water by a membrane that is permeable only to the water. **(a)** The original condition. **(b)** The condition after equilibrium has been reached. **(c)** Representation of the surface of the membrane. The circles are water molecules; the black squares are sugar molecules. **(d)** The condition that would result if a piston with weights were placed in the osmometer tube. For additional discussion see the text.

sucrose (cane sugar) solution, separated from the distilled water by a differentially permeable membrane (cellophane or dialysis tubing) that allows water to penetrate but not sugar. For ease of observation the sugar solution may be colored with a dye, such as Congo red, which will not penetrate through the membrane, or a brown syrup may be used. The concentration of water is greater in the beaker than in the **osmometer,** which means that the free energy of water is higher in the beaker than in the glass tube. Water molecules diffuse more rapidly into the osmometer than they diffuse out, the volume of liquid in the tube increases, and the level of liquid in the capillary rises as shown in Figure 5-3(b).

Note that the water molecules diffuse in both directions but that more of them diffuse inwardly. Net diffusion is then from the beaker into the osmometer. If one could greatly enlarge a portion of the membrane so that molecules were visible, the appearance would be approximately as in Figure 5-3(c). The water molecules penetrate quite readily, as indicated by the small arrows, but the sugar molecules are restricted by the nature of the membrane. These large sugar molecules occupy space at the surface of the membrane, and thus the number of water molecules on the solution side of the membrane is less than that on the water side. Clearly then, more water molecules are in a position to penetrate into the osmometer than are in a position to diffuse outwardly. The rise of the solution level does not continue indefinitely but eventually reaches a definite height.[1] When

---

[1] The approximate height can be calculated and depends upon the concentration of sugar solution used. For those students familiar with chemical principles, the concentration must be in terms of molality. In multiplying the molality (50 per cent sucrose equals about 1.5 molal sucrose) by 22.4 by 33, one arrives at approximately 1100 ft as the height of liquid in the capillary tube. The figures 22.4 and 33 are constants familiar to all chemists; the variable factor is the concentration of sugar that is used. A 1-molal solution has the potential of developing a pressure of 22.4 atmospheres when enclosed in an osmotic system, such as an osmometer, at 0°C. One atmosphere of pressure is capable of holding up a water column 33 ft (10 m) in height.

the level remains constant, the system is at equilibrium. Diffusion is still going on, but now the same number of water molecules is moving out of the osometer as is moving in from the beaker. At equilibrium, the free energy of water on both sides of the membrane must be equal. If this were not true, there would be a net diffusion one way or the other, and an equilibrium would not exist. The question then arises as to how the free energies can be equal even though the concentrations are not equal. Remember that the sugar molecules cannot penetrate the membrane, and thus the water in the osmometer is always less than 100 per cent (which is the concentration of water in the beaker). The explanation arises from an understanding of the influence of external forces on the free energy of water (review Section 5.2) and the development of hydrostatic pressure in the osmometer.

Water has weight. As the solution rises in the capillary tube, more and more weight presses down on the water within the enlarged portion of the osmometer. The weight of this column of water exerts a force (**hydrostatic pressure**), which increases the free energy of water within the osmometer by an amount equal to this external force. With an increase in free energy, more molecules of water move outwardly. Eventually the increased free energy, caused by a buildup of hydrostatic pressure (an external force), reaches the point where it is equal to the free energy of water in the beaker and the numbers of molecules moving in the two directions are equal. The level in the capillary could be maintained at a lower height if a piston and weights were placed in the tube and were pressing down on the solution (as in Figure 5-3[d]). This is also an external force.

If no capillary tube were on the osmometer but only the enlarged portion, the hydrostatic pressure would still develop, but it would not be visible as a column of liquid. This is basically the situation found in plant cells. In a mature cell the cytoplasm consists of thin layers adjacent to the cell wall and surrounding the vacuoles. Most of the water entering plant cells passes through the cytoplasm and into the vacuole. Though really two differentially permeable membranes are involved, the entire layer of cytoplasm can be thought of as functioning like a single membrane in this case. As the water concentration increases within the vacuole, the developing hydrostatic pressure forces the cytoplasm out against the cell walls. Since such walls are rather rigid, their resistance to expansion is in effect a pressure against the cytoplasm and vacuolar contents. It is this **wall pressure** that increases the outward diffusion of water molecules, until equilibrium is reached and net diffusion is zero.

Osmotic Pressure and Turgor Pressure

Hydrostatic pressure against cell walls results in the **turgid** (firm) condition of plant cells and even of an entire structure, such as leaves and nonwoody stems. The maximum amount of pressure that can be developed in a solution separated from pure water by a rigid membrane permeable only to water is termed the **osmotic pressure**. Living cells are not immersed in distilled water, except under laboratory conditions, and the theoretical maximum pressure never develops. The actual pressure exerted by the protoplast against the cell wall is the **turgor pressure**, which is always less than the osmotic pressure unless the cell is in distilled water. The difference between these two pressures is similar to that of a car having a maximum speed of 100 miles per hour and actually being driven at 50 miles per hour. Turgor pressure and wall pressure are equal but are exerted in opposite directions. Just as the protoplast is pressing against the wall with a certain force, so must the wall be exerting a similar force against the protoplast. The crispness of lettuce leaves in a salad depends upon the turgidity of their cells. The longer the exposure to evaporation, the more water is lost, and turgor pressure decreases until the lettuce becomes limp and unappetizing.

In addition to maintaining nonwoody plant parts in a firm condition and keeping leaves expanded (as opposed to limp and drooping), turgor pressure is also one of the factors involved in cell growth or enlargement. When cells first develop, the walls are plastic (i.e., nonrigid). As water is absorbed and turgor pressure increases, the cell walls expand or are stretched. Additional wall material is synthesized by the protoplast. The wall eventually becomes more rigid and further growth ceases. Without turgor pressure such cell enlargement does not occur.

The term **osmotic potential** is frequently used to indicate the potential capacity of a solution just as the horsepower rating of a car engine indicates its capabilities even when the motor is turned off. If the osmotic potential of a solution is 10 bars, this amount of pressure will develop only if this solution is separated from pure water by a membrane (as indicated previously). However, a similar solution in a beaker is referred to as having an osmotic potential of 10 bars, really meaning that it is capable of developing that much pressure under the specified conditions. If this solution, enclosed within a rigid membrane that is permeable only to water, is placed in a container of pure water, an osmotic pressure develops as water diffuses in. At equilibrium, 10 bars of osmotic pressure will have developed within the enclosed area. It might be noted that the free energy of water in the solution within the membrane is equal to the free energy of the pure water outside (or the system would not be at equilibrium). Because of the solute (e.g., sugar) in the solution, the concentration of water within the membrane, however, is *less* than the 100 percent (pure) water that is outside.

Pressure is measured in various units such as pounds per square inch (lb/in²), millimeters of mercury (mm Hg), atmospheres (atm = 14.7 lb/in²), and bars. We will use *bar,* a metric unit, which equals 14.5 lb/in², 750 mm Hg, or 0.987 atm.

5.4 Water Potential

The movement of water, as in osmosis, cannot be accurately explained in terms of differences in concentration. For example, as shown in the previous section, equilibrium was reached in the osmometer even though the concentrations on opposite sides of the membrane were not equal. Thus, care was used to discuss water movements in terms of energy concepts. This is also advantageous because it conforms to the terminology utilized by chemists and physicists (in thermodynamics).

Diffusion, as we have seen, is dependent upon the random kinetic motion of particles. In terms of water, we can say that movement occurs because the activity of water molecules is greater in one area than in the other. For our purposes we may consider this activity to be the free energy of the water molecules, and net diffusion occurs down an energy gradient. We are really concerned with the energy difference between one system and another (e.g., between the external solution and the solution within the cell). The absolute value for the free energy of water in a solution is difficult to measure. The *difference* in free energy can be determined using pure water as the standard of reference. The *difference* between the free energy of water molecules in pure water and the free energy of water in any other system (e.g., water in a solution or in a cell) is termed the **water potential** of that system. If the water potentials differ in various parts of a system, water will tend to move to the region where the water potential is lowest (i.e., down the energy gradient). This is similar to the transfer of heat from a region of higher to one of lower temperature.

Pure water, at atmospheric pressure, has a water potential of zero. (Remember: pure water is the standard of reference.) The presence of solute particles reduces the activity (or free energy) of water (by decreasing the concentration of water molecules) and thus decreases the water potential; it has a negative value since it is decreasing from zero. Surface forces that adsorb or bind water molecules (e.g., to surfaces of soil particles) decrease the activity of water and also lower the water potential. A physical pressure or force increases the activity of water and thus increases the water potential. The free energy of water in a cell or a solution is usually less than that of pure water. Therefore, the water potential of a cell or a solution is usually a negative number. When water is entering a cell, it is moving from a region of higher water potential (*less negative*) to a region of lower (*more negative*) water potential.

Now let us turn to the osmometer of the previous section and discuss the water movements in terms of water potentials. The presence of sugar in the osmometer lowers the interior water potential below that of the exterior water (which is zero in this case). The movement of water is always toward the region of lower water potential, which corresponds to the area of lower water activity or free energy. The hydrostatic pressure that builds up within the osmometer, as water enters, gradually increases the water potential until it is eventually equal to that outside. Water molecules are now diffusing equally in both directions; equilibrium has been reached, even though the concentrations are not equal.

The Greek letter psi (ψ) is used as a symbol for **water potential.** The symbol ψ_s is used to indicate the **osmotic potential;** that is, the amount by which the water potential is reduced as a result of the presence of solute. The more solute that is present, the greater the reduction. The reference standard is pure water at a water potential (ψ) equal to zero. A *reduction* from this value is a negative number. Thus, osmotic potentials are negative values. The surface to which water molecules are adsorbed is termed a **matrix;** the **matric potential** (ψ_m) refers to that component of water potential influenced by the presence of a matrix. Matric potentials are also negative values; they reduce the water potential of a system. For the most part, matric potentials are important in soils, especially at low water content, but are not considered to be significant in osmosis. It is possible that matric potentials are important in cells containing relatively large quantities of colloidal materials. The symbol ψ_p refers to the **pressure potential,** which is usually positive and increases the water potential in the system which is subjected to that pressure. The positive pressure operating in plant cells is the wall pressure or turgor pressure, in an osmometer it is the buildup of hydrostatic pressure.

The relationships among these various potentials can be expressed as:

$$\psi = \psi_s + \psi_p + \psi_m$$

Remember that ψ_s and ψ_m are negative quantities, while ψ_p is usually positive or zero. In discussing cellular movements of water we can ignore the ψ_m component. Note that the water potential and the osmotic potential are equal if there is no imposed pressure, as with a solution in a beaker or in a limp cell having no turgor (i.e., flaccid). The osmotic potential is determined by the amount of solute. Thus, the osmotic potential gives an indication of the total **osmotic concentration** within the solution or cell; that is, how much solute is dissolved in the water of that system.

If a plant cell were placed in distilled water, there would be a net movement of water into the cell because the presence of solute (e.g., sugars) results in a lower water potential within the cell. As water enters the cell, turgor pressure builds up and the resultant pressure effectively increases the water potential within the cell. When this water potential reaches zero, the cell is in equilibrium with the water outside and there is no further net movement of water. The cell is also fully turgid. Starting with a limp cell having an osmotic potential of -10 bars, the situation may be shown mathematically as:

	Water potential	=	Osmotic potential	+	Pressure potential
	ψ	=	ψ_s	+	ψ_p
Original values:	-10 bars	=	-10 bars	+	0 bars
Final values:	0 bars	=	-10 bars	+	10 bars

In order to simplify the mathematics, we assume that the cell volume does not change and that there is no dilution of materials. Note that the osmotic potential of a cell or a solution indicates the amount of pressure that must be exerted on that system to bring its water potential up to zero, the ψ of pure water.

Water movements into and out of cells, and between cells, are determined by water potential differences between the areas in question. The water potentials tend to equilibrate, not the osmotic potentials or osmotic concentrations. At equilibrium, the water potentials are equal on the opposite sides of the membrane. The osmotic concentrations may or may not be equal, and usually are not, because of the many dissolved materials found in plant cells—sugars, inorganic salts, organic salts, organic acids, and so forth.

Plant cells are not exposed to distilled water, of course, except under laboratory conditions. Figure 5-4 indicates the condition that results when a plant cell having an osmotic potential of -10 bars, because of the various materials dissolved in the water of the vacuole, is placed in a solution having an osmotic potential of -2 bars. The external solution usually is the soil solution with its dissolved minerals, and the plant cell could be one of the root cells. The cell is in a rather limp condition, has no turgor pressure, and the osmotic potential and water potential are equal (-10 bars). Remember that the presence of solute lowers the free energy, or

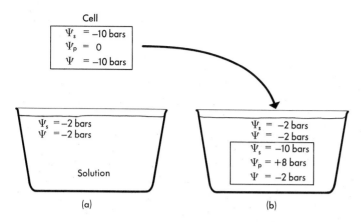

Figure 5-4. Diagram to indicate the relationship of water potential, osmotic potential, and turgor pressure. **(a)** The original condition. **(b)** The condition at equilibrium.

Cell

$\Psi_s = -10$ bars
$\Psi_p = 0$
$\Psi = -10$ bars

$\Psi_s = -2$ bars
$\Psi = -2$ bars

Solution

(a)

$\Psi_s = -2$ bars
$\Psi = -2$ bars
$\Psi_s = -10$ bars
$\Psi_p = +8$ bars
$\Psi = -2$ bars

(b)

activity, and thus the water potential; so we have negative numbers. Only when there is no turgor does the water potential equal the osmotic potential. The external solution is not enclosed, and its ψ_s and ψ are equal (-2 bars). There is a net movement of water into the cell because its water potential is lower than (more negative than) that outside. As indicated previously, water entering the cell brings about an increase in its ψ_p. This increases the water potential within the cell (to a less negative value). When the ψ inside the cell is the same as that outside (-2 bars), there is no further net movement of water into the cell; the same number of water molecules diffuse out as diffuse in—equilibrium has been reached. When there is a relatively large amount of external solution, the concentration outside will not change noticeably. Similarly, the very small amount of water that enters the cell and develops a turgor pressure will not significantly change the concentration within the cell. The osmotic potentials within and without may thus be considered to be the same as before. The original condition is shown in Figure 5-4(a); the equilibrium condition is indicated in Figure 5-4(b). Note that the concentrations, as indicated by the osmotic potentials, are not equal at equilibrium although the net diffusion of water is zero.

The ability of a cell to obtain (or absorb) water is determined by its water potential. In the example just discussed, the cell in the final equilibrium condition has a $\psi = -2$ bars. This cell will absorb water from any solution or cell whose water potential is greater than (less negative than) -2 bars. It will lose water to any system that has a water potential less than (more negative than) -2 bars. The use of negative values is sometimes confusing to beginning students. It might help to think of a thermometer and temperatures below zero; the *greater* the minus number, the *lower* the temperature. This is what happens with water potential values.

The movement of water from cell to cell is influenced in the same manner as just discussed. If the amount of material dissolved in a cell increases, possibly as a result of food (sugar) production or a loss of water by evaporation, the osmotic potential and water potential become more negative. For example, as shown below, the solute concentration of a cell doubles and this results in a change of the osmotic potential from -10 bars to -20 bars. The water content has not changed, so the turgor pressure is still $+8$ bars. Therefore, the water potential changes from

−2 bars to −12 bars; the ability of the cell to absorb water is now much greater than it was before the change in solute content.

	ψ	ψ_s	ψ_p
Initially	− 2 bars	− 10 bars	+ 8 bars
Final	− 12 bars	− 20 bars	+ 8 bars

Initially the cell would absorb water only from cells or solutions with a water potential above (less negative than) −2 bars. After the change, the cell will absorb water from any systems with water potentials above − 12 bars. In vascular plants the water absorbed by root cells moves from cell to cell because of differences in water potentials. Such water potentials frequently vary as solutes are used (as during respiration) or produced (as by photosynthesis).

Up to this point we have indicated that the cell walls are rigid. They are, however, elastic to some degree, and a small increase in volume does occur when a limp cell absorbs water and becomes turgid. With such an increase in volume, there is a slight dilution of the vacuolar contents, thus changing the ψ_s of the cell contents. In most cases this slight change can be ignored. As the turgidity of a cell increases, its ψ_p approaches the value of its ψ_s (which represents the maximum pressure that may develop). This maximum pressure would be reached, however, only if the cell were placed in pure (distilled) water. In actuality the turgor pressure of a cell is somewhat less than this possible maximum, depending upon the external environment (solution) with which the cell comes into equilibrium.

5.5 Entrance of Materials into Cells

Many materials, besides water, enter plant cells: mineral salts from the soil solution, and carbon dioxide and oxygen from the air through tiny pores in the leaves or into root cells from the soil atmosphere. These substances are essential to the normal growth and development of a plant.

Passive Absorption

The movement into or out of a cell is in many instances caused simply by differences in activity or free energy. As the cell uses a certain substance, the concentration of that material within the cell decreases, and more of it diffuses in from the environment if the external concentration is now higher than the internal. In the case of solutes the difference in concentrations is really the only factor that influences free energy, whereas the movement of water is somewhat complicated by the development of turgor pressure. The term **passive absorption** is used to refer to the movement of material into cells as a result of diffusion.

Active Absorption

More and more evidence has indicated that the molecules and ions of many substances continue to diffuse into plant cells even when their concentrations are greater within the cell than they are outside the cell. In other words, the movement is in a direction opposite to that which would be expected on the basis of diffusion phenomena. This accumulation against a concentration gradient is termed **active absorption.** It occurs commonly in root cells of higher plants, but an outstanding example is the accumulation of iodine from sea water by kelp.[2] This iodine, at least 1000 times as concentrated as that in the sea, can be extracted for commercial use.

The accumulation of ions and molecules in the vacuoles of cells is attributed to the energy of respiration being used to hold these materials against a gradient. Oxygen and sugars must be available to the cell, and respiratory inhibitors have been shown to prevent this process. Most mineral absorption quite likely is of this type. In fact, if cells are deprived of oxygen, not only are they unable to accumulate ions but they will actually lose much of what they had previously accumulated. Active absorption appears to be associated with protoplasmic membranes in general. There is selective transport by all intracellular membranes, resulting in compartmentalization of the cytoplasm.

The suggestion that appears to be most logical in explaining the transport of substances across a membrane is that a carrier system operates within the membrane itself. In this view a carrier molecule specific for a particular substance combines with that material on one side of the membrane, moves to the other side where the bond is broken, and then returns again to the first side where another molecule or ion can be picked up. In some way this mobile carrier requires energy in order to function. The exact mechanism is as yet not known.

It is quite likely that an enzyme specific for bond formation is at the outer surface of the membrane, and a second enzyme at the inner surface is responsible for rupturing the bond between the carrier and the material being carried. It is the *carrier complex* (i.e., the carrier molecule combined with the molecule or ions it is carrying) that is free to move from the outer to the inner surface of the membrane, probably as a result of the concentration gradient that would exist. In a similar fashion, after the bond ruptures at the inner surface, the carrier is free to diffuse back to the outer surface. The distance that the carrier complex must traverse is not very great; membranes are relatively thin structures. Therefore, it is also possible that a shift in molecular configuration of the carrier molecule may transfer the active attachment site from one side of the membrane to the other without actual diffusion being involved. This in effect would move the attached molecule from the outside to the inside surface where it would be released.

Figure 5-5 is a diagrammatic representation of the carrier concept. Evidence at present indicates that ions are absorbed in this fashion and that there is considerable specificity with regard to the carrier. In other words, different binding sites

---

[2] Kelp is a general term referring to the giant brown seaweeds (Division Phaeophyta) found growing in the oceans, as along the Pacific Coast of the United States.

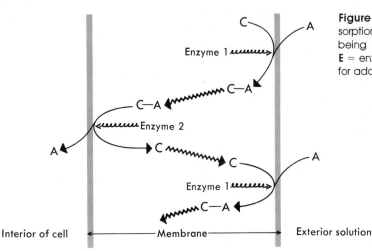

Figure 5-5. The carrier concept of active absorption. **C** = Carrier; **A** = molecule or ion being transported through the membrane; **E** = enzyme; **C-A** = carrier complex. See text for additional discussion.

Interior of cell ←——— Membrane ———→ Exterior solution

on the same carrier or different carriers exist for the various ions that are absorbed.

Electron micrographs often reveal small vesicles (or sacs) pinched off from the plasma membrane. Upon separation from the surface membrane, these vesicles are in effect very tiny vacuoles that contain small amounts of material that had been located at the external surface of the membrane. This process of **pinocytosis** appears to be another method by which substances penetrate membranes. Although the vesicle membrane has never actually been seen to disappear, it has been found that secondary vesicles form and pinch off from the primary one. It may be that pinocytosis produces many small vacuoles where both passive and active transport through the membranes may take place more readily. Certainly such a process would provide a tremendous increase in the surface membranes through which materials may penetrate. This again indicates the importance of the surface-to-volume ratio (see Section 3.10).

Plasmolysis

As has been mentioned, when water enters a cell becomes turgid. If a cell is placed in a highly concentrated sugar or salt solution, water diffuses out of the cell (why?) and the protoplast shrinks away from (or collapses away from) the cell wall as turgor pressure is lost. The vacuoles, which contain most of the water of the cell, shrink and may even disappear completely. Since the cell wall is freely permeable, the space between the protoplast and the cell wall will contain the bathing solution. This condition is termed **plasmolysis** (Figure 5-6), and the cell will die if left in that solution. However, if the cell is placed in distilled water or a dilute solution soon enough, recovery will take place since a higher water potential now exists outside the cell and movement of water will be inward. The "burning" of plants after spraying with insecticides or fungicides is actually a result of plasmolysis of leaf cells because of the high concentration of the spray residues. The brown discoloration is an indication of dead cells. Sometimes the excessive addi-

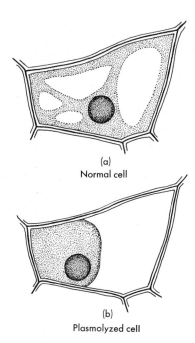

Figure 5-6. **(a)** Normal cell. **(b)** The same cell in a plasmolyzed condition after it has been placed in a concentrated sugar or salt solution.

(a)
Normal cell

(b)
Plasmolyzed cell

tion of chemical fertilizers to the soil may lead to the death of root cells and even the entire plant.

The effect of plasmolysis on living cells is utilized in the salting of meat and fish. The excess salt prevents the growth of decay organisms by plasmolyzing the cells of such molds and bacteria. In a similar manner, jams and jellies are preserved because of the high sugar concentrations, which prevent the growth of molds. Undesirable plants can be eliminated by applying salt to the soil at the base of the plant. Although this method may be expensive, it is suitable when one does not wish to dig up the unwanted plant, as with grass growing up through paved driveways. The recent development of arsenical plant-killing solutions and plant growth regulators, such as 2,4-D, has replaced the plasmolytic killing of plants in most instances.

Summary

1. Net movement of a substance is always from a region of its greater free energy to one of lesser free energy. The rate of diffusion of materials into and out of living cells is primarily influenced by the concentration gradient, by external forces, and by the size of the molecules that are diffusing.

2. All the membranes of a living cell are differentially permeable. Some materials penetrate such membranes much more rapidly than do others, and some molecules cannot penetrate the membrane at all.

3. Osmosis is the movement of water through a differentially permeable membrane. As water penetrates, a pressure builds up within the cell. This turgor

pressure is important in the growth of cells and in the general support of a plant. The osmotic movement of water into or out of plant cells depends upon the differences in water potentials of the two areas in question. Water potential refers basically to the ability of a system to absorb water and is given in negative values.

4. Materials enter cells passively as a result of diffusion, or they may be absorbed actively as a result of the expenditure of energy by the cell. Carrier molecules are probably involved in active transport. Pinocytosis is the pinching off of small vesicles from the plasma membrane.

5. If a cell is surrounded by a solution that is more concentrated than that within the cell, a net movement of water out of the cell results, and the cell plasmolyzes.

Review Topics and Questions

1. Explain in detail how you could determine whether the cytoplasmic membranes or the cell walls act as differentially permeable structures.

2. A row of three living cells is immersed in a 2 per cent salt solution. Cell *A*, which is in contact with cell *B*, contains 2 per cent sugar and 1 per cent salt; cell *B*, which is in contact with cell *C*, contains 3 per cent sugar and 3 per cent salt; cell *C* contains 1 per cent sugar and 2 per cent salt. The differentially permeable membranes of these cells are freely permeable to water and salt but are impermeable to sugar. Assume that only passive absorption occurs, and that turgor pressure is zero originally (i.e., before immersion).

   (a). Explain the direction of initial net movement of all of the materials among the cells and between the cells and the external solution.

   (b) Explain why such net movement finally ceases; in other words, explain what brings about the equilibrium condition.

3. A living plant cell contains 5 per cent sugar and the following salts: 1 per cent nitrates, 2 per cent sulfates, and 1 per cent bicarbonates. This cell is immersed in a solution containing the following salts: 2 per cent nitrates, 2 per cent sulfates, 2 per cent phosphates, and 1 per cent chlorides. The cell membranes are freely permeable to all salts and water but are impermeable to sugar. Assume that only passive absorption occurs, and that turgor pressure is zero originally (i.e., before immersion).

   (a) Explain in detail the reasons for the net movement, or lack of it, of all materials when the cell is first placed in the solution.

   (b) Explain why such movement eventually ceases.

4. Explain why molds and bacteria usually do not grow upon jams, jellies, and salted meats.

5. Human red blood cells do not have walls. Explain what happens to such cells if a drop of blood is placed in distilled water.

6. Living cells are stripped from an onion bulb and placed in a drop of distilled water. They are examined with the aid of a microscope.

   (a) Describe the general appearance of one of these cells as to its turgidity and the location of the protoplast with respect to the cell wall.

(b) Describe the appearance of the same cell five minutes after the distilled water is replaced with a 10 per cent salt (NaCl) solution, and explain any difference in appearance from that in (a) above.

(c) Describe the appearance of the same cell after five hours in the 10 per cent salt solution and explain any difference in appearance from (a) or (b) above. (Assume no toxicity.)

7. Frequently after heavy applications of insecticides, yellowish areas appear on the leaves of plants that have been sprayed. Microscopic examinations reveals that cells in the yellow areas have died. Since the insect spray is not poisonous to plants, explain the death of the leaf cells.

8. Explain the relationship among turgor pressure, wall pressure, and osmotic pressure.

9. Consider an isolated healthy cell bathed in a nutrient solution. Describe and explain the effect each of the following changes will have upon the net ability of the cell to absorb water from the surrounding solution:
(a) an increase in the turgor pressure of the cell;
(b) an increase in the concentration of solutes in the external solution;
(c) an increase in the concentration of solutes in the cell sap.

10. A home-owner decided to triple the amount of fertilizer above that indicated on the label. After adding the dry fertilizer, he sprinkled one half of his lawn for 30 to 40 minutes and then moved the sprinkler to the other half. He went to the movie later, forgetting that the water was still turned on. As a result, the second half of the lawn was thoroughly soaked for over six hours. Several days later, the home-owner noticed that the grass that had received a light soaking was dying, while the rest of the lawn was lush. Explain these results.

11. If a crystal (or small piece) of potassium permanganate (violet or purple in color) is placed on a layer of clear agar (or gelatin), the agar gradually becomes violet in color, the color being more intense closer to the crystal.
(a) Explain in detail this observation.
(b) Discuss how this general process is important to living cells.

12. A 5 per cent sugar solution is enclosed within an inelastic unbreakable membrane that is permeable to water but impermeable to sugar. Explain in detail what will happen if this enclosed sugar solution is placed in distilled water.

13. Explain why the water potential of a cell is a clearer indication of its ability to absorb water than is a knowledge of its osmotic (solute) concentration or its osmotic potential.

14. Differentiate between passive and active absorption. Discuss a possible mechanism of active absorption.

Suggested Readings

BIDWELL, R. G. S., *Plant Physiology* (New York, Macmillan, Inc., 1974).

GREULACH, V. A., *Plant Function and Structure* (New York, Macmillan, Inc., 1973).

The Plant

Before the various types of plants are considered in any detail, the flowering plant will be the example for a discussion of the basic structure and function of plants. Most processes, especially those occuring on the cellular level, are basically alike, whether they take place in a flowering plant or in a one-celled alga. Diffusion and photosynthesis, for example, may be studied in a variety of plants with equal ease and similar results. Differences arise, for the most part, in examining the overall metabolism of plants that exist in greatly differing environments; land plants have the problems of water loss and conduction of materials, which are trifling or not existent in aquatic plants. Also, the energy-yielding processes of plant cells proceed along one pathway if oxygen is present (respiration) and along a different pathway if oxygen is absent (fermentation). The similarities in structure and function are numerous and will be discussed first in a general manner. The differences will be emphasized during the discussion of individual groups of plants as they fit into specific niches of the plant kingdom examined from an evolutionary point of view.

Even though no structures or functions of an entire plant are isolated from all others, to separate individual portions or processes for the purpose of logical discussion and understanding is frequently helpful. After individual parts are clear, they can be placed in juxtaposition as one fits pieces of a jigsaw puzzle together to form a finished picture. One of the difficult problems faced during biological investigations is whether or not the one variable factor is influencing a process directly or indirectly by its influence on other processes. The interrelationships of processes and of structure and function must be kept in mind at all times. Many separations for discussive purposes are purely a matter of convenience and may not appear too logical when viewed on the basis of an entire plant. However, this type of approach, coupled with a final pulling together of the isolated parts, appears to be the best way of treating something as complicated as plant life.

6.1 Vegetative Structures

In general, one may lump together as **vegetative structures** those structures of the plant that are concerned with growth and development, i.e., root, stem, and leaf. These parts of the plant are not directly concerned with sexual reproduction. Figure 6-1 represents the various parts of a flowering plant in diagrammatic form and should be referred to for an understanding of the placement and relationships of the portions under discussion.

Definitions, as they are encountered, refer to the great majority of plants. There are occasional instances where plant parts do not seem to fit such definitions. Comparison with related plants, however, usually clarifies the situation. The investigator realizes that structural modifications exist, and that careful study is necessary to understand relationships. Sometimes, of course, a definition may have to be revised as a result of such studies.

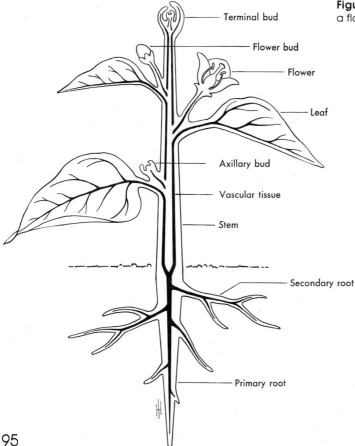

- Terminal bud
- Flower bud
- Flower
- Leaf
- Axillary bud
- Vascular tissue
- Stem
- Secondary root
- Primary root

Figure 6-1. Diagram of the principal parts of a flowering plant.

Root

That part of the plant axis that is typically nongreen and found beneath the surface of the soil is termed the **root.** This greatly branched structure's primary function is the absorption of water and minerals from the soil. The root, however, has secondary functions. Since those materials that are absorbed are used by all cells of the plant, conduction obviously must take place through the root before such utilization can occur. In addition to conduction, many roots contain much stored food materials, and thus storage must be indicated as another function of this part of the plant. The branching of roots and their penetration within the soil serve to anchor the plant as well as the soil, a significant factor in any area subjected to windstorms and heavy rainstorms. A more detailed discussion of roots will appear later (see Chapter 9).

Some plants, such as various orchids, have aerial roots. These plants grow on other plants, such as in the fork between the trunk and a large branch of a tree. They usually obtain essential minerals from decaying organic matter that collects around their roots. The roots function as do underground roots, and anatomical studies indicate typical root structures.

Stem

The **stem** is the continuation of the plant's axis typically found above the soil surface. No sharp line of demarcation, but a gradual merging of one into the other, exists between root and stem. The stem branches in a variety of ways, resulting in the more or less characteristic form associated with different plants (Figure 6-2), and may be woody or nonwoody. Regardless of appearances, the stem functions mainly in conducting water and minerals from the root to other parts of the plant and in conducting food materials from the leaves to the rest of the plant. Vascular, or conducting, tissues are well developed. Stems are involved in the production and support of leaves. Many stems also store food materials and may even manufacture foods if they are green (chlorophyllous). Some stems develop underground, as is true of the white potato; the tuber is actually an enlarged portion of

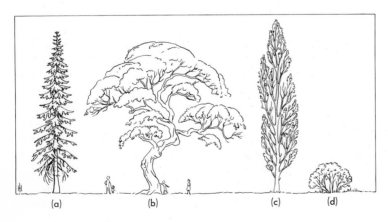

Figure 6-2. Characteristic growth formation of various woody plants. **(a)** Pine (*Pinus*). **(b)** Oak (*Quercus*). **(c)** Lombardy poplar (*Populus nigra* var. *italica*). **(d)** Typical bush or shrub.

(a)   (b)   (c)   (d)

an underground stem. Studies of its structure indicate quite clearly that it is a stem and not a root. The structural arrangement of tissues and their functions will be discussed in Chapters 7 and 8.

Leaf

A **leaf,** an outgrowth of the stem, is usually flat and thin, needlelike, or scalelike. It is green or, if it is some other color (such as red), the chlorophyll is merely masked by an abundance of accessory pigments. The function of a leaf is closely correlated with the presence of chlorophyll. This is the material that enables leaf cells to utilize light energy for the production of food from simple inorganic materials (carbon dioxide and water), a process termed **photosynthesis.** Since food for all living organisms is dependent upon this process, considerable discussion will be devoted to it later (see Chapter 12). Some leaves also store considerable quantities of food materials.

6.2 Sexual Reproductive Structures

Those parts of the plant that are concerned with sexual reproduction and the production of seeds are termed the **reproductive structures.** In the more complex vascular plants these are **flowers** and **cones,** which are basically continuations of the stem with specialized structures comparable to modified leaves and branch systems. The seeds that are produced germinate and grow into new individuals similar in type to the parents. The differences between flowers and cones, the manner in which seeds are produced, and the development of fruits from flower parts will be discussed in subsequent chapters (see Chapters 26 and 27).

6.3 Life-Span

Once a seed has germinated, the growth and development of the plant are influenced by both the environment and the inherited characteristics of that particular type of plant. Factors of the environment would include: (1) water supply, (2) temperature, (3) supply of minerals in the soil, (4) light, (5) oxygen and carbon dioxide supply, and (6) parasites or herbivores. However, even if all these factors are conducive to growth, not all plants will grow indefinitely. The length of the life-span will depend upon the type of plant: (1) **annuals** grow for one season only, producing their seeds and then dying (e.g., beans, some grasses); (2) **biennials** grow vegetatively during the first season and do not produce seeds until the second year, after which they die (e.g., lettuce, carrot, cabbage); and (3) **perennials** grow for several to very many years, producing a new crop of seeds each year after the first few years (e.g., trees, shrubs). These last include plants that may be hundreds or thousands of years old; a *Sequoiadendron giganteum* in California has been estimated to be probably 3000 years old. It was already a giant at the birth of Christ, and unless it is attacked by parasites, destroyed by fire, or blown down, there is no reason why some hundreds of years cannot be added to its age. As mentioned previously (Section 1.3), one of the characteristics of plants is their continued growth throughout their life-span. Each year this tree, as do others, is increasing in size as well as in age. Who can say when all this will cease?

6.4 Tissues

In large multicellular organisms diffusionary processes are much too slow to enable materials to penetrate to all cells as rapidly as they are needed. For example, water is subject to evaporation from the above-ground portions of a plant when they are exposed to an unsaturated atmosphere. In a plant of any large size such water could not be replaced rapidly enough to prevent desiccation if diffusion alone were responsible for water movement. The interior cells of such a plant would probably be deficient in food supply if such material first diffused through the outer layers of cells where it would also be used. Light could not readily penetrate to the interior cells and they would not be able to carry on photosynthesis; neither could carbon dioxide diffuse to such cells. It was fortunate indeed that over exceedingly long periods of time, some plants happened to evolve that contained variously elongated cells where conduction of materials could occur fairly rapidly—a simple kind of conducting tissue. Such plants would have an advantage over neighboring ones and would tend to survive more readily. Plants with conducting tissues, especially if comprised of rows of elongated cells, could survive in drier terrestrial environments than were previously usable. In a similar manner, again over long periods of time, many other types of structures happened to evolve. If they were beneficial, such as a simple root for absorption, such plants would have advantages in a competitive environment and grow more readily than others. Plants with less advantageous structures would not grow as well, would tend to have fewer offspring, and would eventually die out. The more competitive plants would gradually occupy the area relinquished by those that died.

Basically then, evolution implies many different kinds of chance modifications on which environmental factors would act in a selective manner. A differentiation of function enabled the evolution of large, multicellular organisms. Such differentiation of function is correlated, of course, with a differentiation of structure: conducting cells, food-manufacturing cells, storage cells, and reproductive cells are variously modified and frequently are organized as tissues. Generally, then, **tissues** are considered to be groups of cells that perform essentially the same function and are commonly of similar structure; they are organized into a functional and structural unit. They are subdivided into meristematic and permanent types.

It should be borne in mind that millions of years and billions of plants were involved from the time that living organisms originated to the appearance of large, multitissued plants. One must also remember that the various changes that arose during the evolution of plants were of a random nature. Some changes were beneficial, many were not. Some changes happened to provide the plants with a survival advantage. These advantageous changes were then inherited by succeeding generations.

Meristematic Tissue

The continued growth of plants throughout their lives depends upon the activity of **meristematic tissues.** In such tissues the cells are actively dividing, and new cells are continually being produced. There is no differentiation yet of cells, one cell being much the same as any other cell of the tissue. Most of the cells of the meristematic regions are destined, usually after one or more divisions, to enlarge and develop into variously more or less specialized cells. Some of the cells, however, remain meristematic and divide repeatedly. These are the **initials,** and they serve to perpetuate the meristems. **Apical meristems** consist of cells that are basically isodiametric (i.e., having equal diameters), usually appearing cubical, and are located at the tips of both roots and stems. These immature cells have thin walls with no secondary layers, a dense protoplast, and vacuoles are nonexistent or very small. A sample of this tissue (usually a root tip) is used for observing mitosis in the laboratory; usually 10–15 per cent of the cells are found in various stages of division. The **vascular cambium** is located as a thin cylindrical sheath between the bark and the wood.[1] It is somewhat difficult to examine because of the tissues bordering on each side, but it is readily detected in cross sections of stems and roots. It consists of two types of thin-walled, highly vacuolated cells: (1) vertically elongated cells with tapering ends, the **fusiform initials;** and (2) nearly isodiametric, somewhat horizontally elongated, relatively small cells, the **ray initials.** Both these cells undergo divisions producing many new cells, as will be discussed more fully in the next chapter. In many plants a **cork cambium** develops. Living cells of the area become meristematic and produce additional cells that develop into cork. If a plant is damaged, protective cork layers may form in this manner.

In monocotyledonous plants, especially the grasses, a tissue at the base of each leaf usually remains meristematic. Such tissue is termed an **intercalary meristem.** The rapid elongation of such plants as wheat, barley, and grasses is in large part caused by these meristems. The mowing of a lawn does not prevent subsequent leaf growth because the intercalary meristems are below the area of cutting.

Most of the cells produced by the activity of these various meristematic tissues eventually differentiate and become the cells of the permanent tissues. As water is absorbed, the turgor pressure within the cells results in their enlargement. Additional cell wall material is produced, frequently including secondary layers, various modifications of the original cell occur, much new cytoplasmic synthesis occurs, and the cell differentiates, or develops, into a specific type of cell that

---

[1] Bark and wood are general nontechnical terms that have been used for convenience. In mature woody stems, the cambium is located between phloem and xylem. The inner part of the "bark" consists of phloem; the "wood" consists of xylem. (See Chapter 7.)

forms a part of one of the tissues comprising the plant body. In most cases no further divisions occur in such mature cells.

Permanent Tissue

In these tissues the cells are stable and no longer dividing. Although derived from meristematic regions, these cells have considerable structural and physiological modifications. Each type of permanent tissue is composed of specifically differentiated mature cells that make possible an efficient division of labor. For example, the cells of conducting tissues are usually quite elongated, whereas storage cells are short and bulky. Only through the development of a variety of tissues having various specific functions are large organisms capable of existing, especially on land. The number and complexity of the many functions occurring in higher plants could not possibly be carried on efficiently by a single type of cell or tissue.

The protoplast of a cell depends for its existence upon materials obtained from the environment. Even if it can manufacture its own food, as in a photosynthetic cell, it must obtain carbon dioxide, water, and light. These materials enter through surfaces and must be moved in some way through the protoplast. The larger the structure, the more difficult it is for inner areas to obtain such essentials. The volume increases more rapidly than the surface area when size increases, and (therefore) the surface-to-volume ratio decreases. The difficulties involved are more serious with land plants than with those existing in water. In land plants, for example, water is continually being lost from the aboveground surfaces and must be replaced. This becomes impossible without specialization—specialization in terms of absorption, transport, and mechanisms that tend to reduce water loss. Large organisms, thus, can exist *only* because of specialization. No one cell or tissue could possibly undertake all the various functions necessary for a land existence. All cells require food, oxygen, water, and minerals. Some of these essentials are basically supplied through the atmosphere (carbon dioxide for food manufacture, for example) and others from the soil; a spatial distance is involved. Add reproduction to these functions, and the situation becomes even more difficult. Specialization in multicellular organisms is basically a division of labor, differentiations in which particular structures carry out certain functions efficiently (but not all functions). Cells that efficiently transport water are not capable of food production. This is why cellular and tissue differentiation is important. Certain types of cells manufacture food, others transport food, some store food, still others are developed as reproductive structures. The whole organism, thus, is capable of existing as an entity with various types of tissues cooperating to make this possible. In order to comprehend this situation fully, one should have an understanding of the tissue structure of plants as well as their functions.

The permanent tissues are divided into simple and complex types. The **simple permanent tissues** are composed of cells that are structurally and functionally alike for the most part, while the **complex permanent tissues** are composed of several kinds of cells that differ in structure. In the latter type of tissue, the cells are involved in a group of interrelated activities in which one function is usually dominant.

1. Simple permanent tissue. Figure 6-3 consists of diagrams of the kinds of cells that are involved in the formation of simple permanent tissues, while Figure 6-4 shows photomicrographs of certain of these cells.

EPIDERMIS (Figure 6-3[a] and 6-4[a]). This tissue is usually one cell in thickness and forms the surface layers of leaves, flowers, and young stems and roots. The outer walls of epidermal cells are covered and somewhat impregnated with **cutin,** a fatty material secreted by the protoplast. The continuous layer of cutin, which is termed the **cuticle,** is fairly impervious to water and greatly retards the loss of water from a plant. The conservation of moisture and the protection derived from the fairly thick walls of the epidermal cells are important functions of the epidermis.

In many regions of the epidermis, especially on leaves and green stems, are found tiny pores, or **stomata.** Each **stoma** is really the space between two adjacent, kidney-shaped **guard cells.** The number of stomata per unit area of epidermis varies with the location as well as with the kind of plant. They are usually most numerous on leaf surfaces, and frequently more are found on the lower surface than on the upper surface of a leaf. Such openings provide easy access for the diffusion of gases into or out of the leaf, a situation that is essential for food manufacture. Stomata may be associated with other epidermal cells, termed *subsidiary cells,* that differ in shape from ordinary epidermal cells.

The epidermis often bears small, unicellular or multicellular appendages called **trichomes,** which are frequently slender hairs. Some trichomes may be glandular, secreting a sticky material or an irritant. The functions of trichomes are not clear although protection from predators or reduction of evaporation are possibilities. A dense mat of hairs, stiff and bristly hairs, or hairs containing irritants could provide some protection from animals, especially insects. The importance of hairs with regard to evaporation will be discussed in Section 15.5

PARENCHYMA (Figure 6-3[b] and 6-4[b]). The cells of the **parenchyma** are usually more or less spherical, although the shape may be distorted by the pressure of surrounding cells and may appear rather angular or box-shaped. The walls are thin, and the vacuoles are quite large. Various materials are stored in the parenchyma tissues, which are found throughout the plant. If chloroplasts are present, these cells are capable of carrying on photosynthesis under suitable conditions. Cells of this type are capable of becoming meristematic under certain conditions, such as wounding or injury, and form a cork cambium. This is extremely important for recovery from adverse conditions.

SCLERENCHYMA (Figure 6-3[c] and 6-4[c]). The cells of this tissue characteristically have extremely thick secondary walls and are dead at maturity. Obviously they are alive as they develop, but the thick walls become impregnated with such materials as lignins, and the protoplast subsequently dies. The gritty texture of

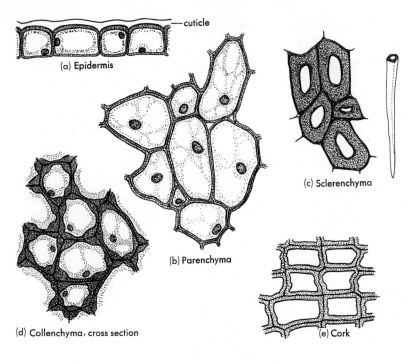

(a) Epidermis

cuticle

(b) Parenchyma

(c) Sclerenchyma

(d) Collenchyma, cross section

(e) Cork

Figure 6-3. Cells of simple permanent tissues.

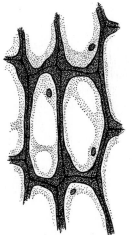

Collenchyma, longitudinal section

many pears is a result of the presence of **stone cells,** or **sclereids,** which are scleren-chyma-type cells. Sclereids are variable in shape, often branched, and also found in seed coats and shells of nuts. Elongated, slender sclerenchymatous cells are termed **fibers,** which commonly occur in strands or bundles. Strength, mechanical support, and protection are the main functions of sclerenchyma tissue. Fibers, obtained from various plants, are twisted together to manufacture thread or yarn from which fabrics and ropes can be produced.

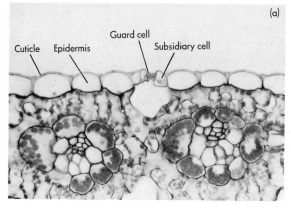

(a)

Cuticle   Epidermis   Guard cell   Subsidiary cell

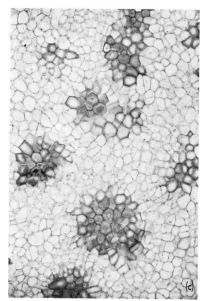

Figure 6-4. Photomicrographs of cell types **(a)** Epidermis, *Zea mays* leaf (×200). **(b)** Collenchyma and parenchyma cells, young *Sambucus* stem (×375). **(c)** Clusters of stone cells, young pear (*Pyrus communis*) fruit (×150). (Courtesy of Robert Gill.)

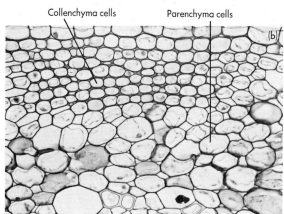

Collenchyma cells   Parenchyma cells   (b)

(c)

COLLENCHYMA (Figure 6-3[d] and 6-4[b]). The cells of this tissue are somewhat elongated, and the primary walls are irregularly thickened, usually at the corners where the cells meet. This strengthening and supporting tissue is frequently found in young stems near the epidermis and associated with the veins (conducting tissues) of leaves.

CORK (Figure 6-3[e]). This tissue is composed of cells that have walls that are impregnated with **suberin,** a fatty material. The cells are dead at maturity and form a rather waterproof layer of tissue. Cork is found as the outer layers of stems and roots of woody plants and also as the protective layers that form when a plant is damaged. Water conservation and protection, then, are the functions of cork. In the case of the cork oak (*Quercus suber*), such a profuse formation of cork occurs that this tissue is removed and used commercially (e.g., cork stoppers, life preservers, insulation). A cork cambium, which develops from parenchymatous cells, produces new cork, which may again be stripped from the tree in another three or four years.

2. Complex permanent tissue. The **xylem** and **phloem,** which form the vascular or conducting tissues, are complex permanent tissues. The specific kinds of cells that comprise these two tissues will be discussed in the chapter on stems (Chapter 7) so that the location of cells and tissues can be clarified. Since these tissues are somewhat complicated, they can be studied more readily—this is especially true of microscopic examination—in conjunction with their location, and then the individual cells can be examined in detail.

Summary

1. The root, stem, and leaf are vegetative structures and not directly concerned with sexual reproduction. The root is that part of the plant axis that is below ground, while the stem is the part typically found above the soil. The former absorbs water and minerals; the latter conducts material throughout the plant. The leaf is an outgrowth of the stem and carries on photosynthesis.

2. The sexual reproductive structures are flowers and cones.

3. Annual plants grow for one season only, biennials produce seeds during the second year and then die, and perennials grow for many years.

4. A tissue is a group of cells that perform essentially the same function and commonly are of similar structure. In meristematic tissues the cells are frequently dividing, as in the apical meristem, vascular cambium, and cork cambium. Permanent tissues are those in which the cells are no longer dividing.

5. Simple permanent tissues are composed of cells that are structurally and functionally alike: epidermis, parenchyma, sclerenchyma, collenchyma, and cork. The xylem and phloem are complex permanent tissues in which several kinds of cells are found.

Review Topics and Questions

1. You are given a slide that contains thin sections, transverse and longitudinal, of a tissue. Discuss in detail the characteristics that you would have to look for in order to determine what kind of simple permanent tissue had been used in preparing this slide for observation through a microscope.

2. You are given a slide that contains thin sections, transverse and longitudinal, of a tissue. Discuss in detail the characteristics that you would have to look for in order to determine whether this was a meristematic tissue.

3. Discuss in detail the changes that occur when a cell differentiates into a fiber. What are the functions of fibers?

4. Describe the difference between annual, biennial, and perennial plants.

5. Discuss the advantages that obtain to a multicellular organism as compared with unicellular or colonial organisms.

6. What is the importance of structure to the function of a cell?

Suggested Readings

CUTTER, E. G., *Plant Anatomy: Experiment and Interpretation, Part I* (Reading, Mass., Addison-Wesley, 1969).

ESAU, K., *Anatomy of Seed Plants* (New York, Wiley, 1960).

FAHN, A., *Plant Anatomy,* 2nd ed. (Elmsford, N.Y., Pergamon Press, 1974).

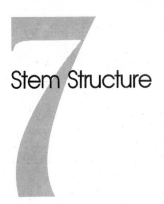

Stem Structure

The stem is the pathway whereby foods, which are produced in leaves, and minerals and water, which are absorbed by roots, are transported throughout the plant. Some of these materials nourish the cells of the stem, some are stored there, and the rest merely pass through on their way to other cells and tissues where they are utilized or stored. In addition to the **translocation** of materials, many stems are also involved in food manufacture, vegetative reproduction, and support. Each of these functions will be discussed in some detail after a discussion of stem structure.

7.1 External Structure

An examination of the various kinds of plants that can be found on any campus will indicate that some plant stems are quite woody, while others are not. One can separate plants into two general groups on this basis: those that have herbaceous stems and those that have woody stems. In both types leaves and buds are present at specific locations along the stem, each point of attachment being termed a **node** and the distance between nodes being called an **internode.** The length of an individual stem depends upon the growth of the internode and varies with the type of plant as well as with various environmental conditions (e.g., temperature, light, soil fertility).

Herbaceous Stems

The stems of this group are generally soft and green and have very little or no tough woody tissue. There is little growth in diameter, and the plants are usually short-lived. The outer surface consists of a thin epidermis in which stomata are present. The green color, of course, is caused by the presence of chlorophyll and indicates the food-manufacturing ability of such stems. The support of the leaves depends upon collenchyma, sclerenchyma, and the turgid condition of individual cells. Examples of plants with herbaceous stems are the common garden peas and beans, common grasses, clover, alfalfa, wheat (*Triticum* sp.), and corn (*Zea mays*).

Woody Stems

In contrast, **woody stems** are hard, thick, and long-lived. The outer surface of the older stems is rough and covered with cork, the common bark of trees and shrubs. In this rough surface are raised areas, or **lenticels,** which are really openings beneath which the cells are loosely arranged with many intercellular spaces. Gaseous exchange can take place through these openings. The bulk of the stem consists of tough woody tissue. A young woody stem may contain chlorophyll and carry on photosynthesis for a short period, but as the diameter increases and cork forms, this ability is lost. In fact, young stems are all herbaceous in appearance at first, and the woody characteristics develop as the stems become older. The increase in diameter of such stems results mainly from the production of wood and cork. Lumber is obtained by cutting the woody tissue of trees into usable sizes and shapes. The difference between a tree and a shrub is merely one of growth-form rather than any intrinsic difference in the internal structure. In a **tree,** the stem (or trunk) rises some distance above the ground before branching occurs, while in a **shrub,** several stems of rather equal size usually arise at or close to ground level (see Figure 6-2).

Buds

New stems and their leaves develop from **buds.** Figure 7-1 indicates a typical arrangement, with a **terminal bud** at the tip of the twig and **axillary** (or **lateral**) **buds** occurring at regular intervals along the stem. The latter type is always located in the angle made by the stem and the leaf stalk. Since the drawing is that of a buckeye (*Aesculus*) twig in its winter condition, after the leaves have fallen, the location of leaves can be determined by the presence of **leaf scars** just below each axillary bud. Since the conducting tissue of the stem is continuous with that of the leaf, this tissue also ruptures when the leaf falls, and **vascular bundle scars** are visible in each leaf scar. In woody plants the more delicate inner structures are often protected by the tough, hard **bud scales,** which drop off as the new stem develops. (The terminal bud scale scars of previous years are visible on the diagram.) These scales basically are modified leaves in which the food-manufacturing ability has been superseded by a protective function. In herbaceous plants no bud scales are present. Since these plants are generally short-lived, the survival value of protective bud scales is not the significant factor that it is for long-lived woody plants, especially those faced with the problem of severe winter conditions. In this case, it seems likely that those plants in which bud scales evolved would tend to survive much more readily than plants without such protection. Many buds are dormant for a period after they have formed. In this condition they are rather resistant and readily survive adverse environmental conditions, such as the cold temperatures

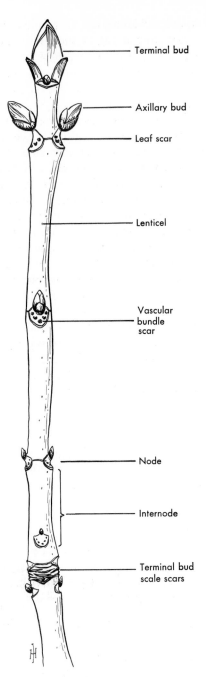

Figure 7-1. Dormant twig of California buckeye (*Aesculus californica*).

Terminal bud

Axillary bud

Leaf scar

Lenticel

Vascular
bundle
scar

Node

Internode

Terminal bud
scale scars

of winter. (See also Section 29.11.) After dormancy is broken by the low temperatures of winter and with the advent of spring (a suitable environment) buds unfold and develop into branching systems (stems and leaves), flowers, or both, depending upon the type of bud.

If a bud is sectioned longitudinally while dormant, the similarity in the three types is seen. All buds possess immature leaves, immature buds in the axils of these leaves, meristematic tissue, nodes, and internodes (Figure 7-2). Careful examination will usually make it possible to identify developing vascular tissue. A bud, then, may be regarded as a very much compressed, compact, undeveloped section of a stem. The term **leaf primordia** refers to small projections that will develop as leaves when the bud unfolds but do not have the characteristic leaf shape as yet. These peglike structures arise by repeated divisions of cells at one locus in the peripheral region near the basal portion of the apical meristem. Continued growth and divisions of these cells produce a bulging of the surface and then the peglike (or fingerlike) projection. Later, as the leaf primordium is developing, axillary buds originate in the area that represents the axil of the developing leaf. Because the axillary buds develop somewhat later, leaf primordia may be distinguishable apparently without such buds. However, in a later stage of development these buds would also be apparent. The unfolding and expansion of a bud result from the divisions of cells of the apical meristem and leaf primordia, and the subsequent enlargement and differentiation of these cells. As this occurs, the internodes elongate greatly. The expansion of axillary buds results in the characteristic branching systems of plants.

Although the shoot apex is dependent upon the shoot for water, minerals, and food (sugars from the photosynthesizing leaves), it is surprisingly autonomous. The shoot apex from various plants, and devoid of all leaf primordia, has been grown on a nutrient culture medium. Under these conditions, it is able to give rise to an entire plant with both shoots and roots. The culture medium, of course, was

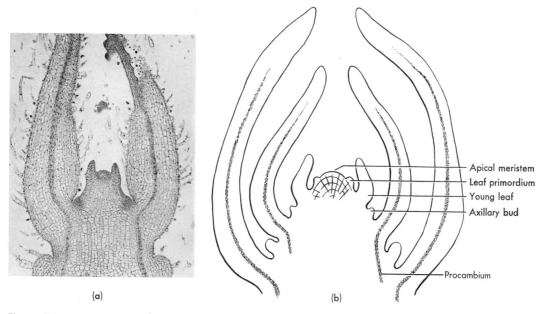

(a)　　　　　(b)

Apical meristem
Leaf primordium
Young leaf
Axillary bud

Procambium

Figure 7-2. (a) Photomicrograph of a *Coleus* stem tip. Two leaf primordia are present, one at each side of the apical meristem. **(b)** Diagram of a terminal bud. (Photo courtesy of General Biological Supply House, Chicago.)

the source of those materials normally provided by the plant itself. In other experiments the apex of a plant was divided into halves by a single, median cut. The two halves continued their development and reorganized into two complete apices, each of which produced a shoot. It is the shoot apex that is important in determining the organization of tissues and structures.

The arrangement of buds and leaves is termed **opposite** if two are at a node, **alternate** if only one is present, and **whorled** if three or more are at a node. The growth of axillary buds results in a different type of growth-form for each of the three arrangements.

In many plants buds may arise at places other than leaf axils and have no connection with the apical meristem. They may develop on roots, stems, or leaves and give rise to new shoots. Such buds are termed **adventitious buds,** and a vascular connection is established between the bud and the parent structure by differentiation from the bud toward the existing vascular system. If a sweet potato (actually a root) is positioned with the lower portion immersed in a container of water, adventitious buds develop at the upper end and give rise to leafy shoots. This method is frequently used to demonstrate adventitious buds.

7.2 Internal Structure of a Woody Stem

A **transverse** (cross) **section** of the woody stem exposes to view a variety of tissues, which are considered to be **primary tissues** if they develop from an apical meristem and **secondary tissues** if they arise from a cambium. The cells at the tips of stems are young, immature, and undifferentiated; so the section to be examined must be cut at a distance from the end of any stem in order for the plane of section to pass through mature tissues (see Figure 7-3). Low magnification allows the identification of tissues, whereas higher magnification exposes individual cells to observation.

Sections through the apical meristem region expose to view a dome-shaped mass of isodiametric cells in which there is no differentiation, each cell looking much like every other cell. As these cells become older, most of them will develop into the cells of one of the primary permanent tissues while others will continue to divide and produce more cells, their growth resulting in the upward (or outward) extension of the stem tip. Rather than waiting for such stems to mature, we examine tissue that is older already and farther down the stem. In other words, Figure 7-3(b) represents a transverse section through cells that were once part of the apical meristem but now have begun to differentiate into specific permanent tissues. This usually occurs merely a few millimeters below the apical meristem, and we can recognize three fairly distinct regions.

The **protoderm** is the outermost layer of cells and will eventually develop into the epidermis as they continue differentiating. The cells of the **procambium** appear as isolated groups of cells arranged in a circular fashion. In a longitudinal view these cells are seen to be much longer than neighboring cells and form strands that can be recognized as the beginning of vascular tissue. Sometimes a continuous cylinder of procambium is formed, but in any event these cells give rise to the

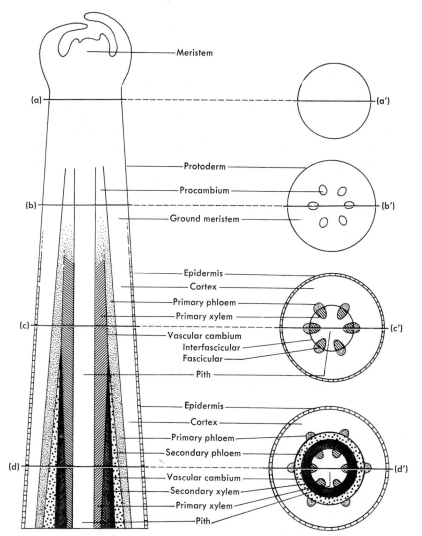

Figure 7-3. Diagram of longitudinal section of a woody stem tip and diagrams of cross sections at different levels, as indicated. Individual cells are not shown; regions and tissues are indicated. The older tissues are the ones farther from the apex.

Labels in figure:
Meristem

(a) — (a')

Protoderm
Procambium
(b) — Ground meristem — (b')

Epidermis
Cortex
Primary phloem
Primary xylem
(c) — Vascular cambium — (c')
Interfascicular
Fascicular
Pith

Epidermis
Cortex
Primary phloem
Secondary phloem
(d) — Vascular cambium — (d')
Secondary xylem
Primary xylem
Pith

primary vascular (conducting) tissues as they mature. The rest of the area shown in Figure 7-3(b) comprises the **ground meristem,** the cells of which eventually form the pith and the cortex. These three distinguishable groups of cells (i.e., the protoderm, procambium, and ground meristem) represent the forerunners of the primary permanent tissues. As they mature and differentiate, they gradually develop into the various tissues as indicated.

In most plants, where secondary growth is involved, the vascular cambium arises in part from the procambium and in part from parenchyma cells between the developing vascular bundles. These two parts are termed the **fascicular cambium** and the **interfascicular cambium,** respectively. The vascular cambium becomes a complete cylinder and produces continuous cylinders of secondary vascular tissues. (See Figure 7-3[c] and 7-3[d].) The differentiation of primary permanent tissues from meristematic tissues usually occurs rather early in the first season's growth. Although most of the secondary tissues are produced during subsequent years, some do develop by the end of the first season of growth. It should be noted that the stem is elongating, and that new primary tissues are produced at the tips of stems during subsequent years. In the older portions of the stem (area toward the base and away from the tip), only secondary tissues are produced. Conduction in older stems is by way of secondary tissues. As materials are transported toward the younger portions of the stem tip, they move quite readily from the secondary to the primary vascular tissues, which in effect form a continuous conducting system. For water and other materials to reach the cells of the apical meristem some diffusionary movements must take place through the nonvascular (undifferentiated) cells at the very apex of the stem.

Primary Tissues and Regions

The very young woody stem consists of primary tissues alone, and secondary tissues are produced later, sometimes not until near the end of the first year's growth or until subsequent years. All the cells that comprise primary tissues were produced by cell division in the apical meristem, after which cell enlargement and differentiation resulted in such cells becoming part of one of the mature (permanent) tissues. The tissues and regions will be discussed in order from the outermost to the central area of the stem (see Figure 7-3).

1. Epidermis. As mentioned previously (see Section 6.4 and Figure 6-3[a]), the epidermis forms the surface protective layer. It is usually one cell in thickness, the cell walls are frequently rather thick, and the outer wall is covered with a layer of cutin. As will be discussed later, the epidermis is usually not present in older stems.

2. Cortex. Just under the epidermis is a region containing primarily parenchyma cells, the **cortex.** As indicated previously (see Section 6.4 and Figure 6-3[b]), the cells are basically spherical, although sometimes slightly elongated, with thin walls and large vacuoles. The amount of this storage region varies with the kind of plant, but it is usually many cells in thickness. One may find collenchyma and sclerenchyma cells in the cortex, in which case an additional strenghtening and supporting factor is present. Water, minerals, and foods pass through by diffusion from cell to cell. The cortex frequently is not present in older stems (see later).

3. Primary phloem (Figure 7-4). This is one of the complex permanent tissues that were mentioned in the previous chapter. Several kinds of cells are present in the phloem, which enables the efficient conduction of organic (food) materials from place to place. In addition there are secondary functions of the phloem as a result of the presence of certain supporting and storage cells. The principal conducting cells are the **sieve tube elements,** vertically elongated cylindrical cells, the side and end walls of which contain numerous thin areas (the **sieve areas**). The

Figure 7-4. **(a)** Diagrams of phloem cells. The sieve tube element and fiber are rather elongated cells; the central portion has been omitted in part. **(b)** Photomicrograph of phloem, longitudinal section, *Cucurbita pepo* stem (×300). (Photograph courtesy of Robert Gill.)

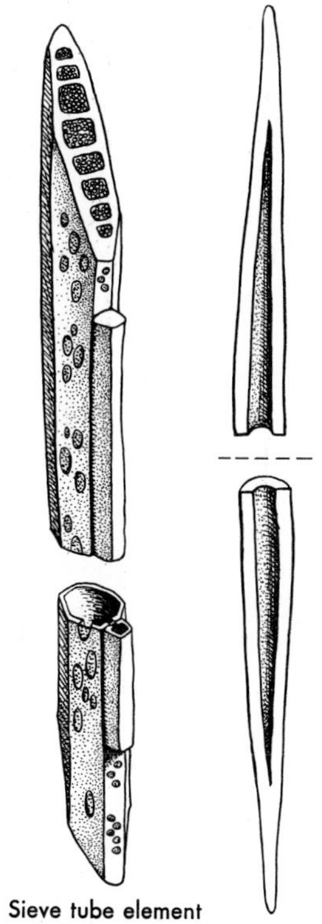

Parenchyma cell

Ray cell

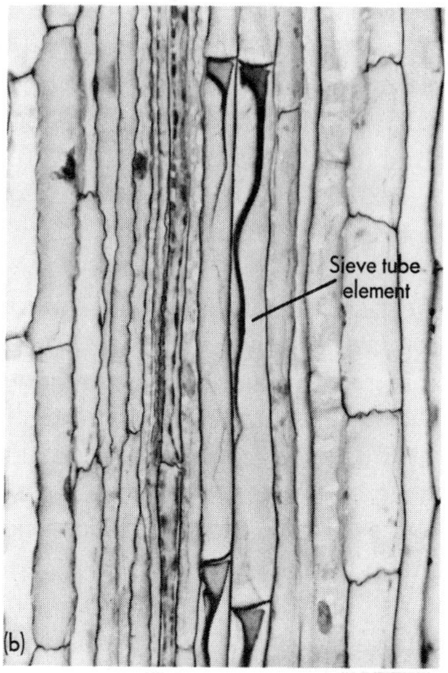

Sieve tube element

Sieve tube element and companion cell

Fiber

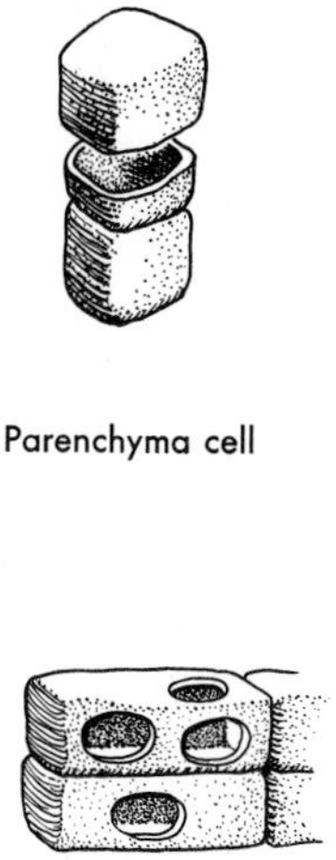

(b)

protoplasts of adjoining cells are connected by strands that penetrate through the **sieve pores** of the sieve areas. These areas in the end walls are usually more complex and termed **sieve plates.** This contiguity undoubtedly facilitates movement of materials through the **sieve tubes,** which are composed of numbers of sieve tube elements arranged end to end. In cone-bearing plants the sieve cells do not form conspicuous vertical series (or sieve tubes) and the sieve areas are rather uniform on all walls (i.e., no specialized sieve plates). The exact mechanism of phloem transport cannot yet be explained, but the rates of movement are much more rapid than the rate of diffusion alone. When mature, the cytoplasm in the sieve tube element is not distinct from vacuoles as in other living cells, and the nucleus has disappeared. Abutting each of these conducting cells (in the flowering plants, but not in the cone-bearing plants) is a somewhat shorter, narrower, vertically elongated **companion cell.** The nucleus of the latter cell is present throughout its life. The exact function of the companion cell is not known, but it may have something to do with both conduction and storage (there are pits in the walls between the sieve tube element and the companion cell). The **phloem parenchyma cells,** scattered through the tissue, are similar to such cells discussed previously and are mainly storage cells. In some plants **phloem fibers,** strenghtening and supporting cells, are present. These cells are narrow, vertically elongated cells with very thick lignified walls and a small **lumen** (the cell cavity). The protoplast often disap-

113

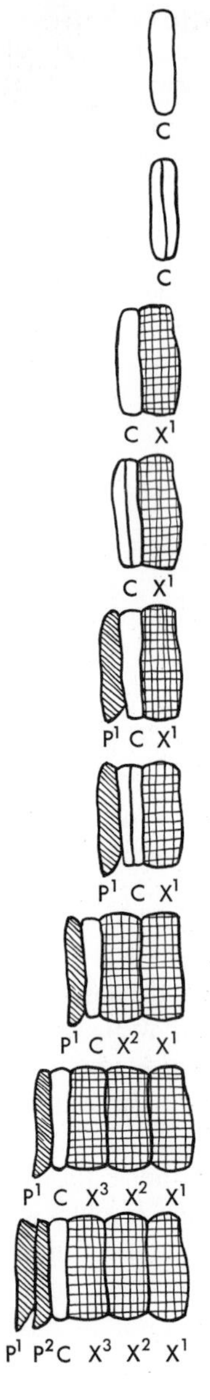

Figure 7-5. Cell (C) of the vascular cambium dividing with the resulting cells differentiating into secondary xylem cells (X^1, X^2, X^3) or secondary phloem cells (P^1, P^2). The central region of the stem is to the right while the exterior is to the left.

pears as fibers mature. The phloem fibers of the flax and hemp plants are made into linen and rope, respectively. The manner in which such fibers are separated from other cells will be discussed in Section 20.6

4. Vascular cambium. Meristematic cells, derived from the procambium or the ground meristem, form a narrow (one or two cells in thickness) cylindrical sheath of tissue, the **vascular cambium,** immediately internal to the phloem (see also Section 6.4). As these thin-walled cells divide, those cells that mature toward the outer part of the stem develop into various cells of the secondary phloem, while those that mature toward the inner part of the stem develop into cells that form the secondary xylem (Figure 7-5). Usually, more xylem than phloem is produced, and as a stem continues growth in diameter, a greater and greater proportion is xylem, or **wood.** The tissues external to the cambium constitute the **bark,** although this term is usually used only after secondary tissues and cork have formed. Note that the cambial layer is retained in that one of the two cells that are formed at each division remains meristematic as part of the vascular cambium. The vascular cambium and other regions external to the xylem are pushed outward by the production of secondary xylem. The entire development results in an increase in the diameter of the stem. Longitudinal divisions at right angles to those shown in Figure 7-5 produce additional cambial cells and allow for the increased growth in circumference of the vascular cambium and secondary vascular tissue.

5. Primary xylem (Figure 7-6). This is the second of the complex permanent tissues that were mentioned in the previous chapter. Here again one finds several kinds of cells and primary and secondary functions. The conduction of water and minerals occurs mainly through tracheids and vessel elements of the xylem. The **tracheids** are lignified, thick-walled, elongated, and tapering cells, which are dead at maturity. Primary pit fields are located in the walls, and spiral or ring thickenings may add to the strength of the cell (see Figure 4-4). The hollow cavity (lumen) of each cell, the pits that are present, and the overlapping ends of tracheids in stems and roots enable water and minerals to be transported very rapidly in a vertical direction. The tracheid's value in providing strength and support to the plant is related to the various wall thickenings, the length of the cells, and the overlapping and grouped arrangement of the cells. **Vessel elements** function in transport, as do tracheids, but are more efficient in that they are much larger in diameter, many pits are present in the thick lignified walls, and the end walls as well as the protoplast have disappeared when the cells are mature (see Figure 7-7). In addition to this, vessel elements are arranged end to end, like barrels on top of each other, forming long vertical tubes, the **vessels,** which may be several feet in length. These are the main conducting structures in the xylem of flowering plants but they are not found in cone-bearing plants, such as pine and spruce. The tracheids, on the other hand, are present in all xylem tissue. The pattern in which the secondary

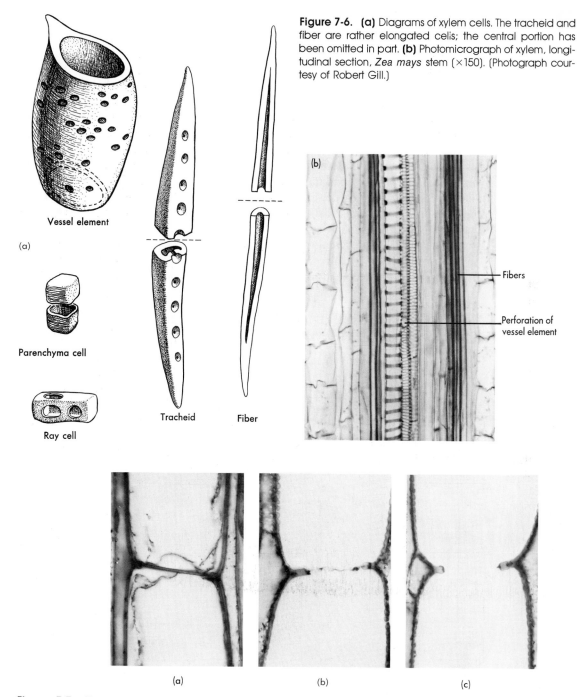

Figure 7-6. **(a)** Diagrams of xylem cells. The tracheid and fiber are rather elongated cells; the central portion has been omitted in part. **(b)** Photomicrograph of xylem, longitudinal section, *Zea mays* stem (×150). (Photograph courtesy of Robert Gill.)

Vessel element

(a)

Parenchyma cell

Ray cell

Tracheid

Fiber

(b)

Fibers

Perforation of vessel element

(a)

(b)

(c)

Figure 7-7. Photomicrographs of longitudinal sections showing stages in the development of a vessel in *Zea mays*. **(a)** End wall between vessel elements intact. This wall is primary in nature, but the secondary thickenings of the longitudinal walls are visible on the sides. The protoplast is present. **(b)** End wall is broken down; remnants are still visible. Note pits in side walls. Protoplast gone. **(c)** End wall is gone, although the rim persists; no protoplast. (Courtesy of Katherine Esau.)

wall is deposited in vessel elements undergoes the same general variations as in tracheids (Figure 7-8). **Xylem parenchyma cells** and **xylem fibers** are similar to the parenchyma cells and fibers found in the phloem, although fibers may be more numerous in the xylem. In conifers (e.g., *Pinus*) typical fibers are rare. In these plants the xylem is quite homogeneous, consisting mainly of tracheids.

During the long period that vascular plants have been on earth, evolutionary modification of the tracheid proceeded in two directions and gave rise to the vessel element and the xylem fiber. These two cells have assumed the tracheid's functions and have largely replaced the tracheid in the most highly evolved vascular plants. In the vessel the conducting aspect has been emphasized, especially in view of the perforations and the end-to-end positioning. In the fibers, on the other hand, the supportive features of the tracheid have been intensified. The fiber is longer; more slender and tapering, has greater wall thickening, and more extensive overlapping of cells. These two derivatives of the tracheid together perform the dual functions of the original tracheid but do so much more effectively.

6. Pith. The central portion, or core, of the stem consists of parenchyma-type storage cells.

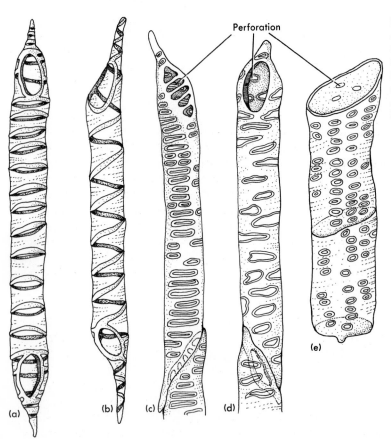

Perforation

Figure 7-8. Vessel elements. **(a)** Annular, **(b)** spiral, **(c)** scalariform, **(d)** reticulate, **(e)** pitted. Vessel elements are perforated at the ends. (From *Plants, An Introduction to Modern Botany,* by Greulach & Adams, 1975, Wiley, N.Y., reprinted by permission.)

(a) (b) (c) (d) (e)

Secondary Tissues

The presence of the cambium in woody stems is responsible for the tremendous growth in diameter of which they are capable (observe any old tree). The tissues that develop from the vascular cambium or cork cambium constitute the **secondary tissues,** which reinforce or replace certain primary tissues. The cells of the cambium divide in two planes (radially and tangentially), thus producing an increase in circumference as well as in diameter. No epidermis, cortex, or pith is formed by the cambium.

1. Secondary phloem. With one difference, the cells that comprise the secondary phloem are very similar to those of the primary phloem as far as structure and function are concerned. The difference lies mainly in their mode of origin. All the cells of the phloem that have been mentioned are vertically elongated or isodiametric. Two of the difficulties encountered by land plants are conduction along a considerable length of stem and the support of this stem and the leaves, flowers, and fruits that are attached to it. Vertically elongated cells are well suited to these functions. However, as the stem increases in diameter, more horizontal conduction is necessary to supply tissues located internally and externally with regard to the vascular tissue as well as the cells of the vascular tissue itself. Some of the cells of the vascular cambium develop into **phloem ray cells,** parenchymatous cells (not found in primary vascular tissue) that are elongated radially and function quite readily in transporting materials across the stem (Figure 7-4). The relatively large simple pits in the walls of these cells facilitate such conduction from cell to cell. These cells are oriented end to end, forming the **phloem ray,** which is usually several cells deep. Various materials are quite likely also stored in these ray cells.

As mentioned previously (Section 6.4), there are *fusiform initials* and *ray initials* in the vascular cambium. The vascular ray cells are derived from the ray initials, while the other cells of the secondary vascular tissues are derived from the fusiform initials. Each cambial initial produces radial files of cells, some toward the outside (secondary phloem) and some toward the inside (secondary xylem), and the files meet at the initials. During cambial activity, cells are continually dividing and maturing; new cells are being added while the older ones are differentiating. Thus, a fairly wide zone of rather undifferentiated cells may be present, and the initials are difficult to distinguish from their recent derivatives (Figure 7-9). For convenience, therefore, the term *cambium* is usually used to refer to a zone of such cells.

2. Secondary xylem. As is true of the secondary phloem, the cells of the secondary xylem are basically similar to those in the primary xylem but differ with respect to the meristematic tissue from which they develop. There are detailed differences between the cells of the primary and secondary xylem, but it is beyond the scope of this book to discuss them. Here also the exception is the **xylem ray cell** (Figure 7-6), which is similar to ray cells found in the phloem. A **xylem ray** extends to the cambium. The term **vascular ray** is used to denote the radially oriented ray cells extending through the secondary xylem and secondary phloem,

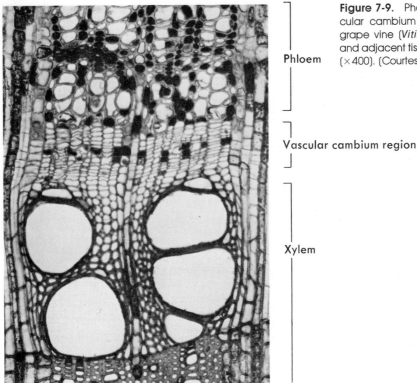

Phloem

Vascular cambium region

Xylem

Figure 7-9. Photomicrograph of active vascular cambium zone. Transverse section of grape vine (*Vitis* sp.) through the cambium and adjacent tissues in second year of growth (×400). (Courtesy of Katherine Esau.)

and thus includes both xylem ray and phloem ray. The derivation of secondary xylem cells from fusiform and ray initials was discussed previously.

Concentric rings that form in the secondary xylem can be seen quite readily whenever a tree is cut down. One ring usually develops each year, and so the term **annual ring** has come to be used, although **growth ring** is more appropriate. The appearance of alternate light and dark bands is caused by differences in the tracheids (and vessels) produced during periods of rapid growth and those produced during periods of slow growth. When growth is rapid, usually during the spring, the cells are larger in diameter and have thinner walls than when growth is slow, as in summer. The sequence of events, then, is rapid growth, slow growth, and no growth, followed by a repetition of this differential process. These events result in large, relatively thin-walled cells gradually merging into smaller, thick-walled cells, which comprise one ring; no distinct line of demarcation appears between the two types of cells. However, since a period of no growth (or very little growth) is followed by a period of rapid growth, a very sharp line can be seen between the thick-walled cells of one year and the relatively[1] thin-walled cells of the subsequent year (Figure 7-12). Viewed without the aid of a microscope (Figure 7-10), the area of thick-walled cells appears dark (much wall, little lumen), whereas the area of thin-walled cells appears lighter (less wall, more lumen). Since the forma-

---

[1] The term *relative* must be emphasized. Remember, these are tracheids. Compared with parenchyma cells, even the relatively "thin-walled" tracheids would be considered to have thick walls.

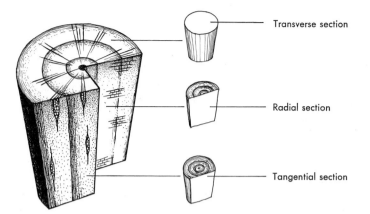

Figure 7-10. Woody tissues of stem shown in three views.

Transverse section

Radial section

Tangential section

tion of rings in the secondary xylem depends upon growth differences, lack of water, damage, or other environmental factors might result in the production of fewer than one ring each year; more than one ring may form if environmental conditions or other factors cause several growth fluctuations during the same year. For this reason the term *growth ring* is preferable.

Figure 7-10 represents a stem cut to show three views of the woody tissues. In the transverse section the growth rings can be seen quite clearly, and the slender xylem rays radiate outward much like the spokes on a wheel. The longitudinal cuts have been made in two planes so that the rings and rays can be seen in surface view and in face view. In a radial section the rays (in surface view) have the appearance of a brick wall, while the growth rings are seen as vertical lines. In a

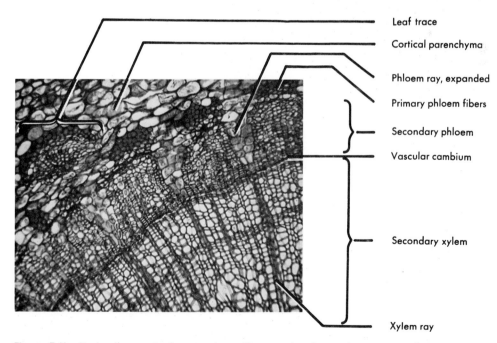

Leaf trace

Cortical parenchyma

Phloem ray, expanded

Primary phloem fibers

Secondary phloem

Vascular cambium

Secondary xylem

Xylem ray

Figure 7-11. Photomicrograph of young stem with secondary tissues, transverse section.

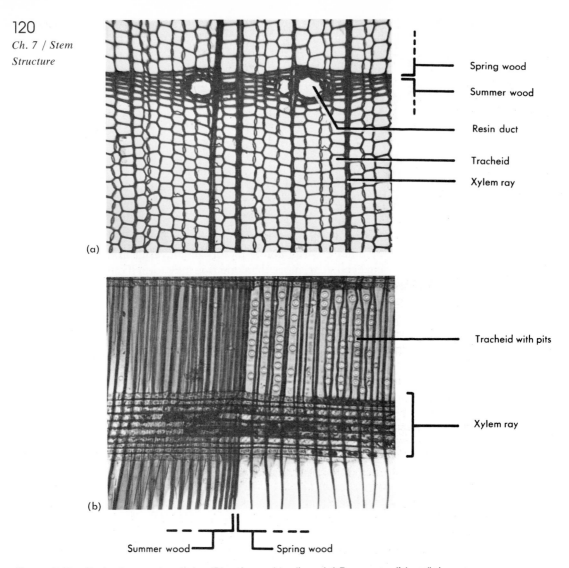

Spring wood

Summer wood

Resin duct

Tracheid

Xylem ray

Tracheid with pits

Xylem ray

(a)

(b)

Summer wood

Spring wood

Figure 7-12. Photomicrographs of pine (*Pinus*) wood sections. **(a)** Transverse; **(b)** radial.

tangential section, however, the rays are seen "head on," appearing somewhat elliptical in shape. The beautiful grains that are present in boards used for cabinet work result from the slanted cuts that are made through the secondary xylem; the growth rings generally appear as vertical or curved lines. The vascular rings of branches from the main trunk add to the intriguing patterns found in wood, and some of the most beautiful cabinet work utilizes wood with large rays in radial sections.

Figure 7-11 is a photomicrograph of a young stem in transverse section showing the early development of secondary tissue; portions of two growth rings are visible in the xylem, and the conducting cells of the primary phloem are crushed be-

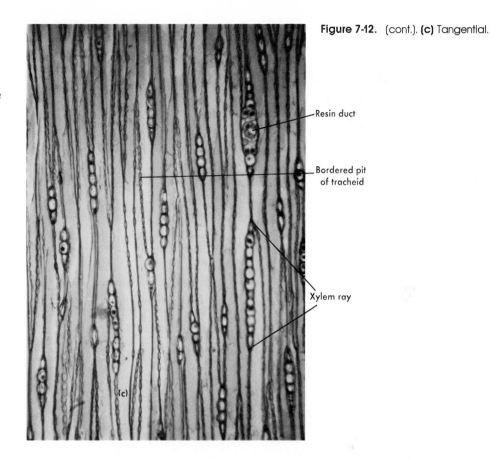

Figure 7-12. (cont.). **(c)** Tangential.

Resin duct

Bordered pit
of tracheid

Xylem ray

(c)

tween the expanding secondary phloem and clusters of primary phloem fibers. Some of the vascular rays expand considerably in the phloem region. At the upper left of the photograph is a small vascular bundle, or **leaf trace,** which is the small extension from the main vascular tissue that extends out into the leaf.

Figure 7-12 (a), (b), (c) are photomicrographs of pine wood (secondary xylem) in transverse, radial, and tangential sections, respectively. The border between two growth rings shows clearly in the transverse and radial sections. The bordered pits of some of the tracheids are shown in face view (Figure 7-12[b]) and in cross section (Figure 7-12[c]); see also Figure 4-5(b). **Resin ducts** or **canals** are scattered through the wood and bark of various cone-bearing trees, notably pine. These interconnected canals are lined with parenchyma cells that secrete resin. This material is collected for commercial use by cutting through the wood and allowing the resin to drip into containers. Resins are used in the preparation of varnishes, incense, and perfumes, and distillation yields turpentine.

Figure 7-13(a), (b), (c) are transverse, radial, and tangential sections, respectively, of oak wood. The vessel elements of the spring wood are frequently very large; they are not in the radial view, which was sectioned through the summer wood. Some of the vascular rays are many cells in thickness; one such **compound ray,** along with smaller ones, is shown in Figure 7-13(c).

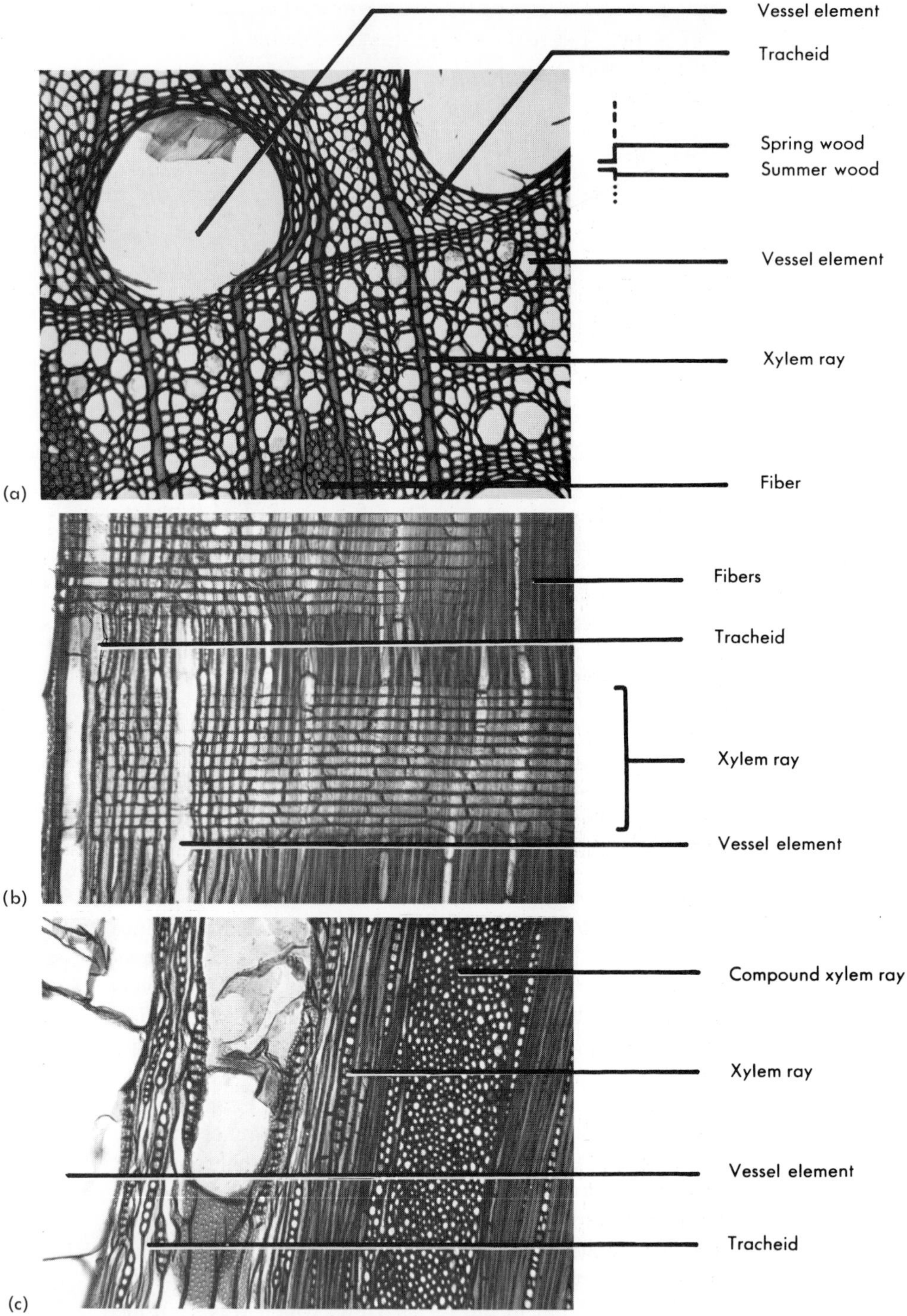

Figure 7-13. Photomicrographs of oak (*Quercus*) wood sections. **(a)** Transverse; **(b)** radial; **(c)** tangential.

3. Cork cambium and cork (Figure 7-14). The cambium that produces xylem and phloem is termed the **vascular cambium** to differentiate it more vividly from the **cork cambium.** As the cells of the cork cambium (or **phellogen**) divide, the outer ones develop into **cork cells,** and the inner ones may develop into a tissue known as the **phelloderm,** in which the cells are of the parenchyma type. Cork (or **phellem**) consists of box-shaped cells, similar to the cork cambium from which they are derived but whose walls are impregnated with suberin. The presence of this fatty material in the cell walls and the lack of intercellular spaces result in cork being relatively impermeable to water and gases. Cork, then, is a protective tissue that replaces the epidermis in woody plants. The protoplasts of cork cells die after suberin is deposited in the walls.

As mentioned previously (Section 6.4), parenchyma cells are capable of becoming meristematic under certain conditions, even though such cells are considered to be portions of permanent tissues. Damage to a tissue frequently results in such meristematic activity and the formation of a cork cambium. This is not an unusual occurrence. In fact, most woody stems have cork as the surface layers by the sec-

Figure 7-14. **(a)** Development of cork and phelloderm from the cork cambium. **(b)** Photomicrograph of *Pelargonium* stem, transverse section, showing early stage of cork development (×75). **(c)** Similar to **(b)**, but a later stage (×75). (Photographs courtesy of Robert Gill.)

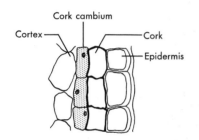
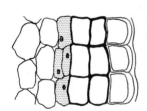
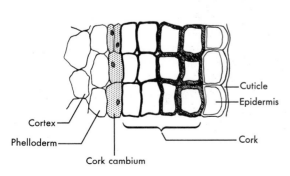

(a)

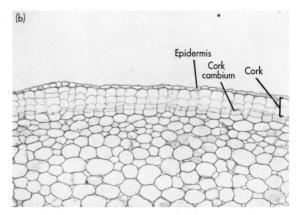

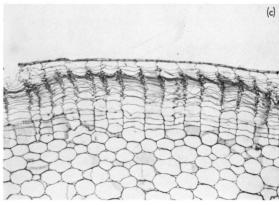

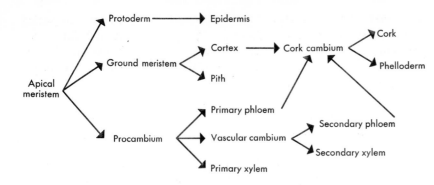

Figure 7-15. Developmental sequence of tissue differentiation in the dicotyledonous stem.

ond year's growth; often cork begins forming before the end of growth during the first year. The production of secondary vascular tissue results in an increase in the diameter of a stem. The epidermis and cortex are permanent tissues and thus can no longer increase in size. These outer tissues rip and tear as more and more vascular tissue develops internally. Cork frequently forms before the damage is apparent. Since the cork is impervious to most materials, any cells external to this tissue die and are sloughed off. In older woody stems all the epidermis, cortex, primary phloem, and even parts of the secondary phloem are lost in this fashion, cork cambium eventually forming from parenchyma cells of the secondary phloem. The so-called "bark," then, consists of secondary phloem, cork cambium, and cork. In roots the cork cambium usually forms from pericycle cells (parenchyma type) at a considerable distance beneath the epidermis (see Section 9.2).

All the cells of the various tissues that have been discussed originally arose as cells of one of the meristematic tissues. As these cells enlarged and differentiated, they eventually developed into cells of one of the permanent tissues. The inherited characteristics and the positioning of such cells determine their eventual structure and function. Cells that were basically quite similar at first may be quite dissimilar when they mature. Figure 7-15 indicates the developmental sequence of tissue differentiation in the dicotyledonous stem. In the herbaceous dicot stem little secondary tissue develops, and in most monocot stems there would be no cambium and therefore no secondary tissue.

7.3 Stem Types

At the beginning of this chapter, herbaceous and woody stems were mentioned. An examination of their tissues indicates that we should actually subdivide herbaceous stems into two groups: **monocotyledonous** and **dicotyledonous.** These two terms will be discussed more fully in Section 27.3, but suffice it to say now that all flowering plants can be termed either monocotyledonous or dicotylendonous. Plants in the former group have one **cotyledon,** or seed leaf, in the seeds, the leaves are typically parallel-veined, and the flower parts are in threes; examples are common grasses, lilies, and corn. The dicotyledons have two cotyledons in each seed, net-veined leaves, and flower parts in fours or fives; common examples are oaks, tomatoes, beans, and pears. Except for the palms, almost all the monocotyledons have herbaceous stems, whereas the dicotyledons may have herbaceous (e.g., tomato, bean) or woody (e.g., oak, pear) stems. Basic internal differences for each group will be considered.

Herbaceous Monocotyledonous Stem

The vascular tissue exists as scattered bundles of xylem and phloem in this type of stem although the bundles may be more numerous at the periphery of the stem. (Figures 7-16[a]; 7-17). Except for the palms and certain other large monocots, which have an anomalous cambium, no cambium, and thus no secondary tissue, is present; the little growth in diameter is dependent upon enlargement of the cells of primary tissues. Even though no definite arrangement of vascular bundles exists —the xylem and phloem never form continuous cylinders of tissues—the xylem is always located on the inner side of a bundle and the phloem to the outer side. The greater part of the stem consists of parenchyma tissue. Usually, however, directly adjacent to the epidermis and frequently surrounding the vascular bundles are sclerenchyma and collenchyma cells, which strengthen and support the stem. In some grasses the centrally located parenchyma cells break down, except at the nodes, and the stem has large hollow areas.

Herbaceous Dicotyledonous Stem

In these stems (Figures 7-16[b] and 7-18) the vascular tissue is also arranged in discrete bundles, but the bundles themselves are arranged in an orderly ring and not scattered. The cambium, which is visible between the xylem and phloem, may be restricted to the individual bundles or may be continuous from bundle to bun-

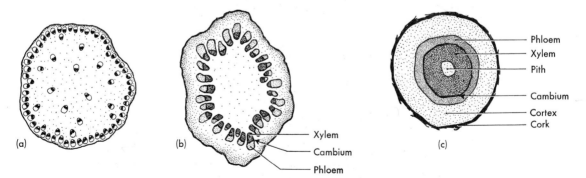

(a)   (b)   (c)

Xylem
Cambium
Phloem

Phloem
Xylem
Pith
Cambium
Cortex
Cork

Figure 7-16. Stem types in transverse section. **(a)** Herbaceous monocotyledonous stem with scattered vascular bundles. **(b)** Herbaceous dicotyledonous stem; the vascular bundles are arranged in an orderly ring fashion. **(c)** Woody dicotyledonous stem; the vascular tissue is arranged in concentric cylinders, which appear as circles in a transverse section. (In this diagram growth rings are not shown; neither is any distinction made between primary and secondary tissues.)

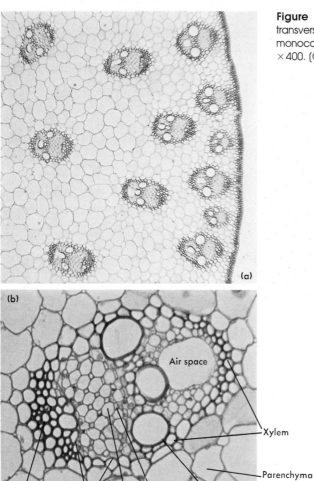

Figure 7-17. Photomicrographs of transverse sections of an herbaceous monocotyledonous stem. **(a)** ×100. **(b)** ×400. (Courtesy of Peter Jankay.)

(a)

(b)

Air space

Xylem

Parenchyma

Fiber   Phloem   Companion cell

Sieve tube element   Vessel element

dle. Whatever the arrangement, secondary tissues are poorly developed, and the stem remains nonwoody. Frequently the cortex may contain many collenchyma cells and the vascular tissue may contain fibers, both of which aid in the support of these stems. The parenchyma cells, between vascular bundles, are continuous with those of the pith and cortex.

Woody Dicotyledonous Stem

This is the type of stem (Figures 7-16[c] and 7-19) that was utilized for the discussion of primary and secondary tissues in earlier sections of this chapter. Although the very young stem has vascular bundles, the conducting tissues of mature woody stems are in the form of concentric cylinders, in which the great

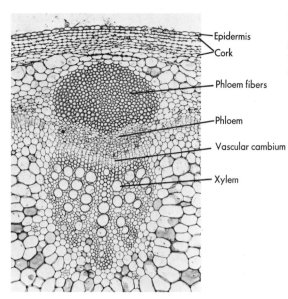

Epidermis

Cork

Phloem fibers

Phloem

Vascular cambium

Xylem

Figure 7-18. Photomicrograph of portion of herbacious dicotyledonous stem, transverse section, *Helianthus* (×100). (Courtesy Carolina Biological Supply Company.)

development of secondary xylem results in the characteristic woody condition. These stems usually have much less pith than the herbaceous dicotyledonous stems.

Summary

1. Herbaceous stems are generally soft and green, have little tough woody tissue, and are short-lived. Woody stems are hard and long-lived. In both types leaves and buds are located at nodes, and the interval between them is the internode.

2. New stems and their leaves develop from terminal or axillary buds. Bud scales are present in woody plants.

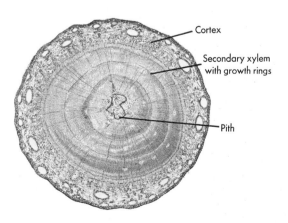

Cortex

Secondary xylem with growth rings

Pith

Figure 7-19. Photomicrograph of woody stem, transverse section, pine (×50). (Courtesy Carolina Biological Supply Company.)

3. Primary tissues develop from apical meristems, and secondary tissues arise from a cambium. A transverse section through a mature woody stem would expose the following tissues and regions: epidermis, cortex, primary phloem, secondary phloem, vascular cambium, secondary xylem, primary xylem, and pith. As additional secondary tissues develop, the outer regions of the stem split and tear, which results in the formation of cork from a cork cambium.

4. The cells that may be found in the phloem are sieve tube elements, companion cells, phloem parenchyma cells, and phloem fibers. In the xylem may be found vessel elements, tracheids, xylem parenchyma cells, and xylem fibers. In secondary xylem and phloem vascular ray cells are also found.

5. The rings that are visible in the trunk, or stem, of a cut tree are formed in woody stems as a result of the differential growth of the secondary xylem.

6. Herbaceous monocotyledonous stems contain vascular tissue that is present as scattered bundles of xylem and phloem. Herbaceous dicotyledonous stems contain similar vascular bundles, but these are arranged in an orderly ring. In the woody dicotyledonous stem, the vascular tissues are in the form of concentric cylinders.

Review Topics and Questions

1. Diagram in cross section a woody stem with a vascular cambium but no secondary tissues. Label all regions or tissues, but do not show individual cells.
2. Diagram the stem of the previous problem after a year's cambial activity. Label all regions or tissues, but do not show individual cells.
3. Diagram in longitudinal (radial) section the stem of the previous problems after six years of growth. Label all regions or tissues, but do not show individual cells.
4. Describe each of the cells that are (or may be) common to both xylem and phloem, and list the main function of each.
5. Describe each of those cells that are located in the xylem or phloem but not in both, indicate in which tissue each is found, and list the main function of each cell.
6. For each of the following, name a plant cell that has that function, describe the structure of the cell, and discuss how such structure is an adaptation for that function: (a) support, (b) conduction of water, (c) food storage, and (d) conduction of food.
7. In terms of internal anatomy, compare a mature herbaceous dicotyledonous stem with a mature herbaceous monocotyledonous stem.
8. A branch was cut from a willow tree that was over three years old. If you were given the branch, explain how you could prove (a) its age at the time of cutting, and (b) the time of year at which it was cut.
9. Explain why tissues outside the cambium of a stem are frequently cracked, torn, and sloughed off.
10. In what ways is a vessel better modified than a series of tracheids for the conduction of water?

Suggested Readings

ESAU, K., *Anatomy of Seed Plants* (New York, Wiley, 1960).

ESAU, K., *Vascular Differentiation in Plants* (New York, Holt, Rinehart & Winston, 1965).

FAHN, A., *Plant Anatomy,* 2nd ed. (Elmsford, N.Y., Pergamon Press, 1974).

WILLIAMS, S., "Wood Structure," *Scientific American* **188,** 64–68 (January 1953).

Stem Function

The preceding chapter was devoted primarily to a discussion of the external and internal structure of the stem. As each structure was described, a specific function correlated with that particular structure was indicated. The correlation between structure and function in storage, supporting, and strengthening tissues or cells is rather straightforward and simple. Storage cells are thin-walled, which enables materials to penetrate readily, and have cytoplasm and large vacuoles in which such materials are stored. The metabolic activity of these cells enables them to accumulate storage products against a concentration gradient or to store products in a nonsoluble form (e.g., starch). Strengthening and supporting cells are vertically elongated and have thick walls and small lumen; sometimes additional ring or spiral wall thickenings are present. However, difficulties arise in finding a similar correlation between structure and the conduction of various materials through a stem.

8.1 Movement of Organic Materials

Experiments have demonstrated quite conclusively that organic materials (e.g., sugars, amino acids, growth regulators) are transported by the phloem. In some of these experiments all tissues external to the cambium are removed in a band around the stem. Though this removes the phloem, the xylem is not disturbed, and water quite readily passes the **ringed** (or **girdled**) area. Any such interruption of the phloem prevents the transport of organic compounds, primarily sugars, from one side of the ring to the other; both upward and downward movement ceases. In control plants—those in the normal condition with intact phloem—movement of sugar occurs quite naturally. Some investigators have even removed sections of the xylem without destroying the phloem (see Figure 8-1) and have found sugar movements to be normal, provided care is taken that the phloem does not dry out. Various experiments have been carried out using tagged or labeled sugar (i.e., sugar containing radioactive carbon atoms), and following the pathways through which such sugar is transported. The rapid vertical movements are found to occur in the sieve tubes of the phloem. As a result of this accumulated information, botanists accept the statement that organic materials move upward (to growing stem tips, for example) and downward (to storage areas in the root, as an example) in the phloem. This is not meant to exclude diffusionary movements from cell to cell, or the movement of very small amounts of organic material in the xylem. The great amount of rapid movement, however, occurs in the phloem.

In the spring of the year sugars are frequently found in the xylem conducting elements of various woody plants. In fact a sugary syrup is obtained from the sugar maple tree by driving metal tubes into the tree trunk to the xylem. The sugar solution slowly drips out into buckets hung on the tubes. These sugars undoubtedly come from the storage cells (parenchyma type) of the pith and xylem. The highest concentrations of sugar are found during the winter and early spring when there is little or no upward translocation. Various investigations have indicated that this sugar is transported down the phloem from the leaves and into the storage areas by way of the vascular rays.

The difficulties that arise in discussing **translocation** (or transport) in the phloem relate to the mechanisms involved, not the tissue. Many measurements have been made of the rate of movement through the phloem, and data indicate a rate that is at least a thousand times the rate that could be accounted for by diffusion alone. For example, it would take approximately 900 days for 1 mg of sucrose, the main organic material transported in the phloem, to diffuse 1 m from a 10 per cent solution. The rates of transport, as actually measured, vary considerably from plant to plant and with the environmental conditions, but they range up to at least 0.5 m per hour and possibly as high as 1 m per hour. If translocation is faster than diffusion, what is the mechanism?

Several suggestions have been made regarding the movement of materials in the phloem. Some plant physiologists consider that a **mass flow** of materials through the sieve tubes occurs as a result of higher turgor pressure on the supplying end of such tubes. Sugar manufactured in the leaf is actively secreted or ''pumped'' into the sieve tubes of the vascular tissue by neighboring parenchymatic cells. In many

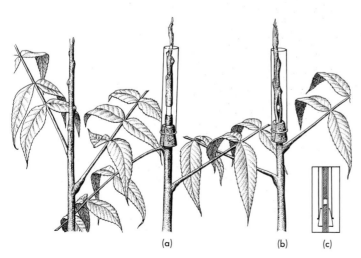

Figure 8-1. Diagrams to indicate one technique of investigating the movement of materials in stems. **(a)** The phloem has been removed by excising the tissues external to the cambium (ringing). **(b)** The xylem has been removed. **(c)** Sectional view of **(b)**. The cut area of the stem is enclosed in a water jacket to keep the exposed tissues from drying in each instance. (Redrawn from *The Translocation of Solutes in Plants*, by O. F. Curtis, New York, McGraw-Hill, copyright 1935 by the publishers and used by permission.)

(a)   (b)   (c)

plant species some of the parenchyma cells are characterized by a greatly increased surface area of the plasma membrane, brought about through the development of irregular ingrowths of the cell wall—the production of plasma membrane material follows the contours of the irregular walls. These **transfer cells** are believed to play an important role in collecting sugars from the food manufacturing cells of the leaf and transferring such solutes to the sieve tubes, a function greatly facilitated by the extensive membrane surface. This *phloem loading* appears to be a mobilization of sugars and is probably an active transport process. Such an increase of solutes decreases the water potential in this region of the sieve tube and water moves in by osmosis from the nearby xylem. The turgor pressure (or hydrostatic pressure) becomes greater at the *source* (a region that supplies sugars to the phloem) than at the *sink* (a region that utilizes sugars that were translocated in the phloem). The difference in water potential between source and sink results in the development of a turgor pressure difference between the two areas. This is the driving force that moves the sugar solution in the sieve tube. At the sink end sugar is removed and utilized or stored. This results in increased water potential and movement of water out of the sieve tube—thus, the mass flow.

An elegant technique demonstrating pressure in sieve tubes utilizes plant-feeding insects. It has been shown that aphids insert their mouthparts (stylets) into a single sieve tube element of young woody branches. The contents of the sieve tube are then forced into the body of the aphid by the turgor pressure of the sieve tube. The insect is killed and the body is cut away, leaving the stylet. Exudation from the stylet continues for a day or more and may be collected. The amount of material collected (which is about 10 per cent sucrose, the primary material carried in the phloem) is frequently comparable to the contents of 100 or more sieve tube elements. This seems to indicate a flow of materials along a considerable length of the sieve tube. Other investigators, however, have shown that respiratory poisons and low temperatures or low oxygen supply reduce or stop translocation. The implication is that sieve tube elements play an active role in transport and not merely a passive role involving turgor pressure, although such activity may simply maintain the concentration or turgor pressure gradients. Mass flow can account for translocation in only one direction at a time, and there are data from various experiments that indicate bidirectional movement in the phloem—as some organic materials are moving out of a leaf, for example, others are moving into that leaf. It is not absolutely certain, however, that the two materials are moving in the same sieve tube; there may be opposite movements in adjacent sieve tubes, and some lateral movement between these sieve tubes. However, it is still somewhat difficult to explain mass flow and the forces involved in such movement with regard to opposing directions in adjacent cells. Also, a mass flow would be rather difficult through sieve plates.

Other plant physiologists consider **cytoplasmic streaming** to be important in transporting materials from one end of a cell to another, followed by diffusion through plasmodesmata. Such flowing movements can be observed in some cells, but not in all, and the rate is apparently too slow. In support of this theory is the fact that conditions that retard or stop cytoplasmic streaming, such as low temperature or low oxygen supply, influence translocation in the same manner. Simultaneous transport in opposite directions could be accommodated quite readily, and diffusion would be over such short distances as not to slow the rate. However, cytoplasmic streaming has not been observed in mature sieve elements.

Another suggestion is that of **activated diffusion,** in which the metabolic activity of living cells speeds up diffusion. Actually, all these suggestions have their proponents and opponents, the data lend themselves to alternative explanations, and no hypothesis has yet knitted the data together in a suitable explanation. Despite the fact that there is evidence against mass flow, as well as for it, most of the investigators of phloem transport consider it to be the most plausible explanation although some modifications may be necessary as a result of continuing research. The simple fact is that the mechanism whereby organic compounds move through the phloem is yet unknown in spite of the tremendous amount of experimentation that has been involved. This is another one of those problems of living organisms that cause us to look with great anticipation to the future.

8.2 Movement of Minerals

Ringing experiments and other types of investigations have not been as conclusive with regard to mineral transport as they have been with regard to sugar movements. The majority of evidence, however, indicates that minerals are, for the most part, transported in the xylem. One of the difficulties in these investigations is that certain minerals are very rapidly utilized in the formation of organic compounds, which then move in the phloem. This is quite likely true of nitrogen and phosphorus. Another difficulty is that salts transfer radially to the phloem and other tissues and may accumulate in living cells. However, experiments using radioactive ions (e.g., potassium, sodium, phosphate, bromide) demonstrated clearly that the xylem is the main pathway for the upward movement of mineral elements from roots to leaves. To prevent lateral movement of the mineral elements longitudinal slits were made in the stem, the bark was carefully pulled away from the wood (but leaving it attached at the ends), and a sheet of paraffined paper was inserted between the phloem and the xylem. Practically all of the tracer element was found in the xylem. Above and below the stripped area, there were considerable amounts of the tracer in both xylem and phloem.

Not all of the mineral elements that are translocated into a given leaf remain in that leaf; others are much less mobile. Much of the export of minerals occurs just prior to leaf fall and takes place by way of the phloem. Once in the stem, these solutes move both upward and downward in the phloem into younger leaves and frequently to apical regions of the stem or root. Such recycling of essential minerals is, of course, advantageous in that such materials are not lost to the plant when the leaf falls. Those minerals left in the falling leaf and the minerals in the rest of the plant body are returned to the soil when these structures decay. In this way essential minerals become available to other plants. Harvesting plant parts interrupts this mineral cycle.

8.3 Movement of Water

That most of the water absorbed is transported throughout the plants by means of the tracheids and vessels in the xylem has been conclusively shown. The difficulty again is in explaining the mechanism involved. Capillary movement of water through the xylem cells could account for a rise of water of only a few feet, even if the smallest diameter of a tracheid is used as a basis for calculation. As indicated previously, the most efficient conducting cells are the large-diametered vessel elements, in which water would rise much less by capillarity than it would in a tracheid. Imbibitional movements within the submicroscopic channels of the cell walls must be ruled out, because water moves in the lumen; plugging the lumen prevents water conduction. For a time, some plant physiologists thought that root pressure forced water upward. Such pressures have been demonstrated in many plants and depend upon the activity of living root cells. However, other plants do not develop root pressure, root pressures are not great enough to force water to the top of tall trees, and water movement through stems continues even though roots have been removed (note almost any cut flowers in the home). Another suggestion was that the loss or utilization of water at the tops of the plant produced a vacuum and that atmospheric pressure pushed water up the hollow tubes of the xylem. Unfortunately, atmospheric pressure could account for only a rise of approximately ten meters (33 ft), and many trees are taller than that.

Strange as the idea may appear at first reading, botanists are now convinced that water is *pulled* up the stem of a plant. Chemists and physicists have demonstrated that water molecules are very strongly attracted to one another; water molecules in thin columns are difficult to separate, and considerable tension (pull) is necessary to accomplish separation. This fact was made use of by plant physiologists in explaining water movements. As water is lost by evaporation from cells at the top of a plant, or as water is used by these cells (e.g., photosynthesis, digestion), more moves into such cells by osmosis from neighboring cells. (Why?) One or more of these cells is next to a tracheid or vessel element from which water is removed. Imbibition likely is involved in water movements through the walls of the tracheid and the adjacent cell. At this point, the **cohesive** strength between water molecules enters our explanation. Since the molecules cling together, as some are removed at the top of the xylem, the rest are pulled upward—just as pulling on the end link of a chain will cause the entire chain to move along. These water columns are under tension, a phenomenon that demonstrates itself by a decrease in the diameter of a tree trunk during times of great water loss and movement. Just as an external force increases water potential, a tension decreases the water potential (review Sections 5.2 and 5.4, if necessary). The result is that the water potential in the tracheids and vessels is less than that of the living root cells, and water moves from the latter to the former. This lowers the water potential in the cells adjacent to the xylem cells, and an osmotic movement of water takes place through living root cells from the soil solution. Note that water movement is osmotic through living cells and cohesive through the nonliving cells (tracheids and vessels). Cohesion is not the basis for all water movement but is really limited to those areas where nonliving cells are involved.

In this discussion of translocation, various mechanisms other than diffusion were emphasized. This should not be interpreted to imply a relatively unimportant role for the process of diffusion. Movements from one parenchymatous cell to another are by diffusion, termed *osmosis* in the case of water. If the distance is short and the gradient steep, as across membranes between adjacent cells, diffusion is adequate. However, the known rates of diffusion are much slower than the known rates of translocation through the xylem and the phloem. Obviously, then, rapid vertical movements over relatively long distances are brought about by other, or additional, mechanisms and not by diffusion alone. Radial (horizontal) movement, as through vascular rays (xylem rays plus phloem rays), is a slow process dependent upon diffusion.

8.4 Specialized Stems

In a number of plants, the stems are greatly modified as to structure and function. Many such modifications are concerned with **vegetative** or **asexual reproduction**—that is, reproduction not involving fusion of **gametes** (sex cells) or the production of seeds. Other structural modifications result in a variety of protective and supporting devices. However, even though the stems may be very different from the "typical," they can be recognized if one looks for nodes (and their attendant buds and leaves) and for the development of branches from the surface. Secondary, or branch, roots originate from internal tissues, and roots do not have nodes.

Runners

The strawberry plant (*Fragaria*) has a long, slender stem that grows horizontally along the surface of the ground. New leaves and roots develop where a node touches the ground, and these are independent plants as soon as the **runner** (or **stolon**) dies (Figure 8-2). In many areas the strawberry is used as a ground cover

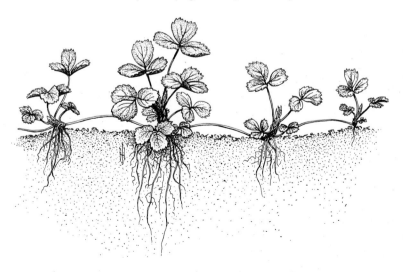

Figure 8-2. Strawberry plant (*Fragaria*). The oldest plant is second from the left. Other plants have developed at intervals along horizontal stems.

because of its rapid reproduction and spreading characteristics. Many plants produce this type of structure. Bermuda grass (*Cynodon dactylon*), for example, has become a very troublesome weed in many areas because of its rapid vegetative reproduction by stolons. If a few plants become established, it may completely overrun an area during the first year of its presence.

Rhizomes

As exemplified by iris and fern plants, a **rhizome** (Figure 8-3[c]) is an underground, horizontal stem, which may be bulky and contain much stored food. Most rhizomes live through the winter, even though the shoots may die, and new shoots develop the next spring from the buds, which are located at nodes. Breaking of the rhizome will not destroy the plant but merely separates the new shoots that are produced. As long as a node with its bud is present, each piece of rhizome is capable of producing a new plant, provided it has sufficient stored food.

Tubers

The **tuber** (Figure 8-3[a,b]) of an Irish potato (*Solanum tuberosum*) is a bulky, short terminal portion of an underground stem. The "eyes" are actually buds at nodes, and the bulky appearance results from the compressed, unexpanded internodes. As long as the piece of tuber contains at least one "eye" when it is planted, shoots and roots will develop, and a new plant results. The food value of potato

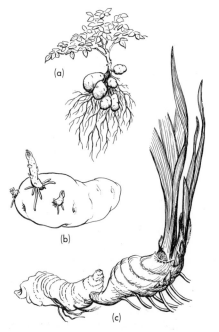

Figure 8-3. Some specialized stems. **(a)** Irish potato plant (*Solanum tuberosum*) with tubers. **(b)** Potato tuber sprouting. **(c)** Iris rhizome sprouted.

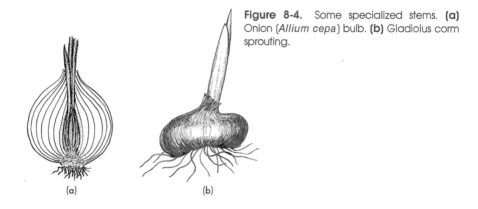

Figure 8-4. Some specialized stems. **(a)** Onion (*Allium cepa*) bulb. **(b)** Gladiolus corm sprouting.

(a)   (b)

tubers for humans results from the enormous numbers of starch grains that are stored in the cortical and pith cells; almost the entire mass of the tuber is storage parenchyma. However, the sweet potato (*Ipomoea batatas*) and yam (*Dioscorea alata*) are storage roots.

Bulbs

Actually, a **bulb** (Figure 8-4[a]), as in the onion (*Allium cepa*) and lily, consists of a very small piece of stem tissue bearing numerous fleshy leaves. Terminal and lateral buds are present, and both may develop into shoots. Frequently, in slicing an onion that has been stored for some length of time, one will notice that the buds have already begun to grow. In this structure the fleshy leaves are the storage organs and contain large amounts of sugar but no starch. The odor and tear-producing characteristics of onions are caused by aromatic oils.

Corms

A **corm** (Figure 8-4[b]), as in gladiolus, consists of a bulky, short, vertical stem, which contains stored foods. The greater part of the structure is stem tissue thus differentiating it from bulbs, which are primarily fleshy leaves.

Spines and Thorns

The protective thorns (Figure 8-5[a]) of rose bushes (*Rosa*) and spines (Figure 8-5[b]) of cacti (*Opuntia*) are examples of stem outgrowths or modified stems. These sturdy, sharp-pointed structures are probably of survival value in decreasing the amount of browsing damage to these plants by herbivores (and probably decrease damage caused by humans also). The spines of some cacti are so dense as to appear as a white hairy covering (e.g., old man cactus, *Cephalocereus senilis*), and this may afford some protection from the sun's rays by reflecting light

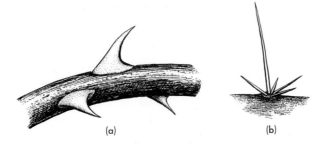

Figure 8-5. **(a)** Thorns of a rose (*Rosa*). **(b)** Spines of a cactus (*Opuntia*).

(a)   (b)

and thus reducing the possible heating effect of the sun. Such cacti can possibly withstand greater sunlight intensity than others not so endowed with enormous numbers of spines. The spines of some plants are modified leaves.

Tendrils

In some plants, such as the grape (*Vitis*), certain of the stems develop as long, thin, coiling structures, which attach the plant to a support (Figure 8-6). Because of unequal rates of growth, the tips of tendrils move slowly back and forth as they elongate. As soon as a tendril comes in contact with a solid object, rapid growth responses result in the tendril coiling about the object. The tendrils of Boston ivy (*Parthenocissus tricuspidata*) end in flattened discs, which adhere to surfaces and enable the plant to "climb" vertical walls. Tendrils may also be modified leaves, but in any case such structures support the photosynthetic areas where they will be exposed to light, a function performed by mechanical tissues (composed of fibers and tracheids mainly) in many plants.

8.5 Grafting and Rooting

Stems are frequently used in the vegetative propagation of desirable varieties of plants. In some instances the cultivated variety, although producing large yields of good quality (a result of plant breeding programs), is more susceptible to root diseases than is the naturally occurring variety. Buds or branches of the susceptible variety may be grafted to the stem of the resistant variety. In **grafting,** the freshly cut surfaces of two stems are bound together firmly so that the two cambial layers are at least in partial contact. **Callus,** consisting of undifferentiated cells, forms from the various parenchymatous cells at the region of the cuts. Differentiation of some of these cells forms a cambium that unites the cambial regions of the two stems. Eventually, differentiation of cells from this continuous cambium layer results in the joining of vascular tissues from one stem to the other. The graft is considered to be successful if this union is sufficiently sturdy to prevent rupture when the binding material is removed. Grafting wax or a similar compound is used to cover the wound surfaces to prevent tissues from drying out and to prevent fungi from entering.

Figure 8-6. Tendril of a grape (*Vitis*) coiled about a twig.

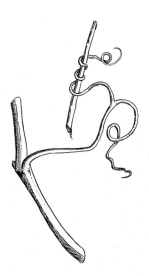

The **scion** is the stem cutting, which is grafted to a rooted portion, the **stock.** Neither the scion nor the stock influences the other's growth habits, except for nutritional materials that may be transported through the graft union. Therefore, to graft several varieties of apples on the same root stock is possible; each cutting then bears a different type of apple. Although plants of two different genera within a family can sometimes be grafted, most successful grafts are between members of the same species. The general growth and tissue differentiation must be quite similar for graft unions to be substantial enough to resist rupture. Figure 8-7 represents several common methods of grafting.

Many citrus (orange, lemon) trees consist of a resistant root stock variety to which other varieties have been grafted. The navel orange is probably the best example of the usefulness of grafting. Because these oranges are seedless, they can be propagated only by grafting. In fact, the entire navel orange industry is dependent upon trees that are all asexual descendants of a single tree found in South

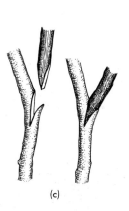

Figure 8-7. Several common methods of grafting. **(a)** Shield budding. **(b)** Whip or tongue grafting. **(c)** Side grafting.

(a)

(b)

(c)

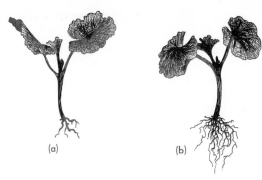

(a)   (b)

America almost 100 years ago. To perpetuate such a desirable variety, it has been necessary to propagate it vegetatively by grafting.

Another method of vegetative propagation that can be used with some kinds of plants is to make stem cuttings and then place the cut ends in moist sand or peat moss. Plants, such as *Coleus,* willow, rose, geranium, and grape, develop adventitious roots (see Section 9.1) at the cut end when handled in this manner. Increased formation of such roots can frequently be obtained by treating the basal portions of cuttings with plant growth regulators (see Section 29.11). Figure 8-8 represents the rooting of cuttings, some of which were treated with naphthalene acetic acid. Some cuttings will not root unless treated in this fashion. The cuttings are usually kept in a shaded area with a humid atmosphere to reduce water loss and thus water requirements. Some of the leaves are also usually removed from the cutting to reduce loss of water. Until roots form, these cuttings are very susceptible to wilting because of poor water absorption.

These two methods of vegetative propagation are frequently useful in maintaining varieties without the changes that so frequently occur as a result of sexual reproduction. The plants that result from grafting and from rooting of cuttings have the same characteristics as the plants from which the stems were taken. Plants that develop from seeds are frequently different from the plant that produced those seeds, as will be explained in Chapter 18. Seeds from an apple tree will always produce apple trees, of course, but the size, shape, quality, or yield of apples may vary from tree to tree even if the latter trees all developed from seeds obtained from a single apple tree or even from a single apple fruit. This is not the case with vegetative reproduction.

8.6 Economic Aspects of Stems

In addition to lumber, wood pulp for paper, and fibers for linen and rope, man has utilized plant stem tissues and their products in many ways. The bark of the cork oak (*Quercus suber*) is useful for stoppers, insulation, life preservers, and padding. Quinine, an alkaloid used in the treatment of malaria, is extracted from *Cinchona* bark. During World War II, when *Cinchona* bark could not be obtained, a synthetic drug, atabrine, replaced quinine to a great extent. For similar reasons, synthetic rubber was used in place of natural rubber from *Hevea brasiliensis* (a tropical tree). The potato tuber has already been mentioned as a food source, but just as important are sugar and molasses from sugar cane (*Saccharum officinalis*). To flavor food, one can use spices from *Cinnamomum zelylanicum* (cinnamon from the bark). Tars and wood alcohol are distilled from hardwood trees, such as oak and hickory, while resins and turpentine are obtained from softwoods, such as yellow pine. In recent years many materials have been developed that utilize cellulose as a base: celluloid, cellophane, rayon, and lacquers (ethyl cellulose). The cellulose is obtained from wood (plant cell walls).

The foregoing is only a very brief indication of the economic uses of stems, usually in the form of tree trunks and branches. Any economic botany text can be consulted for a more detailed discussion of these and other uses of plants and their products.

Summary

1. Organic materials are transported mainly in the phloem, while water and minerals are transported primarily in the xylem. Because of the cohesive strength between water molecules, the columns of water in the xylem are under tension as they are pulled to the tops of the plant.
2. In some plants stems are variously modified as runners, rhizomes, tubers, bulbs, corms, spines, and tendrils.
3. In grafting, the cut surface of the scion is tightly bound to that of the stock so that the new cells that form will tend to form a union.
4. Many stem cuttings will root if the cut portion is placed in moist sand or peat moss. The application of certain plant growth regulators to the basal portion of such cuttings frequently stimulates root initiation.
5. Man has found many uses for plant stems: lumber, wood pulp for paper, wood fibers for cloth and rope, cork, quinine, rubber, food, resins, alcohol, and plastics.

Review Topics and Questions

1. A water molecule, originally in the conducting tissue of a stem, eventually is utilized in the growth of a cell of the apical meristem. Discuss the movement of this water molecule, including the cells concerned and the forces that bring about such movements.
2. Explain why most trees will die if all tissues external to the xylem are removed in a ring around the stem.
3. Describe in detail an experiment that would demonstrate whether or not organic materials are transported in the phloem.
4. Describe in detail an experiment that would demonstrate whether or not root pressure is involved in supplying water to the top of a plant.

5. Describe five types of modified (or specialized) stems and indicate the importance of each to the plant involved.
6. Discuss the importance of grafting to humans.
7. List ten items, commonly used by humans, that are made from plant stems or from parts of plant stems.
8. List five products that humans extract from plant stems.
9. The trunk of a tree is split. In order to save the tree, the owner decides to place either a metal bolt through the trunk or a metal band around the trunk, and asks you for advice. Explain which method he should use.
10. During their honeymoon at a mountain cabin, a couple fastened a wooden bench firmly between two trees. They returned on their fiftieth anniversary. Would they be able to use the bench? Explain, assuming the trees are still living and had grown at the rate of 1 ft per year.

Suggested Readings

BIDWELL, R. G. S., *Plant physiology* (New York, Macmillan, Inc., 1974).

CRAFTS, A. S., *Translocation in Plants* (New York, Holt, Rinehart & Winston, 1961).

GREULACH, V. A., "The Rise of Water in Plants," *Scientific American* **187**, 78–82 (October 1952).

PEEL, A. J., *Transport of Nutrients in Plants* (New York, John Wiley and Sons, Inc., 1974).

Roots

The roots of a plant are the belowground structures through which materials move from the soil to the various parts of the plant. Water and minerals are the primary essential nutrients that the plant obtains from the soil, and these nutrients are either utilized in the roots themselves or are transported to other parts. Note that the root cells are closest to the supply of water and minerals, and the amount of transport out of the roots is influenced by requirements of root tissues themselves. In addition to absorption and conduction, roots serve to anchor the plant, are frequently storage areas, may serve in supporting the plant, and are extremely important in holding soil particles in place (erosion control).

9.1 Gross Structure of Roots

The roots are a continuation of the main axis of the plant, and no sharp line of demarcation exists between what is root and what is stem. When a seed germinates, the root grows down into the soil as the stem grows up and eventually out of the soil. As additional roots are produced, the root system develops as one of two general types: (1) a **diffuse,** or **fibrous, root** system, or (2) a **tap-root** system. In the former, as exemplified by grasses (Figure 9-1[a]), are numerous slender main roots, which are nearly equal in size and which branch profusely. Such a root system has a high surface-to-volume (S/V) ratio and is a rather efficient organ of absorption. (Refer to Section 3.10 for a discussion of the importance of S/V ratios.) In some plants, such as the sweet potato (*Ipomoea batatas*), certain of the larger roots may develop as enlarged storage areas. Because of their root systems, grasses are utilized in preventing **erosion** (the wearing away of soil by wind and water). The small, abundantly branched roots are intertwined throughout the soil particles and tend to anchor both plant and soil in place.

In a tap-root system, such as found in beets, dandelions, and carrots, the primary root grows rapidly and remains the dominant part of the underground structure (Figure 9-1[c]). This main, or tap, root may enlarge considerably as a fleshy storage root containing much stored food materials when the plant is mature.

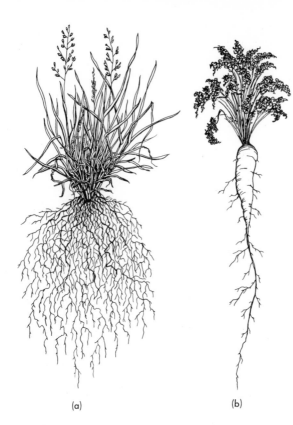

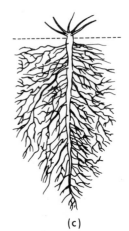

Figure 9-1. (a) Diffuse or fibrous root system of a grass. **(b)** Tap root (fleshy) system of a carrot (*Daucus carota*). **(c)** Dandelion (*Taraxacum officinale*) with nonfleshy tap root.

(a)   (b)   (c)

However, most of the absorption of materials is accomplished by the secondary and tertiary (branch) roots. As indicated in Figure 9-1(b), if a carrot is carefully removed from the soil, the small branching roots are readily seen. Whereas the outer layers of the mature tap root usually consist of cork cells, this is not true of the tiny younger roots through which materials enter the plant.

In many plants **adventitious roots** frequently develop. These are roots that arise from stems and leaves (Figure 9-2). The prop roots of corn are adventitious. They arise from the lower part of the stem and grow out and down into the soil, serving to brace the plant. The rooting of stem cuttings depends upon the development of adventitious roots, as does the rooting of African violet (*Saintpaulia ionantha*) leaf cuttings. The roots that form from tubers and corms, as when white potatoes and gladiolus are planted, are also adventitious. In general, adventitious roots arise from parenchyma cells situated near vascular tissue.

The root system may be shallow, but many are deeply penetrating and very extensive in proportion to the tops. The lateral extent is commonly greater than the spread of the branches (shoots). The primary form and development of the root are governed by hereditary growth characteristics of the species and by the environmental conditions under which the plant is grown. The porosity or compactness of the soil, the moisture and air contents of the soil, and the availability of minerals are all factors that may influence root development. In general, however, root systems are quite extensive. Many lawn grasses have roots that penetrate to at least 30 cm (1 ft). Apple trees frequently have roots that spread laterally for 7 to 10 m (20 to 30 ft) and reach a depth equally as far. Even some garden vegetables develop roots that extend 1 to 2 m (3 to 6 ft) into the soil (e.g., carrots, squash, pumpkin, tomato).

The rate of root growth is usually rather rapid and is also closely correlated with environmental conditions. Under favorable conditions, the rate of root elongation of many plants reaches 5 cm (2 in.) per day. Such growth, and the extensive production of lateral roots, continually serves to bring the absorbing structures of the plant into contact with new soil areas. As a result, new supplies of soil moisture and minerals are available to the plant. In some deep-rooted shrubs and trees, the plant may survive even if the surface layers of the soil are dry, provided the roots have penetrated to underground water sources. Plants in which germination is followed by immediate deep rooting may survive subsequent dry periods. Bur oak (*Quercus macrocarpa*), for example, has been known to produce a root that extended to a depth of about 2 m (6 ft) in one season's growth; the top growth was about 35 cm (14 in.) at this time.

(a) (b)

Figure 9-2. Adventitious roots. **(a)** Prop roots of corn. **(b)** Roots developing from leaves. (Courtesy of Peter Jankay.)

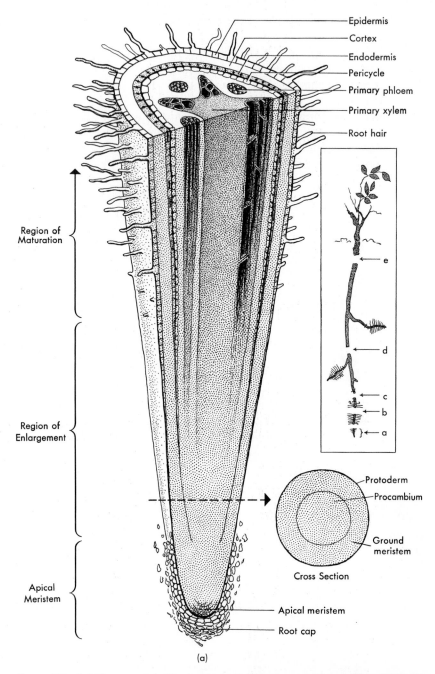

Epidermis
Cortex
Endodermis
Pericycle
Primary phloem
Primary xylem
Root hair

Region of
Maturation

Region of
Enlargement

Apical
Meristem

Protoderm
Procambium
Ground
meristem

Cross Section

Apical meristem

Root cap

(a)

Figure 9-3. (a) Diagram of root tip showing growth regions and tissue differentiation. The enclosed figure indicates the regions at which sections were made. In **(b)** and **(c)** on the facing page, lateral (or secondary) roots are visible.

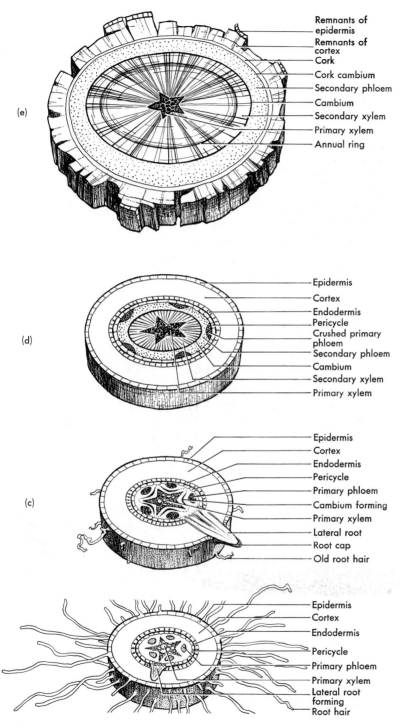

(e)

Remnants of epidermis
Remnants of cortex
Cork
Cork cambium
Secondary phloem
Cambium
Secondary xylem
Primary xylem
Annual ring

(d)

Epidermis
Cortex
Endodermis
Pericycle
Crushed primary phloem
Secondary phloem
Cambium
Secondary xylem
Primary xylem

(c)

Epidermis
Cortex
Endodermis
Pericycle
Primary phloem
Cambium forming
Primary xylem
Lateral root
Root cap
Old root hair

Epidermis
Cortex
Endodermis
Pericycle
Primary phloem
Primary xylem
Lateral root forming
Root hair

(b)

The tremendous extent of some root systems may be emphasized by the follow-ing estimates. By very carefully counting roots and root hairs (the main water-absorbing portion) and by measuring many of them, it has been estimated that a single rye plant had a total of over 13 billion roots and about 15 billion root hairs with a total surface area of over 557 sq m (6,000 sq ft). It was also estimated that new root hairs develop at a rate of more than 100 million per day. This is an enor-mous amount of absorbing surface through which the plant may obtain materials from the soil.

9.2 The Root Tip

The root tip (Figure 9-3) is more elongated than the stem tip and is not enclosed by developing or protective leaves or scales. Its growth regions, therefore, are some-what easier to examine closely.

Apical Meristem and Root Cap

As mentioned previously (see Section 6.4), at the tips of roots and stems are the apical meristems, regions in which the cells are actively dividing so that new cells are continually being produced. The cells are small, thin-walled, and cubical or spherical; they contain a dense protoplast (few, small vacuoles). Protecting this delicate meristematic tissue in the root is a covering that consists of a thimble-shaped mass of cells, the **root cap** (Figure 9-4). As the root grows and the tip ad-vances through the soil, cells of the root cap are sloughed or rubbed off, and new cells are continuously added by divisions of the apical meristem cells directly be-hind it. The cells of the apical meristem become a part of the root cap, enlarge and mature as permanent root tissues, or continue to divide as a part of the apical meristem. In this way, additional tissue cells are produced, while the meristematic area remains fairly constant in size. As in the stem apex, the protoderm, ground meristem, and procambium are distinguishable; but in the root tip the procambium consists of the centrally located cells, and the ground meristem is limited to the cells in the region between the procambium and protoderm.

Area of Cell Enlargement

The new cells that are produced at the distal[1] end of the meristematic region become enlarged as they mature, and this region of cell growth is usually termed the **area of cell enlargement.** Again, remember that no sharp distinction exists be-tween the apical meristem and the area of cell enlargement; one blends into the other. The small vacuoles of the enlarging cell coalesce and expand, an increase in

---

[1] Distal refers to areas away from the apex or, in this case, to the older areas of the meristematic tissue.

Figure 9-4. The root cap area consists of the somewhat loosely arranged cells at the bottom of the photomicrograph. The apical meristem is directly behind the root cap (above in the photograph)—the more densely packed cells ($\times 50$).

protoplasmic contents takes place, and the cell wall stretches. New cellulose molecules are added between those already present, and additional wall layers are secreted. The cells gradually assume separate identities, and indications of tissue development are clearly visible in the older portions of this area.

Area of Cell Maturation or Differentiation

The enlarged cells eventually develop into the various more-or-less specialized cells of the mature tissues. Those cells that have already differentiated can be seen farther back from the root tip. In the **area of cell maturation,** specific types of tissues are clearly visible, and a division of labor here becomes evident: certain tissues store materials, while others transport materials.

As cells divide, enlarge, and differentiate, the root cap and apical meristem regions are forced farther and farther through the soil primarily as a result of the expansion of cells in the area of enlargement. Actually, the extent of the root cap, apical meristem, and cell enlargement areas remains fairly constant; the area of cell maturation is what increases greatly in size. A cell starts its life as part of the apical meristem. As it ages (if it does not become a root cap cell), the cell next exists as a part of the enlargement region and finally as a cell in what is now the

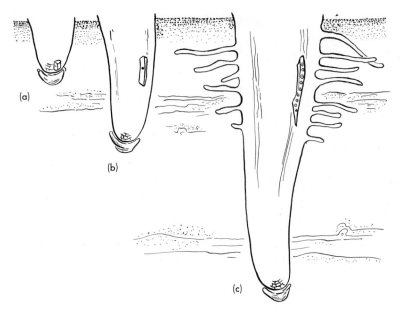

Figure 9-5. Maturation of a cell as the root grows. **(a)** The cell is at first a part of the apical meristem. **(b)** The cell enlarging. **(c)** The cell now matured as a tracheid in the xylem. As this cell develops, the root tip grows downward.

region of maturation. The cell has not changed positions; it has merely changed in structure and function. Figure 9-5 represents this forward, or downward, growth of a root, in which a single cell is shown in its various stages of development. For the sake of simplification, other cells have not been shown, but one must remember that many cells are undergoing a similar maturation.

In a transverse section through the region of mature cells in the root of a woody plant, the various tissues and regions can be diagrammed as in Figure 9-3; outlines of tissues and regions, but not the individual cells, are indicated. Only primary tissues are found at levels b and c, whereas secondary tissues have developed at levels d and e, which represent older tissues.

Epidermis (Figure 9-3). This tissue develops from the protoderm and is similar to the epidermis of the stem except for the presence of root hairs in the youngest portion of the region of maturation. A **root hair** (Figures 9-6 and 9-7) is an elongation of an epidermal cell, which becomes intimately associated with soil particles as it develops. Mucilage and pectic substances of the outer layers of the wall serve to maintain such contact, and root hairs are probably more important in erosion control than are the roots themselves. These hairlike appendages grow very rapidly but have only a transitory existence, dying within a few days. New root hairs are continually being produced, however, as cells enlarge and differentiate. Thus, the general root hair zone remains because the new root hairs replace the dying ones. The elongation of the epidermal cell, as a root hair forms, produces a tremendous increase in surface area with only a slight change in volume; an increase of the surface-to-volume ratio results. Since materials are absorbed through surfaces, the root-hair zone is a very efficient area of the roots for absorption; probably over 90 per cent of all absorption occurs through the root hairs and the root surfaces nearer the tip. The cuticle in this region is usually an exceedingly thin layer.

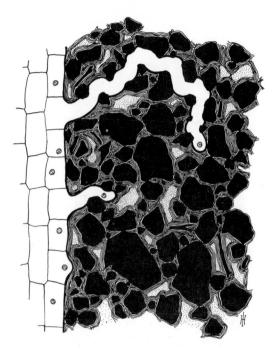

Figure 9-6. Development of root hairs. The soil particles are surrounded by films of water. The stippled areas represent air spaces.

Removing a plant from the soil, as in transplanting, rips off most of the fragile root hairs and root tips and results in a great decrease in the absorption of water from the soil. For this reason, after transplanting, the plant is watered thoroughly and protected from water loss (evaporation). **Pruning,** the removal of plant parts by cutting, aids in decreasing water loss by decreasing evaporating surfaces (leaves are removed).

Cortex (Figure 9-3). This region develops from the ground meristem and is much the same as in stems, except that endodermal layers of cells are more read-

Figure 9-7. Photomicrograph of a root tip with root hairs (×20). (Courtesy of Peter Jankay.)

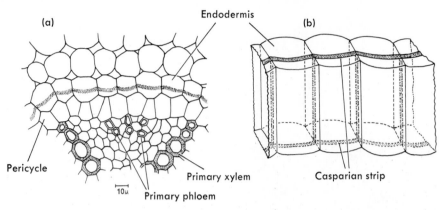

Figure 9-8. Structure of the endodermis. **(a)** Cross section of part of root of morning glory (*Convolvulus arvensis*) showing position of endodermis with regard to xylem and phloem. The endodermis is shown with transverse walls bearing Casparian strips in focus. **(b)** Diagram of three connected endodermal cells oriented as they are in **(a)**; Casparian strip occurs in transverse and radial walls but is absent in tangential walls. (From *Anatomy of Seed Plants*, by Katherine Esau, John Wiley and Sons, 1960, New York, reprinted by permission.)

ily distinguishable in the root. The **endodermis** consists of a sheathlike layer, one cell in thickness, which marks the inner boundary of the cortex. Most of these cells have a bandlike thickening along the radial and transverse walls (Figure 9-8). This **Casparian strip,** a part of the primary layer of the wall, contains suberin and sometimes lignin, or both, and these materials extend into the middle lamella. The cytoplasm of the endodermal cells is firmly attached to the wall at this region. The Casparian strip may form a barrier at which the soil solution is forced to pass through the differentially permeable cytoplasmic membranes rather than along or through the cell walls. In a recent study it was shown that materials, which may penetrate a root by way of the cell walls and intercellular spaces, were stopped at the Casparian strip although they were present in cell walls and spaces of the rest of the cortex. Thus the endodermis exercises some control over the movement of substances into and out of the vascular tissue of the root. In older roots additional wall thickenings may be present. Starch is frequently stored in cortical cells.

Pericycle (Figure 9-3).   The **pericycle** consists of parenchyma-type cells that are located between the endodermis and the vascular tissues. These cells originate from the peripheral cells of the procambium and retain a meristematic capability for a relatively long period of time. All branch roots arise from the cells of the pericycle, force and digest their way out through the cells external to them, and then penetrate out into the soil. Frequently, the root cap is visible before the branch (lateral or secondary) root has completely reached the exterior (Figure 9-9). Tertiary roots are those that form from the pericycle of secondary roots. In older roots, where secondary tissues are formed, the cork cambium and a part of the vascular cambium form from pericyclic cells.

Primary xylem (Figure 9-3).   Although the cells that constitute this tissue originate from the procambium and are similar to those of the stem, the xylem forms

152

(a)

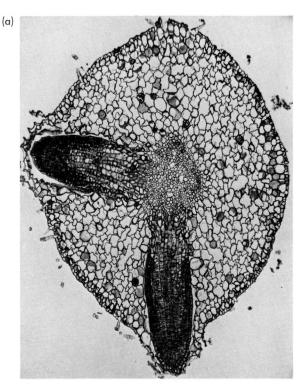

(b)

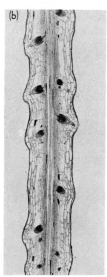

Figure 9-9. (a) Transverse section of a root with developing lateral roots (×50). (Courtesy of General Biological Supply House, Chicago.) **(b)** Longitudinal section of a root with developing lateral roots (×25). (Courtesy of Carolina Biological Supply Co.)

the central core of the root or radiating arms of the vascular tissue—a situation quite different from that in the stem. The general shape of the xylem, as seen in transverse section, is that of a star (Figure 9-10). In most older monocotyledonous roots (Figure 9-11) there is a pith (parenchyma cells) with the primary xylem as radiating arms of the "star," while in woody dicotyledonous roots the central area itself consists of primary xylem. In some herbaceous and a few woody dicotyledonous roots there may be a pith.

What appears as a "star" in a transverse section is actually a fluted column of xylem tissue. Maturation actually begins at the points of the star as the procambium cells in that region differentiate into tracheids or vessel elements, and continues inward, eventually forming a solid column of primary xylem. There is no pith in this type of root. In monocotyledonous roots the development is basically similar except that xylem does not form in the central region, where parenchymatous cells comprise a pith area.

Primary phloem (Figure 9-3).   Located between the projecting arms of the primary xylem are clusters of cells, which constitute the primary phloem. Once again the cells originate from the procambium and are similar to those found in the primary phloem of the stem; only the location of the tissue is somewhat different. In those roots where a pith is present, the primary phloem alternates with the primary xylem.

As the root of a dicotyledonous plant matures, a **vascular cambium** forms from parenchyma cells between the primary xylem and primary phloem and from some

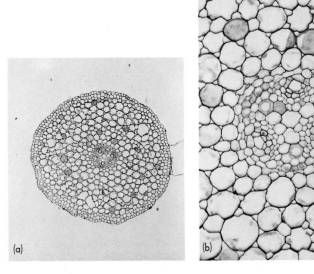

(a)

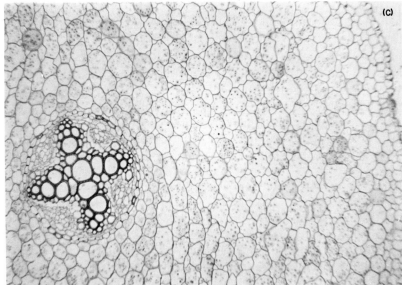

(b)

(c)

Figure 9-10. Photomicrographs of transverse sections of young dicotyledonous root. In **(a)** and **(b)** the cells in the center have not yet matured (as xylem cells). **(a)** ×40. **(b)** ×150. **(c)** ×100. **(d)** ×400 (on page opposite) ([a] and [b] courtesy of Robert Gill; [c] and [d] courtesy of Peter Jankay.)

of the cells of the pericycle. As in the stem, secondary xylem and phloem cells develop from the vascular cambium. Figure 9-3(e) represents a transverse section through a root containing secondary vascular tissues as well as cork, which forms as the increase in diameter tears apart the original outer (primary) tissues. The vascular cambium located on the inner face of the primary phloem begins to function earlier than the part derived from the pericycle. By the formation of secondary xylem opposite the phloem, the cambium is moved outward. Eventually, as the entire vascular cambium becomes functional, the production of secondary tis-

154

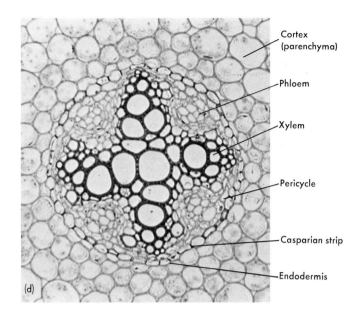

Cortex
(parenchyma)

Phloem

Xylem

Pericycle

Casparian strip

Endodermis

(d)

sues results in the cylindrical arrangement of vascular tissues. As in the stem, the epidermis and cortex are frequently sloughed off and are no longer present in older roots. Such roots absorb little materials because of the cork that now forms the outer layers. This cork develops from a cork cambium that originates from the pericycle; in older roots a cork cambium may arise from parenchyma cells of the secondary phloem. If any cracks or crevices are present in the surface, however,

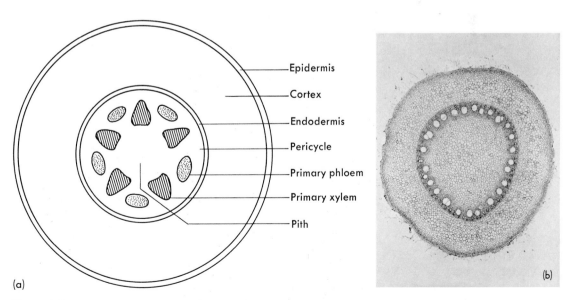

Epidermis

Cortex

Endodermis

Pericycle

Primary phloem

Primary xylem

Pith

(a)

(b)

Figure 9-11. **(a)** Diagram of primary tissues in a monocotyledonous root; transverse section. **(b)** Photomicrograph of a similar root (× 20). (Courtesy of Robert Gill.)

it is quite possible that some absorption occurs through the older roots. Remember that secondary growth does not always occur (as in most monocotyledonous plants).

Figure 9-12 represents a diagrammatic view of primary vascular differentiation in a root of a dicotyledonous plant. The general xylem pattern is actually distinguishable before any of the vascular cells are mature, and the first phloem elements mature before the first xylem elements. The latter tend to mature first in the

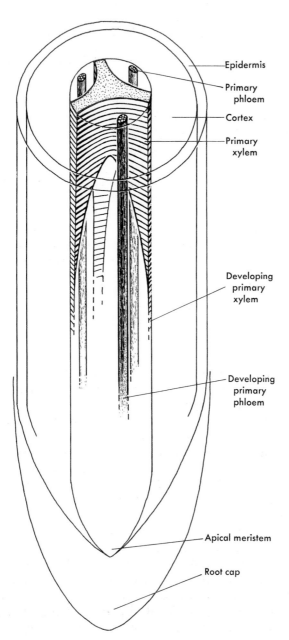

Figure 9-12. Diagram of primary vascular differentiation in a root of pea (*Pisum sativum*). The isolated root was grown in a nutrient medium, but sequence and pattern of differentiation correspond with those observed in roots attached to plants. (Adapted from J. G. Torrey, *Amer. Jour. Bot.* **40,** 525–533, 1953.)

Epidermis

Primary phloem

Cortex

Primary xylem

Developing primary xylem

Developing primary phloem

Apical meristem

Root cap

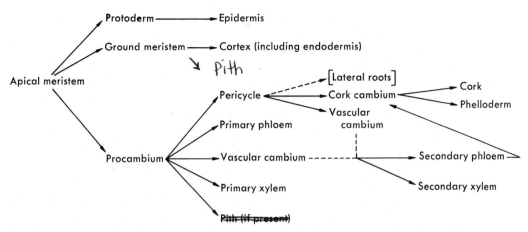

Figure 9-13. Developmental sequence of tissue differentiation in the dicotyledonous root. Lateral roots, although derived from the pericycle, are placed in brackets because they are composed of a variety of tissues.

region of the radiating arms of the developing xylem, and gradually cells toward the inner regions mature until the central portion of the root consists of a solid core of xylem. The sieve elements mature as clusters of cells forming vertical rows between the radiating arms of the xylem. The other tissues of the root are developing also and are indicated in the diagram. Figure 9-13 indicates the developmental sequence of tissue differentiation in the dicotyledonous root (compare with Figure 7-15).

9.3 Absorption

The absorption of water depends upon osmotic forces, as was indicated previously. The student should review Sections 5.3 and 5.4 for a discussion of water potential, wall pressure, and turgor pressure.

Basically, water will be absorbed by the root whenever the water potential within the root cells (usually root hairs) is less than the water potential in the soil solution. This is usually the case, since water is utilized by the various cells of the plant as well as lost to the atmosphere by evaporation, resulting in water movement from the root hairs to other cells. The resultant lowering of turgor pressure, water concentration, and water potential within the root hair brings about further absorption from the soil solution.

The absorption of minerals from the soil is not influenced by the absorption of water, except for the concentration increase in the soil mineral content that results when water is removed. The two processes are independent of each other, just as the absorption of one salt is independent of the absorption of a second salt. Salts may enter a root cell because the membranes are permeable and the concentration of that salt is greater in the soil solution than in the root cell. Utilization of salts within the plant tends continually to decrease the amount of salts dissolved in the root cells. Evidence, however, indicates that **active absorption**—when salts are absorbed against a concentration gradient—is involved in most mineral absorption. This is especially prevalent in the actively metabolizing cells of the root tip. (See Sections 5.5 and 16.3 for additional discussion of this phenomenon.)

Minerals may be absorbed as molecules or as ions. Since these latter particles are charged, they cannot be absorbed alone without upsetting the electrostatic equilibrium between positively and negatively charged particles. When a positively charged ion is absorbed by a cell, either a positively charged ion is lost from the cell (**ion exchange**) or a negatively charged ion must accompany the one that has a positive charge. Respiration occurring in a root cell results in the production of carbon dioxide (CO_2), which forms carbonic acid (H_2CO_3) when dissolved in water. Since the cells contain considerable quantities of water, carbonic acid is present and ionizes to form hydrogen ions (H^+) and bicarbonate ions (HCO_3^-). These ions may then be exchanged for similarly charged ions of the soil solution (e.g., H^+ exchanged for K^+, potassium; HCO_3^- for NO_3^-, nitrate). It is also likely that a direct exchange (**contact exchange**) takes place between ions adsorbed to soil particles and the ions of root cells without the ions of the soil actually entering into the soil solution. The extremely close attachment of soil particles to root hairs makes this possible.

Minerals and water may move freely through the cell walls and intercellular spaces of the root until they reach the Casparian strip of the endodermis. In order to get by this area, the water and solutes must pass through the cytoplasm of the cells of the endodermis. These materials then enter the xylem, and are transported upward through the plant. It has been suggested that there is a gradient of decreasing oxygen and increasing carbon dioxide concentrations from the outer regions of the root to the inner areas. As a result, the living cells closest to the xylem would have a relatively low level of metabolic activity. Metabolic energy is required to accumulate and hold minerals, so these cells would favor the loss of minerals (to the xylem).

9.4 Economic Aspects

The roots of many kinds of plants have been utilized for their food value since the beginning of recorded history and probably earlier. The digging and eating of raw roots of wild plants gradually gave way to the cultivation of plants specifically for roots. The parts are now usually cooked before eating, but many are quite palatable raw. Through plant-breeding programs, plants have been produced that develop tap roots of considerable size and of enormous commercial value. The roots most commonly used for food include beet (*Beta vulgaris*), carrot (*Daucus carota*), radish (*Raphanus sativus*), parsnip (*Pastinaca sativa*), turnip (*Brassica rapa*), sweet potato (*Ipomoea batatas*), and yam (*Dioscorea alata*). The last four are also frequently used as feed for cattle and hogs. The sugar beet is increasing in importance in various countries as a source of sugar, over 5 million tons being produced annually, with the by-products being utilized as cattle fodder, manure, and the molasses for industrial alcohol production. Some plant roots contain drugs, which are frequently utilized: ipecac, an emetic and expectorant, from *Cephaelis ipecacuanha;* rhubarb, for indigestion and as a laxative, from *Rheum officinale;* and ginseng, a stimulant, from *Panax ginseng.*

Summary

1. In a fibrous root system there are numerous slender main roots, which are nearly equal in size, whereas in a tap-root system the primary root remains the dominant part of the underground structure. Adventitious roots are those that arise from stems or leaves.

2. The apical meristem at the end of the root is protected by the root cap. The cells that are formed in the apical meristem eventually enlarge and differentiate into the cells that comprise the various primary tissues of the root or the root cap.

3. The tissues and regions of a mature root consist of epidermis, cortex, pericycle (from which branch roots and cork cambium arise), primary phloem, secondary phloem, vascular cambium, secondary xylem, and primary xylem. In a young root many epidermal cells are greatly elongated as root hairs. In older roots the outer regions are frequently ripped and torn because of the increased diameter that results from the development of secondary tissues. In such cases a cork cambium forms from parenchymatous cells, and cork is rapidly produced.

4. Water is absorbed whenever the water potential is lower within the root cells than in the soil solution. Minerals are absorbed as a result of diffusion along a concentration gradient, ion exchange, contact exchange, and primarily against a concentration gradient by actively metabolizing cells.

5. Roots of many kinds of plants are used as food, in alcohol production, and as a source of various drugs.

Review Topics and Questions

1. Contrast the development of stem branches and root branches.
2. Contrast the roles in plant growth of the apical meristem and the vascular cambium.
3. Diagram a one-year-old woody root in transverse section, and label all structures. It is not necessary to show cells. Secondary tissues should be shown, but no cork has formed.
4. Describe the structural characteristics of roots that increase the efficiency with which they function.
5. You are given an unlabeled microscope slide containing a transverse section of a young woody plant. Explain how you could determine whether the section was made through a root or a stem.
6. As the root tip grows, cells of the root cap are rubbed off by contact with rough soil particles. Explain why a root cap is still present after 20 or 30 years.
7. Since root hair cells do not have a greater ability to absorb water than do other

cells of the root, explain why root hairs are more important in the absorption of water than are other epidermal cells of the root.

8. Would the soil of a cultivated field of beets erode more rapidly or more slowly than the soil of a field of grasses? How would your explanation differ if a field of corn was compared with the field of grasses?

9. Discuss the ways in which minerals may be absorbed by a plant. In what area of the root would most of the active absorption be likely to occur?

10. List ten ways in which plant roots are important to humans.

11. In discussing root tips, one frequently refers to the apical meristem, region of enlargement, and region of differentiation. Explain why these regions cannot be sharply delimited.

12. Compare a root tip with a stem tip, indicating (a) similarities and (b) differences.

Suggested Readings

ESAU, K., *Anatomy of Seed Plants* (New York, Wiley, 1960).

FAHN, A., *Plant Anatomy,* 2nd ed. (Elmsford, N.Y., Pergamon Press, 1974).

Leaves

The food-manufacturing process that occurs mainly in the green leaves of plants will be discussed separately in Chapter 12, while this chapter will be devoted more to the structural aspects of leaves, since a knowledge of structure is essential to an understanding of function. A clear comprehension of why leaves are the centers of photosynthesis (food manufacture) results only from an investigation of the structural arrangement of the tissues concerned.

10.1 External Structure of Leaves

Although tremendous variations occur in the size and shape of leaves, certain basic structures are distinguishable (Figure 10-1). The slender **petiole** (or leaf stalk), present on the leaves of most flowering plants, is a continuation of the stem and contains vascular tissue that is continuous from that of the stem proper to the rest of the leaf. The **blade** is the flattened, expanded portion of the leaf and is usually green in color, because of the presence of many chloroplast-containing cells. Some leaf blades are needlelike, as in pine (*Pinus*), or scalelike, as in cypress (*Cupressus*). Most of the food manufacture occurs in the blades, and much of this food is then conducted through the phloem of the petiole to other parts of the plant. Water and minerals move into the blade through the xylem of the petiole. Some leaves have small, leaflike **stipules** as outgrowths at the base of the petiole. There is also an axillary bud at the base of the petiole in the axil of the leaf. The development of an axillary bud results in the production of secondary branches, flowers, or both, depending upon the type of bud.

In Figures 10-2 and 10-3, examples of variations from this general type are shown. The gigantic *Victoria* leaf and the tiny leaf of *Lemna* are drawn to indicate extremes in size. Stipules are not present in *Azalea* spp., the leaf of *Penstemon* is **sessile** (no petiole is present), and the petiole of a *Cotula* leaf is clasped about the stem. Leaves also vary in the type of leaf margin or edge, as is indicated in Figure 10-4.

The leaves on a stem are usually produced in a definite pattern, and typically

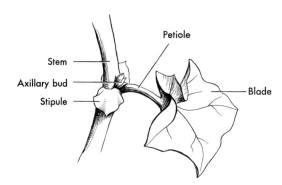

Figure 10-1. The basic parts of a leaf. The ivy geranium (*Pelargonium peltatum*) was used as the model for this drawing.

Stem

Axillary bud

Stipule

Petiole

Blade

with surprisingly little overlapping. Depending upon the species of plant, leaves are alternate, opposite, or whorled, according to whether one, two, or more than two are borne at each node. A general mosaic orientation occurs and each leaf is exposed to the light with a minimum of shading from its neighbors. Leaves have a limited life span; usually only a single growing season, and seldom more than a few years. **Deciduous** plants are those in which the leaves fall off at the end of the growing season, leaving the stems temporarily bare until new leaves are produced at the beginning of the next growing season. The leaves of **evergreen** plants persist

Figure 10-2. Drawing to emphasize the size differences in leaves. The entire *Lemna* plant is shown in the botanist's hand, while only the giant leaf of *Victoria* is visible.

Figure 10-3. Examples of leaf variations. **(a)** The sessile leaf of *Penstemon.* **(b)** The leaf of *Cotula* in which the petiole is clasped around the stem. **(c)** The leaf of *Azalea,* which has no stipules.

(a)

(b)

(c)

at least until new leaves are produced, sometimes for several seasons. The leaves are shed eventually, although not all at one time, and new leaves are produced each year. Most cone-bearing trees, such as pine (*Pinus* sp.), are evergreen as are most flowering trees of moist tropical regions.

Simple and Compound Leaves

The leaf blade may be subdivided into several separate expanded parts, or **leaflets.** Such a leaf is termed **compound,** as distinguished from a **simple** leaf, which has a single expanded portion (Figures 10-5 and 10-6). Each leaflet may consist of

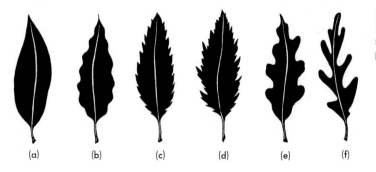

Figure 10-4. Examples of variations in leaf margins. **(a)** Entire; **(b)** undulate; **(c)** serrate; **(d)** double serrate; **(e)** lobed; **(f)** parted.

(a)   (b)   (c)   (d)   (e)   (f)

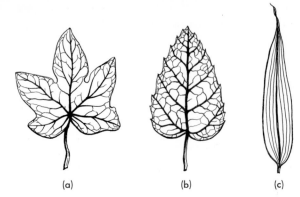

Figure 10-5. Examples of simple leaves. **(a)** Dwarf ivy (*Hedera helix* var.) leaf, palmately net-veined; **(b)** grape (*Vitis*) leaf, pinnately net-veined; **(c)** bamboo (*Bambusa*) leaf with parallel venation (the small interconnecting veins are not shown).

(a)   (b)   (c)

an expanded portion and a short stalk attached to the **rachis,** which is a continuation of the petiole (Figure 10-6). In order to determine whether the structure is a leaflet or a small leaf, the base of the stalk must be examined: buds occur only in the axils of leaf petioles, and stipules are found only at the bases of petioles. Care must be taken because axillary buds may be so small as to escape notice or they may be enclosed by the expanded base of the petiole. Leaves may be **pinnately compound** (Figure 10-6[b]), **palmately compound** (Figure 10-6[a]), or more-than-once compound (Figure 10-6[c]).

Leaf Venation

Vascular tissue extends from the stem through the petiole and into the blade of the leaf, where it forms a network of **veins,** which may be arranged in several ways. **Parallel venation** is a characteristic of most monocotyledonous plants, such as onion (*Allium cepa*), lily (*Lilium*), corn (*Zea mays*), and common grasses. In such plants numerous veins of approximately equal size extend side by side from base to tip of the blade and are interconnected by small and inconspicuous veins (Figure 10-5[c]). **Net venation** is found in dicotyledonous plants, such as oak (*Quercus*), maple (*Acer*), bean (*Phaseolus*), pea (*Pisum*), and sycamore (*Platanus*). In these plants one or more veins are prominent, and the smaller veins form a conspicuous network. If a leaf has one main vein from which the others

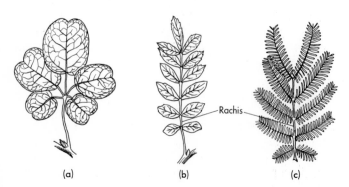

Figure 10-6. Examples of compound leaves. **(a)** *Akebia*, palmately compound; **(b)** Brazilian pepper (*Schinus terebinthifolius*), pinnately compound; **(c)** *Acacia*, double pinnately compound.

Rachis

(a)   (b)   (c)

branch off, it is termed **pinnately** net-veined (Figure 10-5[b]), whereas a **palmately** netveined leaf has several main veins extending from the base (Figure 10-5[a]).

Leaf Development

As mentioned in the discussion of buds (see Section 7.1), leaves develop as lateral outgrowths from the terminal meristems of the stems. A procambial strand differentiates and develops out into the leaf primordium as the latter forms. This strand can be traced back to its attachment to the vascular tissue in the stem below. Primary phloem and xylem differentiate in the procambium; thus, the vascular tissue of the stem is continuous with that of the leaf.

In most leaves growth is rather generalized—that is, growth regions are not localized. Cell divisions in the apex cease rather early, while marginal meristems remain active for a longer period, resulting in the typical expanded form of most leaves. Internal differentiations produce the parenchyma tissue and branching vascular strands, the latter becoming associated with the midrib of the leaf. Variations in the duration of cell divisions and differential cell enlargements cause tensions that result in large intercellular spaces. Most of the leaf expansion is actually a result of enlargement of cells already present. Leaves are determinate (or limited) in growth, as compared to the indeterminate (or unlimited) type of growth found in roots and stems (because of the presence of an apical meristem and sometimes a cambium).

In some monocotyledonous plants, such as the grasses, the basal portion of each leaf remains meristematic, and the leaf continues to grow. Lawn mowers are necessary for this reason—even though the top part of the grass leaf is cut off, this does not stop the leaf from growing, and subsequent mowing is necessary.

10.2 Internal Structure of Leaves

Figure 10-7 is a stereoscopic view of a dicotyledonous leaf, showing the location of various cells and tissues. The leaf blade has both an upper and a lower epidermis, which form a relatively waterproof layer because of the cuticle present on the outer surfaces of the cells and the lack of intercellular spaces. Cutin, a fatty material, is secreted by the epidermal cells and forms this rather impermeable layer; waxes are also found in the cuticle. At intervals are specialized, kidney-shaped epidermal cells, the **guard cells,** between which are pores, or **stomata.** These openings, which may be present in both epidermal layers (although usually more frequent or more numerous in the lower epidermis in dicotyledonous plants), provide a pathway for gaseous exchange between the intercellular spaces within the leaf and the external atmosphere. In darkness, the guard cells are limp, and the stoma is closed; in the presence of light, the guard cells become turgid by absorbing water, and the stoma is open (Figure 10-8). In most plants guard cells are the only cells of the epidermis that contain chloroplasts, an important factor in bringing about their turgor changes, as will be discussed later (Section 15.4). Sub-

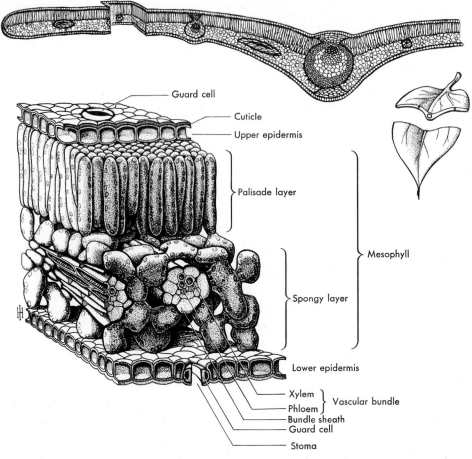

Figure 10-7. Stereoscopic diagram of a dicotyledonous leaf showing the location of various cells and tissues.

sidiary cells, which are somewhat modified epidermal cells, may also be associated with the stomata.

Between the epidermal layers is a parenchyma tissue, the **mesophyll,** usually arranged in two layers in dicotyledonous plants. The layer beneath the upper epidermis is the **palisade layer,** which consists of cylindrical cells whose long axes are at right angles to the epidermis. Although usually only one layer of cells is present, this is influenced by environmental conditions, and many leaves when exposed to the sun tend to have a palisade layer several cells in thickness. Between the palisade layer and the lower epidermis is a mass of loosely arranged, irregularly shaped cells that comprise the **spongy layer.** All cells of the mesophyll contain numerous chloroplasts. The intercellular spaces of the two layers form a continuous pathway through which gas molecules or water vapor molecules may diffuse from atmosphere to cell by way of the stomata, or vice versa.

Extending throughout the mesophyll are strands of vascular tissue, the leaf veins, which ramify and interconnect forming a network within the leaf. Collen-

166

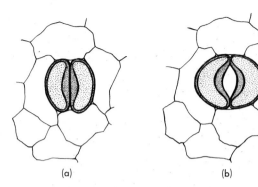

Figure 10-8. **(a)** The two guard cells are relatively limp and the stoma is closed. **(b)** The guard cells are turgid and the stoma is open.

(a)   (b)

chyma is often present on one or both sides of the larger veins, and this provides some mechanical support to the leaf. The number of veins and their profuse branching ensures that the location of any leaf cell is in the vicinity of some vascular tissue. In fact, probably no leaf cell is more than five cell-diameters away from a vein. Such an extensive vascular network, continuous as it is with the vascular tissue of the stem and thus with the root, enables rapid transport of materials into and out of the leaves. The larger vascular bundles of the leaf may be enclosed in one or more layers of compactly arranged cells forming the **bundle sheath.** These cells are usually parenchymatous, although sclerenchyma may be involved in some plants. The cells of the bundle sheath are elongated parallel with the course of the vascular bundle and extend to the end, completely enclosing the terminal tracheids. It is quite likely that the bundle sheath is concerned with conduction; its presence materially increases the contact between mesophyll and the conducting cells, and some of these cells also are modified as *transfer cells* (see Section 8.1).

In many monocotyledonous leaves, such as the grasses, there is no distinct differentiation into palisade and spongy mesophyll, although the cell rows beneath both epidermal layers may be more regularly arranged than the rest of the mesophyll. Such leaves may also develop large amounts of sclerenchyma, which may serve as an important source of commercial fibers in some species. (See Figure 12-4 for a comparison of dicot and monocot leaves.)

Leaf Abscission

Various plant parts (such as leaves, flowers, and fruits) separate from the parent plant during normal growth and as a result of injury, but this phenomenon is most conspicuous as the seasonal defoliation of woody dicotyledonous plants (i.e., deciduous plants). Such **abscission** is the result of the formation of specialized cell layers, the **abscission zone,** from parenchyma cells near the base of the petiole. Two layers usually develop: an abscission, or separation, layer in which structural changes facilitate the separation of the leaf from the stem; and a protective layer in which suberin is deposited in the cell walls (Figure 10-9). The latter occurs on the side toward the stem and protects the surface, which becomes exposed upon leaf fall, from drying out and invasion by parasites.

The abscission zone may form early during leaf development or it may not form

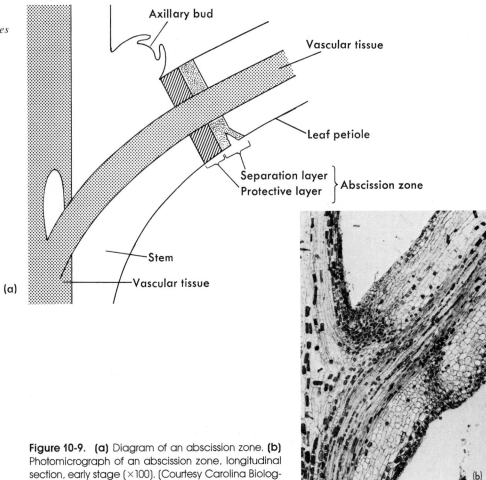

Figure 10-9. **(a)** Diagram of an abscission zone. **(b)** Photomicrograph of an abscission zone, longitudinal section, early stage (×100). (Courtesy Carolina Biological Supply Company.)

until the leaf is mature. The separation of the leaf occurs because the middle lamella dissolves or the cells themselves dissolve. Although some changes may occur in the vascular elements, they do not separate. Wind currents and rain tend to hasten abscission by bringing about leaf movements that facilitate mechanical breakage of the vascular tissue—which is all that is holding the leaf to the stem after separation of the parenchyma cells has occurred. In some cases dead leaves may remain attached to a tree because of the failure of vascular strands to break. In woody plants, cork eventually develops beneath the protective layer.

10.3 Leaf Modifications

The leaves of certain plants are so modified as to lack the appearance of leaves. Just as some stems are modified as supporting tendrils (see Section 8.4), so also

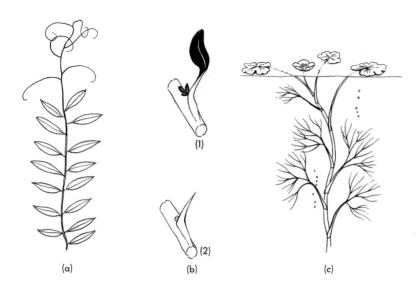

Figure 10-10. Examples of leaf modifications. **(a)** Twining tendrils of *Vicia*; **(b)** thorns of Ocotillo or Jacob's staff (*Fouquieria splendens*): (1) an early stage with the blade present, (2) later stage; **(c)** submerged and exposed leaves of *Ranunculus aquatilis*.

(a)

(1)

(2)

(b)

(c)

may leaves or leaflets function in a similar manner. The twining tendrils of *Vicia* and *Pisum sativum* (peas) are modified leaflets (Figure 10-10[a]). Other leaves may be modified as thorns and spines, as in Ocotillo (*Fouquieria splendens*) (Figure 10-10[b]).

Leaves borne on different parts of the same plant may vary considerably in structure. Leaves exposed to the sun tend to have a thick cuticle, a palisade layer several cells in thickness, and relatively small intercellular spaces in the spongy mesophyll, whereas shaded leaves on the same tree are more likely to have thin cuticles, a palisade layer one cell in thickness, and considerable intercellular spaces. Probably the most important factors involved in such growth are differential water contents and variations in light intensity and quality (wavelengths or colors). The leaves in the sun lose more water, the cells are less turgid, and cell enlargement would tend to be curtailed—remember that such enlargement depends to a great extent upon turgor pressure. With a curtailment in cellular expansion, additional food supplies might be available, for example, during the development of a relatively thick cuticle. The higher light intensities impinging upon exposed leaves would also result in rather high rates of food manufacture (photosynthesis; see Chapter 12). In addition to these effects, various wavelengths (colors) of light are also known to influence significantly the structural development of plants. However, the precise mechanisms involved in the morphological differentiations that have been observed in leaves exposed to the sun and similar leaves not so exposed have not yet been demonstrated.

In some plants the leaves that are produced early or during rapid development are considerably different from those produced later. In Figure 10-11 are diagrammed leaves from the same *Eucalyptus* tree. The "juvenile leaves" were produced on a single shoot that arose from an adventitious bud about 2.4 m (8 ft) up on the trunk. The "adult leaves" were collected from a branch about 7.6 m (25 ft) up the trunk and are typical of most of the leaves produced on that tree. They are slender, slightly curved (sickle-shaped), have a petiole, and are olive-green in

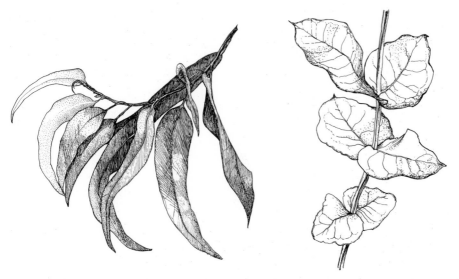

Figure 10-11. Two distinct types of leaves growing on the same *Eucalyptus globulus*. **Left**: Adult leaves; **right**: juvenile leaves.

color. The "juvenile leaves," on the other hand, are ovate, have no petiole (are sessile and clasping), are light bluish-green, and the surfaces are covered with a layer of powdery wax. These latter leaves are not produced on the older branches. They develop only when the tree is young or on adventitious shoots.

Many plants growing in dry areas have very thick, succulent leaves, which store large quantities of water. Water-storage cells, really the parenchymatous cells of the mesophyll, contain copious quantities of hygroscopic[1] mucins, which greatly retard water loss. Examples of these plants are found in many rock gardens of the southern United States: *Sedum* (Figure 10-12), century plant (*Agave*), ice plant (*Mesembryanthemum*), and *Crassula.*

In some aquatic plants, such as *Ranunculus aquatilis,* the leaves that develop submerged in water are quite different in appearance from those that develop above the surface. As indicated in Figure 10-10(c), the former are subdivided into a many-branched structure, while in the latter the blade is unbranched. The finely divided submerged leaf may be advantageous as far as diffusion of materials into the leaf is concerned (i.e., CO_2 for photosynthesis, O_2 for respiration).

10.4 Insectivorous Plants

Much has been written about "carnivorous plants," and these will be discussed in more detail later (Section 14.5), but this phrase is really a misnomer, since these plants do manufacture their own food and only insects have ever been found to be utilized. At least three plants have leaf modifications that enable the plant to par-

170

[1] Hygroscopic materials adsorb water.

Figure 10-12. *Sedum*, a plant with succulent leaves. (Courtesy of Peter Jankay.)

take of insects. Probably the most interesting is Venus' flytrap (*Dionaea musci-pula*) (see Figure 14-8), in which the hinged leaf closes about such things as flies and entraps them. The bodies are digested by enzymes (see Section 11.4) secreted by certain cells of the leaf, and the leaf cells undoubtedly absorb such digested materials. In the sundew (*Drosera*) (see Figure 14-7), certain leaves are basically short stalks with an enlarged end, from which project slender hairs having a sticky knob at the end. Insects stick to these hairs and are digested by enzymes secreted by the hairs. The leaves of the pitcher plant (*Sarracenia*) (see Figure 14-6) are somewhat funnel shaped, with stiff hairs on the inner surface projecting downward. Insects fall into the leaf and drown in the water that partly fills the leaves, since they cannot crawl out past the hairs.

10.5 General Function of Leaves

The most important function of leaves is the manufacture of organic material from carbon dioxide and water, a process (**photosynthesis**) that will be discussed in considerable detail later. Leaves also may be areas in which food or water is stored. The onion bulb (*Allium cepa*) consists mainly of large fleshy leaves in which copious quantities of food are stored (Figure 8-4[a]), while *Sedum* species provide us with an example of leaves that store quantities of water when it is available.

10.6 Economic Aspects of Leaves

The leaves of many plants are utilized as a source of food by humans: cabbage (*Brassica oleracea* var. *capitate*), kale (*Brassica oleracea* var. *acephala*), lettuce (*Lactuca sativa*), spinach (*Spinacia oleracea*), celery (*Apium graveolens*), and rhubarb (*Rheum rhaponticum*). In the last two plants it is mainly the petiole that is usually eaten. The leaves of certain plants are a source of fibers; bowstring hemp (*Sansevieria thyrsiflora*), manila hemp (*Musa textilis*), New Zealand flax (*Phormium tenax*), and sisal (*Agave*). Tea leaves (*Thea sinesis*) are used to make a beverage, while tobacco leaves (*Nicotiana*) are the basis for a billion-dollar industry. A few drugs, condiments, and flavorings are also obtained from leaves: belladonna and atropine from *Atropa belladonna*, purgatives from *Aloe,* cocaine from the coca shrub (*Erythroxylon coca*), digitalis from *Digitalis purpurea,* and witch hazel from *Hamamelis virginiana;* sage from *Salvia officinalis,* basil from *Ocimum basilicum,* thyme from *Thymus vulgaris,* bay leaves from *Laurus nobilis,* and marjoram from *Majorana hortensis;* spearmint from *Mentha spicata,* wintergreen from *Gualtheria procumbens,* and peppermint from *Mentha piperita.*

Summary

1. Leaves arise in shoot apices and are determinate in growth. The basic structures of the leaf are the petiole, the blade, and the stipules. However, leaves vary greatly in size, shape, and as to whether or not a petiole or stipules are present. There is a bud in the axil of the petiole of a leaf.

2. Simple leaves have a single expanded portion, whereas the leaf blade of a compound leaf is subdivided into several parts called leaflets. The vascular tissue in the blade forms a network of veins in which the main veins extend parallel to each other or in which main veins and subsidiary veins form a branching system.

3. Abscission, or leaf fall, results from the separation of specialized cell layers near the base of the petiole.

4. The leaves of various plants are modified as tendrils, thorns, or spines; others are enlarged succulent structures that store copious quantities of water. The leaves of some plants are so modified as to trap insects.

5. Between the epidermal tissues of the leaf is the mesophyll tissue, which may be differentiated into an upper palisade layer and a more loosely arranged spongy layer. The cells of the mesophyll contain numerous chloroplasts and are important in food manufacture. The vascular tissue extends throughout the mesophyll. In the epidermal layers are found numerous stomata through which gaseous exchange occurs.

6. The leaves of some plants are exceedingly important economically: food, tobacco, fibers, tea, drugs, condiments, and flavorings.

Review Topics and Questions

1. Diagram a leaf in cross section and label all structures. Individual cells of the vascular tissue need not be shown.
2. Diagram a leaf as it appears when the section is cut parallel to the epidermis and through the palisade mesophyll. This is a paradermal section. Do not show individual cells of the vascular tissue.
3. Diagram a leaf as it appears when the section is cut parallel to the epidermis and through the spongy mesophyll. Do not show individual cells of the vascular tissue.
4. Why is it possible to find vascular tissue in both transverse and longitudinal views when examining the cross section of a leaf?
5. List seven plants the leaves of which are used as food by animals (including humans).
6. In what ways are leaves used by humans other than as a source of food?

Suggested Readings

ASHBY, E., "Leaf Shape," *Scientific American* **180**, 22–29 (April 1949).
O'BRIEN, T. P., and McCULLY, M. E., *Plant Structure and Development* (New York, Macmillan, Inc., 1969).

11

Energy, Enzymes, and Digestion

11.1 Metabolism

The cell, besides being the basic structural unit of living organisms, is a marvelously complex little "factory" in which a great number of chemical reactions are all going on simultaneously. Some materials in the cell are utilized as "fuel" to supply energy, while others are converted to the more complex components of the cell utilizing this energy. Complex molecules are broken down to simpler ones; small molecules are combined to form more complex ones; protoplasmic constituents "wear out" and are replaced; materials move about within the cell; and in some instances various chemical reactions may result in motility or the production of light (bioluminescence of certain microorganisms). The term **metabolism** refers to all of these various reactions. **Anabolism** includes any metabolic reactions that result in the construction or synthesis of materials, whereas any reactions that involve the breakdown or destruction of substances are termed **catabolism.** Photosynthesis and respiration are examples of anabolic and catabolic reactions, respectively.

Cells are dynamic, not static. Even when cells are not growing or reproducing (dividing), maintenance is a problem. Just as parts of an engine wear out and are replaced, so also are many parts of the cell replaced. The use of nitrogen isotopes has shown a turnover of approximately 50 per cent of the protein content of a rat's liver within a week's time. In other words, half of the protein molecules have been replaced by new ones during that week (similar to the replacing of a fuel pump or a spark plug in an automobile). Such replacement, or maintenance, is undoubtedly typical of all cells, although the rate may vary. Possibly 90 per cent of the living cellular components of an organism are replaced during its lifetime. During senescence, anabolic processes apparently slow down while catabolic ones continue or even accelerate. Death finally results. In plants the leaves die and fall off but the rest of the plant remains viable.

11.2 Bond Energy and Activation Energy

When one speaks of a compound having chemical energy, one is really referring to the bonding forces that hold the component atoms together. Such **bond energy** is usually measured in terms of the energy required to break the chemical bond. Once a bond is broken, the separated atoms are capable of making new bonds and forming new compounds. This is really the basis of chemical reactions.

$$A + B \rightarrow C + D \qquad (1)$$

In the reaction indicated above, bonds in compounds A and B are broken and new bonds are formed, which result in compounds C and D. Not all molecules of compounds A and B will react, however. In any group of similar molecules at a specific temperature, there is a certain average energy possessed by the majority of molecules. In fact, temperature measures the average kinetic energy of a population of molecules. A few of the molecules have higher or lower than average energy, and only the "energy-rich" molecules are able to react. These are the molecules which have sufficient energy to bring about the straining and contorting that weakens internal bonds and makes them susceptible to reaction. The energy above average that is required for a reaction to occur is called the **activation energy**, which may be considered as a barrier to the reaction in that molecules must be raised to this energy level before they will react. Molecules react only when they come in contact. If the mixture is heated, the internal energy of the molecules is increased, and this increases the likelihood of their colliding and reacting; the average energy has been increased and more molecules exceed the activation energy or pass the barrier.

Figure 11-1 indicates energy-distribution curves for a population of molecules at two temperatures. At the lower temperature, only a relatively small percentage of molecules (shown by the stippled area) have energy equal to or greater than the

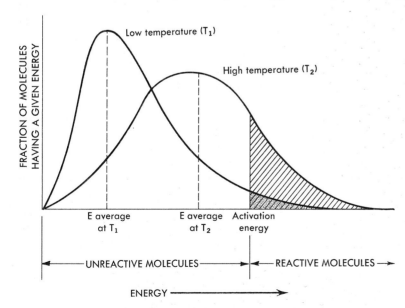

Figure 11-1. Energy-distribution curves for a population of molecules at two temperatures (T_1 = low; T_2 = high). The average energy of the molecules equals the temperature in °K. Only those molecules having energy equal to or greater than the activation energy are reactive; shown by the stippled area at T_1 and the shaded area at T_2.

activation energy and are thus reactive. When the temperature is increased, a much larger fraction of the molecules (shown by the shaded area) have kinetic energies sufficient to be reactive. Heating is an efficient way to increase reaction rates but not in terms of cellular reactions. Proteins, which are such important constituents of the cell protoplast, are rather readily denatured as the temperature rises much above about 45°C. Room temperature (a typical physiological temperature) is considered to be about 22°C, a temperature at which most chemical reactions would occur exceedingly slowly. Then what makes it possible for cellular reactions to occur thousands of times more rapidly than such rates? Our subsequent discussion of enzymes should clarify this (Section 11.4).

11.3 Equilibrium Reactions and Free Energy

In principle all chemical reactions are reversible. In the previous example if A and B react to produce C and D, then C and D can react to produce A and B. The direction and rate of the reaction are determined by the concentrations of the participating molecules.

$$A + B \rightleftharpoons C + D \tag{2}$$

If other factors are kept constant, the greater the amount of A and B added, the more energy-rich molecules will be present, the more frequently will molecular collisions occur between them, and the more rapid will be the reaction. As more C and D are produced, however, more collisions will occur between their molecules and the reaction from right to left will increase. If specific amounts of A and B are placed in contact, reactions will occur that decrease the amount of A and B at the same time that they increase the amount of C and D. Thus, the reaction from left to right slows down as the reaction from right to left increases. Eventually an equilibrium is reached when the rates of the two reactions are the same, and there is no further change in net quantities of the substances; the number of A and B molecules reacting to form C and D is the same as the number of C and D reacting to form A and B. Such reactions will not go to completion unless one of the substances formed is an escaping gas, an insoluble precipitate, or is removed by participation in subsequent reactions.

The amounts of beginning material (for example, A and B) and of end products (C and D) present when equilibrium is reached depend upon the particular reaction and the concentrations involved. If a reaction goes nearly to completion under the specified conditions, there will be little A and B and large amounts of C and D. The reverse is true for a reaction that does not go far to completion (from left to right, that is); only small amounts of C and D will be produced.

In any reversible reaction the direction will depend upon the concentrations of the participating molecules. If large amounts of C and D are added, the direction will be primarily from right to left (toward the production of A and B) until equilibrium is reached. This influence of concentrations on the direction as well as the rate of reactions should be borne in mind throughout subsequent discussions of cellular processes.

Free Energy

The properties of natural systems are such that only a part of their total potential energy is available to do work. This useful or available energy component is termed **free energy.** The free energy change of a chemical reaction can be measured, and this gives the maximum amount of work that the reaction can do. Only when there is a decrease in free energy does the reaction tend to go toward completion. Such reactions are termed **exergonic;** they yield energy. In other words, the total bond energy in A and B of the reaction

$$A + B \rightleftharpoons C + D \qquad - \Delta F \tag{3}$$

is greater than the total bond energy in C and D. This is indicated by the symbol "$-\Delta F$," which means a decrease in free energy and implies that the reaction has an equilibrium point toward a preponderance of C and D. If the total energy of all bonds in A and B is less than in C and D, the reaction as written cannot go from left to right unless energy is supplied; it is an **endergonic** process and requires free energy input. Therefore, we use the symbol "$+\Delta F$," which means an increase of free energy for the reaction as written.

$$A + B \rightleftharpoons C + D \qquad + \Delta F \tag{4}$$

The equilibrium point in reaction (4) would be toward a preponderance of A and B with very little C and D.

We will see later (Section 13.4) that exergonic and endergonic reactions may be coupled or interconnected by a common intermediate so that the former process provides energy to the latter. In such coupled systems the endergonic reaction can be made to take place provided the decrease in free energy of the exergonic process is greater than the gain in free energy needed to cause the endergonic reaction to proceed. For example, in two sequential chemical reactions

$$A + B \rightleftharpoons C + D \qquad +\Delta F \tag{5}$$
$$D + E \rightleftharpoons F + G \qquad -\Delta F \tag{6}$$

the first (5) is endergonic $(+\Delta F)$ and will not go to completion by itself; the equilibrium is toward A and B. The second reaction (6), however, is exergonic $(-\Delta F)$ and the equilibrium is toward F and G. Reaction (6) causes reaction (5) to go toward completion because it removes D, one of the products of (5); D is common to both reactions. The overall reaction is the total of the two and has an overall decrease in free energy.

$$A + B + E \rightleftharpoons C + F + G \qquad -\Delta F \tag{7}$$

D is produced in one reaction but used up in the subsequent reaction and therefore does not appear in the overall scheme. In general, all reactions tend toward an overall decrease in free energy.

There are many reactions in the cell that release energy or make energy available, but there are also many reactions in the cell that require an input of energy (e.g., synthesis of starch, cellulose, fat, and protein). The latter type of reactions can proceed only as a result of having a common intermediate or reactant with the former. As we shall see later, ATP (adenosine triphosphate) is a common reactant of reactions yielding energy (exergonic) and those requiring energy (endergonic).

11.4 Enzymes

Questions that naturally arise when considering the number of reactions occurring within a living cell are whether these reactions occur spontaneously and how they can proceed at physiological temperatures. In many cells sugars are utilized as the building blocks in the synthesis of starch, but sugars dissolved in water and placed in a test tube do not form starch. In many cells the reverse process also may occur, starch being broken down into sugars. This process can be brought about in a test tube, but concentrated acid (usually hydrochloric) and a high temperature are necessary factors that are not present in the cell. In fact, the cell would die if exposed to such concentrated acid or high temperature. Chemical reactions that occur in the laboratory at negligible rates, except possibly at high temperatures, are brought about in living cells at relatively low temperatures because of the presence of molecules called enzymes.

In order to understand the functioning of an enzyme, we might first consider a simple chemical reaction. If hydrogen gas (H_2) and oxygen gas (O_2) are mixed in a chamber, no perceptible reaction occurs. Adding finely ground platinum results in an immediate and vigorous reaction in which water is formed:

$$2H_2 + O_2 \xrightarrow{\text{platinum}} 2H_2O \tag{8}$$

The platinum promotes the reaction without itself being used up; it is a **catalyst.** In the absence of platinum, the hydrogen and oxygen molecules very seldom come into close enough proximity for them to react. When these molecules are adsorbed to the surfaces of the platinum, they are concentrated together in one location, and reactions between molecules occur with great rapidity. Also, stresses and strains probably are established in the adsorbed molecules as a result of the union between the catalyst and these molecules. The molecules, therefore, become more reactive, i.e., are activated. In this condition they undergo changes at a more rapid rate than when present in the free state.

Enzymes are exceedingly large complex organic compounds, which are produced by the living cell and which act as catalysts. They are protein molecules. Additional less complex organic molecules, called **coenzymes,** are frequently necessary for the normal catalytic activity. In some enzyme systems, certain metallic ions (e.g., Mn^{++} and Mg^{++}) are essential for normal activity. The colloidal nature of protein molecules results in the very large surface area per unit of volume, which is characteristic of enzymes. The protein portion of enzyme systems also provides the specificity so typical of enzyme activity. Although many enzymes are specific as to the substrate upon which they act, this specificity resides in many cases in the type of action that is catalyzed rather than the substrate. For example, nicotinamide adenine dinucleotide (NAD) is the coenzyme for various dehydrogenase enzyme systems. In the presence of a dehydrogenase protein, NAD is capable of removing two hydrogen atoms from the substrate molecule. The specific action involving NAD is the removal of hydrogens; the protein portion of this system determines from which specific substrates the hydrogens will be removed. For example, NAD can remove hydrogens from malate (a 4-carbon compound) if malic dehydrogenase (a specific enzyme protein) is present, or the hydrogens can be removed from ketoglutarate (a 5-carbon organic compound) if ketoglutaric dehydrogenase is the protein portion of the enzyme system that is present. Neither the NAD nor the dehydrogenase can function alone; both must be present. The term **apoenzyme** refers to the protein portion of an enzyme system, while coenzyme refers to the nonprotein portion. The latter can usually be separated from the former by dialysis. In this procedure the mixture is placed into a cellophane bag with pores too small to allow the apoenzyme to go through, and the bag is surrounded with a large volume of water. The coenzyme is able to pass through the pores. When this is done, neither portion is functional; they may be recombined, however, and activity returns.

Enzymes are generally sensitive to changes in acidity and are usually most active over a rather narrow range of acidity. Any increase or decrease of acidity beyond the confines of this narrow range results in a decrease of enzyme activity, although the actual pH range that is effective varies with the enzyme (Figure 11-2).

Most enzymes, as well as proteins in general, are inactivated by moderate heat (60° to 70°C) when in the moist condition characteristic of most cells. This is probably a basic explanation of the upper temperature limits of life and also indicates why dry seeds and spores tend to be rather resistant to adverse environmental conditions. In a dried (or dehydrated) condition, enzymes are quite stable. Most of the reactions occurring within a cell proceed at a faster rate with an increase in temperature, up to a maximum beyond which a decrease takes place as enzymes become inactivated. As shown in Figure 11-3, the rate of an enzyme-catalyzed reaction increases about 2 or 3 times for every 10°C increase in temperature until enzyme denaturation becomes more and more pronounced. The numerous weak hydrogen bonds, which hold the enzyme structure in its unique pattern, stretch and finally break as the temperature increases. The rupture of one hydrogen bond makes it easier to rupture the next, and the collapse of the enzyme structure

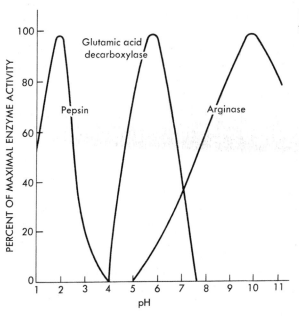

Figure 11-2. Effect of pH on the activity of three enzymes.

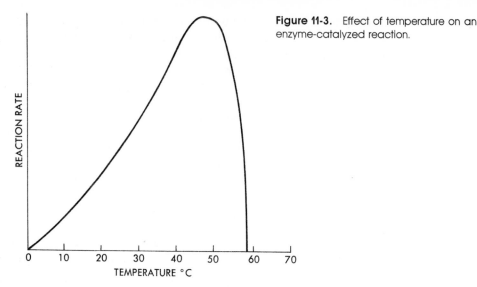

Figure 11-3. Effect of temperature on an enzyme-catalyzed reaction.

REACTION RATE

0   10   20   30   40   50   60   70
TEMPERATURE °C

occurs with increasing rapidity. The loss of catalytic properties typically begins at about 40°C and is usually complete as 60°C is approached.

Enzymes are extremely efficient; minute quantities of them are capable of catalyzing reactions of large numbers of molecules in short periods of time. For example, one molecule of catalase (enzyme) will catalyze the breakdown of approximately 5,000,000 molecules of hydrogen peroxide (to water and oxygen) in one minute. The substrate molecules are adsorbed to the enzyme surface, the reaction takes place, the resultant molecules are released from the surface, and they are replaced by other substrate molecules. The enzyme molecule is used over and over again, since it is not changed by the reactions occurring at its surface. Though the cell contains many different kinds of enzymes, none of them are present in any high concentration. This is in no way detrimental to the cell, since even low concentrations of enzymes bring about reactions of large numbers of substrate molecules.

11.5 Mechanism of Enzyme Action

Any conceptual scheme[1] of enzyme function must be consistent with greatly speeded up reactions occurring at surfaces and without any permanent change of that surface. The present viewpoint is that the surface configuration of the protein portion of the enzymes (the apoenzyme) determines which substrate molecules will be adsorbed. If a coenzyme is an integral part of the enzyme system, the protein must have at least two reactive spots or areas where molecules are adsorbed; the substrate becomes adsorbed in one location, the coenzyme in the second reactive spot, in close proximity to the substrate. Only those molecules that fit the

[1] A conceptual scheme fits a variety of facts into a relatively simply theory.

chemistry made possible by the presence of enzymes. The reactions occurring within a cell would be much too slow without enzymes, and such reactions would not be useful to the cell. The specificity of enzymes also leads to an *orderly* chemistry. Not only are enzymes necessary, but the right enzymes in the right places are necessary to ensure the right reactions. This is where genes enter the picture. As will be discussed more fully later (Section 18.20), protein synthesis is programmed by genes, and enzymes are protein in nature. This indicates the genetic influence in the control of organisms and their reactions.

A general compartmentalization of enzymes within the cells is quite effective as an efficiency factor in enzyme activity. It results in a closer association of enzyme and substrate as well as bringing about an orderly array of enzymes involved in a series of reactions. The substrate molecule can be passed from one enzyme to the next with a bond being ruptured or synthesized at each step. This progression is greatly simplified when there is an orderly orientation of successive enzyme molecules. In the cell the ribosomes, mitochondria, and chloroplasts are cytoplasmic structures that have been shown to contain high concentrations of enzymes. It is also likely that the endoplasmic reticulum of the cytoplasm brings about considerable compartmentalization of enzymes and metabolites. The separation of different reactions would also decrease chances of possible interference between them. Probably most enzymes are structured in some fashion within the protoplasmic framework of the cell (e.g., within organelles or on membranes); the right enzymes in the right places. The development of these structural features is also controlled by genes.

11.6 Vitamins

At this time we might indicate the value of vitamins in the activity of cells. **Vitamins** are organic compounds that are essential in relatively small amounts to organisms but are not utilized as a source of food or energy. A deficiency of one of the vitamins results in abnormal metabolism and may result in characteristic symptoms. Most vitamins participate in enzyme systems as constituents of important coenzymes. For example, nicotinic acid (niacin, one of the B vitamins) is a part of both NAD and NADP, which are coenzymes in the respiratory processes; NADP also functions in photosynthesis. Thiamine (vitamin B_1) is also essential in the respiration of cells. While in man a deficiency of thiamine in the diet results in characteristic symptoms (beriberi), green plants are capable of synthesizing thiamine if they are provided with carbon dioxide, water, minerals, and light. It has been found, however, that thiamine is produced in leaves and transported to the roots. If excised roots are placed in a nutrient solution containing sugar and minerals, growth will occur only when this vitamin is added. In some plants a similar situation obtains with nicotinic acid. Information concerning the importance of vitamins in the nutrition of animals is much more extensive than that concerning their function in plants, primarily because of the ability of green plants to synthesize vitamins. Vitamin deficiency diseases in animals are readily corrected by incorporating in the diet sufficient quantities of vitamin-containing foods (e.g., fresh

Table 11-1 Some important vitamins

Vitamin	Some sources	Role in green plant	Some deficiency symptoms in man
Converted carotene[a] = A	Yellow vegetables, eggs, butter, liver oils	Pigment involved in phototropisms; possibly accessory pigment in photosynthesis	Night blindness
Thiamine = B_1	Yeast, whole grains, meat	Part of enzymes involved in respiration	Beriberi, a disease of the nerves
Riboflavin = B_2	Same as thiamine	Part of enzymes involved in respiration	Skin ailments
Pyridoxine = B_6	Same as thiamine	Part of enzymes involved in synthesis of amino acids	Not certain
Niacin = nicotinic acid	Same as thiamine	Part of enzymes involved in respiration and photosynthesis	Pellagra
Ascorbic acid = C	Fresh fruits (especially citrus) and vegetables	Participates in oxidation-reduction systems	Scurvy
Calciferol = D	Liver oils, eggs, almost all plants	None known	Rickets, defective growth of bones

[a] Beta-carotene + $2H_2O \rightarrow 2$ vitamin A.

fruits and vegetables, cheese, butter). Nongreen plants are usually like animals in their requirements. For example, the fungus *Phycomyces blakesleeanus* cannot synthesize thiamine and must have a supply of this vitamin in its food.

The nomenclature with regard to vitamins is complicated by the frequent use of letters to designate specific vitamins. Such letters (e.g., vitamin A, vitamin B_2 and B_6, vitamin C, and so forth) were used as a convenient method of designating an unknown component of man's diet. As soon as the molecular structure is known with any degree of accuracy, the letter designation should be dropped in favor of a more descriptive term. The situation with regard to B vitamins became even more complex when vitamin B was discovered to be a group of vitamins.

Vitamins in general are obtained from the green plants that form the basic food materials for all organisms. Vitamin A and vitamin D (calciferol) are obtained as precursors from plant material, and then the vitamin itself is synthesized by the animal. Table 11-1 indicates some of the important vitamins and their sources. Vitamins present in such animal products as eggs, butter, and liver oils are derived from plant tissues eaten by the animals. In the case of those vitamins that have a known function in green plants, such as thiamine, riboflavin, and pyridoxine, plants and animals do not differ in their requirements for these vitamins but only in their ability to synthesize the compounds.

11.7 Digestion (Hydrolysis)

Just as complex molecules are continually synthesized from simpler ones in living cells, so are many compounds broken down to simpler ones. **Digestion** is the process whereby complex or nonsoluble molecules are converted to simpler molecules or molecules that are water soluble. Since water is used in this process, it is a type of **hydrolysis.** Some of the more common digestion reactions are indicated below:

$$\text{Complex carbohydrates} \xrightarrow[\text{carbohydrase}]{\text{enzyme}} \text{Sugar molecules} \tag{9}$$

$$\text{Protein} + \text{Water} \xrightarrow[\text{proteinase}]{\text{enzyme}} \text{Amino acid molecules} \tag{10}$$

$$\text{Fats} + 3 \text{ Water} \xrightarrow[\text{lipase}]{\text{enzyme}} 3 \text{ Fatty acid molecules} + \text{Glycerol} \tag{11}$$

In every case, digestion is catalyzed by a specific enzyme that acts upon one substrate but not all. The reactions yield little or no energy and are of no consequence in fulfilling the energy requirements of the cell. The reverse reactions are **syntheses,** which require energy, the energy being made available by respiratory processes of the cell. Any production of more complex materials or molecules from simpler ones can be accomplished only with the expenditure of energy; work is done in rearranging atoms into a more complex structure.

Digestions of one sort or another probably occur in all cells, and some digestions occur outside of the cell. In the latter cases, enzymes are secreted by the cells. If yeast plants are placed in a solution of sucrose (a 12-carbon sugar), sucrase (an enzyme) is secreted, and the sucrose is broken down to two 6-carbon sugars, glucose and fructose. The glucose and fructose molecules are then absorbed and utilized. Such extracellular digestions occur throughout the alimentary canal (mouth, stomach, and intestines) of humans.

When foods are stored, they are usually in the form of nonsoluble materials. Before such foods can be utilized, they must first be converted to a soluble form. For example, when a seed is planted and watered, the first thing that happens (after water is absorbed and swells the seed) is the digestion of starch, fats, and proteins. The resultant sugars, fatty acids, glycerol, and amino acids are then available and are utilized in the subsequent production of new cells, which develop into roots, stems, and leaves. Once the leaves have formed, the plant is no longer dependent upon the foods stored in the seed; photosynthesis can now take place, and foods are produced.

Summary

1. Metabolism refers to all the complex chemical reactions occurring within a cell. Enzymes, produced by the cell, are the catalysts that make possible these various reactions.

2. Enzymes are colloidal proteins that are sensitive to acidity changes and to

moderate heat. Substrate molecules are adsorbed to the enzyme surface, and this orientation makes reactions possible.

3. Vitamins are essential to normal growth and development of all organisms, usually by participating in enzyme systems. Green plants are fortunate in that they are capable of synthesizing vitamins. Animals and many nongreen plants depend upon external sources for vitamins.

4. Useful or available energy is termed *free energy*. Only when the end products of a reaction have less total bond energy than the substrates does the reaction go toward completion—that is, if there is a decrease in free energy.

5. The conversion of complex or nonsoluble materials into simpler or soluble ones is termed *digestion*. It is a type of hydrolysis that occurs in all cells.

Review Topics and Questions

1. Discuss the importance of the fact that enzymes are not used up in the reactions that they catalyze.
2. Enzymes are colloidal proteins. In what way does this characteristic of enzymes influence the activities of living cells? What is the importance of a protein structure to enzyme activity?
3. Discuss the importance of digestion to plants. In what ways do digestions in a flowering plant differ from digestions in a human? In what ways are digestions in these two organisms similar?
4. Explain how you could determine the rate of enzyme activity. Describe an experiment that could be used to test your method.
5. In what ways are the vitamin requirements of plants different from those of animals? In what ways are they similar?

Suggested Readings

DOTY, P., "Proteins," *Scientific American* **197**, 173–178 (September 1957).
FRIEDEN, E., "The Enzyme-Substrate Complex," *Scientific American* **201**, 119–125 (August 1959).
LEHNINGER, A. L., *Bioenergetics* (New York, Benjamin, 1965).

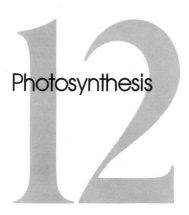

Photosynthesis

12.1 Nutritional Types

In Chapter 10 the structure of a leaf was emphasized, but only a brief mention was made of the functions involved. However, the most important aspect of a study of plants is probably the food-manufacturing process (photosynthesis), which occurs primarily in green leaves. This activity of green cells provides the food and the oxygen for all organisms, including animals and nongreen plants, as well as the requirements of the plant that itself contains such cells. An organism, such as a green plant, that can manufacture its own foods from simple inorganic substances (usually carbon dioxide and water) is termed an **autotroph.** Autotrophs are of two types, **photosynthetic** and **chemosynthetic,** which differ in that the former utilize radiant (light) energy and the latter utilize energy derived from the oxidation of inorganic compounds (see Chapter 20, especially Section 20.6). Those organisms that cannot manufacture their own food are **heterotrophs. Parasites** are those heterotrophs that obtain their food from the living cells or tissues of another organism, while **saprophytes** obtain their food from nonliving organic matter (i.e., dead plants and animals, excretions and secretions from plants and animals). Humans are, of course, heterotrophic, just as are all animals and most nongreen plants; some bacteria are chemosynthetic or photosynthetic but not green.

The manufacture of food, mainly sugar, from carbon dioxide and water in the presence of chlorophyll, utilizing light energy and releasing oxygen gas, is termed **photosynthesis.** This process has been under intensive investigation for many years, but the exact mechanisms involved are not yet completely understood. The raw materials and major end products have been known at least since the work of Sachs in the 1860s, but the complexity of the steps between these end points is only now being clarified. Certainly many steps, some of a cyclic nature, are involved. In our somewhat simplified discussion of photosynthesis we will consider what is necessary for the process, the end products, and then the steps that are involved during the conversion of the raw materials into these end products. Throughout the discussion one should bear in mind that the important aspect of the process of photosynthesis is the trapping of light energy in a usable form and the utilization of this energy in the conversion of carbon dioxide to carbohydrates.

The latter are a source of available chemical energy for cell functions. In effect, photosynthesis converts light energy to chemical energy that all cells of the plant can use.

A few bacteria are photosynthetic but do not utilize water or produce oxygen. The process is essentially similar to higher plant photosynthesis, and these bacteria produce their own food. However, the activity of these organisms is insignificantly small in comparison with the green plants.

12.2 Essential Factors in Photosynthesis

Any factor whose presence is required before a reaction proceeds is termed an **essential factor.** This is not a relative term: if any one such factor is absent, the reaction cannot take place. The raw materials, or substrate, of a reaction are essentials, but so also in many instances are certain environmental factors. In photosynthesis the essential factors are carbon dioxide, water, energy (light), pigments, enzymes, carrier molecules, and a suitable temperature.

Raw Materials

Because of the concentration gradient that results from its use as a substrate, carbon dioxide diffuses through the stomata from the atmosphere into the intercellular spaces of the leaf. The content of carbon dioxide in the atmosphere is rather constant at about 0.03 per cent (300 ppm), a seemingly small amount but totaling approximately 2,000,000,000,000 (2×10^{12}) tons in the atmosphere surrounding the earth. This gas is continually being added to the air by the respiration (Chapter 13) of plants and animals, by the decay of organic materials, by the combustion of fuels, by the weathering of rock, and by volcanic activity, although, in fact, the respiration of plants, including microorganisms, supplies more carbon dioxide to the atmosphere than comes from all other sources combined. The oceans of the world contain tremendous amounts of dissolved carbon dioxide, probably more than is found in the atmosphere, and act as a great reservoir of this material. Once in the intercellular spaces, the molecules of carbon dioxide dissolve in the water that saturates the walls of the cells and diffuse into the cytoplasm and eventually to the chloroplasts of these cells (or areas containing chlorophyll in those plants that do not have chloroplasts) where photosynthesis takes place.

Water, the other substrate, is absorbed from the environment by the cells. In vascular plants this absorption is by the roots and water is transported to the leaves through the various cells and tissues as indicated in previous chapters.

Energy

In the synthesis of any material, energy is required in order to convert simple molecules into more complex ones. As the name implies, the energy source in photosynthesis is light energy, with the red and blue wavelengths (colors) being the most effective. The greens and yellows are mainly reflected from and transmitted through the leaf and have much less effect on the process, but do explain why it is that leaves are green. The sun normally is the source of such radiations, but electric lights can also be used. Only those wavelengths that can be absorbed by chloroplast pigments are useful in photosynthesis. The process may be considered as one in which radiant energy is converted to chemical energy and stored in the form of the carbohydrate end product—an energy-storing or energy-absorbing process. Even though only about 2 per cent of the light energy striking a field of plants is actually stored in the resultant sugar molecules, the process is more efficient than this would appear to be: most of the light from the sun is reflected, transmitted, or absorbed as heat, but approximately 30 per cent of what is actually absorbed is converted to chemical energy (primarily as sugar).

Pigments

The presence of pigments enables green plants to absorb light energy and to use this energy in the production of sugars; chlorophyll is the pigment primarily involved in this process in most plants. There are actually several **chlorophylls** (indicated by letters *a, b, c,* and so forth); chlorophylls *a* and *b* are found in the chloroplasts of the higher or more complex plants, whereas chlorophyll *a* plus other chlorophylls are found in certain of the algae (see Chapter 22).

In the higher plants, those upon which our discussion is based, **carotenoid** pigments, **xanthophylls** (yellow) and **carotenes** (yellow-orange), are always present in the chloroplasts along with the chlorophylls. The carotenes may absorb light and pass this excitation (or energy) to chlorophyll. They supplement chlorophylls as light absorbers, but they are not capable of acting as catalysts in photosynthesis. In some of the red and brown algae, however, there is evidence that indicates that some of the associated pigments may be as effective as chlorophyll, or even more so. In all plants, except bacteria, chlorophyll *a* must be present for photosynthesis to occur. Even those plants that have red leaves are found to contain chlorophyll, which is merely masked by the red pigment. Boiling such leaves in water removes the red **anthocyanins** (nonphotosynthetic pigments), which are water soluble and located in the vacuoles of the leaf cells; the green chlorophyll is then visible. The chlorophylls are lipid soluble (or fat soluble) and not water soluble; thus, they are not removed by this treatment. The membranes, which normally prevent the anthocyanins from diffusing out of the vacuoles, are destroyed by the high temperature of the extraction procedure. Cold water could be used to remove the red pigment if the membranes were first damaged or destroyed, as with a poison.

The absorption spectrum for the chlorophylls indicates that blue and red wavelengths are absorbed much more than green, and this correlates with the maximum effectiveness of wavelengths in photosynthesis (Figure 12-1). The rather high rate of photosynthesis in those wavelengths not absorbed much by chlorophyll (note Figure 12-1[a] especially) result from absorptions by other pigments that transmit the light energy to chlorophyll. In *Ulva,* a green alga, the entire plant (thallus) was used in measuring light absorption. As a result there is a much closer

(a)

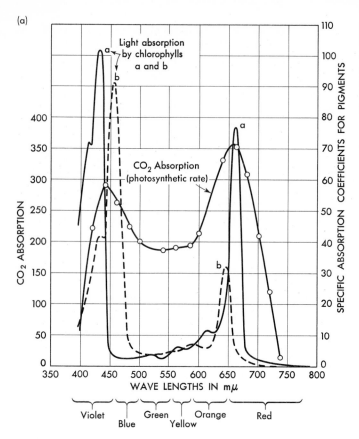

Figure 12-1 Light absorption curves superimposed on the curve of photosynthetic rate as a function of wavelength of light. **(a)** In wheat plants. (From *Introduction to Plant Physiology*, by O. F. Curtis and D. G. Clark, New York, McGraw-Hill, copyright 1950 by the publisher, used by permission.)

(b)

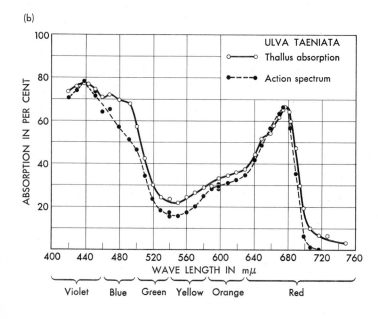

Figure 12-1. **(b)** In *Ulva*, a green alga. (From F. T. Haxo and L. R. Blinks, *J. Gen. Physiol.*, **33**, 389–422, 1950.) See text for further discussion.

correlation between light absorption and photosynthetic rate (action spectrum curve); light is being absorbed by all the pigments that are present.

In most plants the yellowish pigments of the leaves are not noticed because they are masked by the color of the chlorophylls and come into prominence only during the fall of the year. The autumn coloration of many trees results as chlorophyll synthesis decreases while chlorophyll breakdown continues during leaf senescence. (See Section 11.1 regarding maintenance of cell contents.) Eventually the carotenes and xanthophylls are no longer hidden; the leaf color changes from green to yellowish, as in the tulip tree (*Liriodendron tulipifera*), elm (*Ulmus*), aspen (*Populus tremuloides*), and hickory (*Carya ovata*). If the brownish **tannins** are also present, as in some aspens and beeches (*Fagus*), the color may be more golden-yellow. In some trees additional red pigments, anthocyanins, develop, especially if large supplies of carbohydrates are available to the leaves. Red maple (*Acer rubrum*), sumac (*Rhus*), and many oaks (*Quercus*) become quite prominent in the landscape as a result of their dominant red color. As the leaves die, the pigments decompose, and a brownish hue results. The gorgeous hues of northern forests result from these pigment changes that take place near the end of the growing season. Frost does not bring about this coloration. In fact, early frosts tend to decrease the color display, especially with regard to the red pigments, by bringing about the early death of leaves and the rapid appearance of the brown coloration. However, exceptionally brilliant autumn coloration usually occurs if fairly low night temperatures, which retard the removal of food from leaves, are associated with bright sunny days, which favor photosynthesis and sugar production. The brilliance of an autumn display depends upon both climate and the kind of trees that are present. Senescence of the leaf commonly does not affect all parts at an equal rate, nor all leaves equally. Thus, there is considerable variation in colors.

Temperature

Photosynthesis generally takes place over a range of temperature from 5 to 40°C, the rate increasing as the temperature rises, up to approximately 35°C, after which a rapid decline in rate occurs. This decrease in rate is probably caused by enzyme inactivation at the higher temperatures. Since the process is influenced by temperature as well as by light, it is apparent that both chemical and photochemical reactions are involved because the latter are independent of temperature. This will be discussed more fully later. The enzymes and coenzymes that are involved have not all been identified, but the evidence for their presence is quite conclusive. More refined techniques will undoubtedly clarify this situation in the future.

Additional Essential Factors

Present in chloroplasts, along with the pigments already mentioned, are various carrier molecules whose functions have only recently been elucidated. They are important in transferring hydrogen atoms, electrons, and in transferring energy. These activities are discussed in more detail in Section 12.5. The enzymes that enable the numerous individual reactions of the photosynthetic process to occur are also located within the chloroplast. The structure of this plastid is such that the various molecules that participate in photosynthesis are oriented in close proximity one to another, an arrangement that results in a rather efficient mechanism (similar, in a way, to the production line of a factory).

12.3 Materials Produced as a Result of Photosynthesis

Photosynthesis is a complicated process in which a variety of substances may be produced, even though a simple 6-carbon sugar, or **hexose** (a carbohydrate), is the principal end product in the majority of plants. The substances produced quite possibly may vary with the kind of plant as well as with the environmental conditions that influence the process. The hexose that first forms is actually fructose-diphosphate (i.e., fructose with two phosphate groups attached). Fructose is closely related to the more familiar hexose, glucose or grape sugar, and these sugars are readily interconverted. To simplify matters, the term *hexose* will be used during most of the discussion except when more specific details are considered. It should be noted that glucose plus fructose can readily be converted to the common 12-carbon sugar, sucrose or cane sugar. In fact hexoses are really the basic building blocks from which all other more complex molecules are produced. They are also the basic molecules used in respiration, a process from which cells obtain the energy for such syntheses. The importance of photosynthesis is quite obvious.

The first identifiable product of photosynthesis is actually not a hexose but a simpler molecule, and this will be discussed in detail in a subsequent section. At present it is sufficient to indicate that there are at least two basic pathways by which green plants fix carbon dioxide and produce carbohydrates. In the one pathway, frequently termed the *Calvin-Benson cycle* (after the two investigators who elucidated the steps involved), the first product is a *three-carbon* compound called **phosphoglyceric acid** (**PGA**). This material is found after very short periods of photosynthesis, and two molecules of this PGA are readily converted into a hexose molecule. Plants having this pathway are termed C_3 **plants,** or three-carbon plants, to distinguish them from those with the other pathway. In this second type of photosynthetic CO_2 fixation, usually termed the *Hatch-Slack pathway* (after the two investigators who clarified the steps involved), the first product is a *four-carbon* compound called **oxaloacetic acid.** Plants with this pathway are termed C_4 **plants,** or four-carbon plants.

Only the initial steps of carbon fixation differ in these two types of plants, however. The four-carbon compound is broken down enzymatically and yields CO_2, which then participates in the Calvin-Benson cycle. Details are presented later.

The oxygen produced by photosynthesis either is utilized within the cells by respiration or diffuses out of the leaves into the atmosphere via the stomata. Fortunately for the heterotrophic organisms, much more oxygen is produced by green plants than is utilized by them. This excess oxygen (excess as far as the green plant is concerned) is used by man, other animals, and the aerobic nonchlorophyllous plants. Anaerobic, or fermentative, fungi do not utilize oxygen, of course, but are nevertheless dependent upon autotrophic organisms for a supply of food, just as are the aerobic heterotrophs. See Chapter 13 for a further discussion of respiration and fermentation.

12.4 The Overall Photosynthetic Reaction

The equation for photosynthesis was once written as follows:

$$6CO_2 + 6H_2O \xrightarrow[\text{light energy}]{\text{chloroplast}} C_6H_{12}O_6 + 6O_2 \tag{1}$$

This equation suggests that six molecules of carbon dioxide and six molecules of water react in the presence of chloroplasts and light to form one molecule of hexose (a 6-carbon sugar) and six molecules of oxygen. In the early 1940s a series of investigations by Ruben and Kamen resulted in a revision of this basic equation. They utilized isotopes (see Section 3.1) of oxygen to label atoms in the water; until isotopes were used, these oxygen atoms could not be distinguished from those in the carbon dioxide. In a similar manner, radioactive atoms are now used to distinguish, for example, the first carbon atom in hexose from the other five. When the oxygen atoms in the water molecules were labeled, Ruben and Kamen found that these atoms formed the free oxygen that was produced and were not present in the sugar molecule; the oxygen atoms from the carbon dioxide molecules, on the other hand, were found in the hexose. Just how startling this was to a chemist or physiologist can be seen more clearly if we rewrite equation (1) as follows:

$$CO_2 + H_2O \rightarrow [CH_2O] + O_2 \tag{2}$$

The chemical formula CH_2O is a general one for any carbohydrate; equation (2) is really the same as (1), in that carbon dioxide and water react to form a carbohydrate and oxygen. Physiologists used to think that carbon dioxide was split during photosynthesis, the carbon adding to water to form the carbohydrate and the oxygen being liberated to form oxygen gas. This could not be the case, however, since Ruben and Kamen showed that the oxygen gas was derived from water. The dilemma was solved by the understanding that new water is actually formed during photosynthesis; water is used and water is produced:

$$CO_2 + 2H_2O^* \rightarrow [CH_2O] + O_2^* + H_2O \tag{3}$$

The asterisk indicates labeled atoms. The water that is produced is not the same as the water that has been used. Note that the oxygen atoms in the water molecules to the right of the arrow have come from some of the oxygen atoms that were originally present in the carbon dioxide molecules. In order to clarify the situation a bit, we should note that reaction (3) is actually the sum of two types of reactions:

$$2H_2O \xrightarrow{\text{light}} 4[H] + O_2 \tag{3a}$$

$$4[H] + CO_2 \rightarrow CH_2O + H_2O \tag{3b}$$

It is the water molecule that is split in photosynthesis, and the hydrogens are used to convert (actually reduce) carbon dioxide to a carbohydrate. Actually, the symbol 4[H] refers to 4[H$^+$ + e$^-$]; both hydrogen ions and electrons are obtained from water molecules. This will be treated in more detail in the next section.

The following represents photosynthesis in a manner similar to equation (1):

$$6CO_2 + 12H_2O \xrightarrow[\text{light energy}]{\text{chloroplast}} C_6H_{12}O_6 + 6O_2 + 6H_2O \qquad (4)$$

This, then, is a more accurate version of the overall photosynthetic reaction. It is also an excellent example of the importance of being able to label atoms so that the atoms in one reactant can be differentiated from similar atoms in other reactants.

The *energy transfer* aspect of photosynthesis should be emphasized. Sunlight is an enormous source of energy that becomes useful only when it is converted or transferred, to the chemical energy residing in the bond structure of the carbohydrates that are produced in this process. Even photosynthetic cells cannot utilize energy when it is in the form of light, if by "utilization" we mean the use of energy in producing the various constituents of cells. The energy used in synthesizing new materials, or in maintaining existing protoplasmic units, is derived from respiration. In this process various substrates, but especially the carbohydrates produced during photosynthesis, are oxidized, and energy is made available for use by the cell. (This will be discussed more fully in the next chapter.) If substrates other than carbohydrates are utilized in respiration, it should be noted that these substrates are themselves derived from the products of photosynthesis. It can be seen, then, that photosynthesis is the mechanism whereby an energy source is made available.

The discussion in the next two sections is based upon recent evidence and upon general hypotheses that fit most of the data that have been accumulated. However, as additional information is obtained, one should not be too surprised if a revision of at least parts of these sections would be necessary. Similar revisions in various aspects of biology have been made many times in the past as a result of further investigations. Photosynthesis is a complex process, and the difficulties inherent in research on the molecular level should be apparent to everyone. Also, this discussion has been simplified in order to remain within the purview of a general botany text, and refers primarily to the higher plants.

Two types of reactions are involved in photosynthesis: (1) photochemical and (2) enzymatic, or chemical. The photochemical reactions utilize light energy to bring about the decomposition (photolysis) of water whereby hydrogens and electrons are made available for the later enzymatic reduction of carbon dioxide to carbohydrate. In addition, energy is required to bring about this reduction, and a portion of the light energy that is absorbed by the chloroplast pigments is funneled into the formation of high-energy phosphate bonds. Such bonds are extremely important in storing chemical energy, in transferring energy from place to place, and in providing the energy for various syntheses that occur in living cells. Most reactions of a cell in which more complex molecules are built up from simpler ones involve high-energy phosphate bonds. In general, molecules are much more reactive when they contain such bonds, and these phosphate groups are readily transferred from compound to compound. (A more intensive discussion of high-energy phosphate bonds can be found in Chapter 13.) The hydrogens, electrons, and high-energy phosphate bonds from the first steps of photosynthesis are picked up by carrier molecules and are later utilized in enzymatic reactions that may take place in the dark; although these latter reactions are influenced by temperature, they do not require light. The *light reactions* require light and are those in which hydrogens, electrons, and high-energy phosphate bonds are made available. The use of these materials to reduce carbon dioxide to carbohydrates can take place in either light *or* in the dark; light is not required and, thus, the steps involved are termed the *dark reactions*. These latter reactions produce carbon compounds that may be stored, transported, or used to synthesize the basic organic compounds necessary to living systems.

12.5 Light Reactions

Although this discussion is somewhat detailed, it has been simplified and yet presents a useful scheme consistent with current observations and data. However, one should bear in mind that there is not universal agreement among workers in the photosynthesis field. There are occasional conflicting results and differences in interpretation, but the material as presented appears to be the most accepted general mechanism.

In the late 1950s and early 1960s Robert Emerson and his colleagues made rather precise measurements of the yield of photosynthesis at different wavelengths over the visible spectrum. They found that photosynthesis was less efficient in wavelengths greater than 680 nm than would be expected from the absorption spectra of the chlorophylls. This is an area of the spectrum occupied by the red absorption band of chlorophyll *a*. The efficiency could be restored, however, by a simultaneous application of a shorter wavelength. The effect of the two superimposed beams of light on the rate of photosynthesis exceeded the sum effect of both beams used separately—a *photosynthetic enhancement*. It became increasingly apparent that photosynthesis requires the cooperation of two distinct photochemical processes; wavelengths of light shorter than 680 nm affect both processes, while longer wavelengths affect only one.

Numerous spectral analyses of chlorophyll *a in vivo* demonstrated that there are several forms of this pigment in the chloroplast, each absorbing a maximum amount of light at a slightly different wavelength from the others. One form has a maximum absorption at 673 nm and is termed *chlorophyll a 673 (Chl a 673)*; while *chlorophyll a 683 (Chl a 683)* absorbs maximally at 683 nm. Another form of chlorophyll *a* is found in much smaller amounts than either *Chl a 673* or *Chl a 683,* and is termed *P 700* because it has an absorption maximum at 700 nm. These light-harvesting pigments occur in groups, known as **photosynthetic units,** containing 200 to 400 collaborating chlorophyll molecules that can absorb light and funnel this excitation energy on to a reactive molecule. *P 700* has been calculated to exist in the chloroplast at a concentration of 1 or 2 molecules per 300 chlorophyll molecules. It is currently thought that this form of chlorophyll acts as a photochemical reaction center, utilizing the light energy that has been harvested by the other pigment molecules.

The two photochemical processes that drive photosynthesis have specific groups of pigments, respectively. Light energy for **Photosystem I** is collected by *Chl a 683* and *P 700,* while *Chl a 673* and *chlorophyll b* function in **Photosystem II.** Carotenoids of different types appear to be present in both systems, but this situation is not yet clear. In any event, light energy trapped in Photosystem II activates *Chl a 673;* that is, an electron of the chlorophyll molecule is raised to a higher orbital or energy level (see Section 3.2). If this electron merely fell back to its original position, the ground state, it would lose its abosrbed energy as fluorescence and heat. This does not occur in the chloroplast. Instead the high-energy electron leaves the activated chlorophyll molecule and is passed along a series of electron-carrier molecules. The exact pathways of the electrons are not known for certain,

Figure 12-2. A simplified overall schematic diagram of photosynthesis.

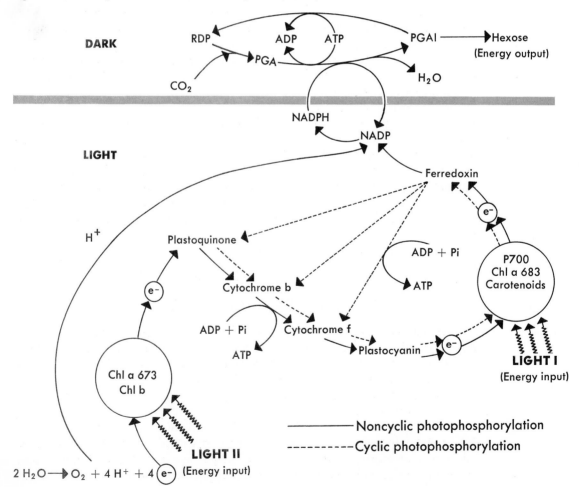

but the most generally accepted one appears to be as presented. Light energy in Photosystem II is also responsible for splitting water into protons (H^+), electrons (e^-), and oxygen gas. These are the electrons that replace those that have been removed from the activated chlorophyll molecules; and activation may again occur.

As shown in Figure 12-2, the high-energy electrons pass from *Chl a 673* to plastoquinone (although there may be a primary electron acceptor functioning before plastoquinone), to one or more cytochromes (iron-containing compounds), to plastocyanin, to the Photosystem I reaction center or *P 700*. As these electrons are transferred along a series of electron acceptors down an energy gradient some of the released energy is used in the formation of high-energy phosphate bonds. In effect, then, some of the light energy is conserved as chemical energy by the conversion of **adenosine diphosphate (ADP)** to **adenosine triphosphate (ATP)**, a molecule to which the high-energy phosphate has been added and that later participates in the reduction of carbon dioxide. The added phosphate comes from inorganic phosphate available in the cell. ATP may be considered to be an energy carrier. It is an available source of energy that enables the cell to carry out "uphill" reactions—that is, reactions against an energy gradient such as the reduction of carbon dioxide to carbohydrate. More energy resides in the chemical bonds of the product (carbohydrate) than in the substrate (CO_2) as a result of this input of energy; actually light energy converted to chemical energy. The cell may utilize this chemical energy later in that respiratory reactions utilize carbohydrates and yield ATP (see Chapter 13).

Light energy harvested by Photosystem I boosts electrons from *P 700* to a higher energy level, which enables their transfer to another primary electron acceptor, ferredoxin. There is some evidence that suggests the existence of an intermediate between ferredoxin and *P 700*. In any case, this second input of light energy raises the energy level of electrons sufficiently to have them accepted eventually by **nicotinamide adenine dinucleotide phosphate (NADP)**, which also picks up the protons (H^+) from water, thus being converted (reduced) to **NADPH,** a powerful reducing compound. The latter molecule eventually participates in the reduction of carbon dioxide.

The reactions involved are as follows:

$$2H_2O \rightarrow 2H^+ + 2OH^- \tag{5}$$
$$\underline{2OH^- \rightarrow 2e^- + H_2O + \tfrac{1}{2}O_2} \tag{6}$$
$$\text{Sum:} \quad H_2O \rightarrow 2H^+ + 2e^- + \tfrac{1}{2}O_2 \tag{7}$$

The net consumption of H_2O is therefore two molecules per molecule of O_2 produced; reaction (7) represents the sum of (5) and (6). Oxygen diffuses away from the chloroplasts; it may be used by the cell (in respiration), or it may diffuse to other cells or entirely out of the plant. The protons (H^+) and electrons (e^-) are used in the production of NADPH:

$$NADP + H^+ + e^- \rightarrow NADPH \tag{8}$$

Although phosphorylation is accomplished when ADP is converted to ATP, the electrons that are removed from the chlorophyll molecules do not return to those molecules but go to NADP; they are replaced by electrons obtained from water.

We use the term **noncyclic photophosphorylation** for this chain of events wherein ATP is produced and electrons are donated by water molecules. It has not yet been established how many ATP molecules are produced during these electron transfers although three ATP are required for the reduction of one molecule of carbon dioxide.

A second type of photophosphorylation has also been found to occur in chloroplasts. In this instance electrons are not derived from water; they are emitted by the reaction center, after excitation by the absorption of light energy, and return to the chlorophyll molecule after passing through a series of acceptors. The term **cyclic photophosphorylation** is used to indicate that the electron returns or cycles back to its original position in the chlorophyll molecule. As the high-energy electron proceeds along this circular pathway, some of its energy is used to bring about the formation of high-energy phosphate bonds (ATP); some of the light energy is conserved as chemical energy. The electron pathway is not known for certain and several possibilities are shown in Figure 12-2 (the dashed arrows). There is also disagreement as to what extent cyclic electron transport occurs normally. If insufficient ATP molecules are formed during noncyclic photophosphorylation, it is possible that the cyclic process may be important in providing essential ATP.

The various enzymes and carrier systems that participate in the light reactions are considered to be closely associated with chlorophyll in the lamellae portion of the chloroplast, especially in the grana. Various of the carotenoid pigments are similarly located. These pigments absorb light energy, and transmit this excitation to the chlorophyll molecule. Thus, all of the complicated machinery for the light reactions of photosynthesis is available in one small area. Such an arrangement may have a high degree of efficiency in that small packets of energy are all focused to one location; energy transfers and various reactions occur within a confined area over short distances. This tends to insure sufficient energy to favor rapid generation of ATP and NADPH even at low light intensities.

The light reactions set the stage for the dark reactions. The former result in the production of ATP, NADPH, and molecular O_2. The oxygen is not used in the rest of the photosynthetic process. The other two products participate in the subsequent dark reactions. Both cyclic and noncyclic photophosphorylation are shown in Figure 12-2. Note that there is an energy input (i.e., light) and that part of this energy is converted to the chemical energy residing in the high-energy phosphate bonds of ATP. Part of this light energy is also used to raise electrons to a sufficiently high level to participate in the reduction of NADP to NADPH.

12.6 Dark Reactions

In the second stage of photosynthesis the products of the light reactions are used to incorporate carbon from CO_2 into organic molecules. These "dark reactions," which take place in the stroma, usually occur in the light and continue only very briefly after a plant is placed in the dark because NADPH and ATP produced in the light reactions are consumed very rapidly. Experiments have shown, however, that they can proceed in the dark as long as NADPH and ATP are available.

As the carbon fixation process was being clarified, it was assumed that there was a single pathway. This idea was accepted for quite a few years, but now it is known that there are at least two pathways, plus a modification of one; other pathways may be discovered in the future.

The Three-Carbon (C_3) Pathway

The mechanism of carbon dioxide fixation worked out by Melvin Calvin and his colleagues, and frequently termed the **Calvin-Benson cycle,** starts when carbon dioxide is added to a pre-existing 5-carbon sugar phosphate (**ribulose diphosphate, RuDP**), forming an unstable 6-carbon compound, which immediately splits into two molecules of **phosphoglyceric acid (PGA),** the first identifiable product. This acid is a phosphorus-containing organic compound with three carbon atoms; thus the term C_3 **pathway.** It is at this stage that some of the products of the light reactions begin to participate, and where energy (as ATP) is required. NADPH and ATP bring about the reduction of PGA to another 3-carbon compound termed **phosphoglyceraldehyde (PGAl),** with water also forming in the reaction:

$$PGA + NADPH + ATP \rightarrow PGAl + NADP + ADP + H_2O \qquad (9)$$

The NADP and ADP are again available for accepting hydrogens and electrons of high-energy phosphate bonds in the light reactions. PGAl rapidly forms an equilibrium mixture with dihydroxyacetone phosphate (also a 3-carbon compound), so it is simpler to consider that PGA has been reduced to a **triose phosphate** (= a three-carbon compound with a phosphate group attached). Some plant physiologists consider this triose phosphate to be the product of photosynthesis because it is the first substance produced with the use of materials from the light reaction. However, it is not a triose that is drained out of the cycle, as will be discussed here.

The many triose molecules produced in this fashion participate in a complex series of interconnected reactions in which 3-C, 4-C, 5-C, 6-C, and 7-carbon compounds are interconverted. From these diverse compounds hexose (6-C) is funneled off in considerable amounts. Because hexose accumulates to a greater extent than the other compounds, it is more readily observed on chemical analysis, and early investigations (with less sensitive methods than present ones) found mainly glucose (a common hexose).

During these interconversions, **ribulose** is also formed. Phosphorylation occurs, using ATP from the light reactions, and RuDP is then available to accept CO_2 (Figure 12-3). In effect this series of complex reactions is cyclic; RuDP is used up and re-formed, and hexose is drawn off the cycle. Since glucose is readily converted to other materials, amino acids, organic acids, and other compounds appear quite rapidly in the cell during photosynthesis. A simplified version of the dark reactions could be represented as follows:

$$CO_2 + NADPH \qquad \qquad ATP$$

$$[CH_2O] + H_2O + NADP \qquad ADP \qquad \qquad (10)$$

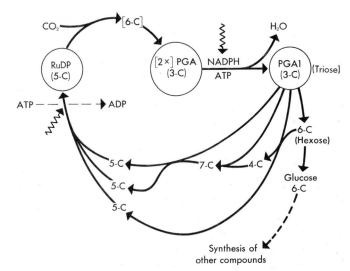

Figure 12-3. A schematic representation showing the Calvin-Benson pathway of carbon in photosynthesis. Each 6-carbon (6-C) compound splits into two molecules of PGA. There would be many CO_2 molecules following this pathway and, thus, many PGA would be produced. Energy input is represented by ⌇⌇⌇.

This indicates the conversion of carbon dioxide to a carbohydrate, $[CH_2O]$ representing hexose primarily. The complex cyclic interconversions have been omitted, and only the beginning materials and end products are shown.

In this whole series of reactions encompassing photosynthesis several materials are changed from one form to another and then back to the original—they are not used up. This is true of chlorophyll, NADP, and ADP. Carbon dioxide and water, however, are used up as hexose and oxygen are produced. Thus, even though reactions number (3) and number (4) indicate the overall photosynthetic process, Figure 12-2 represents the process more accurately. Recent investigations indicate that varying the conditions under which plants are grown may greatly influence the production of fats and proteins. This results from the rapid transformation of simple carbohydrates into more complex compounds.

The Four-Carbon (C_4) Pathway

By the late 1960s, and after the C_3 pathway had been well established for ten years or so, it became quite clear that a C_4 pathway existed. Working with sugar cane, M. D. Hatch and C. R. Slack found that CO_2 reacted with **phosphoenol pyruvic acid (PEP),** a three-carbon compound, and that the product was oxaloacetic acid (4-C) which is readily converted to other four-carbon acids, especially malic. This first photosynthetic product, of course, suggested the name **four-carbon pathway;** also frequently termed the **Hatch-Slack pathway.** The 4-carbon compounds are broken down enzymatically to yield carbon dioxide and pyruvic acid. The CO_2 is accepted by RuDP of the C_3 cycle, and pyruvic is converted to PEP with the utilization of ATP. Thus, only the initial steps of carbon fixation of C_4 plants differ from those of C_3 plants. The C_4 pathway has now been found in various tropical grasses, including *Zea mays* and *Sorghum,* in sedges (*Cyperus*), and in several dicotyledonous plants.

The leaf anatomy of C_4 plants differs from that of C_3 plants (Figure 12-4). In the latter, the chloroplasts are basically similar in appearance and are present throughout the mesophyll tissue of the leaf; thus, CO_2 fixation takes place in each mesophyll cell. In C_4 plants there are prominent bundle sheath cells that contain chloroplasts that differ from those in the mesophyll cells; that is, the chloroplasts in the bundle sheath cells contain numerous starch grains and poorly developed grana. The mesophyll cell chloroplasts usually lack starch grains and contain grana. The fixation of CO_2 by way of PEP occurs in these mesophyll cells. The 4-carbon acids are transported to the bundle sheath cells, where they release CO_2 to the C_3 cycle; the pyruvic formed in this reaction is exported to mesophyll cell chloroplasts, where it is phosphorylated to PEP. Thus, there is a spatial separation of biochemical events; the PEP system in the mesophyll cells and the RuDP system in the bundle sheath cells. The end result, as in the C_3 pathway, is the production of carbohydrate from CO_2 and H_2O.

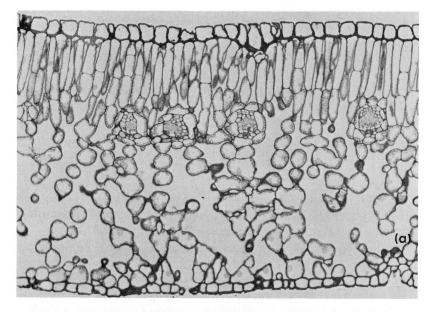

Figure 12-4. (a) Leaf of C_3 plant (*Ligustrum*), cross section ($\times 50$) (Courtesy Carolina Biological Supply Company). **(b)** Leaf of C_4 plant (*Zea mays*; corn), cross section ($\times 100$) (Courtesy of Robert Gill).

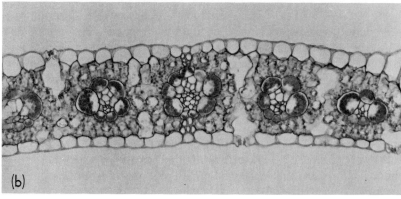

Crassulacean Acid Metabolism (CAM)

Another group of plants, especially succulents, utilize a C_4 pathway, but the biochemical events are separated *temporally* rather than spatially. The leaves of plants belonging to the Crassulacean family, thus **Crassulacean acid metabolism or CAM,** have long been known to display a diurnal pattern of organic acid formation. These acids, particularly malic, accumulate in the leaves at night and decrease during the day; the resulting shift in pH is easily detected (i.e., low pH at night and higher pH during the day). Similar patterns of acid content have been found in other groups of plants (e.g., members of the cactus, orchid, and pineapple families).

The diurnal pattern of organic acid content in CAM plants is accompanied by a diurnal pattern of stomatal opening and closing. The stomata are open at night, CO_2 diffuses into the leaf, and is fixed through the PEP system. In the light the stomata close, which is quite a different situation from other plants. The enzymatic breakdown of malic during the day provides a supply of CO_2 that is fixed by the C_3 pathway (i.e., accepted by RuDP). Thus, CAM is basically a C_4 type of photosynthesis wherein the various biochemical events occur in the mesophyll cells although at different times; the PEP system and the RuDP system are separated by time.

Figure 12-5 shows a comparison of C_4 photosynthesis and CAM. The pathways are quite similar in terms of overall results.

12.7 Photorespiration

As a result of many studies regarding plant productivity, it was found that much of the photosynthate is rapidly oxidized to carbon dioxide and water without the production of metabolically useful energy. Under conditions of low CO_2 and high O_2 concentrations, which are the usual atmospheric conditions, glycolic acid is an intermediate product of photosynthesis in the C_3 cycle; and this 2-carbon compound is the substrate that is oxidized. As much as 20–40 per cent of the photosynthetically fixed carbon may be reoxidized to CO_2 during this process of **photorespiration,** which occurs in small microbodies called **peroxisomes** located adjacent to the chloroplasts. No ATP is formed in this process, in contrast to respiration that occurs in the mitochondria (see next chapter), and the energy is merely dissipated as heat. Thus, the net productivity of the plant is decreased; that is, the amount of organic carbon or dry matter accumulated over a period of time is lower than it would be if photorespiration did not occur.

Investigations have shown that respiratory CO_2 production and respiratory O_2 consumption in chlorophyllous tissues of most species of higher plants proceeds at higher rates in the light than in the dark. In only a small number of plants are the rates of respiration the same in the light and dark; and these are the C_4 plants (e.g., sugar cane, maize, sorghum, crab grass). Illuminated leaves of C_4 plants synthesize only negligible amounts of glycolic acid. Also, any photorespiration in these

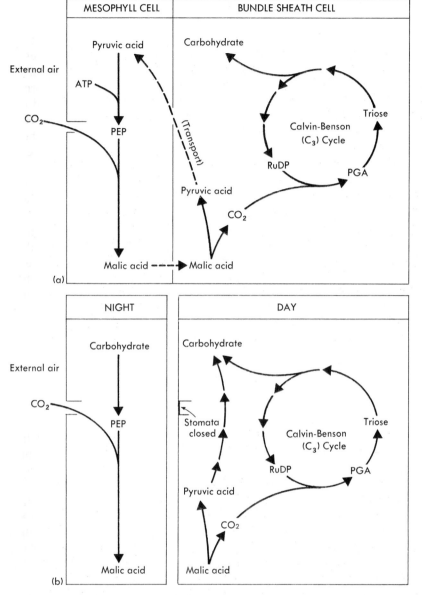

Figure 12-5. Comparison of C_4 photosynthesis and crassulacean acid metabolism (CAM). **(a)** C_4 reactions are spatially separated (i.e., in different cells) but occur simultaneously in light. **(b)** CAM reactions are temporally separated (i.e., by time) but occur within the same cell. From "Comparative Studies on Photosynthesis in Higher Plants" by Olle Björkman, Chapter I in *Photophysiology* Vol. VIII, edit. by A. C. Giese, 1973, Academic Press, N.Y., by permission.

leaves would occur in the bundle sheath cells where the C_3 cycle (Calvin-Benson cycle) is located. The CO_2 would then diffuse out through the mesophyll area where the highly efficient C_4 (PEP) system is located. It appears quite likely that most of this CO_2 would then be re-fixed and not be free to diffuse out of the leaf. In any event photorespiration, which appears to be one of the main factors decreasing net productivity in C_3 plants, is not significant in C_4 plants and this is one factor responsible for the high productivity of the latter plants.

12.8 Selective Advantages of the C_4 and CAM Pathways

C_4 plants apparently evolved primarily in the tropics, areas of extremes in light, temperature, and dryness. Investigations have demonstrated that C_4 plants utilize much greater quantities of CO_2 at high light intensities and high temperatures than do C_3 plants. This is a result of the fact that the enzyme (PEP carboxylase) catalyzing the uptake of CO_2 by PEP has a much greater affinity, or attraction, for CO_2 than does the enzyme RuDP carboxylase of the C_3 system. Although the photosynthetic rates are comparable at low light intensities and low temperatures, as these factors increase the C_4 plants become increasingly superior because the C_3 plants are hampered by their carboxylase system. The former plants efficiently absorb CO_2 (PEP system), which is then deposited in relatively high concentrations at the site of the C_3 pathway. This build-up of CO_2 concentration in the bundle sheath cells speeds up the C_3 system there, and the overall rate of photosynthesis is relatively high.

CO_2 diffuses into the leaves through open stomata, but this also allows the loss of water from moist cells. Such diffusion of water out through the open stomata can be very harmful to the plant if cells and tissues become too desiccated. Because of their generally higher rates of photsynthesis, C_4 plants fix much more carbon dioxide than C_3 plants for the given period of time that the stomata are open. At higher light intensities and temperatures, where water loss would be especially severe, this difference between C_3 and C_4 plants could be particularly significant and advantageous to the latter. That is, C_4 plants could fix sufficient carbon for growth and development in a relatively short period of time, after which the closure of stomata would conserve moisture without hampering the plant. The C_3 plants, on the other hand, because of lower carbon fixation rates, would grow poorly in the same situation.

CAM plants also have an advantage over C_3 plants in certain environmental situations—as in hot and dry habitats. The closure of stomata during the day greatly retards water loss in CAM plants, but photosynthesis is still possible because of the CO_2 fixation that takes place at night (PEP system) when the stomata are open. By accumulating organic acids at night, much more CO_2 can be fixed or converted into carbohydrate than would otherwise be possible with this pattern of stomatal opening and closing.

As indicated in the previous section, the loss of carbon in the form of CO_2 as a result of photorespiration is significantly greater from C_3 plants than from C_4 types. Thus, even if the rates of CO_2 fixation were equal, more carbon would remain in the C_4 plants beause of their lower rates of photorespiration or because of their efficient recovery of the CO_2. Investigators are attempting to breed for low rates of photorespiration in crop plants to increase yield. The success of this program will depend upon further studies of photorespiration and its significance to overall plant growth and development.

The C_4 pathway is of more recent origin and undoubtedly arose many times in the course of evolution. Although primarily in tropical grasses, it has been found in both monocotyledons and dicotyledons, and some plant groups (genera) have both C_3 and C_4 types. In a general way plant distributions reflect these evolution-

205

Sec. 12.9 / The
Use of
Radioactive
Isotopes and
Chromatography

ary patterns in the development of various photosynthetic pathways. CAM plants are well adapted to areas with hot days and cool nights that do not have freezing temperatures. C_4 plants are best adapted to hot, dry, or wet tropical climates with warm nights. C_3 plants are well adapted to cool, wet climates but are abundant in all climates.

12.9 The Use of Radioactive Isotopes and Chromatography

It would have been impossible to determine the complex pathway of carbon in photosynthesis without the use of radioactive isotopes and the technique of paper chromatography. These methods are of great importance in the study of other aspects of metabolism as well as for investigations of photosynthesis, but we will use the latter process as an example in order to clarify the principles involved.

Calvin and his colleagues were able to elucidate the pathway of carbon in photosynthesis by the use of radioactive carbon (^{14}C) in carbon dioxide. If the period of illumination was relatively long, 60 seconds, for example, many compounds contained ^{14}C (e.g., organic acids, amino acids) but most was found in phosphorylated sugars. When the period of illumination was shortened, it was possible to ascertain the order in which various intermediates were formed. With a photosynthetic period of two seconds, the predominant labeled product was found to be phosphoglyceric acid. Increasing the period in a series of experiments enabled these investigators to determine the general order in which labeled compounds appeared.

The technique was elegant. Unicellular aquatic green plants (e.g., *Chlorella*, an alga) were grown in a mineral nutrient solution and exposed to light to allow for photosynthesis. Carbon dioxide with ^{14}C was then injected into the solution. At the chosen interval, two seconds after the ^{14}C was injected, for example, the flask was emptied into boiling alcohol to stop photosynthesis and to extract any materials containing ^{14}C. The extract was chromatographed and radioautographs were made. This is accomplished by placing the concentrated extract at the corner (termed the origin) of a large piece of filter paper (about 46×57 cm) and drying in a stream of warm air. The edge of the filter paper is placed in a trough with the origin and most of the paper hanging down below (Figure 12-6). A solvent placed in the trough moves by capillarity along the paper (much as water is taken up by a blotter), dissolving the compounds in the extract and carrying them along. When the solvent reaches the lower end of the paper, the paper is removed and dried. Compounds are variously soluble in the solvent and adsorbed to differing degrees by the filter paper. Thus, a partial separation occurs. The paper is then rotated 90 degrees, the edge is placed in a trough with a second solvent, and the chromatogram allowed to develop as before. This results in a final separation—a two-dimensional chromatogram. The filter paper is then placed on unexposed x-ray film in the dark. Later, when the film is developed, darkened (exposed) areas indicate the position of labeled products. Known compounds are chromatographed in the same manner. The positions of known compounds can be used as a "map" to

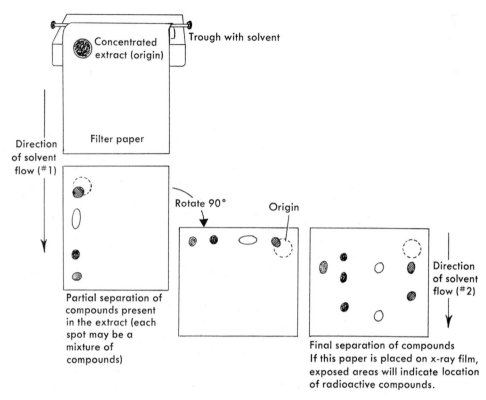

Figure 12-6. Two-dimensional paper chromatography. See text for additional discussion.

determine the identity of "spots" found when the extract is used. It is also possible to cut out areas of paper that correspond to the dark spots on the film and use direct chemical analysis.

Chemical analysis and radioactivity measurements of the compounds separated by paper chromatography made it possible to determine the manner in which the compounds were formed. Such procedures could detect the location of the radioactive carbon atoms; in other words, which carbon atoms of the organic compound first became labeled as a result of the incorporation of $^{14}CO_2$. For example, phosphoglyceric acid and then phosphoglyceraldehyde were found to be labeled in the number one carbon atom of the three carbons that are present in these compounds, as shown in Figure 12-7 (only the carbon atoms are shown, and the asterisk indicates radioactivity). The hexose, which was obtained later, had radioactivity in the two central carbon atoms. This indicated that two triose (3-carbon) molecules joined end to end in forming the hexose.

Because of the many processes or reactions taking place within a living cell, many compounds different from the ones that were originally present in the cell are produced. Some of these latter compounds are transitory stages, or intermediates, in a long series of steps during which the original material is converted to a final product. Organic compounds undergoing such changes contain varying numbers of carbon, hydrogen, and oxygen atoms, as well as smaller numbers of other

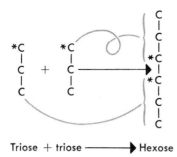

Figure 12-7. Diagram to indicate labeling of carbon atoms when two 3-carbon compounds react end to end in forming a 6-carbon compound (the hexose).

Triose + triose ⟶ Hexose

atoms. Ordinarily, one carbon atom cannot be distinguished from another, or one hydrogen atom from another. However, as mentioned previously, certain atoms in a molecule can be labeled by using isotopes, which may be radioactive. In this manner a physiologist can frequently determine which of a series of intermediates has been formed first; in other words, which compound has the greatest concentration of isotopic atoms at a particular time. Also possible in many instances is the determination of how a compound was broken down or put together by analyzing for the location of the specific isotope.

During the last 10 or 20 years, many radioactive atoms have been produced and used in this fashion, and their production is one of the most important applications of the cyclotron for biologists. The replacement within a cell of one protein molecule by another similar protein molecule could not be detected without such labeling. Radioactive materials are detected quite readily with a Geiger-Muller counter or by the fact that the radiations will expose photographic film. Such studies have resulted in tremendous advances in our knowledge of cellular functions and will undoubtedly lead to additional information in the future. However, many difficulties are inherent in this type of investigation—rapid conversions of one material into another, extremely small amounts of materials involved, techniques frequently injurious to cells and tissues—and many processes occurring in cells are still not well understood or explained. In fact, data are frequently hard to interpret because of the difficulties involved in the methods used. As methods improve, more and more valuable information will be obtained.

12.10 The Fate of the Sugar Molecules

That simple sugars provide the basic building blocks for all the complex molecules that constitute a living cell, its secretions and its excretions, cannot be emphasized too strongly. Hexose and probably other compounds formed during the carbon cycle of photosynthesis undergo many chemical transformations, which result in the formation of starch, cellulose, fats, proteins, vitamins, hormones, and many other compounds. Some of these materials are structural components of the cell (e.g., cellulose, proteins), some are storage products (e.g., starch, proteins, fats), and others are necessary for the normal metabolism of the cell (e.g., vitamins, hormones, enzymes). Hexose frequently is converted to sucrose and trans-

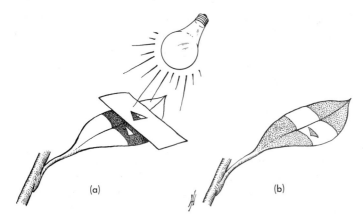

Figure 12-8. **(a)** A cover shades a portion of the leaf. Actually the cover is placed directly on the leaf, but in the diagram it has been shown in a raised position to emphasize the shading effect. **(b)** The stippled area indicates a positive starch test. Note that the covered area is devoid of starch.

(a)

(b)

ported from the mesophyll cells of the leaf to other, nonchlorophyllous cells, where such reactions then take place after the sucrose is digested to hexoses. In addition to hexose being used as a basis for syntheses, many of the sugar molecules are utilized as a source of energy in the respiratory processes of cells. Respiration is that process which occurs in all living cells and makes energy available for cellular functions. This energy-yielding process will be discussed in more detail in Chapter 13. In effect the energy provided by photosynthesis is being released by respiration and used to bring about various other syntheses.

The rapid conversion of sugar to starch, a storage product, frequently enables physiologists to utilize the convenient iodine test for starch as an indication of the occurrence of photosynthesis. In many leaves excess sugars (excess in that more sugars are produced during photosynthesis than are used by the leaf) are rapidly converted to starch, which yields a blue-black color when treated with iodine-potassium iodide solution. If photosynthesis is prevented in half of the leaf by covering this portion with black paper, only the other half of the leaf will yield a positive starch test (Figure 12-8). The chlorophyll is usually first removed by boiling the leaf in alcohol so that the green color does not interfere with the test. In some leaves (e.g., onion, lily) starch is not formed, and this test could not be used. In fact, it must be clearly understood to be an indication of photosynthesis and not really a test for the end product of the reaction, which is hexose. To test for starch is merely more convenient than to test for sugar, and the test is valid as long as its limitations are understood.

12.11 Limiting and Retarding Factors

Photosynthesis is the first metabolic (chemical) process of the plant that has been mentioned in any detail, and this is an excellent opportunity to utilize a process to point out examples of limiting and retarding factors. Remember, however, that these concepts or principles are valid for all processes.

A **limiting factor** is any essential whose deficiency slows down the rate of a reaction. As the amount of this material is increased, the rate of the process will increase. For example, during the night, light is limiting the rate of photosynthesis. As the sun rises and the light intensity increases, the rate of the process increases, up to the point where something else becomes limiting or retarding (see the following paragraph). A similar situation may prevail with any of the essential factors, even though light is most often limiting under natural conditions. It is certainly possible that on a bright summer day at noon carbon dioxide may be a limiting factor; or temperature in evergreens during the winter. The influence of two factors is indicated in Figure 12-9. As the concentration of carbon dioxide is increased, photosynthesis increases up to a certain rate; beyond this the rate remains constant, even though a further increase in the amount of carbon dioxide is available to the green cells (curve ABC). The point at which the curve levels off indicates that carbon dioxide is no longer limiting but that some other factor now is. In this instance it is a low light intensity. If the light intensity is increased to a medium level, the curve assumes the shape ADE; at a high light intensity, the curve is AFG. In each case, the rate increases directly with an increase in carbon dioxide concentration until light becomes limiting: beyond points B, D, and F, respectively, there is sufficient carbon dioxide but not enough light. The solid-line curves are theoretical. Under field conditions, no abrupt change in shape as is indicated at B, D, and F takes place, but rather a gradual transition from an increasing rate to a constant rate, as indicated by the broken lines. This discrepancy results because no single factor is limiting or retarding the process for the plant as a whole or even in all cells of a single leaf. What usually happens is that the rate of photosynthesis may be limited by a deficiency of carbon dioxide in some cells and by a low light intensity in other cells of the same leaf, especially if any portions of the leaf are shaded.

A **retarding factor** is any factor whose presence in excess slows the rate of a reaction. The accumulation of end products of a reaction will tend to slow the reaction; for example, as sugar accumulates in the afternoon the rate of photosynthesis decreases. Poisons may also be considered as retarding factors. The rate of photosynthesis increases as temperature rises, up to approximately 35°C, after which a rapid drop in the rate occurs; temperature was first a limiting factor and then a retarding factor.

Throughout a full day under general field conditions, first one and then another limiting and/or retarding factor is the principal agent governing the rate of photosynthesis, and more than one may be operative at any given instant. This is espe-

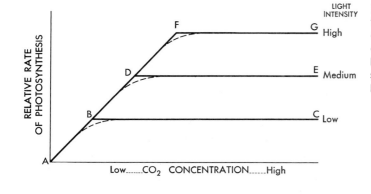

Figure 12-9. A series of curves to indicate the concept of limiting factors. The slope of curve ABDF demonstrates the effect of increasing CO_2 when this factor is limiting. The horizontal portions of the curves demonstrate the lack of effect of CO_2 when light is limiting. (For further discussion see the text.)

cially so if the photosynthetic rate of the entire plant is being considered. For example, CO_2 may be limiting (or sugar may be retarding) in a leaf exposed to sunlight on a cloudless summer day at noon, while photosynthesis in a shaded leaf on that plant may be limited by light intensity at the same time. Usually light is the primary limiting factor in the morning, carbon dioxide around noon, and light again in the evening and certainly at night. The accumulation of sugar (retarding factor) aids in bringing about the decrease in the afternoon, while temperature may be too low in the morning and too high in the afternoon. Obviously, therefore, in discussing the influence of a particular factor on the rate of a process, the assumption must be made that this one factor is either limiting or retarding the process. If this were not so, a variation in that factor would have no influence on the rate of the process. Why?

Water, even though essential to photosynthesis, is probably seldom directly a limiting factor. Most cells that carry on the food-manufacturing process contain 80 to 95 per cent water. Before water could become so deficient as not to be available for photosynthesis, the normal activities of the cell would have been disrupted. The decreased rate of photosynthesis observed when water becomes deficient in the soil is usually caused by a lack of carbon dioxide (limiting) or an increased concentration of sugar (retarding). When water content is low, the leaves wilt, the stomata close (Figure 10-8), and carbon dioxide cannot diffuse into the leaf. The decrease in water content of mesophyll cells (which is what brings about the wilted condition) also results in an increased sugar concentration of those cells, even though the total amount of sugar does not increase.

12.12 Efficiency of Leaf Structure

In the more complex plants, such as the flowering ones, almost all of the food-manufacturing process is restricted to the leaves whose structure is quite conducive to the carrying on of photosynthesis. They are usually broad, flat, and thin, which ensures the exposure of all cells to light, especially since many leaves are oriented with the expanded surface at right angles to the incident rays. Those cells near the lower epidermis receive light, which penetrates through the relatively thin tissues above. The large number of stomata and extensive intercellular (air) spaces of the leaf enable carbon dioxide molecules to diffuse from the atmosphere to all photosynthesizing cells. Such cells (the mesophyll tissues) contain abundant chloroplasts with their chlorophyll pigments and other components, which are so necessary to the process. To these characteristics of a leaf must be added the numerous branching veins that carry water and minerals to the mesophyll and sugars away from this tissue. Cells are not very far removed from the transporting tissues, and materials essential to the normal functioning of these cells are readily obtained. For example, this would include not only water, but also chloride and manganese ions, which are necessary for the photolysis of water (Photosystem II). And then, the rapid transport of sugar away from the mesophyll cells would tend to reduce any retarding effect that might occur as photosynthesis progresses.

12.13 Plant Productivity

There are vast areas of the earth populated by enormous numbers of green plants of many types. It is not possible, of course, to determine the amount of carbon actually fixed (i.e., converted from CO_2 to organic carbon such as sugars and other carbohydrates, fats, proteins, and so forth) by these plants during the course of a year. However, as knowledge of plant activities increased and investigators became aware of rates and numbers of plants involved, it became more and more apparent that tremendous quantities of carbon are processed annually through green plants. The development of more accurate and sensitive techniques, more adequate procedures, and the burgeoning interest of ever more botanists eventually led to estimates that became more and more acceptable. It should be clear that fluctuations in environmental conditions have a great influence on photosynthesis and plant growth, and that the values presented in any of these estimates are basically averages as determined by the best evidence presently available.

Table 12-1 presents estimates made recently by Robert H. Whittaker, although the data have been condensed and simplified from the details presented in his extensive table. The values in column three are rate measurements that indicate the average amount of carbon fixed per unit area per year; such values are commonly referred to as *productivity* estimates. These are really net productivity values; that is, the amount of carbon assimilated into plant material after the loss of carbon by respiratory processes. Probably 25 to 30 per cent of the carbon fixed in photosynthesis is lost through respiration. The values in column four indicate the total amount of carbon fixed annually by the particular ecosystem involved. The term **ecosystem** refers to a community of organisms and its environment in an area.

It is obvious that productivity varies widely in different plant communities, and

Table 12-1 Amounts of carbon fixed annually by green plants

Ecosystem	Area ($10^6 km^2$)[a]	Productivity ($g/m^2/yr$)	Total production ($10^9 t/yr$)[b]
Forests	57	1450	79.9
Grasslands	24	790	18.9
Cultivated lands	14	650	9.1
Deserts	18	90	1.6
Tundra and alpine	8	140	1.1
Fresh waters	4	1100	4.5
Extremes (rock, sand, ice)	24	3	0.07
Total continental	149	770	115
Oceans	361	150	55
Total earth	510	330	170

[a] $km^2 \times 0.3861 = $ sq. miles
[b] t = metric ton = 1000 kg = 1.1023 English short tons = about 2200 lbs
Source: Modified from *Communities and Ecosystems*, 2nd ed., by Robert H. Whittaker, copyright 1975, Macmillan, Inc., New York.

most of the differences are the result of environmental factors: temperature, water supply, soil minerals, and so on. Desert plants are limited by high temperatures and lack of moisture, while plants in the tundra are limited by low temperatures. Tropical forests tend to have higher productivity than forests in a temperate climate primarily because of the temperature difference. Agricultural practices also influence productivity. In general the efficiency of energy conversion in plants—the ratio of energy stored as organic substance to the amount of energy available from the sun—has been found to range from 0.1 to 1.0 per cent for native vegetation or under ordinary agricultural practices because of limiting environmental factors. These values can be increased by 3 per cent to 10 per cent under conditions of intensive agriculture. In any event, the total amount of carbon fixed annually is at least on the order of 170 billion metric tons, which is equivalent to about 187 billion English short tons (one metric ton = 1000 kg = 2200 lbs).

Summary

1. An autotrophic organism can manufacture its own foods from simple inorganic substances, whereas a heterotrophic organism is dependent upon an external supply of food. Photosynthetic autotrophs utilize light energy, while the chemosynthetic autotrophs obtain energy from chemical reactions. Parasites are those heterotrophs that obtain food from living cells or tissues, while saprophytes utilize nonliving organic matter.

2. Photosynthesis is the manufacture of hexose from carbon dioxide and water by chloroplasts utilizing light energy and releasing oxygen gas. The reactions of photosynthesis that require light result in the production of high-energy phosphate bonds and in the splitting of water, which provides hydrogens and electrons. The subsequent reactions (which may take place in the light or dark) utilize the hydrogens, electrons, and high-energy phosphates to reduce carbon dioxide molecules to carbohydrates (assimilation of CO_2).

3. The first identifiable product of CO_2 fixation is either a 3-carbon or a 4-carbon compound; the C_4 pathway is more efficient. Many succulents are CAM plants in which CO_2 fixation involves a 4-carbon molecule.

4. The sugar molecules produced during photosynthesis are utilized in the various syntheses of the plant, are respired, or are stored.

5. A limiting factor is any essential whose deficiency slows down the rate of a reaction. A retarding factor is any factor whose presence in excess slows the rate of a reaction. Under natural conditions, a process within a plant can be influenced by both limiting and retarding factors simultaneously.

6. The structure of the leaf makes it the most efficient part of the plant for photosynthesis.

7. All living things, except for a few groups of bacteria, depend upon photosynthesis for their existence. Besides being the sole source of oxygen, this process is the only significant food-producing mechanism on the earth. It is estimated that green plants annually convert at least 170 billion metric tons of carbon into organic matter.

Review Topics and Questions

1. Describe in detail what you would have to do to determine whether or not a red-leaved plant contains chlorophyll and carries on photosynthesis.
2. In discussing the overall photosynthetic reaction, one usually states that carbon dioxide and water in the presence of light and chloroplasts are converted to carbohydrates, oxygen, and newly constituted water.
   (a) Explain why the word *chloroplast* should be used rather than *chlorophyll.* Does the use of the former term imply that chlorophyll is not essential to the process?
   (b) Explain why one says that water is both used and produced in the process.
   (c) Explain how a plant might possibly carry on photosynthesis without oxygen gas being liberated into the atmosphere.
3. A labeled atom of carbon originally is found as part of a carbon dioxide molecule in the atmosphere over a field of potato plants and eventually is found as part of a starch molecule in a potato tuber. Trace in logical sequence the movement of this atom, indicating the tissues through which it moves, the kind of chemical substances in which it occurs, the processes to which it is subjected, and the energy relations of these processes. Your discussion should include those forces that bring about movement and changes.
4. In what way are radioactive atoms or isotopes useful to a study of metabolism?

Suggested Readings

BASSHAM, J. A., "The Path of Carbon in Photosynthesis," *Scientific American* **206,** 88–100 (June 1962).

BJORKMAN, O., and J. BERRY, "High Efficiency Photosynthesis," *Scientific American* **229,** 80–93 (October 1973).

LEHNINGER, A. L., *Bioenergetics,* 2nd ed. (New York, Benjamin, 1971).

LEVINE, R. P., "The Mechanism of Photosynthesis," *Scientific American* **221,** 58–70 (December 1969).

RABINOWITCH, E. I., "Photosynthesis," *Scientific American* **179,** 24–35 (August 1948).

RABINOWITCH, E. I., and GOVINDJEE, "The Role of Chlorophyll in Photosynthesis," *Scientific American* **213,** 74–83 (July 1965).

13

Synthesis and Respiration

13.1 Syntheses

Whenever a cell enlarges, or when a cell divides into two that then enlarge, additional protoplasmic materials are produced in addition to larger walls or new walls. The reproduction or doubling of chromosomes is an excellent example of a cell manufacturing new living materials from the various nonliving substances found within the cell. The total of all such processes whereby living material is produced constitutes **assimilation,** whereas the production of any type of cellular matter, living or nonliving, is generally termed a **synthesis.** Of course, many kinds of syntheses occur during the development of a cell. Various kinds of carbohydrates (including cellulose), fats, and proteins (including enzymes) are produced, as well as vitamins, hormones, aromatic oils, and many, many other substances. In addition to making its own food, the green plant is capable of synthesizing all its necessary complex organic compounds from this food plus the minerals it absorbs from the soil. Obviously, much work is carried out by individual cells, and much energy is expended. The production of any more complex molecules from simpler ones entails an expenditure of energy, just as a man expends energy in making a fireplace from a pile of bricks and some concrete. Energy and the source of energy thus are important factors in an understanding of cellular function.

13.2 Energy and the Cell

Energy (see Section 3.5) may be thought of as the ability to do work. The work of the cell may include syntheses, maintenance, active absorption, movement, luminescence, and possibly other functions, and in every case such energy is obtained from a complex series of chemical reactions occurring within the cell. During such reactions, energy is made available in a stepwise fashion, similar in a way to the slow unwinding of a mainspring that enables the watch mechanism to keep time.

The potential (chemical) energy of a food molecule is converted to available energy, which the cell uses to rearrange the atoms of simple molecules to form the various arrangements characteristic of more complex molecules of the cell. In this way the protoplast, cell walls, and various other materials are produced. Even cells that are not growing or reproducing still require energy for the ordinary maintanence of normal cell function. Various molecules are rather unstable and are replaced by new, although similar, molecules. Most component molecules of a cell probably are replaced at one time or another during the lifetime of the organism.

Of the many reactions occurring in a cell, some produce energy and others require it. The energy obtained from the former processes is utilized to run the latter, which otherwise would not occur. Energy has not been created; the cell is merely making use of that energy stored in chemical compounds as a result of the original synthesis of the compounds. All organic materials contain potential energy. This potential energy of organic compounds is a result of the conversion of light energy to "chemical" energy during photosynthesis. Such materials may be burned, producing heat energy, or they may be utilized by living organisms, producing both heat and the energy used to form cellular components or to carry on cell activities. During these conversions of energy, a decrease of useful available energy always occurs, since some useless heat is always produced. This is, of course, not to be considered a loss of energy, because heat itself is a form of energy—energy can be neither created nor destroyed, but it can be, and is, readily transformed from one form to another. Heat is energy that dissipates away from the organism and cannot be used by it.

13.3 Respiration

Note that breathing and respiration are not synonymous. **Respiration** refers to the series of complex oxidation-reduction reactions whereby living cells obtain energy through the breakdown of organic material; some of the intermediate (or partial breakdown) materials can be utilized for various syntheses. **Breathing** is merely an exchange of gases, which, in the human and in a few other animals, takes place in the lungs. All living cells respire, an important feature of this process being the energy that is made available.

Although oxidations occur during cellular respiration, this does not necessarily mean that oxygen participates in the reactions that occur. As mentioned previously (Section 3.7), oxidation means that electrons have been removed and this may occur along with protons as hydrogen atoms; but oxygen may not be involved as the hydrogen or electron acceptor. We will discuss both types of respiration, but the emphasis will be on the one involving oxygen because the more complex organisms (such as the flowering plant) depend upon such respiration.

Aerobic Respiration

Respiration that proceeds in the presence of abundant gaseous oxygen is termed **aerobic respiration** or, sometimes, merely respiration. The oxygen is reduced (to

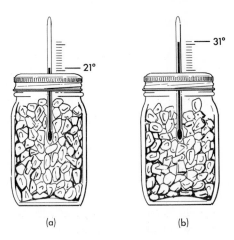

Figure 13-1. Heat production during respiration. **(a)** Sterile dead seeds. **(b)** Germinating seeds. Note the much higher temperature of **(b)**.

water) as the substrate, usually glucose (a hexose), is oxidized. The overall reaction may be indicated as follows:

$$\text{Glucose} + \text{Oxygen} \xrightarrow{\text{enzymes}} \text{Carbon dioxide} + \text{Water} + \text{Energy}$$

The amount of energy produced when a glucose molecule is oxidized to carbon dioxide and water is the same as the amount stored in that molecule during photosynthesis (see Section 12.4). When glucose is oxidized in a cell, some of this energy is in the form of heat, which is wasted as far as the cell is concerned. The water is usually retained within the cell, while the carbon dioxide diffuses out of the cell or is used in photosynthesis. In most organisms this "metabolic water" is of no significance because of the large quantities of water obtained in other ways (e.g., absorption). In the case of the kangaroo rat of the desert, however, this appears to be the only source of water. A broken arrow appears in the overall equation to emphasize that many steps are involved, the complexity of which will be discussed later.

The amount of heat produced during respiration may be considerable, especially in densely packed storage tissues. Figure 13-1 indicates the condition that prevails when vacuum jars are tightly packed with seeds; lot (a) consists of germinated but dead seeds (treated with formaldehyde or some other poison to kill any microorganisms that might be present), and lot (b) consists of germinating (actively respiring) seeds. The temperature in lot (a) is equal to the room temperature (21°C), while the temperature of lot (b) is considerably higher (31°C). Under certain conditions, as in wheat storage bins where the seeds have become moistened or in barns where moist hay has been stored, respiratory heat may result in such high temperatures that the material bursts into flame (**spontaneous combustion**). The presence of bacteria and other microorganisms undoubtedly adds to the respiratory activity and subsequent heat production.

Although the substance most frequently used as a substrate in respiration is glucose, other materials may be utilized. Fats and proteins may be digested, the products of such reactions then serving as respiratory substrates. This is true in the case of storage organs, such as seeds, and in any cell during extreme starvation.

Anaerobic Respiration

Many microorganisms are capable of carrying on oxidation-reduction reactions in the absence of gaseous oxygen. Electrons and H^+ are transferred from one molecule to another, but oxygen is not the final acceptor. This is termed **anaerobic respiration,** or **fermentation,** and the energy released from the substrate is a result of molecular rearrangements. The substrate is only partially oxidized, and the end products are not carbon dioxide and water but various organic compounds and sometimes carbon dioxide. As with aerobic respiration, fermentation results in the production of energy, some of which is then available to the cell. As a matter of convenience, the terms *respiration* and *fermentation* are usually used to designate aerobic and anaerobic respiration, respectively. A fermentation of great economic importance is carried on by yeast cells and results in the production of ethyl alcohol. This is commonly termed simply alcoholic fermentation and is utilized in the manufacture of industrial alcohol, wine, and beer.

Comparison Between Aerobic and Anaerobic Respiration

In general, the substrate is the same—glucose, a simple 6-carbon sugar or hexose. A series of reactions results in the production of a relatively small amount of energy, as a glucose molecule is converted to two molecules of **pyruvic acid,** a 3-carbon compound. These steps in respiration, which result in the production of a key intermediate compound (pyruvic acid), are collectively termed **glycolysis** and occur in the cytoplasm. Oxygen is not utilized in glycolysis, and the pathway of glucose degradation is the same up to this point in both types of respiration. The processes diverge in the reactions that occur after the formation of pyruvic acid, as indicated schematically in Figure 13-2.

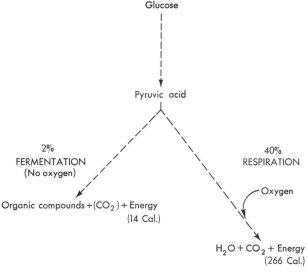

Figure 13-2. A schematic comparison of respiration and fermentation. In alcoholic fermentation (ethanol production) the caloric value of the usable energy produced would be as indicated. The two values for energy made available are presented for comparative purposes. CO_2 is produced in many fermentations but not in all. The energy values are per mole of glucose.

Aerobically, the pyruvic acid molecules are further changed by means of a cyclic pathway in which carbon dioxide is given off at intervals and where hydrogens and electrons are removed, also at intervals. This portion of respiration takes place in the mitochondria. The loss of hydrogens and electrons is a type of oxidation: large amounts of energy are made available, and oxygen is reduced to water as it accepts the H^+ and electrons. As a result, the pyruvic acid molecules are eventually converted to carbon dioxide and water, and the energy of the original glucose molecule is converted to usable energy or heat.

Anaerobically, the pyruvic acid molecules are converted to various organic compounds, such as alcohol and organic acids, with or without the production of carbon dioxide. The amount of energy produced is much less than that made available in aerobic respiration: at least 20 times as much energy is made available aerobically as is made available anaerobically. In fermentation the energy obtained is that made available during the glycolytic reactions. The difference in available energy results because pyruvic acid is not completely oxidized to carbon dioxide and water in fermentation reactions, and much potential energy still resides in the alcohols, organic acids, and other materials that are produced. Some organisms (bacteria) can respire these materials and thus make use of such residual energy. Various alcohols, such as ethanol, isopropanol, butanol, and glycerol; various organic acids, such as formic, acetic, propionic, butyric, lactic, and succinic; and acetone may be produced. The products that result from the fermentation of glucose depend upon the organism involved, and more than one organic compound may result from the activity of a single type of organism.

A mole of glucose yields 673 Cal of free energy when oxidized. (Cal = kcal = 1000 calories.) Much of this energy is converted to heat at various points in the respiratory pathways and is not available to or usable by the plant. In homothermic animals, such as man, it is useful in maintaining a stable body temperature, of course. In any event some of the energy released during respiration is sufficient to bring about the phosphorylation of ADP to ATP, and this is a form of chemical bond energy that is readily usable in various kinds of metabolic work. Fewer ATP molecules are generated during fermentations than during respiration, and this accounts for the considerably less energy available through the former reactions. Estimates of the free energy released when the terminal phosphate bond of ATP is hydrolyzed vary from 7 to 12 Cal/mole. Respiration yields 38 ATP and fermentation yields two; thus, about 266 Cal vs 14 Cal. Even respiration is only about 40 per cent efficient as an energy conversion process, but that is certainly better than fermentation.

The cells of vascular plants carry on aerobic respiration. However, at times an environmental situation results in anaerobic conditions and these cells can survive for the short periods of time during which alcoholic fermentation proceeds. For example, the flooding of a soil could result in oxygen not being available to root cells. Plants vary somewhat in their susceptibility to such a situation, but the root cells in most plants would not survive more than a few days. Some plants have evolved adaptations that enable survival, but these will be discussed later (see Section 14.2). It is probable that death results from the build-up of toxic amounts of fermentation products (e.g., alcohols) or the lack of sufficient ATP molecules for normal cellular activity.

13.4 Energy Transfer

Whenever more complex molecules are produced from simpler ones by rearrangements of component atoms and by various types of combinations, energy is utilized. Except in photosynthesis, such energy is obtained from the respiratory activities of the cell involved. It is beyond the scope of an elementary text to delve into the details of respiration and energy transfers, but some few brief comments are pertinent to an understanding of the great importance of this process.

In the process of photosynthesis light energy is converted to the chemical potential energy of sugar molecules (or other organic compounds). The latter form of energy is made available to the cell in small amounts at various intervals as glucose is broken down (oxidized) in a stepwise fashion through a whole series of enzyme-controlled reactions. Investigations with many kinds of cells and tissues have revealed that certain organic substances and phosphate groups are involved in making available to the cell that energy which is present in substrate (food) molecules. The most important such agent is **adenosine triphosphate (ATP)**.

$$\boxed{A} - \boxed{P} \sim \boxed{P} + \boxed{P} \longrightarrow \boxed{A} - \boxed{P} \sim \boxed{P} \sim \boxed{P}$$

As the name indicates, it is composed of adenosine plus three phosphate groups, the last two of which are attached by "high-energy bonds." In certain of the oxidative steps of respiration (or fermentation), some of the liberated energy is utilized in the formation of these "high-energy phosphate bonds" of ATP. Inorganic phosphate is used, and adenosine *di*phosphate (ADP) is converted to adenosine *tri*phosphate (ATP). The curved bond ($\sim$) indicates a high-energy condition, ADP having a lower energy value than ATP. The ATP molecule that has been formed can now diffuse throughout the cell to the various areas where energy is required, as in the synthesis of proteins, for example. Energy-requiring processes are accompanied by the breakdown of ATP to ADP. With this conversion of ATP to ADP the terminal high-energy phosphate group is transferred to another molecule, thus increasing the energy content of the latter. As a result of this energy increase, such molecules are activated and are now capable of participating in various reactions or syntheses.

The formation of ATP allows a slow or gradual utilization of the energy contained in the substrate molecule (usually glucose), in contrast to the rapid energy release of burning or an explosion. The gradual energy release results in the normal functioning of the mechanism. The cell can also build up a backlog of immediately available energy in the form of an ATP reservoir.

One of the best understood syntheses, and one which we can use as an example, is the formation of starch from glucose. Molecules of glucose as such are not utilized. The reaction,

$$\text{Glucose} + \text{Glucose} + \cdots \rightarrow \text{Starch} \quad + \Delta F \qquad (1)$$

is endergonic (i.e., has a positive ΔF; see Section 11.3) and will not proceed spontaneously but requires a supply of free energy. In other words, the product (starch) of the reaction has a higher energy content than the reactants (glucose molecules), and such a reaction cannot proceed unless there is an input of energy.

Within the cell are ATP molecules, the storage form of free energy available to do work in the cell. The terminal high-energy phosphate group of ATP is transferred to glucose, increasing the energy of the glucose molecule.

$$\text{Glucose} + \text{ATP} \longrightarrow \text{Glucose}\!-\!\boxed{\text{P}} + \text{ADP} \qquad -\Delta F \qquad (2)$$

Reaction (2), being exergonic, proceeds readily. Actually, this reaction is the summation of two reactions and indicates how ATP is used to make reactions exergonic. Hydrolysis of ATP is highly exergonic and yields the phosphate, which is transferred to glucose. The reaction of glucose with phosphate is actually endergonic.

$$\text{ATP} + \text{H}_2\text{O} \longrightarrow \text{ADP} + \boxed{\text{P}} \qquad\qquad \Delta F = -7 \text{ Cal} \qquad (2a)$$

$$\text{Glucose} + \boxed{\text{P}} \longrightarrow \text{Glucose}\!-\!\boxed{\text{P}} + \text{H}_2\text{O} \qquad \Delta F = +3 \text{ Cal} \qquad (2b)$$

By coupling a spontaneous reaction (2a) with a nonspontaneous process (2b), it is possible to make the overall reaction take place spontaneously, as in reaction (2). This is true provided the algebraic sum of the free energy changes accompanying the processes is negative. In this case ΔF for reaction (2) would be -4 Cal $(-7 + 3 = -4)$. In effect some of the energy available in ATP has been transferred to the glucose molecule as the latter is converted to glucose-phosphate, which now has sufficient energy to participate in reactions.

Many glucose-phosphates then join to form a starch molecule:

$$\text{Glucose}\!-\!\boxed{\text{P}} + \text{Glucose}\!-\!\boxed{\text{P}} + \cdots \longrightarrow \text{Starch} + n(\text{P}) + n(\text{H}_2\text{O}) \qquad (3)$$

There is no free energy change in this reaction and it takes place without difficulty. In effect, the increased bond energy gained by the glucose molecules in reaction (2) results in the total bond energy of the glucose-phosphates being equivalent to the bond energy of the starch molecule, a situation that does not exist between glucose molecules and starch. The overall reaction (i.e., the sum of [2] and [3]) is exergonic.

$$n(\text{Glucose}) + n(\text{ATP}) \rightarrow \text{Starch} + n(\text{ADP}) + n(\text{P}) + n(\text{H}_2\text{O}) - \Delta F \qquad (4)$$

The exact number of glucose molecules used is not known, but one water molecule and one phosphate group are liberated whenever two glucose-phosphate molecules are joined. The letter "n" in reactions (3) and (4) represents this unknown number, probably on the order of 1000 units. The ADP produced in reaction (2), as a result of ATP losing one high-energy phosphate, may again be converted to ATP by the respiratory mechanism (utilizing inorganic phosphate).

The various syntheses occurring in living cells are thought to take place in this fashion. Adenosine triphosphate (ATP) is an *energy-rich* compound. A molecule that is going to take part in a synthesis is *activated* by the transfer of a phosphate group from ATP. The ATP is changed to ADP, but it can be "recharged" with energy by the respiratory reactions again. When the activated molecules participate in various reactions, the phosphate groups are liberated and are then available for the conversion of ADP to ATP. There are other energy-rich compounds, most of which involve phosphate groups, but ATP is the best known.

13.5 Respiratory Reactions

For ease of discussion, aerobic respiration can be considered to consist of three identifiable sequences which will now be mentioned in a bit more detail: (1) glycolysis, (2) Krebs cycle (or citric acid cycle), and (3) oxidative phosphorylation. The first sequence of reactions occurs in the cytoplasm rather than in specific organelles of the cell, whereas the latter two sequences occur in mitochondria.

Glycolysis (Figure 13-3)

It is interesting to note the great importance of phosphorylated compounds. Not only must glucose molecules be converted to glucose-phosphates before they can be used in syntheses, but such a conversion (or activation) must also take place

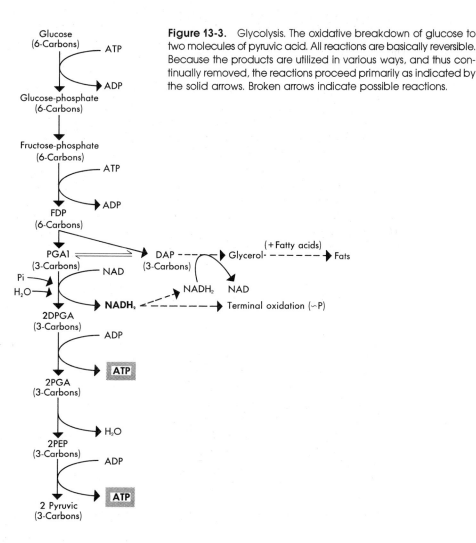

Figure 13-3. Glycolysis. The oxidative breakdown of glucose to two molecules of pyruvic acid. All reactions are basically reversible. Because the products are utilized in various ways, and thus continually removed, the reactions proceed primarily as indicated by the solid arrows. Broken arrows indicate possible reactions.

221

before these molecules can be used in the respiratory reactions. Since ATP is used during phosphorylations, this means that the initial (or primary) steps in respiration actually use energy. The later series of reactions then yield energy, and in much more profuse amounts than the quantity required for those first few steps.

In the initial stages of respiration glucose is phosphorylated by the transfer of a phosphate group from ATP, which is derived from previous respiration.

$$\text{Glucose} + \text{ATP} \rightleftharpoons \text{Glucose}\!-\!\boxed{\text{P}} + \text{ADP} \qquad -\Delta F \qquad (5)$$

This process is exergonic. The **glucose-phosphate** has a higher energy content and is more reactive than glucose. However, because ATP has a much higher energy content than ADP there is an overall decrease in free energy (symbolized by $-\Delta F$) as the reaction proceeds. Internal rearrangement, with no free energy change, converts this sugar-phosphate molecule to **fructose-phosphate,** a slightly different 6-carbon sugar.

$$\text{Glucose}\!-\!\boxed{\text{P}} \rightleftharpoons \text{Fructose}\!-\!\boxed{\text{P}} \qquad (6)$$

The latter compound is further phosphorylated to **fructose-diphosphate (FDP)** by a second ATP molecule.

$$\text{Fructose}\!-\!\boxed{\text{P}} + \text{ATP} \rightleftharpoons \boxed{\text{P}}\!-\!\text{Fructose}\!-\!\boxed{\text{P}} + \text{ADP} \qquad -\Delta F \qquad (7)$$

These processes, and subsequent ones, are catalyzed by enzymes specific for each reaction. As can be seen in reactions (5) and (7), two high-energy phosphate bonds are used up for every glucose molecule that is converted to fructose-diphosphate.

The fructose-diphosphate is enzymatically split into two 3-carbon trioses, **phosphoglyceraldehyde (PGAl)** and **dihydroxyacetone-phosphate (DAP),** with no free energy change. The latter two molecules are interconvertible, but it is the PGAl that participates further in glycolysis. However, as PGAl is used, more DAP converts to PGAl; thus, all the original material from FDP is eventually used.

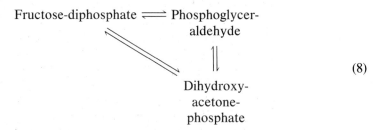

$$\text{Fructose-diphosphate} \rightleftharpoons \text{Phosphoglycer-} \atop \text{aldehyde}$$

$$(8)$$

Dihydroxy-
acetone-
phosphate

Note that PGAl is a familiar compound, one that is also involved in photosynthesis, where it was formed by the energy-requiring reduction of phosphoglyceric acid (PGA). In respiration, however, the first energy-yielding step occurs when inorganic phosphate is incorporated into the phosphoglyceraldehyde molecule concomitant with its oxidation to **diphosphoglyceric acid (DPGA)** by the loss of hydrogens and electrons to **nicotinamide adenine dinucleotide (NAD).** Hydrogens and electrons are removed in pairs from the substrate: two electrons and one hydrogen are transferred to NAD and one hydrogen is released into the medium (i.e., cytoplasm) as a hydrogen ion (H^+). Another hydrogen carrier (NADP) very similar in structure to NAD, differing only in that it has an extra phosphate group, has already been discussed (see Section 12.5). Thus, it should be apparent that a number of reactants and carriers in respiration are the same as, or very similar to, some in photosynthesis.

$$PGAl + NAD + Pi \rightleftharpoons DPGA + NADH + H^+ \qquad -\Delta F \qquad (9)$$

The oxidation, as shown in (9), shifts the internal energy of the molecule so that one of the phosphate groups in the oxidation product is of the high-energy type. As a result, a portion of the energy liberated in this oxidation has been conserved in the formation of a high-energy phosphate bond. If oxygen is available, reduced nicotinamide adenine dinucleotide (NADH) may undergo terminal oxidation in which ADP is converted to ATP. This will be discussed later.

The high-energy phosphate group of diphosphoglyceric acid is transferred to ADP. Because each hexose actually gives rise to two triose (PGAl) molecules, reaction (10) really indicates the manner in which two molecules of ATP are produced for each one molecule of glucose oxidized.

$$DPGA + ADP \rightleftharpoons PGA + ATP \qquad -\Delta F \qquad (10)$$

Two ATP were used up in phosphorylating the original hexose molecule, so the energy gained equals the energy used to his point; not counting the NADH produced and its possible oxidation, of course.

Further internal shifts and the loss of water (dehydration) result in an *intramolecular* oxidation-reduction reaction that converts the remaining phosphate into a high-energy type as **phosphoenolpyruvic acid (PEP)** is formed.

$$PGA \rightleftharpoons PEP + H_2O \qquad -\Delta F \qquad (11)$$

Again, some of the oxidative energy is conserved in a phosphate group, which is then transferred to ADP.

$$PEP + ADP \rightleftharpoons Pyruvic + ATP \qquad -\Delta F \qquad (12)$$

The formation of **pyruvic acid** concludes the glycolytic reactions in which at least a net of two ATP molecules have been gained. Remember that there are two PEP molecules produced from *each* glucose molecule. Reaction (12) indicates that each molecule of PEP provides a high-energy phosphate bond. Thus, some of the energy of the original glucose molecule has been conserved by the production of ATP.

If pyruvic acid were the end product of respiration, this energy-yielding process would soon come to a halt and death would result. This would occur simply because the hydrogen carriers, the NAD molecules, would all be loaded with hydrogens and no longer able to function. There must be a mechanism whereby these molecules relinquish hydrogens and thus become available to bring about the oxidation (by removal of hydrogens) of other substrate molecules. The manner in which this is accomplished during aerobic respiration will be discussed in the next section. In fermentation reactions, the conversion of pyruvic to other organic compounds (see Section 13.3) uses hydrogens from NADH. For example, in alcoholic fermentation, carbon dioxide is removed from pyruvic, forming acetaldehyde, which then accepts hydrogens as it is reduced to ethyl alcohol. This frees NAD and enables it to accept more hydrogens.

$$\text{Pyruvic} \longrightarrow \text{Acetaldehyde} \xrightarrow{\quad \overset{\displaystyle NADH + H^+ \quad NAD}{\displaystyle \smile} \quad} \text{Ethyl alcohol} \quad (13)$$
$$\downarrow$$
$$CO_2$$

It should be noted that fermentations do not produce any additional ATP molecules, and the energy available to the cell resides within the two ATP molecules (per glucose molecule) obtained from glycolysis. Less than 10 per cent of the total available energy is released during this reaction sequence, and only about 2 per cent is retained in the two ATP. Most of the bond energy originally present in the glucose is in the fermentation product, alcohol in this case. In terms of energy availability, or yield, fermentations are rather inefficient.

Krebs Cycle (Figure 13-4)

The further oxidation of pyruvic acid takes place by way of a cyclic series of enzyme-catalyzed reactions occurring in the mitochondria. The enzymes for this **Krebs cycle** or **citric acid cycle,** are located in the soluble matrix of the mitochondrion.

Pyruvic acid is oxidatively decarboxylated; that is, it loses carbon dioxide as well as hydrogens and electrons, and the residual 2-carbon **acetyl** group is accepted by **coenzyme A.** The latter contains pantothenic acid, one of the B vitamins, as a part of its molecular structure. Although the carbon dioxide may diffuse out of the cell, the hydrogens and electrons are accepted by NAD.

$$\text{Pyruvic} + NAD + CoA \rightleftharpoons$$
$$\text{Acetyl} — CoA + NADH + CO_2 + H^+ \qquad -\Delta F \quad (14)$$

The 2-carbon compound can be funneled into the Krebs cycle only in an "activated" form, namely as **acetyl-coenzyme A.** Once again an oxidation has brought about an internal energy shift that resulted in a "high-energy" linkage—in this case, the acetyl group bond to CoA.

In the first reaction of the cycle, the acetyl group is transferred to the 4-carbon **oxaloacetic acid** to form the 6-carbon **citric acid;** coenzyme A is free to accept another acetyl group. In the process a molecule of water is taken up.

$$\text{Acetyl—CoA} + \text{Oxaloacetic} + H_2O \rightarrow \text{Citric} + CoA \qquad -\Delta F \quad (15)$$

The citric acid is oxidized in a series of reactions, by losing hydrogens and electrons to NAD, and **oxalosuccinic acid** is formed.

$$\text{Citric} + NAD \rightarrow \text{Oxalosuccinic} + NADH + H^+ \qquad -\Delta F \quad (16)$$

The latter acid loses carbon dioxide (a decarboxylation) to form the 5-carbon **α-ketoglutaric acid** (alpha-ketoglutaric), a reaction that readily goes to completion since one of the end products (CO_2) is removed by diffusion out of the cell.

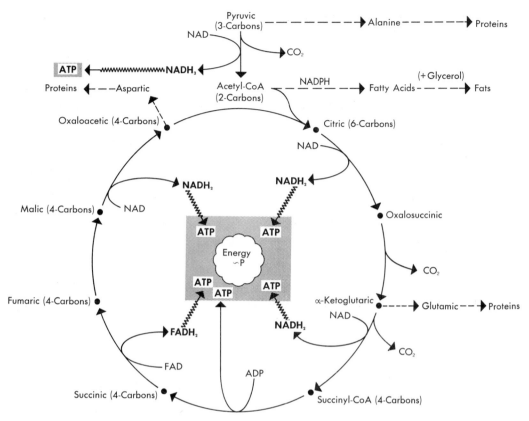

Figure 13-4. Krebs cycle and electron transport. The complete oxidative breakdown of pyruvic acid, which yields most of the energy (ATP) that the cell obtains from respiration. All reactions are basically reversible. Broken arrows indicate possible reactions. Electron transport is indicated by ⤳ (see Figure 13.5 for details).

$$\text{Oxalosuccinic} \rightarrow \alpha\text{-Ketoglutaric} + CO_2 \qquad (17)$$

Another simultaneous oxidation and decarboxylation now occurs, in which coenzyme A also participates, forming **succinyl-coenzyme A** with hydrogens and electrons being accepted by NAD.

$$\alpha\text{-Ketoglutaric} + CoA + NAD \rightarrow$$
$$\text{Succinyl—CoA} + NADH + CO_2 + H^+ \qquad (18)$$

A portion of the oxidative energy is retained in a "high-energy" linkage to coenzyme A. When succinyl-coenzyme A is converted to **succinic acid** (a 4-carbon compound), some of this energy is utilized in converting ADP to ATP (inorganic phosphate is used). Water is again taken up.

$$\text{Succinyl—CoA} + ADP + Pi + H_2O \rightarrow$$
$$\text{Succinic} + ATP + CoA \qquad -\Delta F \qquad (19)$$

225   Succinic is now oxidized to **fumaric acid,** and the hydrogens and electrons are

accepted by **flavin adenine dinucleotide (FAD),** a carrier molecule similar to NAD in its action. One of the components of FAD is **riboflavin,** or vitamin B_2.

$$\text{Succinic} + \text{FAD} \rightarrow \text{Fumaric} + \text{FADH}_2 \qquad -\Delta F \qquad (20)$$

Water is added enzymatically to fumaric resulting in the formation of **malic acid.**

$$\text{Fumaric} + H_2O \rightarrow \text{Malic} \qquad (21)$$

At this point a final oxidation converts malic to oxaloacetic acid with NAD again accepting hydrogens and electrons.

$$\text{Malic} + \text{NAD} \rightarrow \text{Oxaloacetic} + \text{NADH} + H^+ \qquad -\Delta F \qquad (22)$$

These various reactions have brought about a stepwise oxidation and decarboxylation of pyruvic acid during which NAD and FAD molecules have been reduced (by accepting hydrogens and electrons). Energy from the chemical bonds of pyruvic acid has been transferred to the reduced hydrogen acceptors and to the molecule of ATP that is generated when succinyl-CoA is oxidized to succinic acid (reaction [19]). This ATP represents a gain of available and usable energy that may be utilized in various ways by the cell. The further participation of NAD and FAD in reactions that result in the production of ATP will be discussed in the next section.

As can be seen in Figure 13-4, five molecules of hydrogen acceptors (NAD or FAD) are reduced as pyruvic is oxidized, but this molecule can provide only four hydrogens (i.e., pyruvic = $CH_3COCOOH$). The remaining six hydrogens are derived from the three water (H_2O) molecules that participate in the Krebs cycle. Remember that two molecules of pyruvic acid are obtained for each molecule of glucose that is respired. Thus, the Krebs cycle turns twice for each glucose molecule that is utilized, yielding two ATP and ten reduced hydrogen-acceptor molecules.

13.6 Electron Transport and Terminal Oxidations

The NADH and other reduced H-acceptors produced during glycolysis and the Krebs cycle may serve as H-donors in a variety of metabolic processes, some of which will be discussed in Section 13.10. However, most of the usable energy that a cell derives from respiration is obtained when the hydrogens and electrons of these acceptor molecules unite with oxygen to form water in a series of terminal oxidations that are coupled to the generation of high-energy phosphate bonds.

Reduced NAD or FAD (i.e., NADH or $FADH_2$) donate their hydrogens and electrons to a series of carriers and eventually to oxygen. The compounds funneling toward these **terminal oxidations** have a relatively high energy content, about five times higher than ATP. Figure 13-5 represents the mitochondrial system for trapping phosphate bond energy. In effect there is a gradual oxidation of substrate during which the cell traps energy in multiple stages. The carriers involved in electron transport and terminal oxidations are apparently located in an orderly and linear arrangement on the inner invaginated membranes of the mitochondria

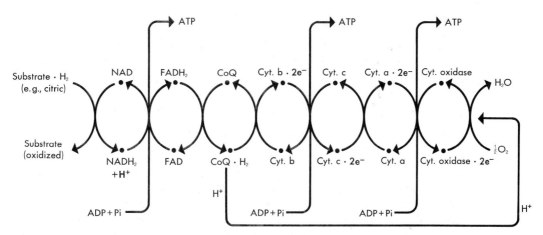

Figure 13-5. Schematic of the respiratory chain of oxidative phosphorylation. Actually the cytochromes transfer one electron at a time; the hydrogens are liberated as H^+ ions in the medium and eventually join with one-half O_2 and the electrons to form H_2O. The reactions in this figure have been simplified.

where they function like a bucket brigade in passing hydrogens and electrons from one to another and eventually to oxygen at the lowest energy level. Each carrier, or acceptor, is at a progressively lower energy level. As the transfers occur down the energy gradient, sufficient energy is released to drive the phosphorylation of ADP to ATP; this is termed **oxidative phosphorylation.** Some of the bond energy is lost as heat, however.

After the hydrogens and electrons are accepted by NAD, they are passed on to flavoproteins (FAD) with some of the energy being stored as ATP; FAD picks up two electrons and one hydrogen from NADH and a second hydrogen from the medium (i.e., cytoplasm). There is further transfer of these hydrogens and electrons to coenzyme Q. The reduced carrier $CoQH_2$ transfers a pair of electrons to a chain of cytochromes (iron-containing carriers) and releases two hydrogen ions (H^+) into the medium. At least two more ATP molecules are formed as electrons pass, one at a time, from one cytochrome to another to lower energy levels. At the terminal end of the cytochrome chain, the electrons, and hydrogen ions (or protons) from the medium, are accepted by oxygen; and water is formed. The removal of electrons from the cytochromes (to oxygen) enables these cytochromes to participate again in electron transport. Without such removal respiration would cease as the cytochromes became loaded with electrons. A similar situation pertains to the other carriers.

The net effect of these reactions is that at least three ATP molecules are produced for each NADH molecule that donates its hydrogens and electrons to the respiratory chain.

$$NADH + H^+ + \tfrac{1}{2}O_2 + 3Pi + 3ADP \rightarrow NAD + 3ATP + H_2O \qquad (23)$$

The $FADH_2$ molecules donate their hydrogens and electrons farther along the respiratory chain, and thus only two ATP molecules are formed. One NADH is produced in reaction (9), another results from reaction (14), and three are produced

**Figure 13-6.** Energy transformations and utilization.

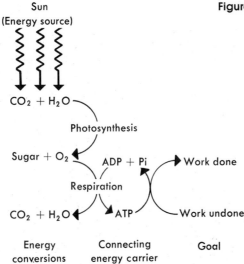

Sun
(Energy source)

$CO_2 + H_2O$

Photosynthesis

Sugar + O_2

$ADP + Pi$ → Work done

Respiration

$CO_2 + H_2O$

ATP

Work undone

| Energy conversions | Connecting energy carrier | Goal |

during the Krebs cycle ([16], [18], and [22]). One molecule of $FADH_2$ is produced during reaction (20) of the Krebs cycle, and reaction (19) produces one ATP. Each such reaction utilizes as its substrate a 3-carbon compound. Since the starting substrate in respiration is the 6-carbon glucose, there are actually two of these 3-carbon compounds and the number of ATP produced during the terminal oxidations of NADH and $FADH_2$ must be multiplied by two. Inclusion of the two ATP gained during the glycolytic reactions then indicates the total number of ATP molecules produced by the aerobic respiration of a single glucose molecule to be 38 ATP, which equals about 40 per cent of the total energy content of glucose. This is not bad for a cell, considering that a gasoline engine is only about 4 per cent efficient.

During photosynthesis, electrons are raised to higher energy levels and their energy is ultimately incorporated into the bonds of carbohydrate molecules. During respiration, these bonds are broken and the electrons pass through a series of reactions that take them from higher energy levels to lower energy levels. Much of the energy released is transformed to high-energy phosphate bonds (ATP). Photosynthesis and respiration combined are means whereby light energy is converted to usable energy. The ATP molecule may be thought of as the connecting link between the source of energy and the work accomplished (see Figure 13-6). The work accomplished would include all of the syntheses of various cellular components and other activities that the cell may undertake (e.g., absorption, conduction, luminescence). It should be noted that some of the intermediate products in respiration may be used by the cell to synthesize some of the complex molecules required for normal functioning. In other words, these intermediates are building blocks for other molecules. This will be discussed in more detail in Section 13.10.

13.7 Summary of Respiration

In the preceding two sections respiratory reactions have been covered in considerable detail. This summary will consider the overall process in a more simplified fashion in order to place greatest emphasis upon the production of available energy (as ATP).

Summary of Glycolysis

If we consider glycolysis as a whole, the conversion of one molecule of glucose to two molecules of pyruvic acid resulted in a net gain of *two* ATP molecules. Two molecules of ATP are used (reactions [5] and [7]) to phosphorylate the hexose sugar, which is activated and made more reactive as a result of the addition of phosphate groups. ATP is converted to ADP by this transfer of phosphate. Two oxidations ([9] + [10]; and [11] + [12]) each yield sufficient energy to generate high-energy phosphate bonds, converting ADP to ATP. A total of four ATP result from these oxidations, and thus a net gain of two overall.

An intermediate 3-carbon compound, dihydroxyacetone-phosphate (DAP), is produced during these reactions and may be used in cellular syntheses rather than being converted to pyruvic. However, assuming pyruvic is produced, an overall, or summary, equation for glycolysis could be shown as follows:

$$\text{Glucose} + 2\text{NAD} + 2\text{ADP} + 2\text{Pi} \rightarrow$$
$$2 \text{ Pyruvic} + 2\text{NADH} + 2\text{ATP} + 4\text{H}^+ \quad (24)$$

If the NADH molecules (of [9]) donate their hydrogens and electrons to oxygen in a series of steps, additional ATP molecules (probably about six per glucose molecule) are formed—a total gain of eight ATP molecules for glycolysis as a whole in aerobic respiration (only two in fermentation). Alternatively, the NADH may serve as hydrogen donors in various reduction processes, as in the reduction of nitrates (NO_3), the formation of glycerol from dihydroxyacetone-phosphate (an intermediate in glycolysis), or the synthesis of fatty acids.

In general, the oxidations serve to concentrate a great deal of the internal energy of the molecule in the form of high-energy phosphate groups. The phosphate groups are then transferred to ADP, forming ATP, and it is the ATP molecule that makes various syntheses possible. There is a decrease in total free energy when phosphate groups are transferred to ADP, but this enables the reaction to take place. Remember that reactions tend toward completion only when there is an overall decrease in free energy (i.e., $-\Delta F$). Note, however, that ATP has a higher energy content than ADP.

Summary of the Krebs Cycle

The oxidation of pyruvic proceeds by a somewhat more complex series of reactions than those of glycolysis, and the number of ATP molecules resulting from the former reactions is almost four times that of the latter: 30 as compared to a maximum of 8. In this simplified summary we will indicate the removal of hydrogens from substrates as oxidations occur, but not the numerous specific reactions. Such details are available in Section 13.5.

Carbon dioxide and hydrogens are removed from pyruvic, and the resulting 2-carbon acetate enters the Krebs cycle proper. Remember that there are two pyruvic produced from each glucose.

$$2 \text{ Pyruvic} + 2\text{CoA} \rightarrow 2 \text{ Acetyl—CoA} + 2\text{CO}_2 + 2(2\text{H}) \qquad (25)$$

The acetyl-coenzyme A enters a cyclic series of reactions, but the overall effect may be indicated as follows:

$$2 \text{ Acetyl—CoA} + 6\text{H}_2\text{O} \rightarrow 2\text{CoA} + 4\text{CO}_2 + 8(2\text{H}) \qquad (26)$$

In summation, these reactions ([25] and [26]) may be shown as follows:

$$2 \text{ Pyruvic} + 6\text{H}_2\text{O} \rightarrow 6\text{CO}_2 + 10(2\text{H}) \qquad (27)$$

Each pair of hydrogens is accepted by NAD or FAD and eventually passed to oxygen (i.e., $\frac{1}{2}\text{O}_2$); water is formed as a result of this transfer of hydrogens. Therefore, five oxygen molecules ($10 \times \frac{1}{2}\text{O}_2$) are used, and ten water molecules are produced for each two pyruvic acid molecules that are oxidized. During these various oxidations, 30 ATP molecules are produced (15 per each pyruvic molecule). We may then indicate the following:

$$2 \text{ Pyruvic} + 6\text{H}_2\text{O} + 5\text{O}_2 \rightarrow 6\text{CO}_2 + 10\text{H}_2\text{O} \qquad (28)$$

Six water molecules are used up and ten are produced. In effect, six water molecules can be canceled out on each side of the reaction.

$$2 \text{ Pyruvic} + 5\text{O}_2 \rightarrow 6\text{CO}_2 + 4\text{H}_2\text{O} \qquad (29)$$

The important facet of the reactions during the oxidation of pyruvic is that much energy is conserved in high-energy phosphate bonds (ATP) and made available for use by the cell, and that some of the intermediate products are available for syntheses by the cell (see Section 13.10).

Overall Respiration

The glycolysis reactions may be shown in a manner similar to the Krebs cycle just preceding.

$$\text{Glucose} \rightarrow 2 \text{ Pyruvic} + 2(2\text{H}) \qquad (30)$$

These hydrogens are also accepted by NAD and may then be transferred to oxygen, forming water and ATP molecules—one oxygen, two water, and six ATP for the two hydrogen-pairs. The reaction may be written:

$$\text{Glucose} + \text{O}_2 \rightarrow 2 \text{ Pyruvic} + 2\text{H}_2\text{O} \qquad (31)$$

We may now combine reactions (29) and (31):

$$\text{Glucose} + 6\text{O}_2 \rightarrow 6\text{CO}_2 + 6\text{H}_2\text{O} \qquad (32)$$

231

Sec. 13.9 /

Factors

Influencing the

Rate of

Respiration

Thirty ATP molecules have been produced as a result of the Krebs cycle and eight during glycolysis; thus, a total of 38 ATP molecules are formed in the overall aerobic respiratory reactions. Probably over 40 per cent of the total energy residing in a glucose molecule eventually appears as ATP. The remainder of the energy is lost to the cell as heat, which merely dissipates away. Respiration produces usable energy (ATP) and building blocks for syntheses. The use of intermediate products as building blocks is discussed later in this chapter.

13.8 Pentose Phosphate Pathway

In addition to the glycolytic pathway there is another oxidative sequence in which a 5-carbon sugar is one of the intermediates as a result of the decarboxylation (i.e., removal of CO_2) from the hexose substrate. We will not delve into the details of this pathway; it is beyond the intent of this text. However, there are several important features of the **pentose phosphate pathway** (also called the *hexose monophosphate shunt*). There are various interconversions that occur, and this pathway is basically the reverse of the C_3-cycle (Calvin-Benson cycle) of photosynthesis (see Section 12.6). NADP is the hydrogen-carrier involved, and NADPH may donate its electrons and proton to NAD. The resultant oxidative phosphorylation can generate ATP, and energy is available to the cell.

The importance of the pentose phosphate pathway actually may be in providing building blocks, which are then used in other syntheses. For example, 5-carbon compounds are essential for nucleic acid synthesis and 4-carbon compounds are used in the synthesis of lignin and various ring compounds such as plastoquinone. In addition, NADPH is the primary source of hydrogens during the synthesis of fatty acids and aromatic compounds and in the reduction of nitrate (NO_3^-) to ammonia (NH_3). The importance of this reduction in the synthesis of amino acids, and then proteins, will be discussed later. In effect, this is again an example of the respiratory pathway playing a part in providing starting materials for syntheses as well as making energy available.

13.9 Factors Influencing the Rate of Respiration

The factors that influence respiration may vary considerably in intensity and thus the rate of the process will fluctuate. The increased availability of food material, such as resulting from the onset of photosynthesis or the hydrolysis (digestion) of storage materials, could increase the rate. This might occur during the transition from night hours to daytime (photosynthesis), or the germination of a seed in which hydrolysis of food reserves is an early event. Oxygen is usually not a limiting factor as long as its concentration is about the 20 per cent or so normally present in the air. As mentioned previously, root cells would be exposed to much

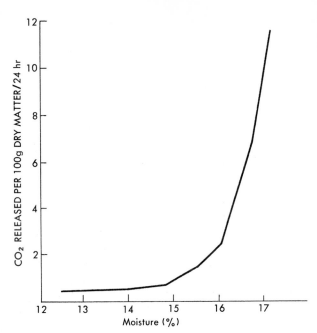

Figure 13-7. Rate of respiration of wheat grains as influenced by moisture content.

lower oxygen concentrations when they are in flooded or water-saturated soils because of the low solubility of oxygen in water. It is also possible that the inner cells of bulky storage tissues (e.g., within a potato tuber) might be faced with oxygen deficiencies. In such situations respiration would very likely be limited by the low concentrations of oxygen.

The most common limiting factor, however, is temperature. In general the rate of respiration increases with temperature up to the point (about 45–50°C) where enzymes are inactivated. In mature and dry seeds and spores water is the limiting factor, and the rate of respiration is very low. As a dry seed absorbs water, which would occur prior to germination, the rate of respiration increases slowly up to a certain water content; a slight further increase in water results in a dramatic increase in respiratory rate (Figure 13-7). This is probably the point at which hydration of enzymes is adequate for active respiration.

13.10 Use of Respiratory Intermediates in Cellular Syntheses

Previously it has been mentioned that the various syntheses of complex molecules that occur in cells utilize ATP as an energy source. In the preceding sections the formation of ATP molecules during the respiratory reactions has been discussed. It should also be noted that some of the intermediate products of these reactions may be utilized in cellular syntheses. One such synthesis was suggested in Figure 13-3.

233

Sec. 13.10 /
Use of
Respiratory
Intermediates
in Cellular
Syntheses

Synthesis of Fats

Dihydroxyacetone-phosphate (DAP), which results from the cleavage of hexose in respiration, is readily reduced to **glycerol.** The hydrogens come from NADH, which forms in various steps of respiration. Three fatty acids may then combine with the glycerol to form a fat molecule.

The fatty acids utilized in the synthesis of fats are also derived from respiratory intermediates. As shown in Figure 13-4, acetyl-coenzyme A molecules may be funneled away from the Krebs cycle with the 2-carbon components (acetyl or acetate) being used in forming **fatty acids.** These acids may be fairly large molecules, the carbon chain usually being 4 to 30 units long and sometimes even longer. The carbon atoms generally occur in even numbers because the chain is built up by the addition of two carbon atoms at a time (from acetyl-coenzyme A). ATP is necessary as the energy source that makes this synthesis possible, and NADPH is the hydrogen source for the reductions that occur as fatty acids are synthesized. Thus, the respiratory process provides the "building blocks" as well as the energy for syntheses.

Fat molecules produced in this fashion comprise part of the structural components of the cell or they may be storage products. Since these reactions are reversible, fats may also be used as substrates in respiration. Upon digestion (hydrolysis), glycerol and fatty acids are produced, which then feed back into the respiratory reactions. Because of the reductions that take place in their synthesis, fatty acids and fats are more highly reduced and have a higher energy content per unit weight than sugar or carbohydrate. A gram of fat contains about twice as many calories as a gram of sugar or starch. This makes fat an excellent storage product, especially in seeds, where space is at a premium.

Protein Synthesis

In Figure 13-4 are shown three positions in which respiratory intermediates may be converted into amino acids. The latter contain nitrogen, which is not found in the former. The plant, however, has absorbed nitrogen salts from the soil, and this is the source of the nitrogen atom. NADH and ATP are required in these conversions to amino acids and eventually to proteins. Once again, various components of the respiratory reactions are utilized in syntheses.

The nitrogen salts absorbed by the plant are primarily in the form of nitrates (NO_3^-) although ammonium (NH_4^+) salts may also be taken up. Nitrate is the most abundant available form of nitrogen in the soil, and ammonia (NH_3) or ammonium ions are somewhat toxic if much accumulation occurs. The nitrates must be reduced before the nitrogen can be utilized in amino acid synthesis because the latter molecules contain an amino group (NH_2). The source of hydrogens and electrons for the reduction is the carrier NADPH, although NADH may participate; both of these carriers obtain hydrogens and electrons during respiratory reactions. Nitrates are reduced to nitrites (NO_2^-) and then to ammonia in a stepwise fashion. The production of ammonia from nitrates is stimulated by light in green leaves, probably because of the production of NADPH.

The primary pathway of amino acid formation is through α-ketoglutaric acid of the Krebs cycle with either ammonia or ammonium ions participating in the reaction along with NADH. However, the amino group is readily transferred to other organic acids, thus yielding a variety of amino acids.

$$\alpha\text{-Ketoglutaric} \xrightarrow[\text{NADH}]{\text{NH}_3 \text{ or NH}_4^+} \text{Glutamic} + H_2O \qquad (33)$$
$$\text{(an amino acid)}$$

After activation by ATP, the various amino acids may link together to form exceedingly large and complex protein molecules containing up to thousands of carbon atoms. Although only 20 amino acids are found in proteins, the variety of proteins that may be produced is almost infinite. This is not really too surprising. After all, the tremendous number of words found in an unabridged dictionary are formed from a mere 26 letters. These are rather short words, whereas proteins are much longer, relatively speaking. Also, one should bear in mind that proteins are three-dimensional, and may contain branches, spirals, and cyclic portions. The patterns that result may be enormously complex, but the functional characteristics of proteins are determined to a great extent by the configuration (or shape) of the molecule.

Proteins not only form the basic structural features of the cell protoplast, some of them function as enzymes. The physiological activities of the cell are dependent upon such molecules. In addition proteins are frequently a storage product, especially in seeds. Upon digestion of proteins, amino acids are formed and these quite readily enter the respiratory reactions; the reactions discussed previously are reversible.

Polysaccharide Synthesis

When many molecules of simple sugars (e.g., hexoses, such as glucose) are combined, a large molecule called a **polysaccharide** is formed. Starch and cellulose are examples of such molecules. Polysaccharides are important structural components of the cell as well as storage products.

As in all syntheses, the molecules must be activated before they can participate in the reactions. This occurs typically as a phosphorylation with ATP as the source of the phosphate group. A glance at Figure 13-3 enables one to see glucose-phosphates and fructose-phosphates, excellent sources of the basic molecules from which more complex ones can be built, but the substrates and energy sources are, once again, obtained from respiration.

Glucose also may be oxidized, hydrogens accepted by NADP, and decarboxylated (loss of CO_2) to form pentoses, 5-carbon sugars. The reactions are a bit complex and will not be detailed in this text. These pentoses may also participate in polysaccharide formation, or in the formation of other large molecules. For example, RNA and DNA both contain a 5-carbon sugar—ribose and deoxyribose, respectively. The importance of DNA has been discussed in Section 4.5; RNA will be discussed later (Section 18.20).

Figure 13-8 summarizes the transformations of glucose that occur in living cells. Some glucose molecules are broken down to carbon dioxide and water, with the

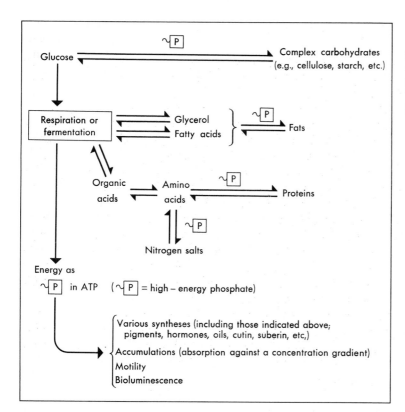

Glucose

Complex carbohydrates
(e.g., cellulose, starch, etc.)

Respiration or fermentation

Glycerol

Fatty acids

Fats

Organic acids

Amino acids

Proteins

Nitrogen salts

Energy as

$\sim\boxed{P}$ in ATP   ($\sim\boxed{P}$ = high – energy phosphate)

Various syntheses (including those indicated above; pigments, hormones, oils, cutin, suberin, etc,)

Accumulations (absorption against a concentration gradient)

Motility

Bioluminescence

Figure 13-8. Synthetic pathways. (Adapted from *Life, An Introduction to Biology*, by G. G. Simpson, C. S. Pittendrigh, and L. H. Tiffany, copyright 1957 by Harcourt, Brace & World, New York, and used with permission.)

concomitant production of energy (in ATP), while other glucose molecules are converted into various complex molecules. The energy necessary to carry out syntheses comes from the ATP reservoir.

Summary

1. The energy that the cell expends in carrying out its complex functions (e.g., syntheses of complex materials, maintenance of the living cell, movement, and accumulation of materials) is obtained from the respiratory activities of the cell. In the more advanced types of plants, gaseous oxygen is utilized during respiration, while in many microorganisms energy is made available by anaerobic fermentation reactions.

2. The initial series of reactions in fermentation (anaerobic) are basically similar to those of respiration (aerobic). Substrates are broken down to pyruvic acid, and the collective steps in this process are termed glycolysis. After the production of pyruvic acid, the two processes diverge. Respiration results in the complete oxidation of the substrate to carbon dioxide and water; fermentation produces a variety of compounds as a result of incomplete oxidation of the substrate.

3. Most of the energy, which is transported about in the cell and transferred

from molecule to molecule as various functions of the cell occur, is found as high-energy phosphate bonds attached to carrier molecules, such as ATP. Some ATP molecules are produced during glycolysis, but the majority of these molecules form as a result of the Krebs-cycle reactions and terminal oxidations occurring in the mitochondria.

4. Various of the intermediate products of the respiratory reactions are utilized in the synthesis of the many complex molecules of the cell.

Review Topics and Questions

1. Not all the substances used in aerobic respiration are broken down completely to carbon dioxide and water. Explain fully what is meant by this statement.
2. In what way is the metabolism of a fungus (a nongreen plant, such as yeast) more like that of yourself than that of a green plant?
3. A man has 2 gm of yeast and divides them into two equal portions. He wishes to use one lot (i.e., 1 gm of yeast plants) for wine production and the other portion as a source of food. (Note: Yeast plants are used in many areas as a food supplement because of their protein and vitamin B content.) His treatment of these two samples of yeast plants would be similar in some ways and different in other ways. Discuss the reasons for each treatment.
4. The rate of respiration is governed by, or influenced by, various factors. Select any one of these, and describe an experiment that would indicate the effect upon respiration of varying this factor.
5. Discuss the various fates that may befall the many sugar molecules that enter the respiratory reactions. (Note: Not all of them end up as carbon dioxide and water.)
6. Discuss the differences among digestion, respiration, and fermentation. Discuss any similarities in these three processes.
7. What becomes of the energy that is liberated during respiration?
8. Compare respiration and photosynthesis.
9. Discuss the importance of adenosine triphosphate (ATP) to living cells.

Suggested Readings

GREEN, D. E., "The Metabolism of Fats," *Scientific American* **190,** 32–47 (January 1954).

GREEN, D. E., "Biological Oxidation," *Scientific American* **199,** 56–62 (July 1958).

GREEN, D. E., "The Synthesis of Fat," *Scientific American* **202,** 46–51 (February 1960).

GREEN, D. E., "The Mitochondrion," *Scientific American* **210,** 63–73 (January 1964).

237
Suggested
Readings

LEHNINGER, A. L., "Energy Transformation in the Cell," *Scientific American* **202,** 102–114 (May 1960).

LEHNINGER, A. L., "How Cells Transform Energy," *Scientific American* **205,** 62–73 (September 1961).

LEHNINGER, A. L., *Bioenergetics,* 2nd ed. (New York, Benjamin, 1971).

SIEKEVITZ, P., "Powerhouse of the Cell," *Scientific American* **197,** 131–140 (July 1957).

STUMPF, P. K., "ATP," *Scientific American* **188,** 84–93 (April 1953).

14 Correlation of Processes, Plant Distribution, and Food Chains

In attempting to simplify the situation that exists in the cell, the processes of photosynthesis and respiration have been discussed individually. Obviously any cell that is carrying on photosynthesis is also carrying on respiration, as well as absorption, digestion, syntheses, and the rest of the various processes characteristic of living cells. A discussion of the correlation between photosynthesis and respiration will aid in an understanding of plant distribution and thus, also, of animal distribution.

14.1 Compensation Points and Plant Distribution

The sugars that are produced in photosynthesis may be used in respiration, involved in various syntheses, or stored. If the rate of respiration is equal to the rate of photosynthesis, all the sugars produced in the latter reaction are used in the former process, and none remain for such syntheses as are necessary to the production of new protoplasm or maintenance of existing protoplasm. That light intensity or temperature at which the rates of photosynthesis and respiration are equal is termed the **compensation point.** A plant cannot grow, or even remain alive for any length of time, under such conditions; it would starve to death.

The growth, or yield, of a plant depends upon food production (photosynthesis) being greater than the food consumption in respiration. If the rate of photosynthesis does not exceed the rate of respiration by a factor of three or more, growth of the plant is not possible. Photosynthesis depends upon light, whereas respiration occurs throughout the 24-hour period. If any sugars are to be available for various syntheses, they must be produced in sufficient amount during daylight hours to provide more than enough substrate for respiration during those hours plus the night hours. Comparative measurements by many investigators indicate that photosynthesis usually averages 8 to 12 times the rate of respiration under natural conditions during daylight hours. Obviously, photosynthesic rates would be low at sunrise and sunset, and much higher during midday. The averages refer to day-long rates.

Temperature and Compensation Points

Figure 14-1 illustrates the relative rates of photosynthesis and respiration of potato leaves during exposure to different temperatures in shade and in full sunlight. The temperature at which the two curves intersect (32°C or 40°C) is the compensation point. Note that the maximum rate of photosynthesis is reached at a lower temperature than the maximum rate of respiration, and that the compensation point is reached at a lower temperature when the light intensity is low. This latter result is caused by the lowering of the photosynthetic rate with the decrease in light intensity. Survival of the potato plant depends upon the average temperature remaining *below* that at which the compensation point is reached. In fact, the maximum yield of most potato varieties is obtained in areas with cool summers. When Irish potato plants were kept continuously at different temperatures, no tubers were formed on the plants kept at 29°C, while large tubers formed at 20°C. The former temperature is so close to the compensation point that the photosynthetically produced sugars were all used in respiration or in general growth, and no sugars were available for the production of storage tissue and products such as starch—thus, no tubers. At 20°C, however, the rate of photosynthesis is considerably greater than that of respiration, and excess sugars are stored in the tubers as starch.

The compensation point varies considerably with the kind of plant. In plotting the rates of photosynthesis and respiration versus temperature, one finds that the shape of the curves is basically similar to that shown in Figure 14.1. The maximum rates and the compensation points, however, may occur at higher or lower temperatures. In those plants that grow in warm climates (e.g., sugar cane, date palm) the compensation point is reached at a relatively high temperature, whereas those growing in colder areas reach the compensation point at a much lower temperature. One of the factors involved in plant distributions is this relationship between respiration and photosynthesis. The potato plant would not survive in an area where date palms are grown, and vice versa. (Why the latter?)

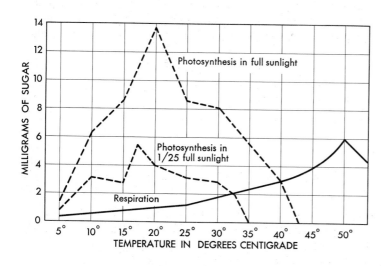

Figure 14-1. Relative rates of photosynthesis and respiration in potato leaves during ten-minute exposures to different temperatures in shade and in full sunlight. Recalculated from data by H. G. Lundegardh. (From *Textbook of Botany*, 1st ed., by E. N. Transeau, H. C. Sampson, and L. H. Tiffany, New York, Harper & Row, 1940.)

240

Ch. 14 /
Correlation of
Processes,
Plant
Distribution,
and Food
Chains

Light and Compensation Points

Various investigators have determined the light intensity at which the rates of photosynthesis and respiration are equal. When the compensation point is referred to in terms of light intensity, one should realize that survival depends upon the light intensity being *greater* than that at the compensation point. This, at first, appears to contradict the previous paragraphs, but careful consideration should clarify the situation. In each case the problem is maintaining photosynthesis at a higher rate than respiration. This may be accomplished by lowering the temperature if it is too high or by increasing the light intensity if it is too low.

Table 14-1 indicates the light intensity at which the compensation point is reached in various plants. Note that this light intensity varies with the kind of plant. This is important in helping to explain plant distribution and plant succession on the basis of light intensities, just as the previous paragraphs utilized similar explanations while dealing with temperature as the environmental factor. For example, the data in this table indicate why a mixed stand of forest trees gradually over a period of years shifts emphases until a climax vegetation is achieved. A **climax vegetation** is one that is stable, the kinds of plants remaining relatively constant as long as no major change in the environment occurs. Such an association, or grouping, of plants may be regarded as consisting of those plants that can compete most advantageously in that environment. For example, if a forest originally consisted of a mixture of pine, red oak, maple, and beech trees, a gradual dying out of the pines and then the red oaks would take place, and the climax forest would consist mainly of beech and maple trees. The pines require the most light, whereas the beeches and maples require the least light, to surpass their respective compensation points, as is shown in Table 14.1. Thus, as seedlings develop and the forest becomes more dense, shading becomes a factor. Eventually the light intensity reaching the seedlings is less than 400 footcandles, and the pines will not

Table 14-1 Compensation points (in terms of light intensity) for various plants (27°C)

Plant	Foot-candles[a]	Plant	Foot-candles[a]
Philodendron	500	Tomato	150
Scotch pine	400	Sunflower	100
Arbor vitae	400	Bean	100
Loblolly pine	300	Beech	100–50
Tobacco	200	Sugar maple	100–50
Soybean	200	Some herbs, ferns,	
Dryopteris	200	and mosses	30
Red oak	200–140	Some algae	1
Cotton	150		

[a] Light intensity is measured in terms of footcandles. One footcandle is an arbitary unit: the light intensity at 1 ft when the source has the brilliancy of an average candle. Full sunlight has an intensity of approximately 10,000 footcandles.

survive. When the shading effect of the forest canopy results in the light intensity being reduced to less than 300 footcandles, the oaks will not survive, and the forest soon consists predominantly of beeches and maples.

Mature pines and oaks may survive for a considerable length of time if their size exposes sufficient numbers of leaves above the more densely shaded areas. Seeds of these trees will germinate, of course, utilizing foods stored within their structures. Once the food reserves are depleted, survival depends upon an excess of photosynthesis over respiration—an impossibility for these plants in the shaded situation. The older pines and oaks may remain viable until damaged, as by fire, lightning, wind, or pathogens. They will not be replaced as would beech and maple. The shift to a beech-maple type of forest is basically a result of competition for light in which certain seedlings have an advantageous position.

Man has been a major factor in bringing about changes in natural climax associations by his great carelessness and wasteful practices. Forest fires, overgrazing, wasteful lumbering methods, and excessive destruction of wildlife are examples of man's destruction of natural resources and its concomitant change in climax associations. The gradual disappearance of forest lands in the eastern portion of the United States and the great Dust Bowl of the Midwest are specific examples of human folly. In Spain, much grassland has been ruined by overgrazing with sheep; with the destruction of native grasses, soil erosion occurred at a tremendous rate, and much of the productive topsoil has been lost. As a result, grasses become established with great difficulty, even if grazing is subsequently prevented. Further examples of man's destructive capacity will be discussed in Sections 15.8 and 31.6.

14.2 Root Environment and Plant Distribution

One of the requirements for the normal growth and development of vascular plants is a supply of oxygen, but some plants do not require as much as others. It is not only the aboveground cell that utilizes oxygen, but also the root cell, and here is where difficulties sometimes arise. The solubility of oxygen in water is relatively low. Thus, when soils are water-saturated or flooded, the amount of oxygen available to root cells is greatly curtailed. If this condition prevailed for any length of time, most plants would not survive. Willow (*Salix serissima*) and bald cypress trees (*Taxodium distichum*), cattails (*Typha latifolia*), some sedges (*Carex* spp.), and rice plants (*Oryza sativa*) are examples of plants that can grow quite readily in soils almost devoid of oxygen.

In some of these plants (e.g., *Typha latifolia*) the internal structure is such that large air channels are continuous from the region of the stomata, which is not submerged, to the root tissues themselves. As a result of this structure, oxygen diffuses from the photosynthetic tissues, or from the outside atmosphere through the stomata, to the various cells that are utilizing oxygen. The root cells are located in an environment with a low external oxygen concentration (wet or flooded soil) but where the internal oxygen concentration is approximately as high as the normal atmosphere. These plants can survive because sufficient oxygen is available to the

242

Ch. 14 /
Correlation of
Processes,
Plant
Distribution,
and Food
Chains

roots. A similar situation probably exists in bald cypress trees, which have peculiar root projections called "knees," which grow above the surface of water or saturated soil and seem to provide a pathway through which air can penetrate to the root cells.

Also, the root cells of some plants (e.g., rice) possibly may be capable of existing under anaerobic conditions. This is a physiological type of adaptation and not clearly understood. Two possibilities appear likely: (1) either the anaerobic respiratory reactions (fermentation) produce more energy than those in other plants, or (2) the root cells have lower energy requirements than similar cells of plants that cannot survive in such an anaerobic environment. Rice seeds will germinate in the absence of oxygen almost as well as if oxygen is available; also, the growth of rice seedlings at low oxygen concentrations is much better than that of most other seedlings. This ability of rice seeds to germinate and grow at oxygen concentrations inhibitory to most other seeds is attributed to the more effective fermentation system of rice as compared with other plants. Presumably it is possible, however, that sufficient oxygen may diffuse from the emerged leaves to the submerged roots even though air channels are not present. Evidence on this point is not clear.

Considering the millions of years that plants have existed upon the earth (at least 350 million for vascular plants alone), it is not surprising that a great variety of structural and physiological differences may have arisen and now exist among plants. Such differences enable some plants to survive in environments that are not conducive to growth of other plants. In this way various environmental niches support some plants and result in the exclusion of others. It is unfortunately true that man has not yet discovered all the explanations for plant distributions.

14.3 Food Chains and Energy

Plant distribution influences animal distribution because of the dependence of the latter upon plants for food. Such food is essential as a source of energy (through respiration) and as a supply of basic building material for the synthesis of various components of cells and tissues. The amount of food that is available to humans depends upon the amount made by green plants in excess of what they themselves use and frequently upon what remains after these plants are consumed by other animals and by nongreen plants. The term **food chain** is used to designate this feeding of one organism upon another although **food web** is probably a better term because of the complex interrelationships that usually exist. At the base (bottom) of every food chain is the green plant, the food producer, and all the other organisms are food consumers. A very simple food chain could be:

$$\text{Corn} \rightarrow \text{Cattle} \rightarrow \text{Man}$$

Corn, rather than grazing grass, is used in this example in order to emphasize the loss of food value that occurs in such a food chain. Because of the respiratory activity of the "middlemen" in the food chain, man can obtain much more food per acre by eating plants directly than by eating the animals that have consumed these plants. Much more food is utilized in the animal as a source of energy than is utilized in the production of body material, especially in the more motile animals where a considerable amount of energy is used in muscular activity. In other words, a relatively large amount of food is eaten in proportion to the gain in weight that results. Consider, for example, the amount of food you the reader consume in a day, a week, or a year, compared to the increase in your weight over the same period of time. Also, much of the body of an animal (a steer in this example) is nonedible as far as man is concerned. The result is that only about 10 per cent of the original food value of the plant is available to a human eating the steer, and this is one of the more efficient animals as far as food use is concerned.

If the yield[1] from 1 acre of corn were utilized by man, approximately 1000 men could be fed for one day, assuming 3000 Calories[2] per day as the normal requirement. If this corn is fed to beef cattle, 125 lb of meat would be available to man, equivalent to the food for 43 men for one day. If the corn is fed to pigs, 273 lb of pork would be available to man, equivalent to the food for 220 men for one day. Either case results in a tremendous loss of the original energy content of the plant food. In general, the following amounts of energy remain when plant food is first converted to meat and before it is utilized by humans: pork 17 per cent, beef 10 percent, fowl 7 per cent, and lamb 3 per cent. A much more efficient utilization of food would result from the direct consumption of plant material. This would eliminate the loss caused by respiratory activity of the intermediate organisms in a food chain, as well as the loss caused by conversion of edible plant products into nonedible animal products.

In those instances where animals feed upon plant material that is not available for human consumption, the conversion to animal products is most advantageous. The supply of food from the great grazing lands of the Midwestern and Western portions of the United States and from many aquatic plants becomes available to humans only after passage through various animals.

It should be borne in mind that this discussion of food chains has been simplified by considering only caloric (energy) intake and disregarding the dietary necessity for such items as fats, proteins, and vitamins. A strict carbohydrate diet could provide sufficient calories but would be sadly lacking in other necessary items. Unfortunately, many peoples in the world are not only obtaining too few calories but they are also obtaining these with a high-carbohydrate—low-protein diet. Thus, the damage of low calories is compounded by the danger of an improper or unbalanced diet. For example, kwashiorkor is a serious disease of children leading to severe mental retardation. It is prevalent in many of the poorer nations and is the result of inadequate amounts of protein in the diet. Prevention and cure are readily obtained by the addition of protein to the food supply. But where is the source of protein in areas where *any* food is hard to come by and where meat is almost never available?

---

[1] These data and calculations are from E. N. Transeau, H. C. Sampson, and L. H. Tiffany, *Textbook of Botany* (New York: Harper & Row, 1940).

[2] Calorie (spelled with a capital C) represents the amount of heat (energy) necessary to increase the temperature of 1 kg of water 1°C. With regard to nutrition, the term Calorie is useful to indicate energy requirements. The Food and Agriculture Organization of the United Nations estimates that the average daily requirement of Calories per person ranges from 2250 to 2710. The average individual daily consumption of Calories was more than 3000 in North America and Western Europe in 1967.

244

*Ch. 14 /
Correlation of
Processes,
Plant
Distribution,
and Food
Chains*

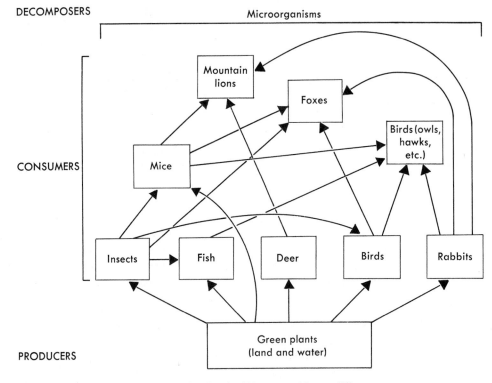

DECOMPOSERS   Microorganisms

CONSUMERS

PRODUCERS

Mountain lions

Foxes

Birds (owls, hawks, etc.)

Mice

Insects   Fish   Deer   Birds   Rabbits

Green plants
(land and water)

Figure 14-2. Interrelationships of a food web. (Where would man fit?)

Food chains in nature are much more complex than the example just given. Figure 14-2 is a more realistic representation of a *food web* (a more descriptive and appropriate term), but it also is simplified. For example, many insects feed upon various animals, and large predatory birds might eat a baby fox if it happened to be available. Many animals take any food that is suitable in size. The important factor is that green plants are always the producers of food; all others are consumers. The decomposers are also really consumers; they decompose (consume, or bring about the decay of) dead organisms of all sorts. In all cases, however, organisms are directly or indirectly dependent upon what the green plants provide. The basic sequence in a food chain consists of:

1. Producer, the green (photosynthetic) plant.
2. Herbivore, or primary consumer.
3. Carnivore, or secondary consumer. There may be several stages of carnivores.
4. Decomposer, or final consumer, the microorganisms.

There is a steep, stepwise decrease in production up the levels, or steps, in a food chain. Only a fraction of the productivity at any level can be harvested by the next higher level (owing to respiration and nonusable material at each level as mentioned previously). Figure 14-3 represents a series of pyramids to emphasize

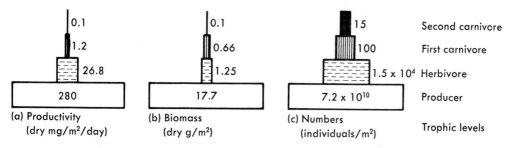

0.1	Second carnivore
1.2	First carnivore
26.8	Herbivore
280	Producer

(a) Productivity
(dry mg/m²/day)

0.1	
0.66	
1.25	
17.7	

(b) Biomass
(dry g/m²)

15	Second carnivore
100	First carnivore
1.5 x 10⁴	Herbivore
7.2 x 10¹⁰	Producer

(c) Numbers
(individuals/m²)

Trophic levels

Figure 14-3. Community pyramids for a shallow experimental pond of low nutrient content. Widths of steps for numbers of organisms are on a logarithmic scale. (From *Communities and Ecosystems*, by R. H. Whittaker, Macmillan, Inc., New York, 1970, by permission.)

this point. In general, productivity, biomass, and numbers decrease up the sequence of levels in a food chain. Pyramids of numbers may sometimes reverse, for example, as thousands of insects feed on a tree. The pyramid of productivity, on the other hand, is a fundamental characteristic of food chains. Decomposers have not been included in the diagrams because they feed at all levels, but the same situation exists with regard to productivity.

Fluctuations in the number of organisms at any level in the food web will have an effect on the other members. If there is a decrease in predator (carnivore) population as a result of disease, this will result in an increase in the number of their prey (herbivores) because fewer are killed. The larger herbivore population might bring about overgrazing and be followed by starvation, which would decrease their numbers. The actual numbers of organisms fluctuate in such fashions. Good rainfall could increase basic food supplies and all organisms increase in number. Drought would have the opposite effect. Such interrelationships, as shown in Figure 14-2, are all too frequently ignored by man. It should be clear that bringing about a change (e.g., killing predators, building houses on productive land) in one portion of a food web is bound to influence the entire complex relationship. One would hope that man is gradually becoming aware of this situation.

14.4 Carbon Compounds and Energy Contents

The interconversion of carbon-containing compounds is an exceedingly important aspect of the study of living organisms. Such compounds constitute the basic structure of all living cells and the energy source for almost all syntheses. The food that an organism consumes consists mainly of carbon-containing compounds of varying degrees of complexity. These are mostly broken down (digested) to simpler molecules, which can then be utilized in the synthesis of those complex components of the living organism or cell. As mentioned already (see Sections 13.1 and 13.2), all syntheses require energy. The most important factor involved in the change of carbon compounds from one type to another is the energy content of each compound. For example, organic carbon compounds, such as sugar, are considered to have a high available energy content—they can be respired (or oxi-

246

*Ch. 14 /
Correlation of
Processes,
Plant
Distribution,
and Food
Chains*

dized) to yield energy. Carbon dioxide, on the other hand, cannot be oxidized to yield energy; it has zero energy content as far as the cell is concerned.

During respiration, organic carbon compounds are converted to carbon dioxide, water, and available energy.

$$\text{Organic carbon} + O_2 \xrightarrow{\text{enzymes}} CO_2 + H_2O + \text{Energy} \tag{1}$$

The energy is utilized or lost as heat (waste), and the carbon compound is now in a form that contains no useful energy (CO_2). Green plants recharge carbon compounds with energy during photosynthesis:

$$CO_2 + H_2O + \text{Energy (light)} \xrightarrow{\text{chloroplast}} \text{Organic carbon} + O_2 \tag{2}$$

This is the only synthesis occurring in living cells that is not dependent upon respiration for its energy. All syntheses utilize energy, and in this case sunlight supplies the energy. In green plants, then, carbon may be circulated from the air as CO_2, into the green plant as various organic compounds, and back to the air as CO_2 once again.

Carbon can be passed quite readily from one organism to another. Animals eat plants and frequently each other. The passage of carbon from animal to plant is

Figure 14-4. A dramatization of certain magazine and newspaper articles.

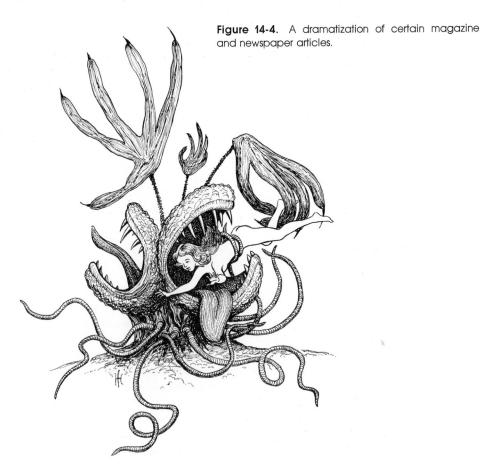

247

*Sec. 14.4 /
Carbon
Compounds
and Energy
Contents*

not so direct, however, in spite of the amazing and fantastic stories in some magazines about carnivorous or man-eating plants (see Figure 14-4). A few insectivorous plants exist, as well as a few that occasionally feed upon nematodes or eel worms (very tiny soil worms, *not* earthworms), but these are not very important. Carbon present as part of an organic compound in an animal usually reaches a plant when the animal respires this carbon compound or when the animal dies and decays. The resultant carbon dioxide is available for photosynthesis.

Even though the concentration of carbon dioxide in the atmosphere is low (about 0.03 per cent), the total amount of carbon dioxide is in excess of 1 million million tons. Wind currents continually bring new supplies of carbon dioxide in contact with green plants; in the case of aquatic plants, the carbon dioxide dissolves in the water, and water currents aid in its distribution.

Of the various sources of the carbon dioxide present in the atmosphere, the most important is undoubtedly the decay of organic material. Remember that such decay is caused by the activity of microorganisms, and thus it is actually respiration and fermentation that continually aid in replenishing the supply of atmospheric carbon dioxide. Other sources of carbon dioxide involve the respiration of any cell or organism even if decay is not involved, burning of fuels (wood, coal, oil), chemical weathering of rocks that contain carbonates, and volcanic action.

Although all organisms eventually die, some of them do not decay completely and, under certain environmental conditions, are converted to peat, coal, or oil. However, this form of carbon is eventually returned to the atmosphere during combustion. One ton of coal produces approximately 3 tons of carbon dioxide when it is burned, and the amount of coal used by various industries is in the thousands of tons per day. No danger of a lack of carbon dioxide exists; in fact, some scientists fear just the opposite. The Industrial Revolution and the subsequent tremendous increase in fuel comsumption may result in an increase in the concentration of carbon dioxide gas in the atmosphere. The suggestion is that such an increase will result in a blanketing effect or a heat trap and that the average temperatures on the earth will rise. Some of the visible light waves, which penetrate through carbon dioxide gas quite readily, are converted to heat when they are absorbed by opaque objects, such as the earth itself, for example. However, heat waves do not penetrate so readily through layers of carbon dioxide. Therefore, when heat is being radiated from the earth to the cooler atmosphere, the carbon dioxide acts as a blanket that decreases the amount of heat lost to the atmosphere; the heat is reflected back to the earth. Whether or not such a heat trap effect will result depends to a great extent upon the amount of carbon dioxide that can be dissolved in the waters of the world. One of the difficulties that could arise would be the melting of the icecaps of Antarctica and Greenland. This would cause a flooding of the earth's coastal lands, where the majority of the world's population now lives. Geophysicists are actively studying various phases of the problem in an attempt to determine whether the carbon dioxide blanket is growing thicker and whether this has any effect on air temperature.

An examination of the preceding paragraphs indicates the existence of a carbon cycle, since carbon dioxide is converted to organic carbon (plant and animal structures) and this material is eventually changed back to carbon dioxide, as is represented in Figure 14-5. The important aspect of the carbon cycle is not the

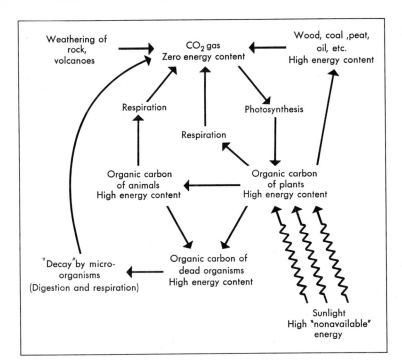

Figure 14-5. Carbon transformations and energy contents.

transfer or movement of carbon-containing compounds but the energy that is stored or liberated as the cycle proceeds. However, no corresponding energy cycle takes place, but a change from zero energy (carbon dioxide) to high available energy (organic carbon compounds) to zero energy (carbon dioxide again). In order for such a situation to exist, a continual input of energy must be maintained, just as a child keeps a hoop rolling by periodically whacking it with a stick—the thrust given by the stick imparts energy to the hoop, which responds by rolling along. The input of energy in the carbon cycle is provided by the sun through green plants. Photosynthesis converts light energy to chemical energy and is the thrust that keeps the cycle turning. The radiant energy of the sun makes life possible; the green plants contribute the means of utilizing such energy. In this figure, sunlight is indicated as being high nonavailable energy. This refers, of course, to the fact that light energy is available only to green plants; it is nonavailable insofar as other organisms are concerned.

Even though we consider carbon dioxide to be on the zero energy level, this does not imply that energy has been lost. One of the basic concepts of physics and chemistry is that energy can be neither created nor destroyed (**The First Law of Thermodynamics**).[3] Energy can be converted from one form to another, however. When gasoline is burned in an automobile, the chemical energy of the gasoline molecules is converted to the kinetic energy of the moving vehicle. As in any process, some of the energy is converted to heat, resulting in a loss of available en-

---

[3] Atomic disintegrations result in the conversion of matter to energy. This merely means that matter and energy are really different forms of the same thing, and we should think of the conservation of both energy and matter.

248

ergy. This heat is lost to the environment, radiated to outer space, and dissipated. Energy is thus lost for use but not destroyed. Remember that during respiration a similar production of heat energy occurs.

An understanding of the interconversion of carbon-containing compounds and the energy content of such materials is essential to an understanding of the importance of plants to all forms of life. The green (chlorophyllous) plant provides food and oxygen, while the nongreen plant (mainly microorganisms) brings about decay processes that make carbon dioxide available for further photosynthesis.

14.5 "Carnivorous" Plants

In the preceding section, a brief comment was made about insectivorous or carnivorous plants, and a somewhat more detailed discussion might clarify the situation. The pitcher plant (*Sarracenia*) (Figure 14-6) has a funnel-shaped leaf, into which small insects may crawl and then drown in the water that collects at the base of the leaf. Small bristles project downward on the inside of the leaf, preventing such insects from crawling out. Digestive enzymes, secreted by the leaf (and possibly bacterial action also), convert the insect body into soluble materials, which are then absorbed. A more interesting insectivorous plant is the sundew (*Drosera*) (Figure 14-7), in which the leaves contain tiny projections that secrete a mucilaginous material. Small insects become entangled in these projections and are digested by enzymes that the leaf cells secrete. In this case also the digestion products are absorbed.

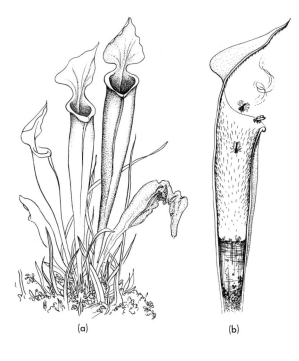

Figure 14-6. The pitcher plant (*Sarracenia*). **(a)** Habit sketch. **(b)** Sectional view showing insects entering and the location of bristles.

(a)   (b)

250

*Ch. 14 /
Correlation of
Processes,
Plant
Distribution,
and Food
Chains*

Figure 14-7. The sundew (*Drosera*). At the bottom left, a fly has become entangled in the small projections from a leaf; the mucilaginous drops prevent its escape.

The most fascinating insectivorous plant, and the one that is probably responsible for the weird magazine articles dealing with carnivorous plants, is the Venus' flytrap (*Dionaea muscipula*) (Figure 14-8). The leaves of this plant are hinged along the midline, with the outer margins bearing strong bristles. The upper surface of the leaf contains short "trigger" hairs, which are sensitive to contact. When an insect alights on the leaf and brushes against these hairs, the leaf snaps shut and the overlapping marginal bristles effectively imprison the insect. Enzymes are secreted by gland cells on the surface of the blade, and the digested material is absorbed. Such an example of a plant with an actual trapping mechanism easily leads the imagination to conjure up an enlarged version capable of enclasping a human. Unfortunately, or fortunately, no carnivorous plants capable of ensnaring even small rodents, much less humans, have ever been found.

As a point of information, the insectivorous plants are green and quite capable of manufacturing food materials. A possible advantage resulting from the ability to

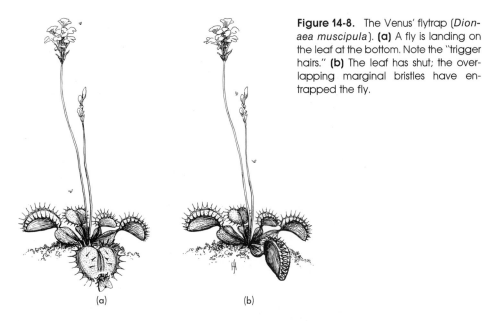

Figure 14-8. The Venus' flytrap (*Dionaea muscipula*). **(a)** A fly is landing on the leaf at the bottom. Note the "trigger hairs." **(b)** The leaf has shut; the overlapping marginal bristles have entrapped the fly.

trap and utilize insects is that such plants may be able to survive in a soil that is low in nitrogen content. Nitrogen could be obtained from the proteins and amino acids of the insect.

The only plants that could even remotely be considered to be carnivorous are a few fungi, and even these organisms obtain the majority of their food from dead organic matter in the soil. However, at least one group of fungi can nicely lasso nematodes. Figure 14-9 indicates the manner in which nematodes are trapped or

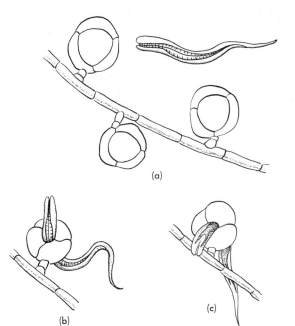

Figure 14-9. Diagram of a nematode (eel worm) being lassoed by a fungus (*Dactylella bembicoides*). **(a)** A nematode approaching one of the rings formed by the hyphae of the fungus. **(b)** The nematode has entered the ring and has been entrapped by the rapid expansion of the ring cells. **(c)** Hyphae of the fungus have entered the body of the nematode. (From J. J. Maio, *Scientific American*, July 1958.)

252

Ch. 14 /
Correlation of
Processes,
Plant
Distribution,
and Food
Chains

lassoed. The three cells that form the ring structure expand greatly when their inner surfaces are touched. Any nematode so unfortunate as to penetrate within the ring is immediately held fast, since the diameter of the ring is greatly restricted when the cells become swollen. The nematode cannot escape, strands of the fungus penetrate the body of the worm, and digestion of the animal takes place. Even if the nematode is vigorous enough to rip the ring from the rest of the fungus filament, this in no way saves the animal. Additional cellular strands, or filaments, are produced by the ring cells, and the result is the same—the nematode is digested and consumed.

These plants are quite interesting and amazing but not nearly the fantastic organisms of various magazine and newspaper articles, which are at best labeled "science fiction" with an emphasis on the "fiction."

Summary

1. The compensation point is that light intensity or temperature at which the rates of photosynthesis and respiration are equal. The plant cannot exist under such conditions, since no food materials are available for maintenance and growth. An understanding of compensation points aids in explaining plant distributions; in order for a plant to survive, photosynthesis must greatly exceed respiration. If a plant has a compensation point at a low temperature, it cannot grow in a warm climate. If a plant has a compensation point at a high light intensity, it cannot grow in a shaded area.

2. Certain plants can survive even though their roots are located where the external atmosphere is deficient in oxygen. In most of these plants, structural modifications make possible an internal movement of oxygen to the roots. In others, however, the root cells appear to be capable of obtaining sufficient energy from fermentation reactions.

3. Whenever one organism feeds upon another, the amount of food value in the second organism is always much less than that in the first because of the respiratory losses that always result. The most efficient use of food, therefore, results when the length of a food chain is kept to a minimum.

4. During respiration, a heat (energy) loss always occurs. Without the continual input of energy from the sun, life would soon cease to exist.

5. Some plants can trap certain kinds of small animals as supplemental sources of food.

Review Topics and Questions

1. A nurseryman in the South obtained a plant from a friend in the North where this plant produces large edible roots. The nurseryman finds that this plant grows very poorly in his region and produces only tiny, thin roots. Explain this situation on the basis of compensation points.

2. A water molecule taken into the root of a plant from the soil solution is consumed in the production of another material after reaching a palisade mesophyll cell of the leaf. The latter compound in turn appears in the root a few hours after its formation in the leaf and undergoes a number of chemical transformations, which result in the production of a new water molecule.

   Describe in detail and in logical sequence the processes, reactions, and tissues involved, from the uptake of the first water molecule to the formation of the second water molecule in the root, including the forces and energy involved in all movements and transformations.

3. Equal numbers of seeds of pine and oak were planted in a forested area. Ten years later many thriving young oak trees in the forested area were observed, but only a few stunted pine trees. Give a possible explanation based upon compensation points.

4. (a) Explain what is meant by a food chain or food web, and give an example that includes yourself.
   (b) Explain why green plants are always the base of any food chain.
   (c) Explain why the last organism gets less and less of the food value of the green plants as the number of organisms involved in the food chain increases.

5. Farmer Muller, in Santa Barbara, California, read that several different varieties of potato were available. In response to his request, friends from all parts of the United States sent him tubers to plant. The tubers were in excellent condition. After being mixed together in a large bin, the tubers were planted, and all plants received the same treatment. Some of these potato plants grew much better than others, and a few plants would not grow at all.
   (a) On the basis of an understanding of compensation points, explain the differences in growth.
   (b) From what sort of climate did the plants that grew poorest originate?
   (c) From what sort of climate did the plants that grew best originate? (Note: The temperature in Santa Barbara is warm during the summer, but not hot.)

6. You have available a group of young birch trees and a group of young alder trees, and you have been given the following experimental evidence:
   (a) At 25°C, the compensation point for birch trees is a light intensity of 100 footcandles.
   (b) At 25°C, the compensation point for alder trees is a light intensity of 50 footcandles.
   (c) Both trees require approximately the same amounts of minerals and water from the soil for a maximum (and equal) rate of growth.
   (d) The area in which these trees are to be planted has plenty of minerals and water, an average temperature of 25°C, and an average light intensity of 150 footcandles.
   Explain in detail which group of trees you should plant in order to obtain a dense forest.

7. Explain why approximately seven times as much land is required to produce food in the form of meat as is required to produce food (of similar caloric value) in the form of grains and vegetables.

254

Ch. 14 /
Correlation of
Processes,
Plant
Distribution,
and Food
Chains

8. "During the entire life of a plant, the quantity of oxygen released in photosynthesis is approximately equal to the quantity of carbon dioxide released in respiration."

 Discuss the fallacy of this statement. (Note: In the photosynthetic production of 1 gm of sugar, the amount of oxygen released *is* equal to the amount of carbon dioxide released in the respiration of 1 gm of sugar.)

9. After a favorable sunny summer season over the North Atlantic, fish are more abundant than after a less sunny summer. Explain.

10. If the source of energy by which green plants live is sunlight, how do they remain alive at night?

11. A plant produces large fruits when grown in the garden but produces only small fruits when grown indoors. Indicate how a knowledge of compensation points helps to explain this phenomenon.

12. A marked atom of carbon, as a part of a molecule of carbon dioxide, occurs in the air above a bean field on May 30. On August 28, this same atom of carbon is found to be part of a molecule of cellulose in the wall of a tracheid in the root of a bean plant.

 Trace in proper sequence the movements of this carbon atom, including all tissues (or regions) and organs through which it moves, all forces involved in its movement, all processes in which it becomes involved, and all changes in energy or chemical form taking place in these processes.

13. Discuss the adaptations that enable some kinds of plants to grow with their roots submerged in water; be sure to include the reasons why such adaptations are successful.

Suggested Readings

BOLIN, B., "The Carbon Cycle," *Scientific American* **223,** 124–132 (September 1970).

PLASS, G. N., "Carbon Dioxide and Climate," *Scientific American* **201,** 41–47 (July 1959).

WHITTAKER, R. H., *Communities and Ecosystems,* 2nd ed. (New York, Macmillan, Inc., 1975).

WOODWELL, G. M., "The Energy Cycle of the Biosphere," *Scientific American* **223,** 64–74 (September 1970).

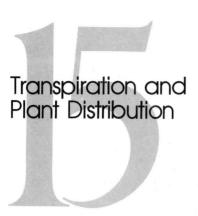

Transpiration and Plant Distribution

15.1 Transpiration

The structure of a typical leaf, including the stomata, has already been discussed. When these pores are open, the diffusion of gases into and out of the leaf occurs quite readily, as in the case of carbon dioxide and oxygen during rapid photosynthetic activity. Before either entering or leaving the cell, such molecules dissolve in the water at the cell surfaces. The enormous number of moist surfaces (the mesophyll cells) exposed within a leaf cause the intercellular spaces to be continually saturated with water vapor. Since the external atmosphere is usually not saturated, a concentration gradient exists, and water molecules diffuse from the leaf (mainly through the stomata) to the environment, where a lower concentration, and hence a lower vapor pressure of water, exists. The cuticle of the plant is not completely impermeable to water, and therefore such diffusion also takes place from various portions of the plant, although to a much lesser extent than through the stomata. This loss of water vapor from a plant is called **transpiration.**

Stomata provide a very effective pathway for diffusion because the rate of diffusion of gases through small pores is more nearly proportional to the perimeters than to their areas. Diffusing molecules tend to accumulate over the pores, and thus the concentration gradient is steeper at the edges than at the center. As a result, diffusion is more rapid at the edge (i.e., perimeter). If a single large pore is compared with numerous small pores whose total area equals that of the large one, it is found that much more diffusion takes place through the small pores. Stomata are tiny; even when wide open they do not comprise more than about three per cent of the total leaf area. However, measurements have shown that water loss through the stomata is usually more than 50 per cent as much as evaporates from a water surface equal in area to the total leaf surface. When the stomata are closed, the loss of water from a plant is much reduced, of course.

Under conditions favoring rapid water absorption and reduced rates of transpiration, liquid water may exude from the tips or edges of many plant leaves. When water uptake exceeds water loss, a hydrostatic pressure builds up in the xylem and water is forced through the intercellular spaces and out of the plant by way of pores (**hydathodes**) situated near vein endings. This exudation of water, or

Figure 15-1. Guttation droplets at tips of leaves. (Courtesy of Peter Jankay.)

guttation, is frequently seen taking place from grass leaves in the spring when relatively cool nights alternate with warm days (Figure 15-1). The rapid accumulation of salts by roots and their consequent buildup in the xylem results in considerable water absorption. Metabolically produced energy is required to accumulate and retain minerals or ions in the root cells. It has been suggested that the oxygen concentration in the deeper cells of the root is lower as a result of the slow rate of diffusion and the utilization of oxygen by the cells nearer the root surface. Thus, respiratory rates tend to be low and insufficient energy is available to retain ions within the deeper cells; as a result, ions leak into the xylem cells. There is, then, an osmotic gradient across the root from the soil solution to the xylem and the living cells of the root provide differentially permeable membranes across which a water potential gradient becomes established. The resultant osmotic movement of water through the root and into the xylem establishes a substantial hydrostatic pressure. The pressure that develops forces water and some solutes out of the leaves. Analysis of guttation water has indicated the presence of nitrates, sulfates, salts of calcium and potassium, and other solutes. The loss of water that results when conditions favor transpiration prevents the development of pressure within the xylem, and guttation does not occur. Losses caused by guttation are usually slight, although some tropical plants may lose as much as 100 ml of water from a single leaf during the night.

15.2 Water Requirement

The quantities of water lost from a plant may be extremely large (Tables 15-1 and 15-2) and are much greater than the amounts actually used by the plant in photo-

Table 15-1 Rates of water loss per day in midsummer

Plant	Water loss (in quarts)[a]	Plant	Water loss (in quarts)[a]
A single plant of corn	3–4	A 12-ft columnar cactus	0.02
A single plant of giant ragweed	6–7	A coconut palm in moist tropics	70–80
A single young 10-ft apple tree	10–20	A date palm in a desert oasis	400–500

[a] One liter is approximately 1.06 liquid quarts.
Source: E. N. Transeau, H. C. Sampson, and L. H. Tiffany, *Textbook of Botany*, 1st ed. (New York: Harper & Row, 1940), by permission of the publishers.

synthesis, digestion, and other metabolic processes. Most plants probably use only about 2 to 4 per cent of the water that is absorbed by the plant: over 95 per cent is lost in transpiration. Obviously, this figure is quite different for many xerophytic (desert) plants, as will be discussed later. Even though most of the water is not used, this large amount is required or the plant will die of desiccation. The **water requirement** refers to the number of pounds of water absorbed by the plant in producing 1 lb of solid material. This figure (see Table 15-3) is obtained by dividing the total quantity (pounds) of water absorbed by the total final dry weight (pounds) of the plant in question. Remember that environmental conditions will greatly influence this ratio by their effect on photosynthesis, respiration, and especially transpiration. In spite of this, the determination of the water requirement of a plant gives a fairly good indication of its distribution, in that desert plants tend to have low requirements and mesophytic plants (e.g., sunflower, bean, tomato, and so forth) have high requirements. Certain exceptions to this statement will be discussed later in the chapter.

In **mesophytic plants,** those that grow in soils containing moderate amounts of moisture, stomatal transpiration accounts for almost all the water vapor that is lost. In some **xerophytic plants,** those that grow in soils having a scanty water supply, there are very few stomata, and the relative amount of water lost through

Table 15-2 Estimated water losses from single plants during a growing season

Plant	Days	Water loss Gallons[a]	Water loss Liters
Tomato	100	30	113
Corn	100	50	188
Sunflower	90	125	470
Giant ragweed	90	140	526
Mature apple tree	188	1,800	6,770
Coconut palm, moist tropics	365	4,200	15,800
Date palm, desert oasis	365	35,000	131,600

[a] One gallon is approximately 3.76 l.
Source: E. N. Transeau, H. C. Sampson, and L. H. Tiffany, *Textbook of Botany*, 1st ed. (New York: Harper & Row, 1940), by permission of the publishers.

Table 15-3   Water requirements for certain plants

Sorghum	250
Corn	350
Red clover	460
Wheat	500
Potato	636
Cucumber	713
Alfalfa	900

Source: Harry J. Fuller and Oswald Tippo, *College Botany*, rev. ed. (New York: Holt, Rinehart & Winston, 1954), by permission of the publishers.

these pores is much less than in the first group of plants. In certain desert plants, then, **cuticular transpiration** becomes more significant than it is in plants growing in other environments. Such transpiration refers to the loss of water vapor directly from the epidermal cells through the cuticle layer, a loss that amounts to only 4 to 10 per cent of the total water vapor lost from mesophytes in general. Such great variations among living organisms must be kept in mind, since most of the basic principles of biology continually refer to the majority rather than to the entirety. The concept of variation is frequently evident but will be covered in detail mainly in the discussion of inheritance and evolution (see Chapters 18 and 28).

15.3 Wilting

The water that is used by or lost from a plant is obtained by absorption from the soil. If the root system can absorb 100 gm of water per hour, and if the plant uses 10 gm per hour, the margin of safety is 90 gm. However, if transpiration exceeds 90 gm per hour, the plant **wilts.** This limp or flaccid condition results because the cells lose water, turgor pressure decreases, and the leaf or even the entire plant loses its rigidity. This is especially true of those plants that do not have much woody supporting tissue, the herbaceous plants. Recovery from this wilted condition occurs when the rate of transpiration again falls below 90 gm per hour.

Plants in a bean field are frequently seen to wilt or droop near midday and then recover turgor as the sun sets and transpirational losses decrease. Such wilting, which occurs as a result of excessive transpiration, is termed **temporary wilting.** There is water in the soil but the root system cannot absorb sufficient quantities to counterbalance the amount lost. When transpiration decreases, the plant recovers as the cells regain turgor because transpiration no longer exceeds absorption. On the other hand, if there is a deficiency of water in the soil, the plant wilts and will not recover even if transpiration is stopped. This is called **permanent wilting.** After all, the plant is still using water even though this may only be a very small amount, and there is simply not enough water to maintain a turgid condition of the cells. When water is added to the soil, the plant will recover if it has not been permanently damaged by being in the wilted condition too long. Wilting, even of the

259

Sec. 15.4 /

Environmental

Factors

Affecting

Transpiration

temporary type, is detrimental to the plant. Water deficiency, which is what causes the wilted condition, reduces the growth of young tissues (cell enlargement and cell division) and adversely affects metabolic processes. Repeated wilting results in stunted growth and a reduced yield of fruit or seed.

One of the primary environmental factors involved in plant growth is water supply, and in many areas of the world, plant production is limited by a deficiency of rainfall. Historical accounts have recorded many instances of famine as a result of periods of drought. This is especially true in areas of marginal rainfall. Any fluctuation of rainfall that results in a decrease in the available water supply has serious consequences with regard to food supplies. This will be discussed in more detail later (Section 15.7).

15.4 Environmental Factors Affecting Transpiration

Environmental conditions may vary considerably even over short periods of time, and these changes result in greatly fluctuating water losses. In general, these changes in transpiration rates result from an effect of the environment upon the vapor pressure of water or upon the stomatal openings.

Temperature

An increase in temperature increases vapor pressure by increasing molecular motion. Since the amount of water vapor within the leaf is usually greater than the amount in the atmosphere, the water vapor pressure in the former area is increased to a greater extent, the vapor pressure gradient steepens, and the rate of diffusion of water molecules out of the leaf increases. If the temperature is high enough, the rate of water loss may exceed the rate of water absorption, and the plant will wilt, as was the case with the bean plants mentioned previously.

The following example will indicate why an increase in temperature steepens the water vapor pressure gradient between leaf and air. We can consider that the internal spaces of the leaf are saturated with water vapor (100 per cent relative humidity), as mentioned previously, and we will assume an external humidity at 60 per cent of saturation (i.e., a relative humidity of 60 per cent). At 20°C, the vapor pressure within the leaf is 17.55 mm of mercury (Hg) and the external vapor pressure is 10.53 mm Hg (60 per cent of 17.55). The difference in vapor pressures, or the gradient, is 7.02 mm Hg. Vapor pressure increases as temperature increases. Therefore, at 30°C, the vapor pressure at saturation (the leaf) is 31.85 mm Hg while that of the air would be 19.11 mm Hg (60 per cent of 31.85, assuming the relative humidity remains at 60 per cent). The difference in vapor pressures is now 12.74 mm Hg, which is nearly double the gradient at 20°C. The rate of transpiration would almost double with a temperature increase from 20° to 30°C even if the external relative humidity did not change. If the external relative humidity decreased, as usually happens when air temperature increases, the gradient would be even steeper. Remember that the intercellular spaces within the leaf remain

Table 15-4 Comparison of vapor pressures within a leaf and that of the air under the specified conditions

	I		II		III		IV	
	Leaf	Air	Leaf	Air	Leaf	Air	Leaf	Air
Temp., °C	20	20	30	30	30	30	30	20
RH, %	100	60	100	60	100	40	100	60
VP, mm Hg	17.55	10.53	31.85	19.11	31.85	12.76	31.85	10.53
VPD, mm Hg		7.02		12.74		19.09		21.32

saturated because of the many moist cell surfaces from which water evaporates. Table 15-4 indicates such a situation if the external relative humidity decreased from 60 to 40 per cent as the temperature rose from 20° to 30°C. The vapor pressure gradient would be 19.09, and water loss would be almost three times as great as at 20°C (compare column III with column I in Table 15-4). In any case, the increased transpiration at higher temperatures is caused primarily by the relatively greater vapor pressure increase within the leaf as compared with that outside. The plant is frequently exposed to such temperature fluctuations as the sun rises and then sets, and air temperature fluctuates as a result of temperature changes in surrounding materials (e.g., soil, rock, buildings).

Light

When plants are exposed to light as the sun rises, the rate of transpiration increases for two reasons. First, the absorption of light by green leaves results in an increased temperature of the leaves. As indicated previously, this causes an increased vapor pressure of water within the leaves and an increased water loss. This is an indirect effect of light. Table 15-4, column IV, illustrates a situation which might occur if a leaf were exposed to direct sunlight on a slightly cool (20°C air temperature) day. Note that the vapor pressure gradient is 21.32 mm Hg and that this would result in a tripling of water loss from that leaf if it had been in indirect light at first so that its temperature was originally the same as that of the air (as in column I).

The second reason involves the direct influence of light on stomatal opening. At night, stomata are partially or wholly closed, and the rate of transpiration is considerably less than during the day, when the stomata are open. The opening of stomata is brought about by the absorption of water and the increased turgidity of the two guard cells, which form the boundary of each stoma (see Figure 10-8), but the exact mechanism involved in these turgor changes is not yet fully understood. We will discuss the suggestions that appear to be most acceptable to plant physiologists at the present time. Investigations are continuing, and one hopes that the difficulties involved in these mechanisms will be resolved.

In the dark, respiration continues and carbon dioxide accumulates. The resultant carbonic acid (H_2CO_3), which forms whenever CO_2 is dissolved in water, causes a slightly acid condition in the guard cells. The sugar-starch equilibrium is

261

Sec. 15.4 /
Environmental
Factors
Affecting
Transpiration

influenced by changes in acidity, and shifts toward the production of starch from the sugars with the increase in acidity. As a result, there is a decrease in the osmotic concentration in these cells. Starch is insoluble and does not contribute to the total amount of solute (i.e., the osmotic concentration). This decrease in the osmotic concentration raises the water potential (see Section 5.4) of the guard cells and they lose water to neighboring cells. As a result of this loss of water, the guard cells become limp (i.e., lose turgor) and the stoma is closed. The guard cells are the only cells of the epidermis that contain chloroplasts. Photosynthesis, which occurs in the light, decreases the CO_2 content (and hence decreases the carbonic acid content), resulting in a decrease in acidity. The sugar-starch equilibrium now shifts in the opposite direction, starch is converted (i.e., hydrolyzed) to sugar, and the osmotic concentration of the guard cells increases. This lowers the water potential, which brings about a movement of water into the guard cells from neighboring leaf cells. The guard cells swell as turgor increases. The walls of the guard cells are much thicker where they are adjacent to each other. The thinner walls bulge considerably, as the guard cells swell, causing the thicker adjacent walls to spread apart. The stoma opens. The procedure is reversed as the sun sets.

Changes in sugar and starch content of guard cells, as indicated above, have been found in many plants. However, some plants do not form starch (onions, for example). Also, in a number of succulents (e.g., members of the Crassulaceae) and some other plants, the stomata are open at night and closed during much of the day. These difficulties have not yet been resolved.

Recent evidence indicates that potassium ions (K^+) are accumulated from adjacent cells by guard cells in the light and lost from the guard cells in the dark. Energy is expended when solutes are accumulated against a concentration gradient, and this would be in the form of adenosine triphosphate. ATP is generated during photosynthesis (i.e., photophosphorylation; see Section 12.5), and this would account for the observation that potassium levels rise in the guard cells as stomata open and fall when stomata close. Such changes would affect osmotic concentrations and water potentials in the same direction as the postulated sugar-starch shifts. It is possible that the stomatal mechanism is complex and may not be the same for all plants, or that both mechanisms operate in leaves. However, the balance of the evidence now seems to favor the potassium ion theory.

Light intensity may fluctuate rather rapidly during the day and bring about many changes in transpiration rates, especially by causing temperature variations in the leaf. Stomata are effective in decreasing vapor loss only after they have closed to 5 to 10 per cent of their full aperture.

External Humidity

One of the basic principles involved in the diffusion of any material is the steepness of the concentration gradient between the two areas in question. Any increase in the atmospheric humidity surrounding a plant would decrease this gradient and thus decrease transpiration. If the atmosphere is saturated with water vapor, transpiration will usually cease. In many green houses the common practice is to water the walls and floors to ensure a high degree of humidity surrounding the plants and thus reduce the water loss from the plants. This reduces the possibility of wilting and results in better growth.

Air Circulation

As water molecules diffuse out through the stomata, they tend to accumulate above these openings, effectively decreasing the concentration gradient by increasing the external humidity. Air movements prevent such accumulation and bring drier air masses in contact with the leaves, thus maintaining high rates of transpiration.

Soil Moisture

If water is not readily available in the soil, the rate of transpiration may exceed the rate of absorption. Such a condition cannot exist for any length of time without wilting of the plant. As was indicated previously, this flaccid (limp) condition is a result of a decrease in turgor pressure of cells as water is lost more rapidly than it is replaced. The stomata, therefore, close when the plant wilts, and transpiration decreases. Remember that closure of stomata also affects CO_2 absorption and photosynthesis. The effect of water stress overrides the effect of other factors. Thus, a water shortage will cause stomata to close regardless of temperature, light, or CO_2 concentration.

15.5 Features That Reduce Water Requirements

A plant that transpires readily must obviously have available to it a considerably greater supply of water than is necessary for a similar plant transpiring at a lower rate. For example, if two plants each utilize 1 gm of water per hour, but the first plant transpires 99 gm per hour while the second transpires 9 gm per hour, the first plant requires ten times as much water as the second: 100 gm per hour vs. 10 gm per hour. Therefore, a low rate of transpiration is especially advantageous for desert plants where rainfall is limited. Remember, however, that modifications that result in low rates of transpiration enable plants to exist in the desert, but that such modifications did not arise in order that the plant could survive in the desert; they are chance occurrences. Many random variations must have evolved in plants during the millions of years that they have existed upon earth. The environment would have acted as a selective agent in that some of these modifications may have enabled a plant to survive under conditions where other plants would die. The features discussed in this section resulted in certain plants being successful "desert dwellers," or at least able to exist in areas of rather sparse water supply.

Any features that result in a decrease of evaporating surfaces will, of course, bring about a lowering of transpiration rates. Such structures are readily found in an examination of desert floras: (1) a reduction in the number of stomata (cacti); (2) reduction in the size (creosote bush, *Larrea tridentata*) or absence (prickly pear, *Opuntia*) of leaves; (3) small leaves that last for only short periods. Ocotillo, or Jacob's-staff (*Fouquieria splendens*), has the appearance, until the rainy season starts, of a dead piece of driftwood stuck upright in the desert soil. Then in a period of days after the rains, hundreds of tiny green leaves appear as if by magic. As soil moisture disappears, these leaves wither, die, and drop off readily. In such a plant, the relatively large evaporating surface, which the leaves represent, is present only during those periods when moisture is available.

Compared to mesophytic plants, many desert plants are found to have few stomata. In others the stomata are sunken in pits, which greatly retard the removal of accumulated water vapor by wind currents (Figure 15-2[a]) and lengthen the path of diffusion. Both factors tend to reduce transpiration from those leaves. Dead hairs on the leaf surface or around the stomata may also interfere with air currents. In some instances hairs may be so numerous that they reflect considerable light and thus maintain a lower leaf temperature than would be the case without such hairs; the Old-man cactus (*Cephalocereus senilis*) is an example. Many of the grasses have rows of specialized bulliform cells in the epidermis. These cells are large and have thin walls. As a result, the bulliform cells lose water, and consequently turgor, more rapidly than other epidermal cells, and the leaf rolls up, a situation diagrammed in Figure 15-2(c). The external humidity of the area enclosed by the rolled leaf soon increases, as transpiration continues and water vapor accumulates. The water vapor pressure gradient between the leaf and this external area decreases. Transpiration, at least from that portion of the leaf surface, decreases. Such rolling also exposes less leaf surface to the heating effect of sunlight.

A thick cuticle is found on many cacti and other succulent plants. In fact, the cuticle of such plants (e.g., the century plant, *Agave*) can sometimes be peeled off like a layer of cellophane. Since the transpiration that takes place through the epidermis is usually a rather high proportion of total transpiration in xerophytes because they so frequently have rather few stomata per cm² of leaf surface, such a thick and continuous cuticle aids in water retention (Figure 15-2[b]).

The root systems of certain desert plants are also important in enabling such plants to survive in relatively dry areas. The deep-penetrating roots of the acacia

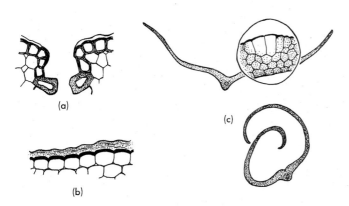

(a)

(b)

(c)

Figure 15-2. Examples of structural features that reduce water requirements. **(a)** A sunken stoma. **(b)** A very thick cuticle. **(c)** At the top, a leaf in its normal position; the enlargement shows the bulliform cells of the epidermis. At the bottom, the leaf is shown after water loss has caused the bulliform cells to lose turgor; the leaf rolls up.

and the mesquite (*Prosopis*) reach submerged water supplies, and these trees survive in spite of relatively high rates of transpiration. Many cacti, on the other hand, have extensive, shallow, fibrous (many-branched) root systems. When correlated with succulent water-storage tissues in which the cells usually contain much colloidal material, this system becomes very advantageous for life in the desert. The rains that occur moisten the upper few inches of soil, where the majority of roots are located, resulting in extremely rapid absorption of water before the soil dries out. This water is then retained within the succulent leaves (as in the ice plant, *Mesembryanthemum*) or stems (many cacti) because of the hydrophilic colloids. Water is adsorbed very strongly to such colloids and evaporates only with great difficulty. If a drop of water is placed in bright sunlight next to a drop of cell sap from the century plant (*Agave*), the drop of water will have evaporated within ten minutes or so, while the drop of cell sap will scarcely have changed because of the hygroscopic colloidal mucilages and gums present in the latter.

15.6 Significance of Transpiration

Many and varied beneficial and harmful effects have been attributed to transpiration. The difficulty in distinguishing between the opposing aspects of this process appears to arise mainly from a misunderstanding of the term, and the student would do well to refer to Section 15.1 at this time.

Lowering of Leaf Temperature

Evaporation of water from a surface lowers the temperature at that surface because of the loss of water molecules of a relatively high kinetic energy; the molecules with the highest kinetic energy are the ones that evaporate first. The decrease in overall kinetic energy of molecules at the surface appears as a decrease in temperature, which is really a measure of kinetic energy. This basic principle has resulted in the statement that transpiration is not only beneficial but actually essential in preventing leaf temperatures in bright sunlight from rising so high as to injure the cells. Calculations based upon known rapid rates of transpiration, however, indicate that this lowering of temperature would amount to probably no more than 2 to 5°C—not a very substantial decrease. Also, those plants in which such a lowering of temperature would be most effective are the desert plants, and the rate of transpiration in some of these plants is so slow as to result in a mere fraction of a degree decrease in temperature. On the other hand, the loss of heat by radiation, conduction, and convection undoubtedly is the major factor that prevents leaf temperatures from greatly exceeding air temperatures. Radiation is especially effective in flat, thin structures such as leaves. It is questionable whether the heat loss by transpiration ever plays an essential role in preventing heat injury to plants. If transpiration stops, for example, there would be an increased loss of heat by radiation and convection as a result of the increased leaf temperature.

Mineral Absorption

Frequently, high rates of transpiration are said to result in high rates of mineral absorption. While most of the minerals are transported in the xylem along with water, our understanding of diffusion and differentially permeable membranes emphasizes the independent absorption of these two materials. Active absorption and diffusion of ions are not influenced by the rate of water absorption and movement. Even the transport of minerals through the xylem is not directly related to rates of transpiration. One might think of the transpiration stream as a rapidly moving belt between the supplying area (the roots) and the receiving area (the leaves), as indicated in Figure 15-3. In this figure, the amount of coal reaching the end of the belt at the right depends upon the speed of addition of coal to the belt and not upon the speed of the belt. Also, movements of minerals tend to follow concentration gradients.Even if transpiration should stop, water would still move through the xylem as a result of the utilization of water in various parts of the plant. The movement would be much slower than when transpiration is active, but the mineral requirements are pretty low in terms of actual amounts. In other words, only a relatively small amount of minerals is needed. Such small amounts are readily supplied even at low rates of transport; and there is at least some movement of minerals in the phloem. There is no evidence of any plant developing mineral deficiency symptoms because of a low rate of transpiration; adequate quantities are transported even if there is no transpiration.

Water Supply

Another misconception concerning the importance of transpiration is embodied in the statement that this process is essential in bringing water to the top of tall trees, that without transpiration water would not be pulled to the tops of trees by the "transpiration stream" and the tree would die. Herein lies a simple basic mis-

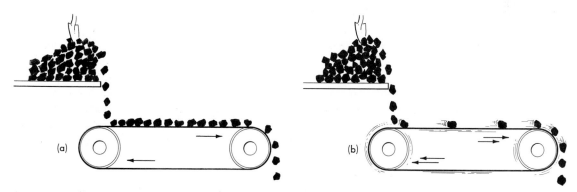

Figure 15-3. Effect of transpiration on the rate of translocation of solutes in the xylem; analogy with a belt delivering coal. **(a)** Belt moving slowly. **(b)** Belt moving rapidly. The amount of coal dropping from the belt at the right is the same in both instances. The rate of delivery depends upon the rate of shoveling and not upon the rate of movement of the belt. (From *Plant Physiology*, by Jacob Levitt, copyright 1954 by Prentice-Hall, Englewood Cliffs, N.J., reprinted by permission.)

understanding. Transpiration is the *loss* of water from a plant, and a loss of water can never increase the moisture content of anything. The fallacy involved in the statement arises from correlating unrelated facts. In other words, while high rates of transpiration will result in the absorption and movement of large amounts of water into and through the plant, it is *not* true that these large amounts of water result in an increase of moisture in the leaves. After all, such water merely replaces that which was lost because of transpiration. Even if transpiration could be stopped entirely, water would still be absorbed by and transported through the plant. Why? Without transpiration the rate of water flow through the plant would be reduced by 95–98 per cent, but the slow flow that remained would still provide water just as fast as it is being used. But it would be *used,* not *lost.* And that is the main point about transpiration—it is a *loss* of water.

Harmful

The significant factor, then, about transpiration is that it is harmful. It is a loss of water, a loss of one of the essentials of life and one of the substrates of an exceedingly important process. (What process?) Because a process occurs in a plant does not ensure that it must be beneficial. With wet cell walls within a leaf and open stomata (both essential for photosynthesis), transpiration must of necessity occur. It is simply a matter of diffusion of water vapor from an area of high concentration (the leaf) to an area of lower concentration (the atmosphere). Danger arises only when the amount of water diffusing out in this fashion exceeds the amount entering the roots. The resultant wilting usually greatly reduces the yield of the plant and may result in death. Water stress may directly reduce photosynthesis because of the effects of dehydration on enzyme activity and membrane permeability, or desiccation may cause stomatal closure, which would result in the rate of photosynthesis being limited by low concentrations of CO_2. Transpiration, in effect, is a necessary evil. Leaves, which are efficient for gas exchange in photosynthesis, cannot help also losing water.

Many of the practices employed by nurserymen and farmers are efforts to reduce transpiration rates and water losses. One of the most important procedures is to remove weeds from any close proximity to the desired plants. Such weeds transpire a vast quantity of water, which is thus unavailable to other plants. The undesirable plants also remove minerals from the soil. The result if weeds are not removed is stunted growth of the crop plants. Nurserymen whitewash greenhouses during the summer as an effective way to reduce light intensity, temperature, and thus transpiration. Frequently, a fine spray of water is used to saturate the atmosphere in the greenhouse and in this way decrease transpiration rates. The resultant saving in time and effort of watering is considerable.

In propagating plants by cuttings or in transplanting from one location to another, every effort must be made to maintain as low a loss of water as possible because of the low rates of absorption that result from the lack of roots or from damaged roots. For this reason, some of the leaves are removed, the plants are frequently covered with glass jars or plastic, and they are always carefully and thoroughly watered. Until a root system is well established in the new area, even fairly low rates of transpiration may result in wilting because of the poor absorption of water. Flaccid (wilted or plasmolyzed) cells are not capable of growth or division. Certainly, a good supply of water is essential for the normal development of the cutting or transplant.

15.7 Plant Distribution and Water Relations

One of the basic concepts in an understanding of plant distribution concerns the interaction of water supply and transpiration. Rainfall, temperature, and light are of primary consideration when discussing regional plant distribution, while soil, topography, and light (again) enter the picture on a more local basis. The influence of temperature and light on compensation points and the resultant effect on plant distribution have already been discussed (see Chapter 14). We should mention at this point that many areas in the world have temperatures too low to sustain any great development of the more complex plants. In fact, unusually late or severe frosts may result in considerable damage to young plants and greatly decreased yields of those plants that survive. Countries such as Canada and the U.S.S.R. have considerable difficulties in growing crop plants because so much of the land is far enough north to cause growing seasons to be rather short—about three months or less between the last severe frosts of the Spring and the earliest frosts of the Fall. One of the many goals of plant breeders is to develop plant varieties with short growing seasons so that they may be used advantageously in such northern climates.

As an indication of the effect of transpiration on plant distribution, one might examine the areas in which apple trees, corn plants, and cactus plants are growing, and compare this with the amount of water necessary to sustain each plant. The average loss of water during a single summer day is approximately as follows:

Apple tree, 10 ft tall	10–20 liters
Corn plant	3–4 liters
Cactus, columnar, 12 ft	0.02 liters

The apple tree thus requires the most water, and the cactus plant requires the least. An examination of the structures of these three plants explains the significant differences in transpiration rates. The leaf of the apple tree is not greatly modified (i.e., it has a thin cuticle, no mucins, many stomata), water losses are high, and thus the tree grows only in those areas where relatively large amounts of moisture are available—that is, in regions of good rainfall, along streams and rivers, or in irrigated areas. The leaf of the corn plant, on the other hand, is somewhat modified and typically grasslike. Fewer stomata, a thicker cuticle, and bulliform (or motor) cells are present in the epidermis. Because of the lower rate of transpiration, the corn plant more readily survives in drier areas than does the apple. The cactus plant is, of course, the most modified of the three: the evaporating or transpiring surface is greatly decreased by the absence of leaves, it has a thick cuticle and few stomata (which are sunken), and the fibrous root system is near the surface and correlated with water-storage tissue. The result is that the

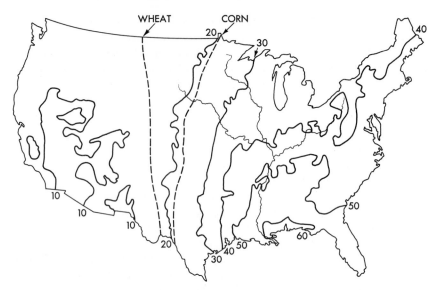

Figure 15-4. The solid lines indicate average annual inches of rainfall in the United States. The westernmost boundaries for wheat growth and for corn growth are shown by dashed lines.

cactus is found in the dry uplands of the desert where there is very little water supply and where neither the apple tree nor the corn plant could survive.

Of course, some exceptions to this correlation between transpiration-rainfall-plant distribution do exist. For example, some desert annuals transpire quite rapidly and yet exist readily in these relatively dry areas. An examination of the life cycles of such plants furnishes an explanation. These small, flowering plants germinate and produce flowers and seeds all within the span of the few weeks when desert showers provide the necessary moisture. Such plants actually escape desiccation as a result of a very short growing season; and they survive through the dry season as resistant dormant seeds. Another example would be the date palm, which frequently loses 400 to 500 liters of water in a single day while growing in a desert. The factor involved in this case is the extensive water supply of the desert oasis (or the irrigation canal). However, apple trees could not grow in this location, even though a sufficient water supply is available. (Review the concept of compensation points with regard to temperature for an explanation.)

Figure 15-4 indicates the average annual inches of rainfall for the United States. Note the gradual decrease in rainfall across the eastern two thirds of the nation as one proceeds from east to west. The greater complexity of the rainfall lines in the western third of the country is because of the vast mountain ranges, which influence rainfall. This effect will be discussed later. The companion Figure 15-5 indicates the very general type of plant growth found in these regions. Plant distribution follows the rainfall belts in a general way.

In the eastern third, one finds broad-leaved plants and deciduous trees (e.g., truck crops, fruit trees, beech, maple), which have but few modifications that result in decreased transpiration. The rainfall in this area is sufficient to support such plants.

In the central area are located the vast grasslands of the United States. The farther west one travels, the less the rainfall, and broad-leaved plants are found only near rivers and streams. The grassland itself changes gradually from the tall (6 to 10 ft) native grasses in the east to moderately tall bunch grasses farther west to the short grasses (less than 2 ft) of the West: the farther west, the shorter the grass. As mentioned previously, grass leaves have lower rates of transpiration than broad leaves; the short grasses have less leaf surface and transpire less than the tall grasses. Consequently, tall grasses require less rainfall than broad-leaved plants but more than short grasses. The grasses are, however, less modified with regard to transpiration than are desert plants.

In the arid regions (the deserts) are found those plants that have the lowest rates of transpiration or that escape desiccation by one means or another. The various cacti, creosote bush (*Larrea tridentata*), Jacob's-staff (*Fouquieria splendens*), and desert annuals are examples of this flora.

The rainfall lines in the western third of the United States are more complex because of the vast mountain ranges in this area (the Rocky Mountains, the Sierra Nevada range, and the coastal mountain range). The moisture-carrying winds blowing in from the Pacific Ocean rise along the slopes of these ranges, moisture condenses as the air is cooled, and rain falls mainly on the western slopes, leaving a rain shadow to the east. By the time the winds reach and pass the Rockies, they are devoid of moisture. In general, then, rainfall is much heavier on the windward side of a mountain range than it is on the leeward side. In fact, if the mountains are relatively high, almost no rain may fall on the leeward side.

Temperature fluctuations run at right angles to rainfall belts, temperatures in the northern hemisphere being higher in the south than in the north. Since temperature greatly influences transpiration, a rainfall that would support certain plants in the north would be insufficient for these plants in the south. For example, in some parts of Mexico, 35 in. of rainfall is insufficient to support anything but desert vegetation because of high transpiration rates. In the southern part of the United

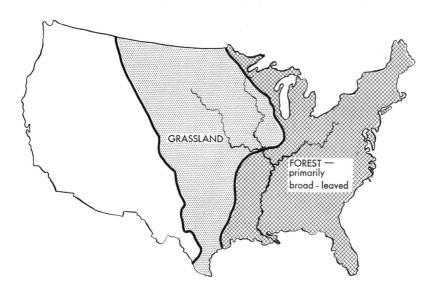

Figure 15-5. The shaded areas indicate general types of plant growth. For further discussion see the text.

GRASSLAND

FOREST —
primarily
broad - leaved

States, this is sufficient for grassland, while trees readily grow with this amount of moisture in the northern part of the country. Mountain areas again complicate the picture because of the influence of altitude on temperature and thus on transpiration.

As discussed more thoroughly in Chapter 17, soil types cause variations in the kinds of plants that can be supported by a specific amount of rainfall. In general, more effective water penetration and aeration occur in sandy soils than in clay soils. Therefore, a rainfall that would make possible the growth of trees in a sandy soil would be sufficient only for grasses in a clay soil. Trees require more water than do the grasses. If much of the rainfall does not penetrate, but runs off as surface flow, relatively little will be available to the plant roots, and this would be particularly detrimental to the trees. Also, tree roots penetrate more deeply than do grass roots. Where water penetrates only the upper regions of the soil, it would be available to grasses but not to trees.

Irrigation systems provided by man can and do enable crop plants to be grown in areas not having a suitable rainfall. This situation exists in the Imperial Valley of California, which receives a meager 10 in. of rainfall per year and develops some of the best crops of lettuce, tomato, and cotton in the country. This is an example of the results that man can achieve; it also emphasizes the importance of a water supply. Table 15-5 indicates the type of response that can be achieved by supplementing rainfall with irrigation even in the more humid regions of the United States. The increases in yield resulting from irrigation are especially noticeable during relatively dry seasons, but even in wet seasons an increased yield does result. The answer to the economic question of whether the additional output will raise farm incomes enough to justify the higher cost frequently depends upon the farmer himself. Benefits and costs vary greatly from farm to farm, and each farmer needs to appraise his own situation. In general, however, the advantages outweigh the disadvantages, and irrigation is continuing to expand. In relatively dry areas, of course, irrigation is necessary if crops are to be grown successfully. In any irrigation procedure, however, there is always the danger of salt buildup in the soil. Irrigation water contains salts, of course. As it flows along canals, evaporation occurs; more water is lost as plants transpire. The salts (minerals) are left behind in ever-greater concentrations and may reach the point where the amounts

Table 15-5   The response of cotton to irrigation in Athens, Georgia

Year	Rainfall (June, July, Aug.)	Irrigation	Seed cotton yields (lb per acre)	
			Unirrigated	Irrigated
1949	12.05 in.	3.00 in.	1155	1286
1950	9.93 in.	1.50 in.	1087	1430
1951	12.37 in.	4.00 in.	2165	2528
1952	8.82 in.	9.44 in.	742	2534
1953	6.25 in.	9.91 in.	934	1731

Source: Water, The Yearbook of Agriculture, 1955 (Washington, D.C.: United States Department of Agriculture), p. 257.

present are toxic to at least some crop plants. This has already happened in the Imperial Valley.

An understanding of the relationship between transpiration rate, water requirement, and rainfall is important to a knowledge of plant distribution, including both native plants and cultivated plants. We are dependent upon plants for food, either directly or indirectly by way of grazing animals, and a knowledge of reasons for plant distribution will enable us to use land and plants more efficiently and with less detrimental effects. Figure 15-4 indicates the distribution of two important crop plants in the United States. In general, the Corn Belt is limited to those areas receiving 25 to 50 in. of rain per year, whereas wheat can be grown with as little as 15 in. of rain. The latter transpires less than the former and thus has a lower water requirement. The effect of a north-south temperature fluctuation is quite distinctly shown in the slant of the line representing the western boundary of wheat growth: about 13 in. of rainfall is sufficient in the North, while 17 in. is required by similar wheat growing in the southern states. An understanding of the reasons for such plant distribution should enable us to decide upon the type of crops to plant in a particular region. Unfortunately, we do not always take advantage of available information.

15.8 The Dust Bowl

An enormous area of the central portion of the United States is now known as the "Dust Bowl" because of the tremendous dust storms prevalent throughout that location. Its size was approximately 12,000,000 acres as of 1958, including large portions of Nebraska, Kansas, Colorado, Oklahoma, New Mexico, and Texas, an area greater than that of Massachusetts and Vermont combined. Dust clouds frequently rise to 20,000 ft, carrying soil as far as the Atlantic Coast, where muddy rain sometimes results. This is a recent development; the seriousness of the situation was first emphasized by the Dust Bowl of the 1930s. This entire area had for generations been lush grazing land. A Dust Bowl did not really exist before the advent of farming, and it never should have developed. It has been created entirely by man through his disregard of an understanding of plant distribution. Even more frightening is the fact that the Dust Bowl of the 1950s was almost twice as large as that of the 1930s, and that the former Dust Bowl was predicted 10 to 15 years beforehand. Let us now examine the causes of this situation.

To grow crops like beans or tomato in the Wheat Belt is obviously impossible. The same reasoning indicates the impossibility of growing wheat in the short-grass regions of the Midwest. However, the figures for annual rainfall refer to average amounts over a period of many years. Seasonal fluctuations occur, so that the wheat frontier can be pushed farther west during the wet cycles, and then the dry seasons wipe back this frontier. In other words, an area that averages 15 in. of rain may receive 17 in. or more for a year or so; 15 in. is received further west then, and wheat can be grown there. However, the average is 15 in. per year, so it stands to reason that there are bound to be years with less than 15 in. if there are years with greater than that—that's what an *average* is all about really. Poor range land with a low rainfall average and low prices tease the farmer into buying such land and growing wheat during the wet years. One drought season, which is bound to come, will wipe out his crop. The danger zone, wherein rainfall fluctuations can ruin crops, includes almost the whole western half of the Wheat Belt. Corn may also be involved in some cases.

This attempt to grow crop plants in unsuitable areas was the cause of dust storms and gave rise to the Dust Bowl. The climate in this area has not changed appreciably for generations. The high winds of the Great Plains have existed for ages without creating dust storms, because the soil has been held in place by the roots of the native short grasses. This native vegetation is plowed under in preparing the soil for wheat and corn production. The failure of these crop plants results in the soil being devoid of vegetation; it dries out and blows away in any heavy wind (see Figures 15-6, 15-7, 15-8, and 15-9).

The wind erosion, which was made possible by man's disregard of known facts about plant distribution, has resulted in the loss of millions of tons of topsoil. Such damage is almost irreparable, since it took 100 to 200 years to create each inch of topsoil. This damage has an exceedingly far-reaching effect, besides the more immediate problem of poverty and migration of the farmers from the Dust Bowl area; it creates a social problem in regions not directly a part of the Dust Bowl, as the West Coast states discovered during the 1930s. At this time, many farmers in the Midwest lost everything and moved to the coast in search of a livelihood. With jobs few and no means of support, they were a further drain on the national economy. All this, of course, was unnecessary.

The solution of the Dust Bowl problem is simple, though drastic, and consists basically of returning most of this area to grazing land. In many areas crops such as sorghum, rather than wheat, also may be planted. The real difficulty involved is

Figure 15-6. An approaching dust storm. Dry topsoil, devoid of vegetation, is being blown about by winds. (Courtesy United States Department of Agriculture, Soil Conservation Service.)

Figure 15-7. The aftermath of dust storms. Plants are almost completely covered by drifting soil. If such plants are not first destroyed by the abrasive action of blown soil particles, they will be smothered and die. (Courtesy United States Department of Agriculture, Soil Conservation Service.)

Figure 15-8. Dust storm damage. The field in the background has been severely damaged by wind erosion of its topsoil. The small grassy area in the foreground is gradually being covered by windblown soil. (Courtesy of the United States Department of Agriculture, Soil Conservation Service.)

Figure 15-9. The western region of the United States showing the tremendous extent of the Dust Bowl area in the 1950s.

to convince people not to plant wheat, which gives them a greater financial return for a short period, but to plant the more stable sorghums and grasses, which are more suited to the area. Obviously, overgrazing would have the same deleterious effect upon the soil as is achieved by attempting to grow wheat. The important thing is to keep some sort of vegetative cover on the soil. This decreases the pounding force of raindrops, increases water penetration and decreases runoff, and the plant roots help to keep the soil in place.

Some areas have adopted land-use ordinances that have been very successful in combating the Dust Bowl problem. Experts of the Soil Conservation Districts indicate the most suitable use of land (e.g., grazing, sorghum, wheat, and so forth), and the land owner is eligible for crop insurance and price supports only if he follows these suggestions. In other words, he is not forced to conform with proper land-use practices; he is just not rewarded for ruining his land if he wants to disregard the suggestions. Many areas throughout the Dust Bowl region already have been at least partially reclaimed by proper land use, and much more should be done to remedy a disgraceful situation.

Summary

1. Transpiration is the loss of water vapor from a plant. Most of this loss occurs through the stomata, although some water vapor is lost directly from the epider-

mal cells. This process is what is really responsible for the tremendous water requirements of most plants, since relatively little of the water that is absorbed is actually utilized by the plant. Without the losses through transpiration, plants would require relatively little water.

2. The rate of transpiration is influenced by those environmental factors that influence evaporation in general: temperature, external humidity, and air circulation. In addition to these factors, soil moisture and various structural features of the plant are important.

3. Transpiration is harmful to the plant. It is a loss of water, and such a loss is always detrimental.

4. The rate at which a plant transpires is the main factor determining the amount of rainfall necessary for that plant to survive. Of course, minor variations as a result of soil conditions and the proximity of bodies of water do occur. Irrigation systems enable plants to grow in areas that do not have a sufficient rainfall, but the gradual buildup of salt concentrations may be harmful.

5. The Dust Bowl of the central portion of the United States is a result of man's disregard for proper land use. Such a situation need never have occurred, and it can still be remedied with appropriate agricultural procedures.

Review Topics and Questions

1. When plants are transplanted, some of the leaves are usually removed (pruning). Explain in detail the advantage of this practice.
2. Explain the functioning of four different structural or physiological features that enable certain plants to exist in desert areas.
3. Explain in detail how humans have been responsible for the development of the Midwestern (U.S.) Dust Bowl. In what ways have fluctuations in annual rainfall been influential? Indicate why such a situation is a nationwide problem and not just a local one.
4. Briefly describe the east-west distribution of apple, corn, and wheat plants in the United States. Explain in detail for each plant the correlation between structure, water requirement, and rainfall.
5. Describe an experiment that would enable you to determine the effect of air temperature on the rate of transpiration.
6. Explain why a plant could wilt even though the soil in which it has been growing is thoroughly watered.
7. A healthy, potted plant is thoroughly watered and then placed in a warm dark room. After three hours, a bright light is turned on for another three-hour interval. Draw a curve plotting the relative rate of transpiration versus time for the six-hour period, assuming that sufficient water was present in the soil at all times. Explain each change in the curve, and discuss the factors that are limiting or retarding during the experiment.
8. Explain the fallacy involved in each of the following statements, and indicate the correct interpretation where pertinent:
   (a) Transpiration supplies the plant with water.

(b) Transpiration keeps the cell walls moist.

(c) Transpiration draws water to the leaves for use in photosynthesis.

(d) Transpiration must be beneficial; otherwise, through evolution and natural selection, the process would have been eliminated.

(e) An increase in transpiration results in an increased absorption of water and therefore increases the turgor of the plant.

9. List those factors that influence transpiration, and discuss the effect upon transpiration of varying each factor individually.

10. Describe in detail an experiment that would demonstrate whether stomata are the main pathways through which water is lost from a leaf.

Suggested Readings

ELLISON, W. D., "Erosion by Raindrop," *Scientific American* **179**, 40–45 (November 1948).

GREULACH, V. A., "The Rise of Water in Plants," *Scientific American* **187**, 78–82 (October 1952).

PENMAN, H. L., "The Water Cycle," *Scientific American* **223**, 98–108 (September 1970).

WENT, F. W., "Climate and Agriculture," *Scientific American* **196**, 82–98 (June 1957).

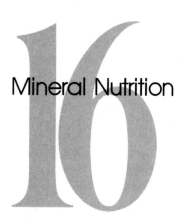

Mineral Nutrition

16.1 Essential Elements

Although plant material can be chemically analyzed for the elements that are present, this does not prove the essentiality of the particular elements that are found, because root cells will absorb any soluble minerals from the soil, whether or not they are essential to the plant. Even toxic materials will be absorbed if they are soluble. The cell membranes are differentially permeable but not selective between two molecules of equal size and solubility, only one of which is toxic. If the material is soluble and the molecules are not too large, it will be absorbed. At least 60 different elements have been found in various kinds of plants, including demonstrable amounts of gold (in *Equisetum*) and silver. Almost all the elements that occur on earth probably can be found in some plant or other. In most instances, no significance can be attributed to the presence of elements that have not been shown to be essential. However, the absorption of the element selenium by plants may be important—not to the plants, but to the animals grazing upon those plants. In certain grassland areas of the western United States (South Dakota and Wyoming), selenium is quite prevalent, and cattle grazing in those regions are sometimes poisoned. The disease is known as "blind staggers" and frequently results in death. Soils containing more than 0.5 ppm of selenium are considered to be potentially dangerous when used for grazing land.

In order to determine the essentiality of an element, one must show that the plant develops abnormally in the absence of that element, that no other element can replace the one in question, and that normal development is obtained when the element is provided to the plant. This is not necessarily the simple task that it may seem. Many elements that are now known to be essential have been discovered only very recently. They are necessary in minute amounts and are frequently found as contaminants in other chemicals, in the glassware, or in the water used for experimental purposes. Technical advances, as in the purification of chemicals, for example, have made possible the compilation of the following list of elements essential for green plants in general:

277   C HOPKNS CaFe Mg MoB CuMnZn Cl

Chemical symbols (see Table 3-1) have been utilized, and the list has been arranged in a special way for ease of remembering. The symbols can be translated into phrases:

"See Hopkin's cafe? Mighty good.
Mob comes in clusters."

Such associations (i.e., chemical symbols with phrases) are useful tools for memorization.

Most of the elements listed in the preceding paragraph are absorbed as minerals from the soil. Carbon, hydrogen, and oxygen are obtained mainly from carbon dioxide and water. Iron, molybdenum, boron, copper, manganese, zinc, and chlorine are frequently called **micronutrients** because they are required in much smaller amounts than are the others (i.e., about 0.1 to 100 ppm). The remaining elements (carbon, hydrogen, oxygen, phosphorus, potassium, nitrogen, sulfur, calcium, and magnesium) are required at more than 1000 ppm and are termed **macronutrients.** This does not mean that any of these elements is "less essential" than others; without any one of them, growth of the plant is abnormal, and death eventually results. Again, let us emphasize that these are the *known* essential elements; the history of mineral nutrition experiments makes plant physiologists realize that other elements may be required in such small amounts that they are almost invariably present as minor contaminants during experimental procedures.

A few plants require, or at least grow better with, additional elements besides those just mentioned. Sodium (Na) may be essential for some blue-green algae, for *Atriplex vesicaria* (bladder saltbush), and for *Halogeton glomeratus;* turnips, beets, and celery grow better when supplied with sodium. Silicon (Si) is certainly essential for diatoms (unicellular marine algae), and may be required by some grasses (at least it is very abundant in grasses). Cobalt (Co) is required by some blue-green algae, at least those that are capable of atmospheric nitrogen fixation (see Section 20.6). These are special cases, however, and these elements (i.e., Na, Si, and Co) have not been shown to be essential for most plants.

16.2 Functions of Essential Elements

The reasons for the essentiality of several elements becomes quite obvious when one examines the molecular structure of the various compounds found in living cells. The main structural components of the protoplast are the protein molecules, which contain carbon, oxygen, hydrogen, nitrogen, and frequently sulfur. Most of the rather complex molecules within a cell contain carbon, hydrogen, and oxygen atoms, but in addition to these atoms, the chlorophyll molecule contains magnesium, the middle lamella consists primarily of calcium pectate, and certain respiratory enzymes contain iron or copper as part of their prosthetic groups. Without any one of these component atoms, gross abnormalities or death results.

Only some elements (C, H, O, N, S, P, Mg, Ca, Fe, Cu, Zn) have been shown to be component parts of cellular molecules. However, evidence has shown that certain minerals are essential for the normal development of a plant because the ions concerned (e.g., Mg^{++}, Mn^{++}) are necessary for the functioning of various enzyme systems. The activity of some enzymes of the glycolytic reactions is dependent upon the presence of magnesium ions, but magnesium may not actually be a part of the enzyme molecule. In other words, magnesium may be functioning quite differently from the iron atom mentioned in the previous paragraph. Ions that are necessary for enzyme activity but have not been shown to be part of the enzyme molecule are termed **activators**. The micronutrients appear to function mainly in this way, but so also do some macronutrients (i.e., K, Mg).

The Role of Specific Elements

A brief consideration of the role of various essential elements and the consequences of their deficiencies may serve to emphasize the importance of soil minerals in plant metabolism. Carbon, hydrogen, and oxygen are not obtained in the form of minerals and will be omitted. Besides, the essentiality of these three elements is obvious.

Phosphorus. The importance of phosphorus in energy metabolism has been emphasized repeatedly in discussions of high-energy phosphate bonds being used to drive chemical reactions and the manner in which such organic phosphate bonds form. Besides ATP, one also finds phosphorus as a component of DNA, RNA, NAD, NADP, and phospholipids. The last-named molecules are important constituents of membranes along with proteins.

In view of phosphorus' many functions within cells, it is not surprising to find that phosphorus deficiency affects all aspects of plant metabolism and growth. Plants are stunted, symptoms appear first in older leaves, which are frequently lost, and soluble carbohydrates tend to accumulate. Leaves of phosphorus-deficient plants are usually darker green than normal ones, anthocyanin formation is enhanced, and necrotic spots appear on leaves, fruits, or other structures.

Potassium. This element appears to have no structural role in plants, but it is an activator for many enzyme systems and has a role in stomatal movements. It is also implicated in protein synthesis in that potassium-deficient plants tend to have relatively high amounts of amino acids and a low protein content.

The first indication of potassium deficiency is a mottled chlorosis (i.e., inhibition of chlorophyll synthesis) of older leaves and spreading to younger ones, followed by necrosis of leaf margins, and a curling of leaves. A widespread blackening of the leaves may also occur.

Nitrogen. Nitrogen is a constituent of proteins (which includes enzymes and membrane components), nucleic acids (DNA and RNA), chlorophyll, growth regulators, and many other important organic compounds within the plant. It is one of the most important structural components of cells.

The most typical nitrogen deficiency symptoms are the marked reduction in growth rate and the development of chlorosis in older leaves and spreading to younger ones. As a result, the plant is stunted and generally yellowish in color as the carotenoid pigments are no longer masked by chlorophyll. Stems and leaf veins may take on a red color as a result of the abnormal production of anthocyanins. This seems to be brought about by the accumulation of sugars that cannot be used in protein synthesis and are then available for anthocyanin synthesis.

Sulfur.   This element is a constituent of three amino acids from which proteins and enzymes in general are synthesized. In addition sulfur is found in vitamins (biotin and thiamine) and in coenzyme A. Sulfur deficiency results in symptoms generally similar to those of nitrogen deficiency except that chlorosis appears first in the younger leaves.

Calcium.   Much of the calcium in plants is found in the middle lamella of cell walls as calcium pectate, the calcium salt of pectic compounds. Meristematic regions are the first to respond to a calcium deficiency with a marked inhibition of bud development and death of root tips frequently occurring. Young leaves become chlorotic, and the tips and margins develop necrotic lesions. Calcium has also been reported to function as an activator of some enzymes.

Magnesium.   This element is an essential part of the chlorophyll molecule, and it is an activator of numerous enzymes (especially those involving phosphate transfers). Magnesium also plays a role in maintaining the integrity of ribosomes. When magnesium is present in deficient amounts, the ribosomes fragment into subunits that are no longer capable of protein synthesis. The characteristic symptom of magnesium deficiency is an interveinal chlorosis of the older leaves; in severe cases the leaves may become whitish as a result of the loss of carotenoids as well as chlorophylls.

Iron.   All of the previous mineral elements are considered to be macronutrients; the rest are micronutrients because they are required in such relatively small amounts. Iron is essential for the synthesis of chlorophyll although it is not part of the molecule, but the specific role is not known. Some of the enzymes and carriers that function in the respiratory and photosynthetic mechanisms of living cells are iron compounds (e.g., cytochromes, ferredoxin). Iron deficiency results in chlorosis that is interveinal and sharply confined to the younger leaves; the principal veins remain green unless deficiency is extreme.

Molybdenum.   Not much is known about the function of molybdenum in plant cells except that it is necessary for nitrogen metabolism, such as the reduction of nitrates to ammonium ions. There must be other function(s), however, because plants require molybdenum even when grown in the presence of ammonium fertilizer. Deficiency of this element brings about a general chlorosis of older leaves and a marginal wilting and necrosis.

Boron.   This element is required in very small amounts, but its function is not known although it has been implicated in carbohydrate transport. Translocation and absorption of sugars are much reduced in the absence of boron, and its addition increases the translocation of products of photosynthesis. The most distinctive symptom of boron deficiency is death of the stem and root apical meristems; young leaves may thicken, blacken, and become necrotic.

Copper. This element is a constituent of certain oxidative enzymes and of plastocyanin, a compound forming part of the photosynthetic electron transport system. Deficiency of copper causes necrosis of leaf tips, which then extends down along the leaf margins.

Manganese. Manganese is required for the activity of many enzymes and is thus involved in respiration (especially the Krebs cycle), fatty acid synthesis, DNA and RNA formation, nitrogen metabolism, photosynthesis (photolysis of water); and it is necessary for normal development of chloroplast membranes. Deficiency of manganese results in interveinal chlorosis accompanied by numerous small necrotic lesions.

Zinc. This element is a constituent of several dehydrogenase enzymes and is an activator of other enzymes. Zinc is also necessary for the production of indoleacetic acid, a growth regulator that is essential for the normal enlargement of cells. It also appears to be involved in protein synthesis. A lack of zinc results in a great reduction in the size of leaves and the length of internodes, resulting in an overall stunted plant. Leaf chlorosis may also occur, which implicates zinc in chlorophyll formation.

Chlorine. Chlorine is essential for the photosynthetic reactions in which water is split and oxygen released, but the mechanism is not understood. Although deficiency probably never occurs in the field, it has been shown by the use of mineral nutrient solutions; leaves wilt, become chlorotic, and frequently turn bronze; roots are stunted.

16.3 Absorption of Minerals

As indicated previously (Section 5.5), absorption of minerals may occur as a result of diffusion, active absorption, or ionic exchange. Recent evidence indicates that the latter two mechanisms are by far the more important. Remember, too, that the growth of roots through the soil continually brings the absorbing surfaces of the plant in contact with additional soil particles and soil solution. Diffusion and associated phenomena were discussed in considerable detail in Chapter 5, so absorption of minerals will now be discussed more from the viewpoint of active absorption or accumulation.

Many plant cells are capable of building up concentrations of salts greater than the concentrations of the same salts in the external solution. This was first demonstrated with freshwater algae but has also been shown to occur in root cells. These accumulated salts are present within the cell in a dissolved state; so this phenomenon is truly an absorption *against* a concentration gradient (i.e., from an area of lower concentration to an area of higher concentration). This is similar to a stone rolling uphill. Work is required to accomplish such a difficult task, which means that energy is expended. Such energy is derived from respiratory reactions.

Salt accumulation occurs primarily in cells that are capable of cell division and growth—cells that are relatively active metabolically. Meristematic cells and cells in the early stages of enlargement are especially active. Thus, we find greatest accumulation near the root apex and decreasing accumulation in those cells farther from the root tip. Respiratory activity of the cells decreases as the distance from the apex increases. Since salt accumulation is dependent upon energy, various conditions conducive to respiration must prevail: adequate aeration (oxygen), sufficient food (sugar, or other substrates), and a favorable temperature. A deficiency of any of these factors has been shown to result in little or no accumulation.

The absorption of mineral ions by a cell always entails transport across a barrier (i.e., the cell membrane). Evidence from various investigations suggests that ATP is the source of energy for the operation of the carrier system postulated as functioning within membranes (see also Section 5.5). This ATP may be supplied by the mitochondria (respiration) or by the chloroplasts (photophosphorylation in photosynthesis). The manner in which ATP functions is not known. It may be that the carrier must be phosphorylated (by accepting phosphate from ATP) before it can bind with the ion that is to be transported.

As has been mentioned before (see Sections 8.2 and 9.3), once the minerals have been absorbed into the root cells, upward transport occurs primarily in the xylem.

16.4 Hydroponics

Because of newspaper and magazine articles (some rather fantastic ones, at that), the average person is aware that plants can be grown in water culture and that soil is not essential. Such soil-less agriculture is frequently termed **hydroponics.** This has been a well-established practice with botanists since at least 1865, when Knop and Sachs published the results of their investigations. In essence the "test-tube farmer" is merely providing water and minerals to the plant—the same materials that the plant would obtain from the soil. Obviously, any plant grown in water culture must be supported in some manner because of the lack of a solid soil foundation.

The physiology of the plant is the same whether it is grown in soil or in a comparable nutrient solution. Care must be taken that the solution contains all the essential elements in prescribed amounts and that the total salt concentration is not so high as to be injurious. Soil-grown plants have the same requirements. Regardless of their location, these plants require light if photosynthesis is to occur and the plants are to grow.

The foregoing paragraph should make clear the fact that many of the claims made for hydroponics are great exaggerations, misrepresentations, or actual falsehoods. Plants cannot be grown in a basement, an attic, or a closet unless light is available in addition to minerals and water. Artificial light can be used instead of sunlight, but to do so at present is certainly not economically feasible. It is also necessary to aerate the mineral nutrient solution or root growth will be curtailed severely because of the low solubility of oxygen in water. The tremendous yields attributed to plants grown by the water-culture method are based upon results from a limited number of plants grown under ideal conditions of light, spacing, temperature, and support. To apply such results on an acre basis without adequate correction for natural shading and other facets of competition is erroneous. When soil-grown and solution-grown plants are maintained under similar conditions, no significant difference in yield or quality is apparent between the two groups.

Mineral-nutrient solution techniques have been used satisfactorily for a wide variety of purposes since their inception in the mid-1800s. One of the most important uses is in determining the essential character of various mineral elements. The chemical composition of such solutions can be arranged quite readily, whereas a soil solution may contain ions that become adsorbed to soil colloids and thus not easily detected or not readily available to the plant. The growth of two similar groups of plants, one growing in a nutrient solution containing mineral X, the other lot growing in a nutrient solution devoid of X, can thus be compared. If the growth of the first lot is better than that of the second, then mineral X can be added to lot number two. If this elicits an increased growth, and no other elements cause similar results, one may consider X to be an essential mineral. This is the general method that has been used to prove that at least certain minerals are essential. Additional precautionary measures should be taken, but a discussion of them is beyond the scope of this text. It should be noted, however, that possible sources of contamination are the chemicals, glassware, and distilled water. For example, sufficient boron actually leaches out of Pyrex glassware to supply the required small amounts for normal plant growth. In using such glassware boron deficiency could not be demonstrated and thus neither could its essentiality. Running distilled water through copper tubing provides sufficient quantities of copper. Polyethylene tubing is a source of molybdenum, cobalt, and possibly other elements. Sulfur may enter leaves, soil, or nutrient solutions by way of sulfur dioxide, which is prevalent in the atmosphere of industrial areas (from sulfur in coal). It would not be at all surprising if newer, more delicate, techniques extend the list of known essential elements.

16.5 "Organic Farming"

We might at this time mention the foolishness of the controversy (amounting almost to a feud in some localities) of "organic farming" versus the use of chemical fertilizers in the growth of crop plants. The "organic farmer" usually maintains that chemicals should not be added to the soil and that compost (partially decomposed organic material), manure, or plowed-under plant parts should be used. The reasoning behind this firm stand appears to be based upon the fact that chemical fertilizers are poisonous when eaten or will at least make a person rather ill. I wonder if steer manure is not in the same category. In actuality, the addition of chemical fertilizers or compost material usually results from the same basic requirement of the plant—the need for minerals.

In most soils, nitrogen is the element that is most often deficient. As plants grow, they absorb nitrogen salts (in the form of ammonium ions or nitrate ions) from the soil. The harvesting of a part or all of the plant results in the removal of nitrogen from the soil. Even more important is the fact that nitrates are very readily leached from the soil because of their high solubility in water. This may result in considerable losses of nitrogen in areas of even moderate rainfall. After continued cultivation, the lack of nitrogen becomes evident through the characteristic symptoms that appear in plants subsequently grown in that soil. A reduced yield always results, and the solution to the problem is merely to add nitrogen. The only difference between "organic farming" and any other method is the *form* in which nitrogen is added. Chemical fertilizers usually contain nitrate or ammonium salts, whereas manures are rich in organic nitrogen compounds (e.g., proteins, amino acids, urea). Nitrogen in this latter form is *not* available to the crop plant. Various microorganisms in the soil convert such organic nitrogen compounds to ammonium salts and nitrate salts (Section 20.6), which are then readily absorbed by the plants growing in that soil. In effect, then, the organic material added to the soil supplies minerals that can be added more rapidly and more easily by the judicious use of chemical fertilizers. The directions given on a bag of fertilizer should be followed rather carefully however. The old idea that "if 1 lb is good, 2 lb will be twice as good" is dangerous. (Review plasmolysis, Section 5.5.)

The real advantage of using organic material rather than chemical fertilizers accrues from the beneficial effects on soil structure, which result from adding the former. Water-holding capacity, aeration, and mineral-holding capacity of soils will be discussed in Chapter 17. It is sufficient to say at this time that the addition of organic matter greatly enhances the advantageous physical characteristics of the soil. In fact, mixing organic matter of various sorts with soil may greatly improve plant growth even if minerals had not been deficient, simply by the improvement of soil structure. The soil is loosened, aeration is promoted, drainage is promoted in clayey soils, and the water-holding ability of sandy soils is increased. Of course, if plants of the legume family (such as clover, alfalfa, and vetch) are plowed under, considerable amounts of nitrogen also are added because these plants usually accumulate rather large quantities of nitrogen. The nitrogen-fixing (conversion of nitrogen gas to organic nitrogen compounds) ability of such plants is discussed in Section 20.6. Organic matter, especially leaves and straw, can also be used as a **mulch** or a covering layer on top of the soil. This decreases the pounding effect of rain, decreases runoff and erosion, and increases percolation of water into the soil. Mulches also retard evaporation of water from the soil as a result of a shading effect and the lowering of soil temperature.

It is an excellent practice to retain grass clippings, leaves, vegetable tops, hedge prunings, and even weeds to form a **compost heap.** These plant materials are usually mixed with phosphorus, potassium, and nitrogen salts to ensure a good supply of minerals. Occasional turning of the heap speeds decay. After several months, the partially decomposed material is in excellent condition for use as a fertilizer or mulch.

The amount and kind of minerals absorbed from the soil varies considerably with the kind of plant. This is one of the reasons that rotation of crops is a good idea. However, some minerals are removed and not returned to the soil, since they are in the product (e.g., fruit, tuber) that is sent to market. At intervals, at least nitrogen, phosphorus, and potassium must be added to the soil. Most of the common commercial fertilizers contain primarily a supply of nitrogen, phosphorus, and potassium (N, P, K), those elements that are most frequently lacking in sufficient amounts in the soil. Most of the other essential minerals are present in such large amounts that they need not be replenished. They are part of the chemical structure of the mineral particles and rock making up the soil. If local conditions warrant, other minerals may also be applied, of course.

Summary

1. Certain elements are known to be essential to the normal growth of green plants, some of these being required in minute amounts. Additional elements may be added to the list as investigations progress.

2. The essential elements may be required because they are the component atoms of structural molecules of the cell, or they may be activators of enzyme systems. In some cases the function of the element is unknown.

3. Although minerals diffuse into plant cells, most absorption of minerals is a result of accumulation or ionic exchange.

4. Most plants can be grown quite readily if their roots are immersed in a mineral nutrient solution. Such solutions are valuable tools in mineral nutrient investigations. They do not, however, have any real advantages over soil if the plants are treated similarly and if no mineral deficiency exists in the soil.

5. If crop plants are to be grown for any length of time on a given area, fertilizer will have to be added to that soil. As the plants are harvested, some of the minerals they have absorbed will be removed with them. Eventually plant growth will be poor unless such minerals are replaced, as by fertilizing.

Review Topics and Questions

1. Describe in detail an experiment that would enable you to determine whether silicon is essential for the growth of bean plants.
2. Describe in detail an experiment that would enable you to determine whether roses are able to absorb nitrate salts through their leaves.
3. If plants in general could absorb minerals through leaf epidermal cells, would this ability be of any value?
4. Plant growth is better when ammonium sulfate (NH_4SO_4) is added to a soil. Give three possible explanations for this increased growth.
5. The following statement appears in a newspaper: "Garden vegetables can be

grown quite readily in a kitchen closet by using mineral nutrient solutions.''
Discuss this statement critically.

6. Plants in a field show abnormal growth. Describe in detail what you should do to determine whether this abnormal growth results from a mineral deficiency.

7. List the essential elements. Indicate, as far as is known, the function of each essential element.

8. Explain why the root system is poorly developed when a plant is grown in a mineral nutrient solution. What could be done to obtain a better root system in such a situation?

Suggested Readings

ANDERSON, A. J., and E. J. UNDERWOOD, ''Trace Element Deserts,'' *Scientific American* **200,** 97–106 (January 1959).

DEEVEY, E. S., JR., ''Mineral Cycles,'' *Scientific American* **223,** 148–158 (September 1970).

McELROY, W. D., and C. P. SWANSON, ''Trace Elements,'' *Scientific American* **188,** 22–25 (January 1953).

OSBORN, F., *Our Plundered Planet* (Boston, Little, Brown, 1948).

PRATT, C. J., ''Chemical Fertilizers,'' *Scientific American* **212,** 62–71 (June 1965).

Soils 17

17.1 Soil

The preceding chapter stated that plants obtain essential minerals from the soil, and this indicates that at least a brief discussion of soils would be of benefit to an understanding of plant growth. Few people do not know what is meant when the word *soil* is used, and yet hardly anyone can give a definition of the word, since all of the connotations implied by such a small word are difficult to put into words. Possibly we can think of **soil** as the very thin uppermost part of the earth's crust, which has developed from the weathering of minerals and contains living organisms and the products of their decay. Such a structure consists of an intimate mixture of mineral materials, organic matter, water, and air in varying proportions. The water and air are in the pore spaces, which are between and within the solid particles. The basic framework of the soil consists, then, mainly of small fragments of mineral matter derived from solid rock (bedrock) by long periods of weathering.

17.2 Origin of Soil Material

The weathering processes may be considered as mechanical disintegrations or as chemical decompositions, and the soil materials that result are designated as sedentary or transported.

Mechanical Forces of Weathering

Within this category we should first consider the effects of temperature fluctuations. Mineral aggregates (the rocks) expand and contract irregularly with increases and decreases in temperature, setting up differential stresses that eventually produce cracks and crevices. In addition to such physical strains are the enormous forces of expansion resulting from water freezing in such cracks. Growing roots also frequently exert a prying effect on rock and may aid in some disintegration. Water, ice (glaciers), and wind are additional agents that bring about abrasive action of rock particles upon each other, with smaller and smaller particles resulting from such grinding action. As a result of these physical or mechanical forces, large solid bedrock material is gradually converted to small rock particles.

Chemical Processes of Weathering

Scarcely has the disintegration of rock material begun than its decomposition takes place. Various hydrolyses result in the formation of soluble materials and frequently the slow development of colloidal clays. If organic matter is present, the carbon dioxide and carbonic acid that result from decay processes greatly increase the rate at which materials dissolve. Through bacterial and fungal action bringing about decay and carbon dioxide production, plant and animal residues may exert a much greater influence on soil production than the mechanical effects of growth and burrowing of the organisms themselves. In many instances bare rock surfaces are gradually decomposed as a result of the activity of lichens, especially in producing organic acids (see Section 21.12). Oxidations, especially of iron-containing rock, result in the weakening and crumbling of the rock. Again, the result of all these forces is to convert bedrock into small particles.

Sedentary or Residual Soil Material

These soils have developed from the disintegration and decomposition of parent material (bedrock) in place, with little or no transportation. There tends to be a rather uniform gradation, with the finer particles on top and the coarser ones below. Most residual soils have undergone long periods of intense weathering and may be thoroughly oxidized and leached, especially in warm, humid climates. In cooler, drier climates, weathering is less drastic; the Great Plains of the United States would be a good example of this type.

Transported Soil Material

Those soils that have been moved to new positions by the mechanical forces of nature are frequently subdivided according to the agencies of transportation and deposition.

Colluvial. In soils of this type, the movement has been caused by **gravity;** talus at the foot of slopes, landslides, and mud flows are examples. The soil material formed in this way is usually coarse, has unfavorable physical and chemical characteristics, and is of little importance agriculturally.

Alluvial. When the agent of transport is running water, deposition occurs when the water slows down; such soils are alluvial deposits. The flood plains along the

Mississippi River and the delta formed at the mouth of this river are excellent examples of this type of soil; so also are the deltas and flood plains of the Ganges (India), Po (Italy), and Nile (Egypt) rivers. These deposits are frequently extremely fertile.

Glacial till. During the Ice Age, starting about one million years ago, northern North America, northern and central Europe, and northern Asia were invaded by a succession of great ice sheets, during which time immense **glaciers** pushed outward (especially southward), receded, and pushed out again several times. There were four such invasions of the central United States with glacial ice covering the area less than half of the time. As the snow and ice thickened, increasing pressure brought about strong lateral thrusts that soon initiated outward movement, and the immense mass was pushed forward. Hills were rounded, valleys were filled, and great masses of rock particles were ground and carried about. When the ice melted away (and the glaciers receded), a mantle of glacial drift remained. Since the interglacial periods were considerably longer than the time during which ice actually overlay the country, plants and animals existed in many areas that were subsequently covered. This resulted in organic debris being mixed in the glacial drift.

Glacial soils are thus quite variable because of the diverse ways by which the debris was laid down (including streams of water issuing from the glaciers), differences in the composition of the original materials, and fluctuations in the grinding action of the ice. In the United States a large area of the northern portions of the Midwest and the Northeast consists of soil deposits associated with glacial ice and the streams flowing from them. Most of these soils are of rather recent development, resulting from the last ice invasion, which disappeared from northern Iowa and central New York only about 10,000 years ago. Not much leaching has occurred, and they are quite well suited for agriculture.

Aeolian deposits. The most important **windblown** material is termed **loess.** Such soils are found in the central and northwestern United States, in northern France, in northern Argentina, and in the great plains of China. They are generally silty in nature and vary considerably with regard to fertility and productivity, depending upon the climatic conditions existent in different parts of the loessial area.

Many kinds of soil may develop from quite similar parent material because of differences in weathering and transport, in topography (e.g., slope, drainage), in organic accumulation, and in climate. Over long periods of time, even slight differences may have profound effects.

17.3 Soil Formation

The disruption of the original silicate minerals by weathering paves the way for clay formation by various hydrolyses, decompositions, and recrystallizations that produce complex aluminum silicates of very small particle size. These colloidal clay particles are platelike and composed of sheetlike molecules held together by various types of bonds, including hydrogen bonds. The result is a cohesive, porous structure with a tremendously large negatively charged surface area to which positively charged particles are adsorbed—characteristics that are exceedingly important in relation to soil fertility.

Living organisms bring about an accumulation of organic matter, which gradually blends with the mineral matter. The resulting complex mixture with its colloidal properties can now be termed soil. A *true soil,* then, cannot be formed without the presence and decay of some organic matter; mere physical and chemical weathering is only part of soil formation. The colloidal particles, both clay and organic matter, are quite gelatinous and add much to soil structure, as will be discussed later. Basically, *soils* are porous aggregates of inorganic particles with an admixture of decaying organic matter in varying amounts. The pore spaces between and within the solid particles contain water, water vapor, or air. The volume composition of a representative soil in optimum condition for plant growth would approximate:

50 per cent pore space: 25 per cent water
                                25 per cent air
50 per cent solid space: 45 per cent mineral material
                                5 per cent organic matter.

The amount of water and air in the pore space would be subject to great fluctuations, of course. During and directly after a rainstorm, many of these spaces may be completely filled with water, most of which gradually drains down through the soil as a result of gravitational forces.

If a section downward through an undisturbed (or virgin) soil is examined, usually a unique and well-defined series of layers are found. These are termed horizons, and they may be divided further into several subhorizons. The uppermost layer, which is usually no more than 25 cm thick, is the **A horizon** and is usually referred to as the **topsoil.** The subhorizons vary from the litter at the very top, to the darkly colored mixture of organic matter and mineral material, and then the lighter colored layer which results from leaching of solutes. Below this surface soil lies the **B horizon** or **subsoil,** usually 50–100 cm in depth, markedly weathered, primarily a mineral soil, including some minerals accumulated from those leaching down from the upper layers, but with little or no organic matter. The subsoil gradually merges with the less-weathered parent rock material. Although the roots of some plants penetrate to the subsoil and the gradual weathering of these layers produces additional available minerals, it is the topsoil that is most advantageous for root development and the growth of various microorganisms. Productivity in terms of plant growth depends primarily upon the top layers. It is for this reason that erosion is so detrimental—it is the topsoil that is lost first. The development of a good soil takes hundreds of years, but the most productive layers may be lost in a day or so as a result of erosion.

The thickness, composition, and texture of the soil in these horizons varies considerably from one type of soil to another and depends upon the environmental conditions under which the soil developed. Climate, vegetation, and soil are intimately related and frequently form predictable patterns. This will be discussed later (see Section 17.7).

17.4 Soil Texture

The mineral particles of the soil vary greatly in size and quantity, with the larger fragments embedded in and coated over with colloidal materials of a rather gelatinous nature. The **texture** of a soil refers to the sizes (e.g., coarse, medium, or fine) of the particles that dominate, and soil particles are arbitrarily classified on the basis of their diameter as follows:

Coarse gravel	Over 5.0 mm
Fine gravel	5.0–2.0 mm
Coarse sand	2.0–0.2 mm
Fine sand	0.2–0.02 mm
Silt	0.02–0.002 mm
Clay	Below 0.002 mm

Soils are named on the basis of a mechanical analysis of the percentage of sand, silt, and clay that they contain. In **sandy soils** the silt and clay particles comprise less than 20 per cent of the material by weight, whereas a **clay soil** contains at least 30 per cent clay particles. The third basic soil group, the **loam** class, is the most important agriculturally and contains about 50 per cent sand and 50 per cent silt plus clay. For example, a typical loam soil would contain 45 per cent sand, 35 per cent silt, 15 per cent clay, and 5 per cent organic matter. Various kinds of sandy, loamy, and clayey soils occur in the field, and these groups grade into each other. This results in such terminology as loamy sands, sandy loams, silty clay loams, sandy clays, and silty clays. For our purpose, however, we will consider only the three broad categories first listed and discuss the importance of soil texture to plants. Remember, for our limited discussion, as far as size of particles is concerned, sand is the largest and clay the smallest.

Penetration of Air and Water into the Soil

The particles that form a soil are not solidly packed together; spaces or channels exist between the individual particles. The larger the particles, the larger the spaces between them, and through such channels penetrate air and water. Sandy soils have excellent penetration characteristics, whereas such movements in clayey soils would occur with difficulty because of the extremely tiny channels that exist (Figure 17-1[a], [b]). Rain falling upon a clay soil is mainly lost as runoff,

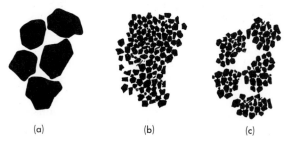

Figure 17-1. Soil particles indicating the variations in space between the particles of different sizes, and the development of a crumb structure. **(a)** Large particles with large spaces. **(b)** Small particles with small spaces. **(c)** Small particles grouped into large aggregates forming a crumb structure.

(a) (b) (c)

291

while the rapid penetration in a coarse (sandy) soil might result in water sinking too rapidly to depths below that occupied by roots. All too frequently, the pounding effect of raindrops packs clay particles so tightly together and so tightly into spaces between the larger particles that it is virtually impossible for water to penetrate a soil with a relatively high clay content. This emphasizes the importance of a loam soil with its generous mixture of sand and clay that results in rather good penetration characteristics without such excessive drainage as occurs in sands. Root growth and penetration are also much easier in sandy soils or loams than in clays because of the larger channels in the former two as compared with the clays.

Water-Holding Capacity

Water is adsorbed on the surfaces of soil particles as well as being held in small channels as a result of capillary action. Besides having a porous structure, most clays are of colloidal size, and this results in a tremendously large surface-to-volume ratio. One gram of clay probably has at least 55 square meters (about 500 sq ft) of surface, which is thousands of times greater than an equivalent weight of sand. Because of this enormous surface and capillarity, much more water is held in a clay soil, once it is penetrated, than in a sandy soil. Capillary water and a great deal of the adsorbed water are available to plants by absorption through root systems. Clays, then, act as a reservoir of water, greatly decreasing percolation (permeation) losses and retaining water, which is later available to plants. The water-holding capacity is lowest in sandy soils, while loams, of course, are in an intermediate position. Too much water may sometimes be retained in the soil. If a high concentration of clay is present, such a large quantity of water may be held by the soil particles as to saturate the tiny channels of the soil. If this condition persisted for any extended length of time, it would be injurious to plant growth, since it would cause a deficiency of oxygen in the root environment. This is similar to what happens when a soil is flooded. The solubility of oxygen in water is very low. Once the small amount of dissolved oxygen is used (for example, during respiration by root cells and microorganisms), anaerobic conditions result. Roots would then grow poorly, if at all, and eventually the tops would suffer.

Fertility

This term as applied to soils refers to the mineral content of the soil. Just as water is adsorbed to the surfaces of soil particles, so also are mineral ions (e.g., K^+, Mg^{++}, Ca^{++}, and so forth) held to these surfaces. In this case, clays also act as storehouses of mineral nutrients, greatly retarding the leaching of minerals out of the soil with the water that drains through. The finer the texture, the greater the mineral-holding capacity and thus the greater the general fertility of that soil. Permanent agriculture on sandy soils is possible only with constant application of fertilizers, usually in the form of chemicals, to replenish the supply of minerals. This is especially true in regions of high rainfall where sandy soils are usually severely leached and infertile. As mentioned in Section 9.3, the respiration of root cells produces carbon dioxide, which forms carbonic acid with water. The dissociation of this acid liberates hydrogen ions (H^+), which are able to displace various positive ions that are adsorbed on the surfaces of the clay particles. Such **ion exchange** (or **cation exchange**) releases ions into the soil water from which they may be absorbed by plant roots.

The mineral content of the parent rock material will influence the fertility of the soil that develops. Granite is largely silica, crumbles coarsely, and forms sandy soils that tend to be relatively infertile because of the small amounts of minerals originally present. Micas and microcline are high in mineral content and tend to develop into rather fertile soils. Limestone, on the other hand, contains rather large amounts of calcium and magnesium and frequently develops into rather alkaline soils that are not very fertile because iron, manganese, and zinc are less soluble (and thus less available to plants) in such soils. Alkaline soils may be improved by adding elemental sulfur or sulfates, which increase acidity (i.e., decrease alkalinity). The sulfur is oxidized to sulfates; hydrolysis of sulfates produces sulfuric acid. If a soil is too acid, ground limestone may be added to decrease the acidity. This may be necessary in some soils because the solubility of iron and manganese is so high in acid soils that toxic concentrations may result. As the acidity is decreased, solubility decreases, there is less soluble trace element, and toxicity decreases. In general mineral elements are most readily available in neutral to slightly acid soils (i.e., pH 6–7). It has also been found that bacteria are most active under neutral or slightly acid conditions; high acidity inhibits them.

As weathering occurs, minerals become available to plants. The plants absorb minerals, grow, die, and decay. Thus, minerals are recycled back to the soil. The organic matter (from the plants) increases the "holding capacity" of the soil and decreases leaching. If this cycle is broken by the removal of nutrients through the harvesting of crops, the soil may deteriorate rapidly. In humid regions considerable leaching occurs and also removes minerals. This is especially true of nitrates (NO_3^-), which are not adsorbed by the negatively charged soil particles. Since this is the form in which nitrogen is primarily absorbed by plants, such loss is very detrimental to plant growth. The practice of adding chemical fertilizers and organic matter to the soil results from such losses.

17.5 Soil Structure

The general characteristics of a soil are determined not only by the texture but also by the **structure,** which refers to the arrangement or grouping of soil particles into aggregates. When the aggregates are rather porous and relatively small, the soil is considered to have a **granular,** or **crumb, structure,** which is characteristic of many soils high in organic matter (humus) and is particularly important in land under cultivation. The soil granule actually consists of a rounded porous mass of mineral particles of varying sizes bound together by the colloidal clays and organic matter contained within them. In effect, small soil particles are lumped together to form larger ones with the result that a crumb soil has many of the advantageous characteristics of a light sandy soil plus those of a heavy clayey soil (Figure 17-1[c]).

Importance of Structure

The relatively low amount of inorganic colloidal materials in sandy soils results in the looseness, good aeration, good drainage, and easy tillage of such soils. However, they have a very low capacity to hold water and minerals, which in turn means low soil moisture and a general lack of fertility. Organic matter will act as a binding material, will greatly increase the water-holding and mineral-holding capacity, and will supply minerals as it decays. Heavy clayey soils, on the other hand, tend to "puddle," especially if cultivated when wet. That is, when such plastic and cohesive soils are cultivated the spaces between particles become greatly reduced. The result is that such puddled soils are very hard and dense when dry, and very sticky and viscous when wet. Although such soils have excellent water-holding and mineral-holding capacity, they are very difficult to work and may become almost impervious to air and water and even roots. As should now be clear, loam soils with their mixtures of sand, silt (intermediate in characteristics between sand and clay), and clay have the advantageous characteristics of both sandy and clayey soils and fewer of the disadvantageous characteristics. The development of a crumb structure enhances the advantages. There are large spaces between the crumbs (or aggregates of soil particles), which provide good penetrating characteristics while the individual small particles of the crumbs retain the excellent water-holding and mineral-holding capacity that is so important with regard to the growth of plants.

The most important factor in the encouragement of granulation is organic matter, especially humus. It not only binds but also expands, making possible the porosity that is characteristic of soil granules. Decaying plant roots are important sources of organic matter, and the maintenance of a soil in sod is an effective means of promoting granulation. The slimes and other viscous products of microorganisms and the filaments of fungi all serve to establish a crumb structure.

Humus

All organic matter in the soil is eventually decayed by the activity of various microorganisms, the bacteria, fungi, and actinomycetes. Cellulose, starch, sugars, proteins, and such compounds are decomposed quite readily, while lignin, oils, fats, resins, mucins, and others are decomposed with difficulty. As a result, the organic material is gradually converted to carbon dioxide, water, simple minerals (e.g., nitrates, sulfates, calcium compounds, phosphates), and a colloidal complex called **humus.** This last product is a result of enzymatic dissolution of the more resistant organic materials, which then tend to unite with proteins and other nitrogenous compounds.

The complex, amorphous, colloidal condition of humus produces a material whose adsorptive characteristics are far in excess of those exhibited by clay. All the advantageous aspects of clays are greatly magnified in this converted organic material. The high porosity and low plasticity of humus tend to alleviate the puddling tendency of clays when the two are mixed. A small amount of humus has physical effects on the soil in excess of its proportionate amount. The addition of organic material to the soil, then, has tremendous advantages: it increases granulation, water-holding capacity, and mineral-holding capacity, and it is a source of minerals.

Because organic material is so extremely important in a soil, plowing under of grass is much preferred to burning. A compost is commonly made of alternate layers of vegetable matter, soil, and manure. The pile should be kept moist, and the addition of minerals will aid in the formation of an effective fertilizer. Bacterial action rapidly decomposes the mass, but occasional turning aids decomposition by increasing aeration (and oxidations). After some months, compost may be spread over the surface or, preferably, disced into the ground. Manures, of course, also are excellent as soil additives. Organic increments are the major sources of nitrogen and frequently phosphorus, in addition to improving the physical characteristics of the soil.

Soil Organisms

The source of the organic material that is converted to humus is the great variety of living organisms found in the soil—their secretions and excretions while living, and their tissues when they die. In some instances portions of the organism, but not the entire individual, may die. The sloughing off of root cap cells, the sloughing off of bark, and the dropping of leaves, fruits, and twigs are good examples of this. Most of the organic matter consists of plants and their products, with a much smaller proportion contributed by animals.

In addition to the roots and underground stems of the vascular plants, many other kinds of plants live in the soil. The most important are the bacteria, followed in importance by the molds. Enormous numbers of bacteria may be present, as many as 5 billion per gram of soil, but they are so small (seldom exceeding 1 μm in diameter, or about 1/25,000 of an inch) that their presence is not apparent except through their activity. Further discussion of bacterial decay processes will be encountered in Chapter 20; mold decay activity is discussed in Chapter 21.

Many animals contribute to the humus of a soil when they die, but in addition to this many of them are more actively involved in the formation of soil structure. Ants, beetles, millipedes, sowbugs, slugs, and snails utilize much undecomposed plant tissue as food, and their feces contribute to humus; further decomposition is carried on by bacteria and molds. Burrowing animals (e.g., gophers, ground squirrels, prairie dogs, moles) may be undesirable as far as agricultural practices are concerned, but their burrows serve to aerate and drain the land, and they may transfer considerable quantities of soil. One of the most important animals in the soil is the earthworm. The number of earthworms varies with the amount of organic material available to them for food; some soils may have over 1 million per acre. Vast quantities of soil (as much as 14,000 kg or about 15 tons per acre) may pass through their bodies during the year, and such turning and passage greatly influence soil structure. The mucus secretions and the grinding action within the digestive tract of the earthworm result in the familiar "casts" that surround the earthworm holes. Such earthworm casts may weigh as much as 8,000 kg or about 8.8 tons per acre. These are examples of the granular condition that results from earthworm activity. The holes left in the soil serve to increase aeration and drainage.

Although any one group of animals is probably unimportant in organic transfers or humus formation, the total mass becomes highly significant even if considered only on the basis of contributing at death to the accumulation of decomposable tissue.

17.6 Soil Moisture

Soils are capable of supporting plant growth only if they contain water in addition to essential minerals. The source of this water is primarily rain and snow, but the amount of water that is actually available to plants is determined basically by the characteristics of the soil. Variations in the clay and organic matter content of the soil have considerable effects upon the water-holding capacity of the soil, as already mentioned. Water reaching the soil by precipitation may be considered in one of the following categories: runoff, gravitational, capillary, and hygroscopic.

Runoff Water

With heavy rains especially, the upper layers of the soil may very rapidly become saturated with water. Frequently the pounding effects of the raindrops tend to compact the soil and clog pores with the finer mineral particles. As a result of these effects, a considerable amount of the subsequent rainfall flows away over the surface of the soil as **runoff water,** which eventually enters streams, rivers, lakes, or oceans. Such water may carry rather large amounts of suspended soil. The steeper the slope down which water is flowing, the more soil is washed away. Vast damage has been done in various parts of the world by such water erosion and the loss of soil particles and soil solutes (minerals) that results. Runoff water does not penetrate the soil and is of very little value in the growth of plants. It is exceedingly important, however, in its harmful aspects as a cause of erosion. Such erosion is a prominent problem on more than half of the croplands in the United States, and similar serious erosion problems are found in many other countries. Historians believe that the fall of ancient empires, such as that of Rome, was accelerated by the deleterious effect on agriculture of the washing away of fertile surface soils.

The presence of vegetation, litter, mulch, or similar material protects the soil surface against the beating of raindrops. The water is scattered as fine spray and infiltrates the soil more readily than in soils not so protected. Plowing and cultivating along the contours interrupt and slow water flow, and allow greater penetration of moisture into the soil. Steep slopes present such an erosional hazard that they are best left in permanent vegetation without cultivation or grazing. Even gentle slopes present an erosional hazard if overgrazing results in destruction of the vegetative cover so that the soil is no longer held in place by plant roots.

Gravitational Water

Not all of the water that penetrates the soil is necessarily available to plants. **Gravitational water** is that which moves downward through a moist soil in response to gravity until it flows away in underground streams or reaches the **water table,** the permanently saturated zone that is present in deeper layers of many soils. The water table may lie within a few feet of the surface or as deep as 600 m (about 2000 ft). Water that percolates through a soil is available only as it passes by plant roots; it usually is below the reach of roots within a few days, unless a series of rains follow one another in rapid succession. The rate at which gravitational water moves depends upon the physical characteristics of the soil; it is relatively fast in sands with their large particles and channels, while it is slow in clays where the small pores and tremendous surfaces hold water rather tightly.

A poorly drained soil may be very unsatisfactory for the growth of plants because gravitational water occupies the larger pore spaces of the soil and excludes air. Roots, as well as various soil organisms, do very poorly under these conditions. Plants vary somewhat in their ability to withstand a reduced oxygen supply to their roots. Some will be injured within a few hours, while others can survive for days. However, except for certain plants that have special adaptations (see Section 14.2), death will usually occur if the soil remains saturated with water. Even with well-drained soils, gravitational water frequently has markedly detrimental effects. Soil nutrients, particularly minerals, dissolve in the water and are **leached** out of the soil. Nitrates, calcium ions, sulfates, and potassium ions are lost to a considerable extent by leaching. The presence of plants reduces such losses because they absorb ions and they introduce organic matter that adsorbs (holds) ions. These absorbed ions are recycled or returned to the soil upon death and decay of the plants. Of course, if plant material is harvested and removed, the ions that have been absorbed will not be returned to the soil. Nitrate presents a special problem because this ion is negatively charged and thus not held to soil particles. Nitrogen is an element that is required in relatively large amounts and it is not a component of parent rock material. The leaching from soils of large amounts of nitrogen in the form of nitrates, therefore, frequently results in a deficiency of nitrogen and symptomatic abnormalities in plant growth.

Capillary Water

As gravitational water drains out of a soil, it leaves behind much moisture in the form of fluid water held to soil surfaces and filling the smaller channels of the soil;

such moisture is termed **capillary water.** This is the principal source of moisture for plants. The attractive forces are not very powerful, and plant roots readily absorb such water. The **soil solution** basically consists of the water and dissolved salts that remain in the soil after free drainage has taken place.

Field capacity is defined as the percentage of water held in a soil after drainage has stopped. It is really a measure of the maximum storage capacity of the soil. The actual amount of water retained depends upon the soil type and is much greater in clayey soils and those containing considerable organic matter than in sandy soils because of the much higher adsorptive capacity of the colloidal clays and humus particles. Typically, at field capacity, a clay soil will have a water content of 30 per cent and a sand soil will contain about 8 per cent. Knowledge concerning the field capacity of a soil can be put to good use in calculating the amount of irrigation water needed to wet that soil to the depth of root penetration. Any water added in excess of the field capacity will simply be lost as drainage or runoff. Water should be added gently and slowly to allow sufficient time for penetration and to hold runoff to a minimum.

Hygroscopic Water

Some of the water is held so tenaciously within soil particles and as thin films on their surfaces, especially with regard to colloidal particles of clay and humus, that it cannot be absorbed by plants; such moisture is termed **hygroscopic water.** The forces of attraction are so great that most of this water is in a nonliquid state and is not biologically or chemically active. It exists in equilibrium with the water vapor in the atmosphere above the soil.

Available and Nonavailable Water

The types of soil water just discussed are not sharply separated one from another, but from a continuous series from water that is not retained by the soil to water that is retained with considerable force or tension. Also, it should be borne in mind that the actual amount of each type of water that is present is determined by the kind of soil. The proportion of organic matter, clay, silt, and sand markedly influences how much water will be retained against the force of gravity and against the forces of root absorption.

During a rainfall and shortly thereafter, some gravitational water is available to plants, but this is usually of rather short duration. Capillary water, though, is readily available for longer periods and the first increments are very easily absorbed by plants. As such absorption takes place and there is less and less water in the soil, however, it becomes increasingly difficult for water to be removed simply because the water layers closer to the surfaces of soil particles are more strongly held; capillary water merges with hygroscopic water. The forces of attraction between soil surfaces and water molecules vary inversely as the square of the distance between them. Therefore, the water molecules close to the surface of a soil particle are held with immensely greater tensions than those found in the center of a channel between particles. Subsequent increments of water become harder and harder to absorb until the plants cannot obtain sufficient water and they wilt. Because of the very steep increases in retention forces with small decreases in water content, plants grown in a particular type of soil tend to wilt at about the same time as soil moisture content decreases.

The **permanent wilting point** is the percentage of water remaining in the soil when a plant wilts and will not recover unless water is added to the soil. This wilting occurs before all the capillary water has been absorbed. In other words, some of the capillary water is held too strongly for plants to absorb it. Again, the actual amount of water retained by a soil at the permanent wilting point will vary with the type of soil; it tends to be low in sandy soils and high in clayey soils.

Available water, then, is primarily capillary water. The amount of available water in a soil is greatest in clayey soils and least in sandy soils; organic matter is even more effective than clay in retaining available water. If sand and clay are both saturated with water and complete drainage occurs, the amount of available water would be on the order of 4 per cent by dry weight for sand and about 15 per cent for clay.

All the water that is theoretically available to plants is not necessarily absorbed by plants. Much evaporation occurs from soil surfaces, and this escaping water vapor is lost as far as the plant is concerned. Although considerable amounts of water may be lost in this fashion, there is very little further loss once the surface layers have dried because little capillary movement of water from the deeper layers takes place. Moisture in the upper layers may be conserved by spreading mulches of grass, leaves, manure, paper, and other materials over the soil surface. Shading lowers soil temperature and decreases evaporation. Actually, much more water is lost from soils by transpiration from plants than by direct evaporation from the soil surface because plant roots tap water supplies in the lower regions of the soil where evaporational losses would be unimportant. The eradication of weeds from a field of crop plants is done primarily to make more moisture available to the desired plants. Weeds do utilize minerals, which are then not available to other plants, but this is usually a minor factor. If weeds are allowed to get too large, a shading effect may be detrimental to other plants of course.

One of the most important factors involved in water availability is root growth. During periods of high transpiration rates, water may be absorbed and removed so rapidly from the soil that the absorbing areas of the root become surrounded by relatively dry soil from which more water cannot be absorbed. There is little capillary movement of water in the soil. Without the addition of water (rain or irrigation), the plant is dependent upon the growth of roots to new areas where moisture is available. (See Section 9.1 for discussion of root growth.) This has given rise to the idea that roots grow in search of moisture, but this is not the case. As roots grow, including the many laterals, the ones that grow best are the ones that project into areas with favorable moisture and aeration. They may extend somewhat into relatively dry soil, but roots will not penetrate even a few inches of soil at the permanent wilting point.

17.7 Important Soil Groups

The natural vegetation of an area is influenced by the climate, particularly humidity and temperature, and so is the soil development of that region. The climate, soil, and plant communities are so intimately related that they frequently form predictable patterns, especially because the plants that are present influence soil formation. In the following discussion we will utilize the new comprehensive soil classification system developed by the staff of the United States Department of Agriculture, although reference will be made to the older terminology, which has been in use for quite a few years and may be more familiar.

Spodosols, previously termed *podzols,* are mineral soils that develop in cold, temperate, humid regions where the native vegetation is predominantly forests; as in the northeastern United States and Southeastern Canada. There is much leaching, the lower part of the A horizon becoming bleached gray in appearance and accumulation occurring in the B horizon. The soils are acid (pH 3 to 5), and rather low in mineral content (i.e., low fertility) as a result of the rather high rainfall (more than 90 cm, or 35 in., per year) and leaching. When properly fertilized, they can become quite productive as in Maine where potatoes are an important crop. Decomposition of organic matter is rather slow, and the A horizon contains a substantial amount of humus in the upper regions.

Mollisols, which include the *chernozems* of the older terminology, develop under prairie vegetation and the grassland soils of the central part of the United States are in this group. The moderate rainfall of 38–76 cm (15–30 in.) per year does not cause much leaching, and these soils are quite fertile. The extensive fibrous root systems of the grasses are responsible for the high humus content, about 10 per cent, and the black color of these soils. The surface horizons generally have a granular structure; and the mildly acid to neutral condition (pH 6.5–7.5) is favorable for the solubility of most essential soil minerals. Because of their fertility, high adsorptive capacity, and ease of cultivation, mollisols are rated among the world's most productive soils although productivity may be curtailed in some areas by low rainfall.

Aridisols are found in very dry climates; they are the soils of deserts and semideserts. Vegetation is sparse and the high temperatures bring about rapid decomposition of what little organic matter is produced. There is very little humus, but salts accumulate as a result of the scanty rainfall (and little leaching) and the soils are quite alkaline. Where irrigation water is available, aridisols can be made quite productive because of the high mineral content. This has been accomplished in parts of the semiarid regions of the western United States. Such high productivity, unfortunately, is not usually retained. Irrigation waters contain salts that add to those in the soil. The high rates of evaporation and transpiration remove much of the water, further concentrating the salts. Cultivation of crop plants becomes more and more difficult and may eventually not be possible at all.

Oxisols, or *latosols,* are highly weathered because of the abundant rainfall and high temperatures of the tropical rain forest areas where these soils develop. Hydrolyses and oxidations are intense and the resultant hydrous oxides of iron give the soil a distinct reddish color. Organic matter decomposes so rapidly under this climatic regime that there is very little humus in the soil. Fertility is poor because of the extensive leaching; and heavy fertilization is required if crop plants are to be grown, especially as a result of the curtailment of recycling of minerals by harvesting plants—less litter and debris remain to decompose and release minerals.

There are other soil groups, of course, but the intent of the discussion was to foster an awareness of relationships among climate, soils, and plants, not to present details of all groups. Further details are available in various soils and ecology texts.

Summary

1. The soils in which plants grow consist of mixtures of mineral materials, organic matter, water, and air, in varying proportions. The small fragments of mineral materials are derived from solid rock by long periods of mechanical and chemical weathering. The organic matter consists of living organisms, their secretions and excretions, and their decay products. Mixing of the various soil components is a result of the activity of wind, water, glaciers, gravity, and the vertical shiftings of the earth's crust. The uppermost layer with its mixture of organic matter and mineral particles is called topsoil.

2. The texture of a soil refers to the sizes of the particles that predominate, and soils are classified on the basis of an arbitrary scale of particle sizes. Since particle sizes vary gradually from one extreme to another, no sharp line of distinction can be drawn; thus a great variety of soil types exists. However, the texture of a soil determines the qualities of air and water penetration, water-holding capacity, and fertility (mineral-holding capacity). In general, penetration of air, water, and roots occurs much more readily through soils in which large particles (sand) dominate. On the other hand, water-holding capacity and fertility characteristics are mainly a result of the small particles (silt and clay) and the organic matter.

3. Soil structure refers to the arrangement or grouping of soil particles into aggregates as a result of the binding characteristics of colloidal clays and organic matter. A soil with a granular or crumb structure has many of the advantages of both sandy and clayey soils.

4. When organic matter is added to a soil, it is gradually converted by microorganisms to carbon dioxide, water, minerals, and a colloidal complex called humus. A small amount of humus has advantageous physical effects on the soil far in excess of its proportionate amount.

5. Basically, a soil that is conducive to the good growth of plants consists of a mixture of sand, silt, clay, and organic matter that exhibits the advantages, and not the disadvantages, of each portion.

6. Water that reaches the soil by precipitation is classified as runoff, gravitational, capillary, or hygroscopic. The primary source of water for plants is capillary water; hygroscopic water is unavailable. The actual amount of water in a soil, available or not available, depends upon the type of soil.

7. There are various soil groups, and of these: the mollisols are the most productive; the aridisols are desert soils; and the oxisols are the extremely leached soils of the moist tropics.

Review Topics and Questions

1. Explain why soil *B* is more productive than soils *A* and *C*, even though they all receive the same amount of rainfall, light, and temperature.

Particles	Soil *A*	Soil *B*	Soil *C*
Sand	80%	50%	5%
Clay	5%	20%	90%
Organic matter	1%	4%	4%

2. Briefly discuss the development of a loam soil.
3. Soil is composed of mineral particles (ranging in size from coarse sand to colloidal) and organic matter in process of decomposition. Discuss how variation in amounts of these constitutents affects each of the following: (a) fertility, (b) water-holding capacity, and (c) aeration.
4. What are the advantages of using compost or manures rather than chemical fertilizers? What are the advantages of using the latter rather than the former?
5. List the agencies that may be involved in transporting soils from place to place.
6. List five animals that contribute to soil formation and indicate how each functions.
7. A clay soil tends to have more available as well as more nonavailable water than a sand. Explain.
8. What is meant by field capacity? By permanent wilting point?

Suggested Readings

BRADY, N. C., *The Nature and Properties of Soils,* 8th ed. (New York, Macmillan, Inc., 1974).

DAUBENMIRE, R. F., *Plants and Environment,* 3rd ed. (New York, Wiley, 1974).

KELLOG, C. E., "Soil," *Scientific American* **183,** 30–39 (July 1950).

18 Inheritance and Variation

The basic vegetative structures and nutritional aspects of the flowering plant have been discussed in previous chapters. Now, before discussing reproduction of any plant group, we will attempt to arrive at an understanding of inheritance and the importance of sexual reproduction in general. Then we will delve into the plant kingdom, covering reproduction in the various groups as we encounter them. This seems to be an advisable means of approaching the plant groups, since reproductive structures form the basis for all modern systems of classification and can also be used to demonstrate evolutionary advances.

18.1 Meiosis

During sexual reproduction, when **gametes** (sex cells, which are egg and sperm in flowering plants) unite, two sets of chromosomes are combined, thus doubling the chromosome number of the fertilized egg over that of the unfertilized egg. It should be noted that the gametes fuse but not the chromosomes; the two sets of chromosomes are located within the newly constituted nucleus of the fertilized egg. In the higher forms of life, the multicellular organism develops as a result of many mitotic divisions starting with the fertilized egg; the resultant cells have the same number of chromosomes as the fertilized egg. (Review Section 4.4 if necessary.) Without any mechanism for reducing the chromosome number, a doubling would occur each generation. This does not happen; the number of chromosomes is usually constant for each kind or species of plant or animal: corn, 20; pine tree, 24; housefly, 12; fruit fly, 8; man, 46. Every corn plant has 20 chromosomes, no matter how many generations are examined. Therefore, at some time after fertilization and before the next fertilization, or once each generation, a reduction in chromosome number is accomplished. The rather interesting type of nuclear division that accomplishes this reduction is **meiosis.**

Note that each example listed in the foregoing paragraph had an even number of chromosomes. This is not a coincidence. Chromosomes occur in pairs in most organisms, and the members of a pair are **homologous chromosomes,** or **homologs.**

Such homologous chromosomes are both alike *and* different. As indicated in Figure 18-1, they are alike in size, shape, and in the characteristics that are controlled, but they may be unlike in the way in which these characteristics are controlled. For example, leaves may be broad or narrow; the characteristic is leaf shape, which is governed by a **gene** at a specific location on the chromosome. In this case, the basic characteristic of leaf shape is determined by two alternate (or contrasting) forms of a gene. Such alternate forms of a gene, occupying the same locus (position) on homologous chromosomes, are termed **alleles.** One member of a pair of homologous chromosomes comes from the female parent (egg), and the other homolog comes from the male parent (sperm). The organism thus is **diploid** (has two sets of chromosomes). Some organisms (such as many algae and fungi) have only one set and are **haploid.** All organisms that reproduce sexually have both haploid and diploid phases, but the emphasis (i.e., the majority of cells of an organism) is usually haploid *or* diploid. A haploid cell has N chromosomes and a diploid cell has $2N$ chromosomes; the N refers to one set of chromosomes (for example, in corn plants $N = 10$ and $2N = 20$; in a pine tree $N = 12$ and $2N = 24$). The dominant form in the more complex plants and animals is diploid: trees, man, ferns, fish, flowering plants, cats, dogs, and so forth. In these organisms the haploid condition occurs in only a few cells or structures.

Meiosis (Figure 18-2) is similar to mitosis (Section 4.4) but differs in some important respects, including the fact that the nucleus divides twice in meiosis while the chromosomes have doubled only once. Duplication of chromosomes occurs during interphase as in mitosis. Much of the first prophase occurs as during mitosis (shortening and thickening of chromosomes, appearance of the spindle, disappearance of the nuclear envelope), but the homologous chromosomes pair in meiosis. This pairing, or **synapsis,** which takes place much like the closing of a zipper, is quite precise along the entire length of the chromosomes. This is an intimate association of homologous chromosomes, but they do not fuse. If part of a chromosome is missing (as occasionally happens), the homolog forms a loop (Figure 18-3[a]); if one part is inverted, the homolog forms a different kind of loop (Figure 18-3[b]). This indicates that the genes must be strongly attracted to one another. The chromosomes have doubled and four chromatids are actually present during synapsis (Figure 18-2[b]).

Two successive nuclear divisions occur in meiosis and Roman numerals are used to distinguish between them; for example, prophase I as compared with prophase II. The chromosomes migrate to the central region of the cell (metaphase I),

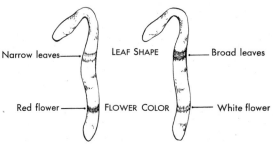

Narrow leaves — LEAF SHAPE — Broad leaves

Red flower — FLOWER COLOR — White flower

Figure 18-1. Homologous chromosomes. They are alike in size, shape, and in the characteristics that are controlled. They may be unlike, as in the diagram, in the way in which these characteristics are controlled. The shading on the chromosomes presents areas where alleles are located; alleles (or genes) cannot be seen.

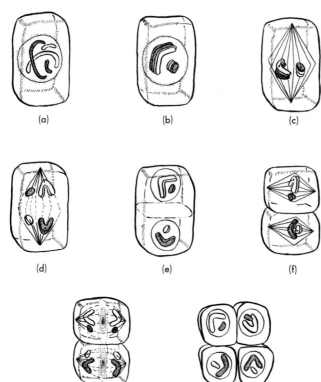

Figure 18-2. Meiosis. **(a)** The diploid condition of a cell (meiocyte); the chromosomes will duplicate before meiosis begins. **(b)** Prophase I; synapsis has occurred and tetrads of chromatids are present. **(c)** Metaphase I. **(d)** Anaphase I; the homologous chromosomes are separating, but not the chromatids. **(e)** Telophase I. **(f)** Metaphase II. **(g)** Anaphase II; chromatids are separating. **(h)** Haploid spores (meiospores) resulting from meiosis.

and then the *homologous chromosomes separate* (anaphase I). The kinetochores (centromeres) have not yet duplicated, and the chromatids do *not* separate. This is very different from mitosis, where the homologous chromosomes do not pair, and where the two groups of chromosomes at anaphase are the product of the separation of the chromatids. In meiosis, during anaphase I, because *homologous chro-*

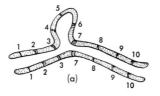
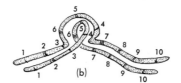

Figure 18-3. Loop formation during synapsis. Double chromatids are not shown. **(a)** The type when part of one chromosome is missing. **(b)** The type when part of one chromosome is inverted.

mosomes separate, the two groups of chromosomes moving to opposite poles of the cell are not similar (Figure 18-2[d]). For example, if the original diploid condition consists of two long chromosomes and two short chromosomes, the groups moving toward the poles each have *one* of the long chromosomes and *one* of the short chromosomes, but never both long or both short. It is strictly a matter of chance as to which long and which short move together, as will be clarified shortly. Diagrams sometimes appear a bit confusing because each chromosome has been duplicated prior to meiosis and now appears as two chromatids. One must keep this in mind and also remember that these chromatids have not separated in anaphase I of meiosis. In mitosis, of course, the chromatids separate at anaphase. The student should compare Figures 18-2 and 4-24.

Telophase I of meiosis is generally similar to telophase in mitosis, although the second nuclear division may occur with only a short interphase or no interphase. In any event there is no DNA synthesis, and thus no chromosome duplication, during this interphase of meiosis. In some cases cytokinesis may not take place until both nuclear divisions have occurred. Remember, homologous chromosomes do not occur together at telophase I; they are at opposite poles of the cell.

Prophase II does not vary significantly from other prophases. During metaphase II (Figure 18-2[f]) the chromosomes usually line up at right angles to that of the first division. The kinetochores have now duplicated, and the *chromatids separate* during anaphase II. The second division of meiosis results in four haploid cells, called **meiospores** (*N*) in plants (Figure 18-2[h]). These spores are quantitatively and qualitatively *different* from the original cell. They have half the number of chromosomes of the original cell, and the spores are not all genetically alike. The diploid cell that undergoes meiosis is more descriptively termed the **meiocyte** (*2N*).

Figure 18-4 represents a more diagrammatic view of meiosis to emphasize the way in which two pairs of chromosomes may line up at metaphase I. The letters on the chromosomes represent genes; the alleles **A** and **a** govern the same characteristic but in contrasting ways (e.g., broad versus narrow leaves), and the same is true of alleles **B** and **b** (red versus white flower color). The alignment of chromosomes is determined purely by chance and may be as at the left in the diagram or as at the right. The haploid meiospores, then, will have a genetic component that is also determined by this chance arrangement. The meiospores may contain large **A**s and **B**s, or small **a**s and **b**s, as diagrammed at the bottom left. The other arrangement results in meiospores that contain large **A** and small **b**, or small **a** and large **B**. Since quite a few cells are undergoing meiosis at the same time, the four possible kinds of meiospores (**AB, ab, Ab,** and **aB**) occur at about equal frequency —like the results of flipping a coin. If three pairs of chromosomes are present, they can line up in four different ways (Figure 18-5), and eight kinds of meiospores can be produced (**ABC, abc, ABc, abC, aBC, Abc, aBc,** and **AbC**). With each additional pair of chromosomes, the number of ways of lining up is doubled. Thus, the number of possible haploid cells is 2^n, where n = the number of pairs of homologous chromosomes. Actually, n = the number of *heterozygous* pairs, but this will be discussed later (Section 18.8). Human cells have 46 chromosomes, or 23 pairs, resulting in 2^{23}, or approximately 10,000,000, possible kinds of haploid cells (2 multiplied by itself 23 times). Remember that even with all the different ways of

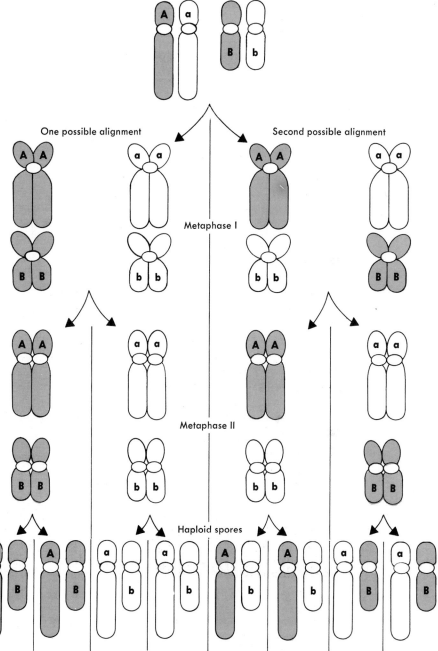

Figure 18-4. The possible alignments of two homologous pairs of chromosomes during meiosis, and the haploid conditions that result. Only chromosomes are shown; no other structures are represented. In this situation there are two possible alignments at metaphase I – the left and right arrangements in the diagram. If the alignment is as at the left, the haploid condition shown in the two lower-left bracketed diagrams is the result. The haploid condition shown in the two groups at the lower right results if the alignment happens to be as indicated at the right in metaphase I. The four possible kinds of haploid cells are produced in approximately equal numbers. Note that the kinetochores have not duplicated by metaphase I; thus, the chromatids do not separate in anaphase I. By metaphase II, the kinetochores have duplicated and the chromatids separate in anaphase II.

(1)	A A a a	A A a a	a a A A	a a A A
	B B b b	B B b b	B B b b	B B b b
	C C c c	c c C C	C C c c	c c C C

(2)	A B C	A B c	a B C	a B c
	and	and	and	and
	a b c	a b C	A b c	A b C

Figure 18-5. Possible alignment of three pairs of chromosomes and the haploid conditions that result. In **(1)** the letters represent the chromatids at metaphase I; in **(2)** the letters represent chromosomes after telophase II. There are eight possible kinds of haploid combinations.

lining up, each haploid cell has a complete set of chromosomes containing one of every kind of gene (one of each letter).

Eventually in plants the meiospores develop into structures that then produce sex cells (gametes—egg or sperm), which are haploid. The manner in which this is accomplished will be mentioned as we discuss the various plant groups individually. On the other hand, in humans and other animals meiosis results directly in the production of egg or sperm, and there is no spore stage comparable to that in plants.

18.2 Significance of Meiosis

During this type of division, the nucleus divides twice, while the chromosomes have doubled only once. In the first division the homologous pairs of chromosomes separate. The chromatids separate in the second division. The products of meiosis are genetically quite different from the original cell. When the many different kinds of haploid gametes recombine during fertilization, the resultant diploid cells may vary greatly from each other and from the originals. This is one of the most important features of sexual reproduction—the production of variants. These variations depend upon random alignments and segregations of chromosomes during meiosis.

18.3 Sexual and Asexual Reproduction

Sexual reproduction involves both meiosis and fusion of gametes. As a result of meiosis, and the resultant distribution of chromosomes (and genes), the gametes and thus the offspring are highly variable. This is very desirable as a survival factor. Some of the offspring may be better suited to the environment than are the parental types, the environment may change and then be more favorable for the offspring, or the offspring may be more capable of existing in a new environment to which they may have migrated. Our whole concept of evolutionary change has, as part of its foundation, the occurrence of variations as a result of sexual reproduction. Sexual reproduction thus presents a selective advantage in that the variability that results in natural populations enables them to adjust to changing environments or to invade new environments—the best adapted of the variants are the best competitors and tend to survive.

Asexual reproduction bypasses both meiosis and fusion of gametes. The new plant is formed directly from a fragment or an extension of the old plant. The rooting of stem cuttings, the production of new plants by Bermuda grass (*Cynodon dactylon*) and strawberry (*Fragaria virginiana*) from runners, and new plants developing from bulbs and corms—all are examples of asexual (nonsexual) reproduction. The resultant plants are genetically like the original, and no variations are possible, except for the possibility of unpredictable changes in the genes (see discussion of mutations, Section 28.11). In the long run—in terms of evolutionary ages—asexual reproduction alone would be detrimental to most plants under natural conditions, because the biggest source of variations is eliminated. Often, of course, desirable varieties of cultivated plants are propagated vegetatively (asexually) to maintain uniformity (e.g., citrus, avocado, roses, and so forth), but this is an artificial situation.

18.4 Genetics

Prior to 1900 nothing was known about chromosomes and genes. Really scientific, carefully devised experiments were not carried out, and this led to the general impression that inheritance worked in a "blending" fashion—that is, the traits of the parents are blended together in the offspring, like mixing colors of paint. Casual observation would indicate that this is true. However, the work of Gregor Mendel, published in 1866, was to blaze a path away from this idea of blending and toward the concept of "particulate inheritance." Interestingly enough, this work was not really understood and was almost completely ignored until about 1900, when the study of chromosomes and genes presented a mechanism that clarified Mendel's ideas. Thus, although Mendel is conceded to have established the foundations of **genetics,** this study of inheritance and variation can be considered to be a new science, because most of the enormous amount of investigations included in this term have been carried on since only about 1900.

Mendel's Experiments

Gregor Mendel (1822–1884) was an Augustinian monk who had actually trained to be a biology teacher. He was interested in growing plants and seeing how hereditary characteristics were transmitted. Others had had this interest also, but Mendel made several fantastically shrewd decisions. First, he thought of studying the inheritance of one character at a time—one variable. He ignored all characteristics except the one in which he was interested. This had not been done before. Mendel kept accurate records of how many times the characteristic appeared in the offspring of selected parents, and he was careful to maintain pollination under control so that all pollen came from one selected parent. Also he had allowed plants to self-pollinate and observed that each produced offspring of its own kind. For example, in the garden pea plant with which he worked, Mendel found that one strain grew to a height of 6 to 7 ft, while the plants of another strain grew only to 1 or 2 ft in height.

Mendel crossed a tall plant, which was a descendant of a long line of tall plants, with a short plant, which was a descendant of a long line of short plants. This was easily accomplished by transferring the pollen, which produces the male gamete (sperm), from the flower of one plant to the pistil of the flower of the other plant; the female gamete (egg) is produced in the pistil. The egg and sperm from the different plants unite during fertilization, to produce the **zygote,** or fertilized egg. The offspring, therefore, will receive genes from both parents by way of the gametes. If the blending idea of inheritance was accurate, the progeny of this mating would all be intermediate in height. In the next section we shall discuss in detail the results that Mendel obtained.

18.5 Monohybrid Cross

When the mating, or cross, is made between parents differing in a single character, it is termed a **monohybrid cross.** One such cross is that under discussion, as originally performed by Mendel. The first-generation offspring, commonly termed the F_1 or *First Filial Generation* (*filial:* pertaining to son or daughter), were *all* tall plants. No intermediate or short plants were produced. Mendel continued by self-fertilizing or self-pollinating the F_1 (most plants have bisexual flowers) and discovered that the F_2 (second generation) contained both tall and short plants but again no plants of intermediate height.

From his results Mendel deduced that both the tall and the short trait must have been present in the F_1, since the F_2 consisted of both types. He also counted the large number of offspring and realized that he was getting approximately three fourths tall plants and one fourth short plants in the F_2 (Figure 18-6). Mendel suggested that these observations could be explained by making two simple assumptions. First, he assumed that each trait (e.g., plant height) is determined by a "factor" and that these factors occur in pairs (e.g., tall versus short). Second, when gametes are formed, the two factors are separated from one another. These conclusions were based upon observations of the results of his experiments plus sound reasoning. Remember that he knew nothing of meiosis, genes, or chromosomes—no one did at that time. We now call these factors **genes,** and they are usually symbolized by the use of letters. **T** can represent tallness, while the contrasting but related characteristic of shortness can be indicated by **t.**

Figure 18-6.

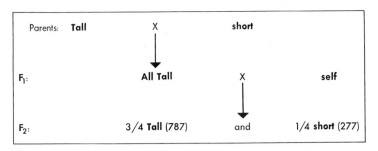

Because the original tall plants always gave rise to tall plants when self-pollin-ated, they must have contained genes **TT**. Remember that the "factors" (genes) occur in pairs. Similar reasoning indicates that the short plants should be desig-nated as **tt**. We now know that these paired genes are on homologous chromo-somes which separate during meiosis (Section 18.1), and each haploid gamete con-tains only one of the genes (Figure 18-7). The F_1 tall plants must have both tall and short factors, or **Tt**, and the F_1 gametes then would be of two kinds (**T** and **t**) be-cause the homologous chromosomes would again separate during meiosis. Since only the tall characteristic shows up when both tall and short factors are present, tallness is said to be **dominant** over short, and the gene for shortness is **recessive.** Only the dominant characteristic is expressed in the appearance of the organism when both alleles are present.

When the F_1 plants are self-fertilized, the eggs and sperm combine at *random,* like flipping two coins at the same time and counting how many times the various combinations appear—two heads, two tails, or one head plus one tail. Each coin will come down heads one half of the time and tails one half of the time; the proba-bility is that for each coin heads (or tails) will show up one chance out of every two. The larger the total number of tosses, the closer one gets to the theoretical 1:1 ratio. If a coin were tossed only four times, three heads and one tail or even four of one kind would not be too surprising. However, for 1000 tosses to result in 900 heads and 100 tails, or even 700 to 300, is extremely unlikely. More probably, the 1000 tosses would come close to a 500:500 ratio; one actual count resulted in a 497:503 ratio. This is, of course, one reason that it was important for Mendel to count large numbers of progeny.

In the foregoing paragraph only one coin at a time was considered. Suppose the two coins are tossed together? What is the probability of *both* landing heads? When two events take place, each having a similar possible result, and if one does not influence the other, the two separate probabilities are multiplied together to find the probability that both will happen simultaneously. In other words, each coin will land heads one half of the times; they will both land heads $\frac{1}{2} \times \frac{1}{2}$, or one fourth of the times. Also, they will both turn up tails one fourth of the times, and one tail plus one head will occur two fourths (or one half) of the times. To verify this figure of one half, toss a penny and a nickel so that one coin can be distin-guished from the other. One fourth of the times, the penny will be heads and the nickel will be tails; another fourth of the times, the penny will be tails and the nickel will be heads ($\frac{1}{2} \times \frac{1}{2} = \frac{1}{4}$). In each of these cases there is one head and one tail: $\frac{1}{4} + \frac{1}{4} = \frac{2}{4}$, or $\frac{1}{2}$.

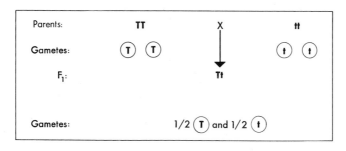

Figure 18-7.

Figure 18-8.

Now let us return to the F_1 tall plants that Mendel caused to be self-fertilized. One half of the sperm contain **T** and one half contain **t**, one half of the eggs contain **T** and one half contain **t**, and the fusion of egg and sperm occurs at random. *Either* one of the two kinds of sperm may fuse with *either* one of the two kinds of eggs (Figure 18-8). As was indicated by the appearance of the F_1 progeny, tallness is dominant to shortness, and any plant containing at least one **T** would be tall. The F_2 zygotes in the above would give rise to plants three fourths of which would be tall ($\frac{1}{4}$ **TT** plus $\frac{2}{4}$ **Tt**) and one fourth of which would be short. This is exactly what Mendel observed. His actual count in one experiment was 787 tall and 277 short, whereas 798:266 would be an exact $\frac{3}{4}:\frac{1}{4}$ (or 3:1) ratio. However, a 3:1 ratio fits these and many other experimental results more accurately than does any other simple ratio. Also, the more offspring, the closer the figures are to the theoretical ratio.

We can now diagram the complete series of crosses, as shown in Figure 18-9. All the tall plants in the F_2 looked alike, but Mendel found that one third of them (**TT**) produced only tall progeny when self-fertilized, while two thirds of them (**Tt**) produced three fourths tall plants and one fourth short plants (just as had the F_1). The short plants (**tt**) produced only short plants when self-fertilized.

Frequent use is made of the Punnett square, or checkerboard, in diagramming crosses between individuals. In this method the kinds of eggs are written across the top line and the kinds of sperm down the first column. The squares are then filled in by carrying the letter symbols down each column and across each row. The two letters in each square thus represent the possible kinds of progeny (F_2) that may result from self-pollinating the F_1 (Figure 18-10).

In order to understand fully the inheritance of characteristics, one must be com-

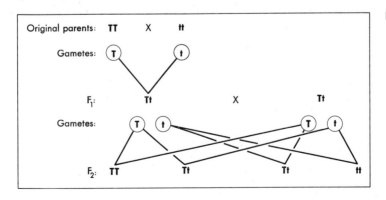

Figure 18-9.

SPERM \ EGG	T	t
T	TT	Tt
t	Tt	tt

Figure 18-10.

pletely familiar with meiosis. The following diagrams represent the diploid condition and the important phases of meiosis that result in the formation of haploid structures in the aforementioned monohybrid cross. Remember that Mendel started with pure strains. The plants were **true breeders;** tall plants always gave rise to tall plants, and short plants always gave rise to short plants. Therefore, one parent plant had **T** on both homologous chromosomes, and the other parent plant had **t** on both homologous chromosomes (Figure 18-11). The figures are diagrammatic, and no attempt is made to show details (e.g., kinetochores, various phases) of meiosis. The student may wish to refer to Figures 18-2 and 18-4.

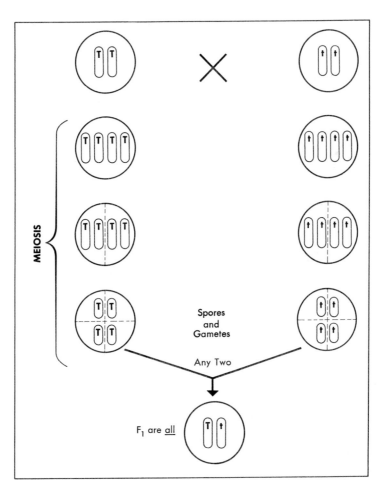

Figure 18-11.

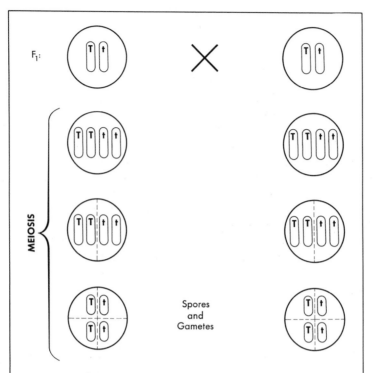

Figure 18-12.

In plants, meiosis results in the production of haploid spores that produce structures that eventually give rise to gametes by mitotic divisions. The gametes and the spores from which they arose are thus genetically alike.

The F_1 plants are then self-pollinated or self-fertilized. Actually, since all of the F_1 plants are alike in this case, they could be cross-pollinated. The F_2 should be the same in either case. Many spores and therefore gametes are produced by each plant, but one half are **T** and one half are **t**. This ratio follows inevitably as a result of meiosis (Figure 18-12). As mentioned previously, either one of the two kinds of sperm may fuse with either of the two kinds of eggs, and this fusion occurs at random (Figure 18-13).

Three fourths of the F_2 generation have at least one **T** gene and are tall (**T** is dominant), while one fourth have both recessive factors (**tt**) and are short. The larger the number of offspring, the closer the actual ratio is to the theoretical one.

Figure 18-13.

F_1 SPERM	F_1 EGG		F_2 ZYGOTES	
T and **T**	$1/2 \times 1/2 =$		$1/4$ **TT**	
T and **t**	$1/2 \times 1/2 = 1/4$			$3/4$ **Tall T–**
t and **T**	$1/2 \times 1/2 = 1/4$		$2/4$ **Tt**	
t and **t**	$1/2 \times 1/2 =$		$1/4$ **tt**	$1/4$ **short tt**

Figure 18-14.

18.6 Test Cross

In appearance a **TT** plant is the same as a **Tt** one in our example. Genes cannot be seen, of course, and it is not possible visually to distinguish between the alleles that we have indicated by the use of the letters **T** and **t**. Therefore, in order to determine the gene makeup, a test cross is frequently made. One of the F_2 tall individuals is crossed with (fertilized by) a short plant. The latter is selected because both genes of the pair must be **t** for the recessive short characteristic to appear. We know, thus, that the short plant must be **tt**.

In the first instance, as shown in Figure 18-14, the tall plant must have had two **T** genes. (Why?) In the second instance the tall plant must have had one **T** gene and one **t** gene. (Why?)

When such a test cross is made utilizing the tall F_2 plants under discussion, one third of them are shown to be **TT** and two thirds are **Tt**. Notice that the determination of the gene complement of the tall plant was possible only by using a plant of known recessive gene content, the short **tt**, in the test cross. We know that the gametes from this short plant can carry *only* the **t** allele. The appearance (tall or short) of the progeny of this test cross will really be determined by the allele present in the gamete of the tall plant. If this allele is **T**, the diploid plant will be tall (i.e., **Tt**). If this allele is **t**, the resultant plant will be short (**tt**).

18.7 Useful Terms

In discussing inheritance and variations, several terms are quite useful. Some of these have already been mentioned. An individual is **homozygous** if the genes in any given pair are alike (**TT** or **tt**), and **heterozygous** if they are not alike (**Tt**). The term **allele** refers to one of the alternate (or contrasting) forms of a gene at a particular locus (location or position) on homologous chromosomes, for example, **T** or **t**. Every organism has more than one pair of genes and may be homozygous for some characteristics and heterozygous for others. The common practice is to term an individual heterozygous (or a **hybrid**) if even only one pair of genes is not alike.

A **dominant** factor is the allele that is expressed and masks the other when both are present. A **recessive** factor, then, is masked and not expressed when both are present. A **Tt** pea plant is tall because the dominant factor, **T**, masks the recessive factor, **t**. The external appearance (tall or short, for example) of an individual is its **phenotype;** the genetic makeup of an individual is termed its **genotype.** Thus, two plants may have the same phenotype (e.g., tall) but different genotypes (**TT** and

Tt). Of course, many alleles show no dominance. For example, some four-o'clock plants (*Mirabilis jalapa*) have white flowers and others have red flowers. If these two plants are crossed, the F_1 progeny all have pink flowers; the F_2 consists of all three types of individuals (i.e., white-flowered, pink-flowered, and red-flowered). This situation, where the F_1 is intermediate in the characteristic, is termed **incomplete dominance, partial dominance,** or **lack of dominance.** However, most cases in which the progeny are intermediate between the parents are the result of multiple genes (Section 18.13).

18.8 Dihybrid Cross

Two or more independently inherited traits may occur in various combinations in an organism if the genes for these traits are on different homologous pairs of chromosomes. Again, we may turn to an experiment by Mendel to discover whether two pairs of contrasting characteristics are inherited independently and whether Mendel's ideas of factors and simple ratios are suitable explanations.

Mendel selected a pea plant that always produced yellow round seed when self-fertilized (or self-pollinated) and one that always produced green wrinkled seed when self-fertilized; they were true breeders. In other words, both plants were homozygous for seed color and for the appearance of the seed coat. He then crossed these two plants by placing the pollen from one in the flower of the other (actually, the pollen was placed on the pistil of the other flower; see Section 27.2 for flower parts) and found that the resultant plants (or F_1) produced seed that were *all* yellow and round. Thus, the yellow and round characteristics were shown to be dominant to green and wrinkled. If **Y** represents the allele that results in the production of a yellow pigment in the seeds, **y** can be used to designate the green characteristic. In a similar fashion, **R** and **r** may be used for round and wrinkled, respectively. In this case, the two allelic genes for seed color were on one pair of homologous chromosomes and the two alleles for type of seed were on a different pair of homologous chromosomes. As shown in Figure 18-15, the parental types were **YYRR** and **yyrr** and the F_1 plants were **YyRr.** The latter were **dihybrids;** they were heterozygous for two characteristics.

The F_1 were allowed to self-pollinate, and Mendel found that four kinds of seeds were produced in the F_2: approximately $\frac{9}{16}$ yellow round, $\frac{3}{16}$ yellow wrinkled, $\frac{3}{16}$ green round, and $\frac{1}{16}$ green wrinkled. Table 18-1 indicates the actual numbers of plants as counted by Mendel in one of his experiments. The "theoretical" num-

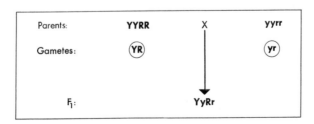

Parents:	**YYRR**	X	**yyrr**
Gametes:	**YR**		**yr**
F_1:		**YyRr**	

Figure 18-15.

Table 18-1 Progeny of a dihybrid cross

Actual number	Character	Ratio	Theoretical number
315	Yellow Round seed	9/16	313
101	Yellow wrinkled seed	3/16	104
108	green Round seed	3/16	104
32	green wrinkled seed	1/16	35
Each character by itself:			
416 (315 + 101)	Yellow seed	12/16 or 3/4	417 (313 + 104)
140 (108 + 32)	green seed	4/16 or 1/4	139 (104 + 35)
423 (315 + 108)	Round seed	12/16 or 3/4	417
133 (101 + 32)	wrinkled seed	4/16 or 1/4	139

bers are those that would have been obtained if the F_2 represented an exact
9:3:3:1 ratio. It is obvious that the actual count was very nearly such a ratio.
Many additional dihybrid crosses by various investigators over the years have
confirmed this ratio. If one now examines each trait (or character) by itself, one
finds that there are $\frac{12}{16}$ yellow and $\frac{4}{16}$ green, $\frac{12}{16}$ round and $\frac{4}{16}$ wrinkled. In both cases
the results (really $\frac{3}{4}$ to $\frac{1}{4}$) are what could be expected from a monohybrid cross with
dominance (see Section 18.5), which demonstrates that single-factor ratios are
maintained during such a dihybrid cross. Obviously, then, the two pairs of genes,
or the chromosomes that contain the genes, segregate independently of one another. This is termed **independent segregation,** or **random assortment.** The way the
Ys separate has nothing to do with the way the **R**s separate, *because they are on
different chromosomes.* However, since homologous chromosomes do separate,
the haploid cell (gamete in this case) cannot contain two **Y**s or two **R**s, but only
one of each. Whether the capital letters or small letters are present in a gamete
(i.e., **YR, Yr, yR,** or **yr**) depends upon the alignment of chromosomes during metaphase **I** of meiosis. Remember that these letters represent alleles. (See Section
18.1 and Figure 18-4 for a more complete explanation.)

Although only four kinds of gametes are involved, they combine entirely at random (see Figure 18-16). The four-letter symbols in the checkerboard thus repre-

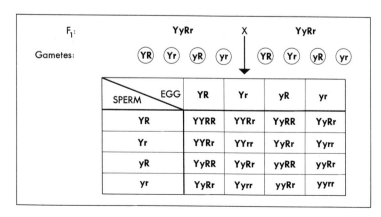

Figure 18-16.

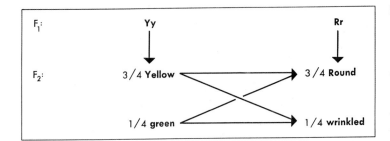

Figure 18-17.

sent the possible kinds of progeny that could be expected in the F_2. The larger the number of progeny, the closer the ratio approaches $9:3:3:1$. This is the phenotypic ratio, whereas the symbols in the Punnett square represent the genotypes. A difference between phenotype and genotype merely means that dominance is involved.

Because the pairs of genes segregate independently, the computation is made simpler by considering each characteristic alone and then combining the ratios. Refer to Section 18.5, and treat each trait in the F_1 as a monohybrid cross. The heterozygous yellow-seeded individual will produce three fourths yellow seeds and one fourth green seeds when self-pollinated; the heterozygous round-seeded individual will produce three fourths round seeds and one fourth wrinkled seeds. This is considering the F_2 in terms of one trait at a time (Figure 18-17).

Now, to determine how many of the F_2 seeds will be *both* yellow *and* round, the two ratios are combined as indicated: $\frac{3}{4}$ yellow $\times \frac{3}{4}$ round equals $\frac{9}{16}$ yellow round seeds. Other possible combinations are also indicated in Figure 18-18.

In the foregoing, only the phenotypes are considered. The genotypes can be determined in a similar fashion by indicating the heterozygous condition of some of the round seeds and yellow seeds. The heterozygous yellow-seeded individual will produce one fourth homozygous yellow-seeded, two fourths (or one half) heterozygous yellow-seeded, and one fourth homozygous green-seeded individuals when self-pollinated. Similar reasoning is used for the heterozygous round-seeded individual (Figure 18-19). In order to determine how many of the F_2 progeny will be homozygous for both traits, one must again combine ratios; $\frac{1}{4}$ **YY** $\times \frac{1}{4}$ **RR** $= \frac{1}{16}$ **YYRR**. Other possible combinations (e.g., $\frac{2}{4}$ **Yy** $\times \frac{2}{4}$ **Rr** $= \frac{4}{16}$ **YyRr**; $\frac{2}{4}$ **Yy** $\times \frac{1}{4}$

F_2:

3/4 **Yellow** x 3/4 **Round**	$= 9/16$ **Yellow Round**
3/4 **Yellow** x 1/4 **wrinkled**	$= 3/16$ **Yellow wrinkled**
1/4 **green** x 3/4 **Round**	$= 3/16$ **green Round**
1/4 **green** x 1/4 **wrinkled**	$= 1/16$ **green wrinkled**

$9:3:3:1$

Figure 18-18.

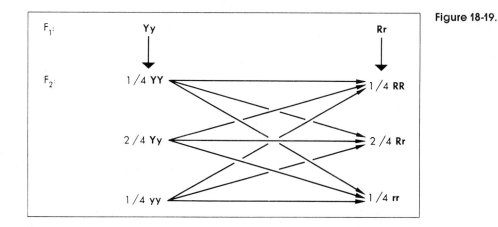

Figure 18-19.

$\mathbf{rr} = \frac{2}{16}$ **Yyrr**) are also indicated in Figure 18-20. The results shown are the same as those obtained by using the Punnett square or checkerboard, which was shown previously.

18.9 Another Test Cross

To show that the F_1 individuals of the previous section are actually heterozygous is easily done by crossing them with the recessive parent (**yyrr**). The resulting progeny (Figure 18-21) are of all four possible combinations of yellow, green, round, and wrinkled, and in equal numbers if the sample is quite large. This indicates that all four traits were present in the individual being tested and that the pairs of genes segregate independently (note the four kinds of gametes produced by the heterozygous individual).

Figure 18-20.

F_2:   1/4 **YY** x 1/4 **RR** = 1/16 **YYRR**   2/4 **Yy** x 1/2 **RR** = 2/16 **YyRR**

1/4 **YY** x 2/4 **Rr** = 2/16 **YYRr**   2/4 **Yy** x 2/4 **Rr** = 4/16 **YyRr**

1/4 **YY** x 1/4 **rr** = 1/16 **YYrr**   2/4 **Yy** x 1/2 **rr** = 2/16 **Yyrr**

1/4 **yy** x 1/4 **RR** = 1/16 **yyRR**

1/4 **yy** x 2/4 **Rr** = 2/16 **yyRr**

1/4 **yy** x 1/4 **rr** = 1/16 **yyrr**

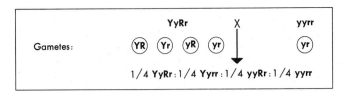

Figure 18-21.

This test cross can also be treated one trait at a time. The heterozygous individual produces gametes one half of which carry the factor for yellow seed, the other half carrying the factor for green seed. Similarly, in the heterozygous individual, one half of the gametes will have a gene governing round seed formation, while the other half will have the gene that results in the production of wrinkled seeds. All the gametes are alike in the recessive individual (Figure 18-22). One half of the progeny are yellow-seeded, one half are green-seeded; one half are round-seeded, one half are wrinkle-seeded. However, we are interested in both characteristics. The possible combinations of characteristics are indicated in Figure 18-23 and are, of course, the same as were obtained previously.

Whether the checkerboard method or the proportion method is utilized depends upon personal preference. Genetics is basically mathematical in approach and is more readily understood if the student understands the chromosomal basis; complete understanding comes with practice in working problems. The reader is strongly urged to attempt all the problems available at the end of this chapter.

18.10 Self-Pollination and Homozygosity

Continued self-pollination increases the proportion of offspring that are homozygous and is the reason that self-pollinated plants tend to be true breeders. Assuming that all parents have an equal number of progeny, with each generation of inbreeding (self-pollination) the proportion of heterozygous individuals is reduced by half. Remember that when a monohybrid is self-pollinated, only one half of the progeny are heterozygous. As our example (Figure 18-24), we shall use one of the original characteristics with which Mendel worked. The F_1 is all heterozygous as a result of crossing a homozygous tall (**TT**) with a homozygous short (**tt**) pea plant. When all the individuals from the F_1 on are allowed to self-pollinate, the heterozygotes form a smaller and smaller proportion of the total population. In the F_2, only half of the individuals are heterozygotes. This drops to one fourth in the F_3 and to one eighth in the F_4. Subsequent proportions would be one sixteenth, one thirty-second, one sixty-fourth, and so on.

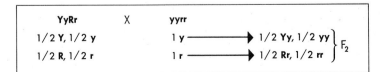

Figure 18-22.

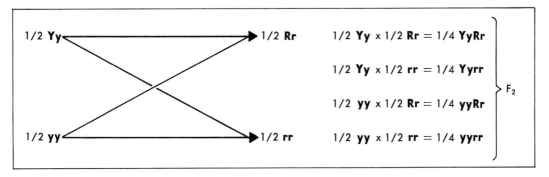

Figure 18-23.

Application of this principle of increased homozygosity is the fundamental reason why brother and sister matings in humans is frowned upon. Everyone carries a few recessive harmful genes. In matings of close relatives, the probability is high that these harmful genes are the same and the homozygous condition is quite likely to occur. Of course, such matings of brother-sister or cousin-cousin were forbidden, or at least looked at askance, before there was any real knowledge of heredity. Such prohibitions were, however, probably a result of observations that these matings resulted in abnormally high numbers of defective children.

It is worth noting that self-pollination (or inbreeding) does not favor an increase in the number of recessive alleles. It merely brings about the homozygous condition that enables such alleles to be detected phenotypically. With reference to homozygosity, the effects of inbreeding are the same for dominants as for recessives. The effect is less spectacular for dominants, however, since they have not been phenotypically submerged.

Inbreeding does not always produce harmful effects. Much of the work of plant and animal breeders involves inbreeding to produce the homozygous condition of advantageous recessive alleles. However, if the recessive allele is deleterious and masked by its dominant allele, inbreeding is much more likely to produce the deleterious characteristic in the homozygous condition to the detriment of the progeny.

Figure 18-24.

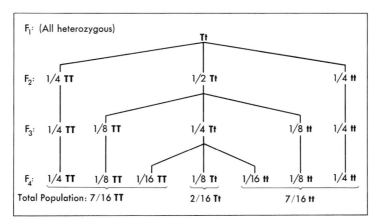

18.11 Mendelian Principles

As a result of the preceding discussions, it is possible to list four basic principles of genetics. These are a direct result of Mendel's work in the middle 1800s and verified repeatedly by many other investigators.

1. **The principle of unit characters.** The inherited characteristics of an organism are controlled by factors, or genes, and these genes occur in pairs.
2. **The principle of dominance.** One gene in a pair may mask, hide, or inhibit the expression of the other gene. Dominance does not occur between all gene pairs (or alleles).
3. **The principle of segregation.** The members of pairs of genes separate prior to gamete formation (i.e., during meiosis), and only one gene of each pair is present in a gamete.
4. **The principle of independent assortment.** The members of different pairs of genes on different chromosomes are distributed to the gametes independently of one another and are combined at random during fertilization.

18.12 Lack of Dominance

One instance in which the F_1 plants are intermediate in a characteristic, as compared with the parents, has already been mentioned (Section 18.9). The diagrams of a one-factor cross (in Figure 18-25) and a two-factor cross may help to clarify this situation where there is a lack of dominance. In the first example the phenotypic and genotypic ratios are identical, because every genotype is expressed in the phenotype. If it is not clear how the F_2 ratio is obtained, the student should refer to Figure 18-8 or 18-9, and substitute **R** for **T** and **r** for **t**.

In the example shown in Figure 18-26(a), dominance is expressed in one pair of genes but not in the other. Each characteristic can be treated separately, and then

Figure 18-25.

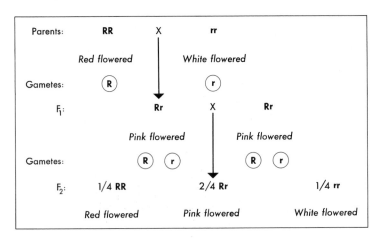

Figure 18-26.

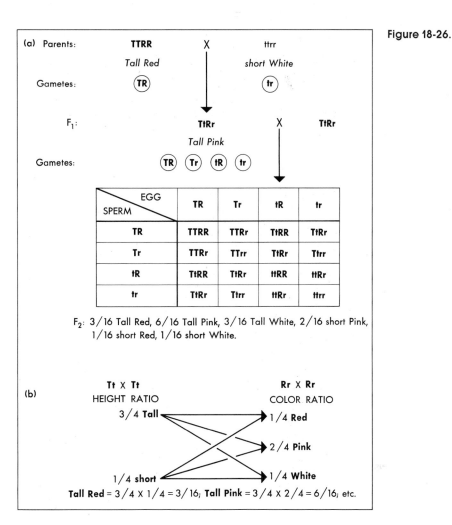

(a) Parents: TTRR   X   ttrr

Tall Red   short White

Gametes: (TR)   (tr)

F₁: TtRr   X   TtRr

Tall Pink

Gametes: (TR) (Tr) (tR) (tr)

SPERM \ EGG	TR	Tr	tR	tr
TR	TTRR	TTRr	TtRR	TtRr
Tr	TTRr	TTrr	TtRr	Ttrr
tR	TtRR	TtRr	ttRR	ttRr
tr	TtRr	Ttrr	ttRr	ttrr

F_2: 3/16 Tall Red, 6/16 Tall Pink, 3/16 Tall White, 2/16 short Pink, 1/16 short Red, 1/16 short White.

(b)

Tt X Tt   Rr X Rr
HEIGHT RATIO   COLOR RATIO

3/4 Tall   1/4 Red

2/4 Pink

1/4 short   1/4 White

Tall Red = 3/4 X 1/4 = 3/16; **Tall Pink** = 3/4 X 2/4 = 6/16; etc.

the various combinations of both traits can be determined (Figure 18-26[b]). The results are the same as were obtained with the checkerboard method.

In order to eliminate the possibility of a misunderstanding when there is a lack of dominance, geneticists frequently use lower case letters with numeral subscripts to designate alleles. In our example, this would result in a red-flowered plant being designated as r_1r_1, white-flowered as r_2r_2, and pink-flowered as r_1r_2. The use of upper case letters is then restricted to indicate dominant alleles.

18.13 Multiple Alleles

In the discussion thus far, we have assumed that a particular locus on a chromosome is occupied by either of two alleles. There are instances, however, in which a given locus may bear any one of a series of alleles, and the diploid individual possesses any two genes of the series. When there are more than two kinds of alleles for a given locus, we speak of them as **multiple alleles.**

In many kinds of plants, including clover, tobacco, cherry, and evening primrose, multiple alleles determine compatibility in sexual reproduction. This was first worked out in tobacco plants. Some plants are self-sterile in that they fail to produce seed when self-pollinated. This is due to a series of alleles, usually designated as S_1, S_2, S_3, and so forth. A plant may be hybrid (heterozygous) for any two alleles (e.g., S_1S_2, or S_2S_3, or S_1S_3, and so forth). Pollen containing either kind of allele present in a given plant cannot grow down the floral parts of that plant to the area where the eggs are located. Thus, if the plant is S_1S_2, pollen containing either S_1 or S_2 cannot grow in that plant (even if the pollen is from another plant). If the S_1S_2 plant is crossed with (pollinated by) an S_3S_4 plant, fertilization could occur. The pollen would be S_3 *or* S_4; the offspring would be S_1S_3, S_1S_4, S_2S_3, or S_2S_4 (since the eggs would be S_1 *or* S_2. If the cross were $S_1S_2 \times S_1S_3$ (male parent), only the S_3 pollen would be functional. The value of self-sterility is that it causes cross-fertilization and thus tends to increase the number of variations in the offspring. The greater the possible variations, the greater the possible survival value in different environments or in changing environments.

Sudden, unpredictable changes in an existing gene occasionally occur. Such **mutations** are inherited, of course, if they are not so deleterious as to result in death of the organism. In this way new alleles arise bringing about the multiple allelic condition. In some plants 20 or more alleles have been identified. Remember, however, that the diploid organism can have only two alleles of the series, one on each of the homologous chromosomes; in the haploid condition only one of the alleles would be present.

18.14 Multiple Genes

Up to this point, all the characters that have been discussed are those that can be sharply or distinctly divided into clearly differentiated classes (e.g., the pea plant was tall or short, with no in-between sizes). However, many characters vary quantitatively in that gradations are continuous from one extreme to the other. Such characteristics are somewhat more difficult to study and are beyond the scope of this text. They will be given a very brief discussion at this time to alert the student to the complexities of genetical studies.

Mendel's genius lay in the choosing of sharply distinguished contrasting characters. Before this, various individuals had found that the F_1 was more or less intermediate between two extreme parents and that the F_2 was composed of all gradations from one extreme to the other. This was the basis of the old idea of "blending." Not until after simple Mendelian inheritance was understood was an explanation of this quantitative gradation possible. If several sets of genes are involved in the expression of a single characteristic, and if no dominance exists between the alleles, then they will be cumulative in their effect. We speak of **multiple genes** or **factors** when several sets of alleles produce more or less equal and cumulative effects on the same character.

In this type of heredity no *obvious* ratios may be found. This is quite different

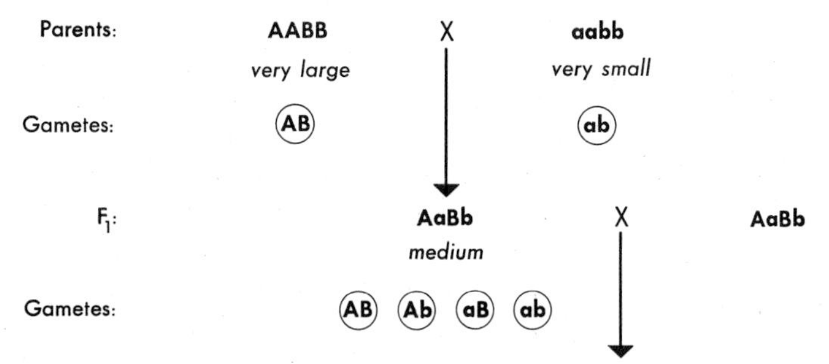

Figure 18-27.

Parents:	AABB	X	aabb
	very large		*very small*

Gametes: (AB)   (ab)

F_1:   AaBb   X   AaBb
   medium

Gametes: (AB) (Ab) (aB) (ab)

EGG / SPERM	AB	Ab	aB	ab
AB	AABB *very large*	AABb *large*	AaBB *large*	AaBb *medium*
Ab	AABb *large*	AAbb *medium*	AaBb *medium*	Aabb *small*
aB	AaBB *large*	AaBb *medium*	aaBB *medium*	aaBb *small*
ab	AaBb *medium*	Aabb *small*	aaBb *small*	aabb *very small*

from the incomplete dominance discussed in Section 18.12. For example, one investigator found in a cross between a plant having very large flowers and one having very small flowers that the flowers of the F_1 were medium in size and that the F_2 had flowers varying in size from very small to very large. Moreover, about 1 out of 16 of the F_2 was as small as the one original parent and 1 out of 16 was as large as the other original parent plant. To a geneticist this immediately suggests that two factors are involved, in which the double dominant and double recessive each occur about once in every 16 individuals (refer to a dihybrid cross, Section 18.8). If we assume two pairs of genes, **A** and **a**, **B** and **b**, the cross may be diagrammed as in Figure 18-27. Thus, even where "blending" appears to occur, Mendel's basic principles are upheld.

Other such examples of multiple genes involve more than two pairs. The complexity increases, but the basic principles remain the same. Although two pairs of genes produce a $1:4:6:4:1$ phenotypic ratio (as in Figure 18-27), many pairs of genes (which is more frequently the case) produce ratios that are harder to discern because of the number of classes and the slight differences between classes; but they *are* there.

18.15 Lethal Genes

The classic genetic ratios may be modified by genes whose effect is sufficiently drastic to kill the bearers of certain genotypes. These are **lethal genes,** whose effect may be produced at almost any time between fusion of egg and sperm (i.e., the zygote stage) and a late stage in life.

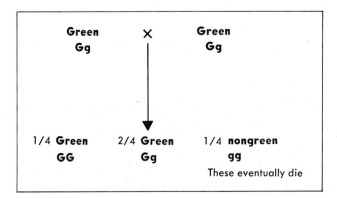

Figure 18-28.

In corn (*Zea mays*) a gene (**G**) for normal chlorophyll production is dominant to its allele **g**. Plants with the genotype **GG** or **Gg** contain chlorophyll and are photosynthetic, while **gg** plants produce no chlorophyll. If two heterozygos plants are crossed, the seedlings show the classic 3:1 phenotypic ratio of a monohybrid cross (Figure 18-28). Germination and early seedling development depend upon food stored in the grain. In normal green plants sufficient green tissue develops before all the food reserve is exhausted, and the plants are capable of food production (photosynthesis). The nongreen plants are not capable of photosynthesis and die after the food reserves of the grain are depleted. Thus, after two weeks or so, the initial 3:1 phenotypic ratio becomes a 3:0 or a 1:0 ratio (i.e., all the living progeny are green). The genotypic ratio is converted from the usual 1:2:1 to 1:2 upon death of the nongreen plants. If the lethal gene had exerted an effect in the zygote stage (or at any time before emergence of the seedling), the observed phenotype would be all green, and the genotypic ratio would be 1:2 without the classic ratios appearing at all.

18.16 Linkage

In studying independent assortment, Mendel experimented with characteristics that were governed by genes located on separate chromosomes. However, all genes are not on different chromosomes; many chromosomes include hundreds or thousands of genes. Two genes on the same chromosome are said to be **linked.** They do not usually separate independently but remain together during meiosis.

If two pairs of genes on different chromosomes are involved, four possible kinds of gametes are produced by a dihybrid individual (see Section 18.8). If these genes are closely linked, however, a similar dihybrid individual can produce only two possible kinds of gametes. As shown in Figure 18-29, there will also be fewer variations in the progeny, resulting from self-fertilization, if the genes are linked. In the unlinked situation any one of four possible kinds of sperm can fuse with any one of four possible kinds of egg; thus producing nine possible genotypes in the offspring. When linkage is involved, however, only one of two possible kinds of

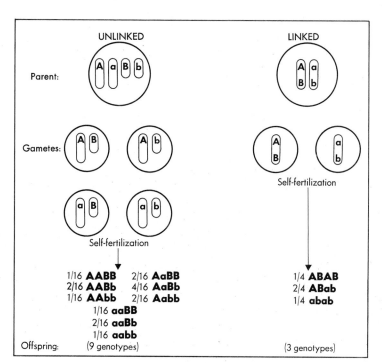

Figure 18-29.

sperm can fuse with any one of two possible kinds of egg; and there are only three possible genotypes in the progeny. Review Figures 18-11, 18-16, 18-19, and 18-20 if the situation with regard to these offspring is not clear.

Crossing-over

Many investigators noticed that linked genes sometimes become unlinked or change linkage, and the frequency with which this occurs depends upon the distance separating the two genes on a chromosome. The explanation for this behavior was found upon careful examination of meiosis (Section 18.1). In prophase I, during synapsis, the chromatids of homologous chromosomes become intertwined and frequently exchange parts at one location. Figure 18-30 represents this exchange, or **crossing-over.** Note that only two of the four chromatids are involved in this exchange. As a result, all four chromatids of the tetrad are different, and the resultant haploid cells will be of four kinds instead of two, as would be true if crossing-over did not occur. Without crossing-over, chromosomes (1) and (2) of Figure 18-30 would be alike, and chromosomes (3) and (4) would be alike. Just one pair of chromatids crossing over doubles the possible kinds of haploid cells (i.e., gametes). The letters represent genes located on the chromosome. In the four chromosomes located in the four haploid cells that are produced as a result of meiosis, genes **ACB** are still linked in one gamete and **acb** are linked in another. In the other two gametes, however, the linkage is now **ACb** and **acB,** respectively, as a result of crossing-over.

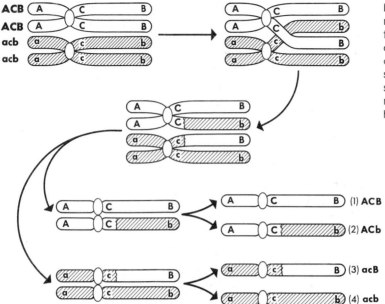

ACB, ACB, acb, acb

(1) ACB
(2) ACb
(3) acB
(4) acb

Figure 18-30. Crossing-over during meiosis. In the first division of meiosis the kinetochores have not doubled and thus the chromatids do not separate; the homologous chromosomes separate. In the second division the chromatids separate. The numbers indicate the four types of haploid cells that result.

Crossing-over is a normal occurrence during synapsis, and breakage may occur at any point on the chromosome where two chromatids happen to be overlapping. As a result, crossing-over may occur simultaneously at several locations on a single chromosome, resulting in an almost infinite variety of gene recombinations. The longer the chromosome, the greater the possibility of more than one crossover taking place. Also, the farther apart that two genes are on a chromosome, the more likely that crossing-over will occur, separating them. Variations in gametes, and thus in progeny, result from crossing-over as well as from the random alignment of chromosomes during meiosis. The possibilities for variations are thus enormously increased. Such variations in living organisms provide the material for evolutionary change.

In order to understand more fully the significance of linkage let us take a hypothetical example. What would have been the progeny in Mendel's classic dihybrid cross (see Figure 18-16) if genes **Y** and **R** had been linked on one homologous chromosome (**y** would then be linked to **r** on the other homologous chromosome)? If no crossing-over occurred (i.e., the genes were very close together on the chromosomes), then the dihybrid would produce only two kinds of gametes rather than four. The progeny would have a phenotypic ratio of 3:1 and a genotypic ratio of 1:2:1 (Figure 18-31), very different from that actually obtained by Mendel. Note that six genotypes of the classic dihybrid cross would not appear in the offspring if complete linkage was involved.

Many examples of linkage have now been studied by geneticists. If crossing-over occurs, some recombinations of genes would occur. In our example, in the previous paragraph, the gametes would be either **YR** or **yr** if the genes were so close together on the chromosome that crossing-over could not take place between them—similar to genes **A** and **C** (and their alleles) in Figure 18-30. Note that

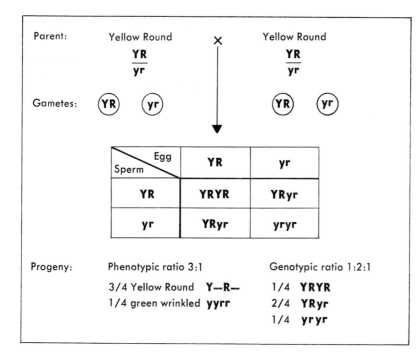

Parent: Yellow Round   ×   Yellow Round

$$\frac{YR}{yr} \qquad \frac{YR}{yr}$$

Gametes: (YR) (yr)   (YR) (yr)

Sperm \ Egg	YR	yr
YR	YRYR	YRyr
yr	YRyr	yryr

Progeny:

Phenotypic ratio 3:1

3/4 Yellow Round **Y—R—**
1/4 green wrinkled **yyrr**

Genotypic ratio 1:2:1

1/4 **YRYR**
2/4 **YRyr**
1/4 **yryr**

Figure 18-31. The conventional method of indicating linked genes has been used. $\frac{YR}{yr}$ implies that **Y** and **R** are linked on one homologous chromosome, and **y** and **r** are linked on the other homologous chromosome. The pair of homologous chromosomes of each parent separate at anaphase I of meiosis.

these two genes remain linked; with regard to these genes there are only two kinds of haploid cells after meiosis, **AC** and **ac**. However, if the two genes were far enough apart on the chromosome so that some crossing-over could take place between them, then some **Yr** and **yR** gametes could be produced—similar to genes **A** and **B** (and their alleles) in Figure 18-30. With regard to these genes, there are four kinds of haploid cells after meiosis: **AB** and **ab**, which represent the situation where there is no crossing-over of chromatids; **Ab** and **aB**, which represent the result of crossing-over of chromatids. **Ab** would correspond to **Yr** of our example; **aB** would correspond to **yR**. The per cent of crossing-over depends upon the distance apart of the two genes in question (**Y** and **R**; **y** and **r**); the farther apart, the more likely that crossing-over will take place between the genes.

The student should diagram a dihybrid cross (utilizing the letters **AaBb**) in three ways: (1) with the genes on different pairs of homologous chromosomes ($2N = 4$ chromosomes), (2) with the genes linked on one pair of homologous chromosomes ($2N = 2$ chromosomes) and no crossing-over, and (3) with linkage plus crossing-over.

18.17 Extranuclear Inheritance

The existence of genes located in chromosomes and controlling phenotypes in a predictable fashion has been conclusively demonstrated. However, there are a few instances in which a given character is transmitted via the cytoplasm and largely independent of chromosome-borne genes. This is usually referred to as **extranuclear inheritance, cytoplasmic inheritance,** or **extrachromosomal inheritance.**

In certain varieties of the four-o'clock (*Mirabilis jalapa*), for example, branches may be green, variegated (patches of green and nongreen tissue), or white (no chlorophyll). The phenotypes of the offspring are determined by the female (egg-producing) parent. Seeds borne on white branches produce only white offspring; those borne on green branches produce green plants; and those borne on variegated branches segregate in irregular ratios of white, green, and variegated (Table 18-2). The type of pollen is not important.

This type of inheritance is considered to be caused by the transfer of proplastids in the cytoplasm of the egg cell. The proplastids reproduce and can mutate (or change). Plastids do contain DNA and, therefore, presumably store genetic information. The proplastids produce plastids as the plant develops, and if the plastids are defective as to chlorophyll synthesis, the cells containing such plastids are white (nongreen). The sperm contains very little cytoplasm and, thus, carries few if any proplastids. Egg cells produced in flowers on a white area carry only white plastid primordia, and the progeny are white. Normal proplastids are carried by eggs produced in flowers on green areas. Egg cells from flowers in a variegated area may contain either normal or defective proplastids or both. After fertilization, such cells (now zygotes) undergo division and give rise to new plants (the progeny). If both types of proplastids are present, some cells of the new plant fortuitously receive a majority of green plastids and some receive larger numbers of white plastids, resulting in variegated plants. Remember that the division of the cytoplasm following mitosis is not a quantitatively exact process; patches of green and nongreen (white) tissues result.

As was mentioned previously (Section 4.2), both the plastids and the mitochondria are self-replicating organelles that contain DNA, and so carry at least some of their own hereditary potentialities. Extranuclear inheritance is really a minor factor, however. The nucleus and the genetic factors it contains are of much greater significance in the total biological inheritance of plants.

Table 18-2 Progeny of a variegated four-o'clock

Branch type from which pollen obtained	Branch type of pollinated flower	Progeny grown from seed
White	White	White
	Green	Green
	Variegated	White, green, and variegated
Green	White	White
	Green	Green
	Variegated	White, green, and variegated
Variegated	White	White
	Green	Green
	Variegated	White, green, and variegated

18.18 Environment and Gene Expression

The phenotypic response of gene action is a result of the coordinated effect of the entire set of genes: a single gene may affect many characteristics, or each expressed characteristic may be influenced by many genes. Each gene probably governs one basic reaction, but the characteristics that are expressed as the phenotype result from many such reactions. Therefore, to state that every characteristic of an organism is governed by one gene is incorrect. In some cases we are fortunate enough to find that one gene has an obvious effect on the mature organism and thus is studied readily. This is not, however, the situation that always exists.

The genes of an organism determine its *potentialities,* but the realization of these potentialities depends on the environment in which the genes perform their function. The genotype may express itself differently in different environments, giving rise to modifications of the phenotype. The genetic material provides the messages and directions for phenotypic growth, but development also depends upon various environmental factors (e.g., light, temperature, moisture, and mineral supply). For example, with most plants if the seeds are germinated in the dark, chlorophyll will not develop in the seedlings. If these seedlings are exposed to light before they die, chlorophyll develops rapidly. The genes for chlorophyll production are present, but light is also essential for such production. Also, if magnesium is lacking in the soil, chlorophyll formation will not occur because the chlorophyll molecule has, as one of its component parts, an atom of magnesium. The potentiality for chlorophyll formation is present in the genes, but an essential atom is missing and the molecule cannot be synthesized.

The time of flowering in many plants is influenced by the duration of light, and in others by temperature. Water supply is another factor that significantly influences plant growth, severe stunting being an easily recognizable effect of water deficiency. Various other examples could be mentioned. It is important to understand that it is the genotype that is inherited and that the environment influences the way in which this genotype is expressed (as the phenotype).

18.19 Plant Breeding

Long before Mendel's time man had been seeking to improve the plants and animals that he found useful. This was accomplished primarily by selecting the "best" progeny in each generation. One should be aware that the term *best* refers to an organism that has certain characteristics that man finds useful for his purposes. Such plants (or animals) are not necessarily best suited for competition with other organisms without man's intervention. In fact, most domesticated plants do rather poorly without man's pampering in terms of weed control, insect control, fertilizing, and so forth.

Such **selection** over many generations has produced many desirable food plants and plants maintained for their aesthetic qualities. This method, however, does not change genes or introduce new genes. It results in the isolation of certain characteristics and produces homozygosity. Selection within a pure line (i.e., one that is homozygous) is ineffective. Thus, there is a limit to the effectiveness of selection. In addition, before genetics was developed as a science, selection was based only on the phenotype or appearance, and the basis for selection was not always clear. Early plant breeders could not distinguish between genetic and environmental factors, and many errors resulted. Records usually were not kept, especially if results were undesirable. There was really no concerted effort to organize and evaluate various results. This was not strange in view of the lack of knowledge concerning the inheritance of factors, as so clearly demonstrated later by Mendel and others. It is interesting to note, however, that many crop plants were developed, presumably by selection, by prehistoric man before any science was known.

With an understanding of the genetic control of plant development came the real advance in the production of plants that best serve man's needs. Plant breeders now utilize their knowledge of genetics to develop varieties that carry those genes considered to be most advantageous. For example, plants that are resistant to certain diseases but are low in yield may be crossed with plants that have a high yield although susceptible to disease. Through such genetic recombinations resistant plants of high yield may be obtained. Even if the factors for yield and disease susceptibility are linked, we now realize that crossing-over may provide us with the combination that we desire. New gene combinations are brought about by chromosome assortment and crossing-over. It may take many generations to achieve a desired result, but the breeder now has a scientific basis for anticipating, predicting, evaluating, and planning a breeding program.

The plant breeder, in effect, is making practical applications of his scientific knowledge. An understanding of genetics is an exceedingly useful tool, which enables man to improve yield (quantity and quality), to develop disease-resistant plants, to produce varieties that may be grown in many environments not suitable to the native (or wild) type, and so forth. Many complexities are involved, of course, in that characteristics are frequently determined by multiple genes, multiple alleles, or closely linked genes; lethals may be present; or dominance may mask results. In spite of such difficulties, the situation is considerably improved over a "hit-or-miss" type of selection, and the possibility of not achieving the desired result because that particular characteristic may not show up in the phenotype of the first few generations. Selection is still a part of plant breeding but the *reasons* for making such selections are now based upon genetic principles. The development of the science of genetics, thus, resulted in greatly improved varieties of crop plants.

Hybrid Corn

One of the best examples of what can be accomplished by plant breeding is the development of hybrid corn. Corn, or maize (*Zea mays*), has been in cultivation for many centuries, certainly long before it was "discovered" (for the Old World) by members of Columbus' expedition in 1492. In fact, there is evidence that corn was a crop plant more than 5000 years ago. The earliest method to improve corn was selection; grain was collected from the largest and most vigorous plants and used for subsequent planting. Considerable improvement resulted, but eventually the yield stabilized and no further increase was obtained. The corn plant produces separate male and female flowers and is almost 100 per cent cross-pollinated. Corn varieties are, thus, heterozygous. Selecting the "best" grain still resulted in wide variations among the progeny because of gene recombinations.

In order to counteract this difficulty resulting from heterozygosity, pure lines of corn were established by self-pollinating for about seven generations (see Section 18.10). Such inbreeding actually decreased vigor by increasing the number of recessive, deleterious genes in the homozygous condition. The effects of many of these genes had been concealed by their dominant alleles. However, selection during inbreeding eliminated less desirable strains and retained the best homozygous strains with the desired traits. When two such pure lines of corn were crossed, the resultant hybrid seed produced plants that were taller, more vigorous, and bore larger ears than their inbred parents; the crop yield was uniform, much greater than the inbred lines, and usually better than the varieties from which the inbred lines were derived. The uniformity of yield was especially important. One difficulty remained, however. The inbred strains used as seed-parents produced only small ears bearing small numbers of grains. The hybrid seed is necessarily produced on one of these parents, i.e., the small number of grains represents the hybrid. As a result of this low yield and the cost of the hand labor required to insure the specified crosses, hybrid seed obtained by this "single-cross" method was rather expensive.

The "double-cross" method of producing hybrid seed was then introduced. This involved four inbred strains, two crossings, and required two seasons before seed was produced that the farmer could plant to give the commercial crop (Figure 18-32). The desired hybrid seed is produced in greatly increased numbers by this

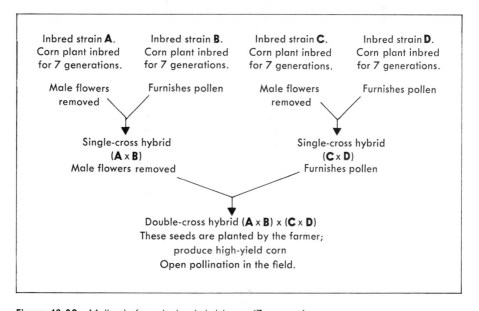

Figure 18-32. Method of producing hybrid corn (*Zea mays*).

means. Although seed from the first cross is produced on small ears borne by small inbred plants, seed from the second cross is produced on large ears borne by large "single-cross" plants.

The double-cross technique is the one used today in the production of the majority of corn grown in the United States. This has doubled the average yield from about 25 bushels per acre to about 60 bushels per acre. Many pure inbred lines have been produced, and from them many different kinds of hybrids with various advantageous characteristics (e.g., disease resistance and climatic adaptation) in addition to high yield per acre. The farmer must buy new hybrid seed each year to retain these desirable qualities. Remember that the corn plant is normally cross-pollinated. However, corn from a field planted with hybrid seed is produced by open pollination of *closely related* plants, resulting in large amounts of self-pollination and increased homozygosity of the grain obtained during the third year (the farmer's yield). If such grain were used for subsequent crops, yield would be quite variable and reduced (i.e., some large ears, some small; all kinds of gradations). The uniformity of hybrid corn results from crossing inbred lines that are typically homozygous to obtain the heterozygous condition. There is now a highly organized industry supplying hybrid seed tailored to the various climatic, soil, and disease conditions prevailing in the area where the crop will be grown.

Hybrid Vigor

The increased growth of the F_1 plants that results from crossing two inbred lines is known as **hybrid vigor** or **heterosis.** It occurs in many plants besides corn. The basis of hybrid vigor is not clearly understood. It may be a result of combining favorable dominant genes with unfavorable recessive alleles in the hybrid. Each inbred line may be weak because of homozygous detrimental recessive genes, and it is not likely that similar recessives are found in the two lines selected for producing hybrid corn. Thus, the dominants from one parent cover the recessives of the other, and the hybrid is more vigorous than either. If many genes are involved in vigor and many of them closely linked, it would be exceedingly difficult to obtain homozygous dominant strains. Attempts to obtain such strains have not been successful.

It is also possible that hybrid vigor may result from an interaction between alleles that is superior to either of the homozygous conditions. The implication is that alleles do separate things, and the sum of their products is superior to the single products produced by either allele in the homozygous state. There is some evidence to support this view, but it is not yet possible to state conclusively what is the mechanism for hybrid vigor.

18.20 Gene Function

Specific locations on chromosomes have long been known to govern certain characteristics of organisms. These loci are termed **genes,** and as the chromosome duplicates, so also are each of the genes duplicated. We are not yet certain whether a chromosome consists of a single exceedingly long DNA molecule combined with protein to form a nucleoprotein, or whether separate DNA molecules with protein link together to form the chromosome. In either case DNA is what determines development, and DNA replication determines the transmission of hereditary characteristics. A *gene* is considered to be a segment (or portion) of a DNA molecule.

All the cells of the more complex multicellular organisms are derived from a single diploid cell, the fertilized egg or zygote.[1] All such cells of an organism, therefore, contain the same complement of chromosomes and genes as a result of mitosis. Obviously there are various kinds of cells in the organism; so cell differentiation must occur. The development of an organism, then, is an orderly sequence of cellular differentiations that transform the fertilized egg into a multicellular adult form. Our subsequent discussion will delve into the method by which genetic information is used to determine differentiation and development, especially in view of the identical genetic complement of these cells.

From the experimental data available, the most logical explanation concerning gene function involves the influence of genes (and DNA) on protein synthesis. The enzymes, which govern all of the metabolic reactions of the cells, are proteins; thus, the kinds of proteins that are produced determine the activities and development of an organism. Genes govern the activities that can occur in a cell, but remember that many times the environment influences what actually does occur. For example, a green plant is capable of carrying on photosynthesis, provided that carbon dioxide and other essentials are available to it, but, as indicated in the chapter on mineral nutrition, if certain minerals are not available chlorophyll will not develop. Variations in environmental conditions may affect plant development. For example, as will be discussed more thoroughly in Section 29.12, day length (which varies during the growing season) is important in the flowering of many plants. The capability to flower, as determined by genetic information, is present but the day length determines whether or not this capability will be expressed. Thus, a plant that requires short days for flowering will not flower if exposed to long days. When one speaks of genes governing the development of an organism, one really means the potential development; for such development to occur, the environment must also be suitable.

Protein Synthesis

More and more evidence has accumulated that indicates that the deoxyribonucleic acid (DNA) molecules of the chromosomes act as templates that determine the order in which the nucleotides of ribonucleic acid (RNA) are arranged. There are three types of the latter produced in the nucleus, but all move to the cytoplasm. The small **transfer RNA (tRNA)** molecules carry out their function in the cytoplasm, while the larger **messenger RNA (mRNA)** and **ribosomal RNA (rRNA)**

[1] In many of the algae, fungi, and bryophytes there are multicellular structures composed of haploid cells and these will be considered later as we discuss those plant groups individually.

migrate to the ribosomes, some of which are on the endoplasmic reticulum. The rRNA is part of the structure of the ribosome, along with proteins.

The nucleotides of RNA are quite similar to those of DNA (Section 4.5), except that the sugar is ribose and **uracil (U)** substitutes for thymine as one of the nitrogenous bases. The base sequence of an RNA molecule is determined by the DNA through a process of base pairing between bases on one of the DNA strands and bases of the ribonucleotides that are within the nucleus. The hydrogen bonds, which hold the parallel strands of the DNA together, are relatively weak bonds, and the helix comes apart quite readily. Hydrogen bonds will form between adenine (A) and uracil (U), thymine (T) and adenine (A), and guanine (G) and cytosine (C); no thymine is present in RNA and no uracil is present in DNA. For example, if the base sequence in DNA is C—T—A—C—G . . . , the complementary sequence in RNA would form as G—A—U—G—C . . . (see Figure 18-33). The complementary nucleotides are now in position for the ribose of one to bond to the phosphate of the next, until the chain of nucleotides forms an RNA molecule, which consists of a single strand rather than two parallel strands as found in DNA. This first portion of the overall information transfer from DNA into protein synthesis is termed transcription. The message of DNA (e.g., C, T, A, C, G) is transcribed into a similar complementary system of the RNAs (G, A, U, G, C in this example). The transfer, messenger, and ribosomal RNAs are each complementary to specific regions on the DNA. In effect, then, **transcription** is the conversion of genetic information present in a sequence of bases in DNA to a sequence of bases in RNA.

In the next process, **translation,** the coded information of the DNA programs the synthesis of specific polypeptides by means of the RNAs that have been produced; the genetic information is translated from a nucleotide code to an amino acid code. A **polypeptide** is a compound made up of amino acid residues joined by peptide bonds (Figure 18-34); a *protein* may consist of one or more specific polypeptide chains. All three RNAs are functional in this final translation of information, but it is mRNA that specifies the amino acid sequence of the polypeptide end product. The rRNAs function as a structural component of the ribosomes. The mRNAs, which also migrate to the ribosomes, contain a series of triplet **codons;** that is, a sequence of three bases (e.g., GCA, UUG, and so forth) each of which is specific for a particular amino acid. The order in which these codons are arranged along the mRNA determines the order of specific amino acids in a polypeptide, and therefore the chemical identity and structure of the protein.

Although the type and sequence of amino acids in a polypeptide are determined by the nucleotide sequences of mRNA, amino acids do not join directly to the mRNA strand. The selection of amino acids from the cellular pool is brought about by the activity tRNAs. Amino acids are first activated by ATP and are then attached to tRNA. There is a specific tRNA for each of the 20 amino acids that are used to form the myriad of protein molecules that are essential to the normal development and functioning of a cell. As each tRNA picks up its particular activated amino acid, the combined molecules move toward a ribosome where the mRNA is located.

Somewhere along its nucleotide chain, each tRNA is thought to contain a specific **anticodon** consisting of three bases (e.g., CGU, AAC). The three bases of an

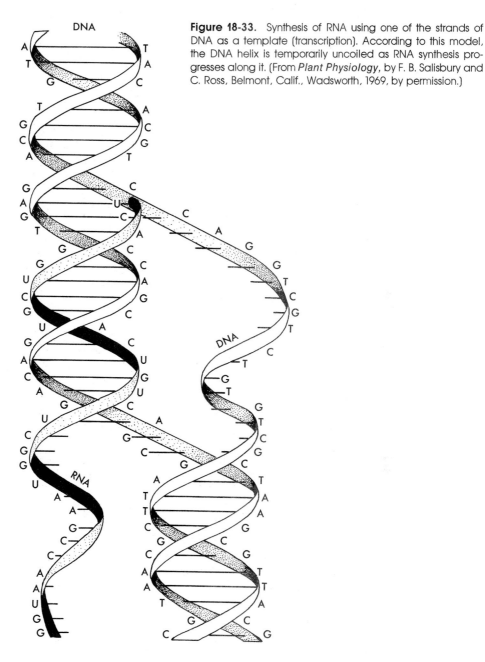

Figure 18-33. Synthesis of RNA using one of the strands of DNA as a template (transcription). According to this model, the DNA helix is temporarily uncoiled as RNA synthesis progresses along it. (From *Plant Physiology*, by F. B. Salisbury and C. Ross, Belmont, Calif., Wadsworth, 1969, by permission.)

anticodon are complementary to the three bases of a specific codon on an mRNA. The anticodons pair with the respective complementary codons (A to U, G to C, and so forth), and thus the proper sequence of tRNAs is lined up along the mRNA (see Figure 18-35). As the tRNAs become associated next to each other on an mRNA, their attached amino acids become enzymatically linked together one after another to form increasingly longer chains of amino acids (polypeptides)

GLYCINE

$$H-N-C-C-OH$$

with H, H on top and O (double bond), H below N

ALANINE

SERINE

Figure 18-34. Amino acids (e.g., glycine, alanine, serine) are linked by a peptide bond (CONH) with the loss of water (HOH). A polypeptide is thus a series of amino acid residues joined by peptide bonds and having an amino group (NH_2) at one end and a carboxyl group (COOH) at the other. The example shown is a tripeptide, formed by the condensation of three amino acids.

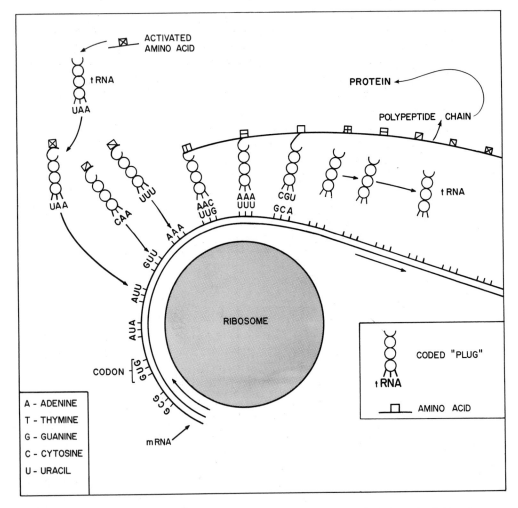

ACTIVATED AMINO ACID

t RNA

UAA

PROTEIN

POLYPEPTIDE CHAIN

t RNA

UAA

UUU

CAA

AAA

AAC UUG

AAA UUU

CGU

GC A

GUU

AUU

AUA

RIBOSOME

CODED "PLUG"

t RNA

GUG

CODON

t RNA

AMINO ACID

GCG

mRNA

A - ADENINE
T - THYMINE
G - GUANINE
C - CYTOSINE
U - URACIL

Figure 18-35. Diagrammatic representation of a possible mechanism of protein synthesis. Transfer RNA is represented as a twisted strand with a bond to an amino acid at one end and a coded three-nucleotide section. The mRNA strand is diagrammed to show only the nucleotides composing condons. See text for further discussion. (Adapted from *The Principles of General Biology*, by M. S. Gardiner and S. C. Flemister, copyright 1967 by Macmillan, Inc., New York, used with permission.)

and, eventually, a protein molecule. As each amino acid links to the next, it detaches from its tRNA, which then moves away from the mRNA and can accept another similar amino acid. When the chain of amino acids is completed, it is then released as a protein molecule.

It appears that a number of ribosomes are simultaneously associated with the mRNA strand; such a complex is termed a **polysome** or **polyribosome.** Either the mRNA moves along the ribosomes, or vice versa, bringing new codons into position, and thus also positioning the new amino acid that has been brought to that codon by the tRNA. In this way an amino acid is added each time a ribosome passes a group of three mRNA nucleotides (the codon). The ribosomes control the attachment of amino acids on the elongating polypeptide chain. Specific codons probably determine when the amino acid chain is of the correct length and is released from the ribosome. Several ribosomes are usually moving along the mRNA and are simultaneously involved in protein formation; the farther along the mRNA, the longer would be the polypeptide chain. Each cell, of course, is constantly forming several different kinds of proteins using different mRNA molecules coded by separate genes.

Specific enzymes are involved in each of the processes just described: the transcription of RNAs, the attaching of an amino acid to its tRNA, and the incorporation of individual amino acids into a protein.

Transcription and translation preserve the genetic information by the sequences of bases and complementarity. The genetic information is not used up but is retained in the nucleus (chromosomes) as DNA. Thus the genetic information is transmitted from generation to generation in the form of the chromosomes that contain DNA.

Differentiation and Development

The preceding discussion has indicated how the genetic information of a cell is translated into the production of proteins. Some of these proteins form structural components of the cell; others are the enzymes responsible for the patterning of a cell's activities. The various processes that may occur within a cell depend upon the presence of enzymes specific for each such process. The result of these activities is the general development of the organism. Multicellular organisms develop from an initial single cell that undergoes mitosis followed by similar divisions of the subsequent cells; in the more complex plants, the initial cell is the zygote or fertilized egg. The precise doubling and sorting out of chromosomes during mitosis result in each cell of the plant having identical chromosomes and genes. And yet, some cells differentiate as storage cells, some as conducting cells, others as photosynthetic cells, and so forth. Apparently, some mechanism exists in the cell to turn genes "on" or "off" at different times and/or in different environments. The manner in which transcription ("on") and repression ("off") are controlled will determine what kind of cell is formed and how it functions (i.e., both structural and functional characteristics are controlled in this manner).

Because the genetic material within the various cells of an organism is similar, the question of differentiation and development was very confusing. It was not until the brilliant work of Jacob and Monod, published in 1961, that the broad out-lines for genetic control of such development was envisioned. The following dis-cussion is based upon their work.

Jacob and Monod have suggested that there are three types of genes functioning as a general unit. An organized group, called an **operon,** consists of an operator gene adjacent to one or more structural genes. The **operator gene** acts as a switch, either blocking the structural genes or permitting them to function. The **structural genes** are responsible for the synthesis of mRNA, which then carries the informa-tion for the synthesis of specific polypeptides. The activity of a **regulator gene,** located elsewhere on the chromosome, determines whether or not the operator gene will permit transcription of the structural genes in its operon. The regulator gene is responsible for the synthesis of an mRNA that results in the production of a protein, known as a **repressor,** which blocks the action of the operator gene and thus shuts off transcription of the entire operon. If the repressor reacts with an appropriate substance (an **inducer**), diffusing into the cell or produced within the cell, derepression occurs (i.e., the repressor is inactivated). This permits action of the operator gene, which in turn enables the structural genes to produce mRNA. It appears that one mRNA is produced per operon and that several polypeptide chains are coded by each such long mRNA molecule; some codons appear to sig-nal chain termination or spaces between polypeptides. It is possible that specific spatial orientations of several proteins may be established in this fashion and re-sult in organized sequences of reactions. This suggested flow of information in the cell is diagrammed in Figure 18-36.

It is also possible that the regulator gene is responsible for the production of a protein, an **activator,** which *allows* the operator gene, and thus the structural genes, to function. An appropriate substance, perhaps a high concentration of end products, may react with the material from the regulator gene and thus block ac-tion of the operator gene and its operon. When the concentration of end products drops, the operon again functions. This type of feedback inhibition tends to pre-vent the cell from using energy or material in the production of substances already present in adequate amounts.

It has also been suggested that two operons may interact in regulating gene ac-tivity. In this model (Figure 18-37) each operon contains a regulator gene as well as structural genes. The regulator gene of each operon produces a repressor that affects the operator gene of the other operon. In this way regulator gene I blocks operator gene II, and regulator gene II blocks the activity of operator gene I. The two repressors are inactivated by different inducers; inducer I inactivates repres-sor I, and inducer II inactivates repressor II. The first inducer to which this sys-tem is exposed will determine which set of structural genes will function. By such a mechanism, cell differences may be established. The introduction of a particular inducer may cause a particular set of genes to begin functioning and repress the activity of other genes. Other cells with similar genetic composition but subject to other inducers develop differently.

The mechanisms as postulated would permit rapid control of cellular activity without disruption of the metabolic machinery. Differential concentrations of sub-strates, end products, oxygen, growth regulators, or even carbon dioxide might exert varying influences on cellular metabolism and control cell differentiation.

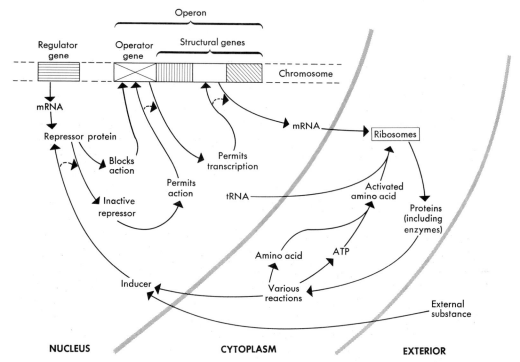

Figure 18-36. Diagram indicating the possible flow of information in a cell.

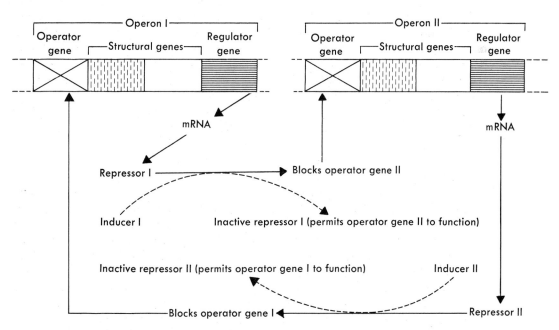

Figure 18-37. Possible method of mutual control of gene activity in two operons. See text for additional discussion.

For example, cells toward the interior of a tissue would tend to be exposed to a situation quite different from surface cells. It is also possible that the products from one cell may influence the activities of a neighboring cell. Although the genetic material is similar, regulation of gene activity could vary from cell to cell. As a result, cells may function and develop differently from one another. All such interactions could readily affect cell differentiation and the general development of an organism.

Although the operon concept is well substantiated in at least some bacteria, other systems may function in higher organisms. It has been suggested, and supported by recent evidence, that the histone constituents of chromosomes play a role in controlling gene action. **Histones** are proteins that form complexes with DNA, and this appears to repress RNA synthesis. If the histones are removed, a marked increase in RNA synthesis occurs. The possible role of histones in the regulation of gene activity is still hypothetical, but future investigations may clarify the situation.

Summary of Gene Function

The DNA of the genes (or chromosomes) determines the coding of the transfer RNA, messenger RNA, and ribosomal RNA molecules. This in turn determines which amino acids will combine with the tRNA. The coding of the messenger RNA is of paramount importance because this determines the order of alignment of the tRNA molecules with their attached amino acids and thus determines the orientation of amino acids with respect to one another. The alignment of amino acids (i.e., their sequence) determines the specific polypeptide and protein that is formed. Basically, *one gene controls one polypeptide,* and a protein may consist of one or more polypeptides. Since enzymes, which are proteins, govern the metabolic activities of the cell, the genes effectively determine what these activities will be. Various mechanisms function to bring about variations in cellular activity by turning on or turning off reactions, or at least changing their rates. The actual appearance and functioning of an organism result from the cooperative activity of individual cells.

Summary

1. Meiosis is a nuclear division that occurs at some time prior to gamete formation and results in a diploid cell giving rise to haploid cells. Not only does a reduction in chromosome number occur, but also a separation of the homologous pairs of chromosomes. The number of different kinds of haploid cells that are produced depends upon the random alignment of chromosomes at metaphase of meiosis.

2. The DNA (deoxyribonucleic acid) molecules act as templates that determine the configuration of the ribonucleic acid (RNA) molecules. The latter are involved in protein synthesis.

3. Four basic principles are involved in genetical studies: (a) unit characters,

(b) dominance, (c) segregation, and (d) independent assortment. Dominance, however, does not occur between all gene pairs (alleles).

4. Genes that are located on the same chromosome are said to be linked. Such genes may change linkage by a process known as crossing-over in which chromatids exchange parts.

5. Characteristics of organisms may be determined by many genes or by a series of alleles, and some genes may be lethal.

6. Genes determine the potentiality of an organism, but the environment influences the manner in which the genotype is expressed.

7. Man makes use of his knowledge of genetics to improve plants for his use.

Review Topics and Questions

1. Briefly define the following: diploid, haploid, allele, synapsis, homozygous, heterozygous, hybrid, phenotype, genotype, linkage, and crossing-over.
2. Assuming that the diploid number is four, diagram a cell in the following stages of division:
   (a) Early prophase of meiosis   (c) Metaphase I of meiosis
   (b) Metaphase of mitosis   (d) Metaphase II of meiosis
3. Describe the manner in which genes are presumed to function.
4. Briefly explain the meaning of each of the four basic Mendelian principles: (a) unit characters, (b) dominance, (c) segregation, and (d) independent assortment.
5. Genes **A**, **B**, and **C** are located on the same long chromosome; their alleles are similarly located on the homologous chromosome. Assuming that the diploid cell is heterozygous for all factors, list the types of gametes that may be produced (a) if crossing-over occurs between genes **A** and **B**, (b) if crossing-over occurs between genes **b** and **c**, (c) if crossing-over does not occur, and (d) if crossing-over occurs between A and B *plus* crossing-over between **B** and **C** (i.e., double crossing-over).
6. Why does linkage present a problem to plant breeders who are trying to develop new kinds of plants?
7. Explain the significance in the life of plants of each of the following: (a) synapsis, (b) mitosis, (c) sexual fusion, and (d) meiosis.
8. You have a red-flowered snapdragon that is resistant to a rust disease, and a white-flowered snapdragon that is susceptible to this rust disease.
   If white flowers and disease resistance are both recessive characters, describe step by step what you would have to do to produce a true-breeding strain that is white-flowered and disease-resistant. Assume no linkage.
9. Prior to the time of Mendel, crosses between phenotypically different organisms had been performed with the result that the progeny were intermediate between the two parents. The conclusion was that the progeny resulted from a blending of the characteristics of the two parents.
   Mendel came to the conclusion that each parent contributed factors to the

progeny and that the factors from one parent were inherited independently of the factors from the other parent. He did not agree that any blending occurred. Discuss the type of information that led Mendel to this conclusion.

10. In Jimsonweed purple flower color (**P**) is dominant over white (**p**) and spiny pods (**S**) over smooth (**s**). These pairs of alleles are not linked.

    (a) A Jimsonweed of genotype **PPss** is crossed with one of genotype **ppSS**. What will be the genotype and phenotype of the F_1 generation? If the plants of the F_1 generation are self-fertilized, what will be the phenotypes of the F_2, and in what ratio will these phenotypes occur?

    (b) Two Jimsonweeds with white flowers and smooth pods are crossed with one another. What will be the genotype and phenotype of the F_1 generation?

11. In corn plants one gene controls the height of the plant. Normal height (**N**) is dominant over dwarf (**n**). A normal corn plant is crossed with a dwarf plant. Of the progeny 247 are normal and 264 are dwarf. What are the genotypes of the two parent plants?

12. Briefly explain how the following influence the possible number of different kinds of gametes produced by an individual: (a) linkage and (b) crossing-over.

13. Plant *Y* is always self-pollinated (self-fertilized) and bears red flowers; flower color is governed or controlled by one gene pair. In one fertilized egg a single gene changed, by chance, to a recessive white-flower characteristic.

    (a) What kind (color) of flowers will be found on the plant *Z* developing from this fertilized egg?

    (b) If this plant *Z* is self-fertilized, what will be the genotypes and phenotypes of its progeny? Be sure to include ratios.

14. In peas red flowers (**R**) are dominant over white flowers (**r**). A pea plant with white flowers is crossed with one that has red flowers. Of the progeny 147 have white flowers and 161 have red flowers. What are the genotypes of each parent and of each type of progeny?

15. A strain (*X*) of tomato characterized by the ability to grow at low temperatures carries a gene (**t**) for this character that is recessive. Strain *X*, however, has no resistance to a serious wilt disease. A second strain (*Y*) is characterized by resistance to the wilt disease; this resistance is the result of a recessive gene (**r**). A certain area in Wyoming is characterized by low temperatures during the growing season and also by the prevalence of the wilt disease; the common tomato cannot be grown there for both reasons.

    You are given (1) strain *X*, homozygous for ability to grow at low temperature and for lack of resistance to wilt, and (2) strain *Y*, homozygous for resistance to wilt and for lack of ability to grow at low temperature. You wish to produce strain *Z*, which will grow successfully in the Wyoming area. By means of diagrams (or checkerboards), indicate the necessary crosses and point out the genotype of strain *Z*. Assume no linkage.

16. A flower grower finds in his nursery a plant with white flowers (recessive to the usual red), which he wishes to propagate from seed. However, the plant carries on the opposite end of the same long chromosome a recessive gene for dwarfness, which is undesirable.

Explain how the grower can obtain a tall, white-flowered plant that will breed true for both of these characters. Describe all steps in the process.

17. A plant-breeding program in one of our agricultural experiment stations resulted in the production of two varieties of melon. Variety X is homozygous for disease susceptibility (**S**), for large fruit (**L**), and for smooth fruit (**w**). Variety Y is homozygous for disease resistance (**s**), for large fruit (**L**), and for wrinkled fruit (**W**). The genes governing disease resistance (or susceptibility) and size of fruit have been determined to be at opposite ends of the same long chromosome.

   Explain how the plant breeder should continue his program in order to obtain a plant that is homozygous for disease resistance, for large fruit, and for smooth fruit.

18. Starting with one plant that is homozygous for smooth leaves (**M**) and heterozygous for fused petals (**F**), and a second plant that is heterozygous for smooth leaves and homozygous for fused petals, explain how a plant breeder could obtain a plant that has hairy leaves (the recessive allele of smooth) and separate petals (the recessive allele of fused).

19. A tall red-flowered plant (heterozygous for both characters) is crossed with a dwarf white-flowered plant (both characters recessive). The F_1 consisted of 1008 tall red-flowered plants and 988 dwarf white-flowered plants. Explain these results.

20. A rancher had one peach tree that produced much larger fruit than the others but had such a poor root system that the tree did not grow very well. His other trees, which had excellent root systems, produced small fruit. Since fruit size and root system each is dependent upon a single factor respectively, and since small size and poor root system are dominant, explain how the rancher would (or could) produce trees with a large fruit and a good root system.

21. On one long chromosome of a potato occurs a dominant gene (**G**) that causes giant tubers but is lethal (causes death of the plant) if it is homozygous. On the same individual chromosome (but at the other end) is a second dominant gene (**B**) that causes a bitter flavor and is undesirable.

   You wish a strain of potato with giant tubers and a pleasant flavor. Explain how you can obtain the desired strain, assuming that the strain that is presently available is heterozygous for flavor.

22. An onion has, on opposite ends of a long chromosome, gene (**w**) (recessive) for white bulbs and gene (**s**) (recessive) for small bulbs. Another plant of the same species carries the corresponding genes (**W**) (dominant) for nonwhite bulbs and (**S**) (dominant) for large bulbs. Both these plants are homozygous.

   How can we obtain from these plants a true-breeding strain of onion with large white bulbs? Describe the steps and processes necessary to arrive at such a strain.

23. In violets the gene that bears the potentiality for blue petals is dominant over that for yellow petals, and the gene that bears the potentiality for large flowers is dominant over that for small flowers.

   A plant with yellow petals and large flowers was crossed with one that had blue petals and small flowers. The F_1 generation was composed of

47 plants with yellow petals and large flowers;

45 plants with yellow petals and small flowers;

50 plants with blue petals and large flowers;

46 plants with blue petals and small flowers.

What were the genotypes of the parents? Show the method you used to determine these genotypes.

24. A plant is heterozygous for yellow flowers (red is recessive) and for hairy leaves (smooth is recessive). This plant is self-fertilized, and the offspring (or progeny) consist of

72 plants with yellow flowers and smooth leaves;

143 plants with yellow flowers and hairy leaves;

68 plants with red flowers and hairy leaves.

Explain the genetical basis for such results, including the reason why no red-flowered, smooth-leaved individuals were produced.

25. By adjusting the environment, an ordinarily late-flowering, rust-resistant plant was crossed with an early-flowering, rust-susceptible plant. All the 200 plants of the F_1 generation were early-flowering and rust-resistant. In the F_2 generation the progeny were as follows:

302 early-flowering, rust-susceptible;

584 early-flowering, rust-resistant;

287 late-flowering, rust-resistant.

Let E = early-flowering, e = late-flowering, R = rust-resistant, and r = rust-susceptible.

(a) Diagram, showing only the chromosomes and genes concerned, a diploid cell of each of the plants crossed initially and a diploid cell of the F_1 generation. Show as many chromosomes as necessary to indicate the distribution of the genes.

(b) Give the phenotypic ratio that would be expected if the early-flowering, rust-resistant plants of the F_2 were self-fertilized.

Suggested Readings

BEADLE, G., and M. BEADLE, *The Language of Life* (Garden City, N.Y., Anchor Books, Doubleday, 1967).

BURNS, G. W., *An Introduction to Heredity* (New York, Macmillan, Inc., 1969).

CRICK, F. H. C., "The Genetic Code III," *Scientific American* **215,** 55–62 (October 1966).

HURWITZ, J., and J. J. FURTH, "Messenger RNA," *Scientific American* **206,** 41–49 (February 1962).

MANGELSDORF, P. C., "Hybrid Corn," *Scientific American* **185,** 39–47 (August 1951).

MERRELL, D., *Introduction to Genetics* (New York, Norton, 1975).

PTASHNE, M., and W. GILBERT, "Genetic Repressors," *Scientific American* **222,** 36–44 (June 1970).

STRICKBERGER, M. W., *Genetics* (New York, Macmillan, Inc., 1969).

YANOFSKY, C., "Gene Structure and Protein Structure," *Scientific American* **216,** 80–94 (May 1967).

Plant Classification

19.1 Classification

The plants and animals of the world have been of interest to mankind for many thousands of years. During that time much information has been accumulated, and gradually man realized that disseminating such knowledge is vital to a thorough understanding of living organisms. The more data that were accumulated, the more important became the need to apprise others of the work done by various individuals. In order to discuss plants and animals, they must be given names, preferably in some logical way, or else the 350,000 species of plants known today would just be a gigantic hodgepodge of confusion.

With this in mind, **taxonomists** (those biologists especially interested in plant classification and relationships) have developed a system of categories within which any plant may be placed and given a name. Many such systems have been developed, with many changes being made as botanists learned more about plants, and with many changes undoubtedly yet to come. The system used in this text is one that has been adopted by many botanists, but one that is subject to change, just as are all systems of classification. There are, of course, several other systems in current use. Each interested botanist must make a decision as to which system he thinks is most suitable. It is also likely that new ideas and new data and new interpretations will cause him to revise his concepts and accept a different system at some later date. There is no *one official* classification.

In general, the higher the category, the more individuals in it and the fewer the number of categories at the same level. Conversely, the lower the category, the fewer individuals in it and the larger the number of categories at that level. The following outline of the position of the "cork oak" will serve to illustrate the method of grouping organisms.

Kingdom Plantae
Division Antho*phyta*
Class Dicotyledon*ae*
Order Fag*ales*
Family Fag*aceae*
Genus *Quercus*
Species *suber*

The **scientific name** consists of the generic and specific names—*Quercus suber* in this case—and the other groupings need not be used. A number of closely related species are combined into a larger unit, the genus; a number of closely related genera likewise are combined to form a family; families combine to form an order; and so on to the all-inclusive kingdom. In the classification used in this text it is suggested that division names end in the suffix *-phyta* or *-mycota,* classes end in *-ae* or *-eae* or *-mycetes,* orders in *-ales,* and families in *-aceae.* The names of genera and species have diverse endings, and in some cases long-time usage has prevented the complete adoption of the endings here suggested for the other groupings; long-established custom sometimes antedates the present rules of botanical nomenclature. Each group in any system of classification, regardless of rank, is referred to as a **taxon** (plural, **taxa**). Rules governing the application of botanical names to plants and their taxa are embodied in the *International Code of Botanical Nomenclature,* which is revised at International Botanical Congresses that are held every five years. Revisions are necessary because all schemes of classification are established by humans and are not necessarily natural entities, and there are bound to be differences of opinion as to which system best shows natural relationships.

Taxonomic groupings are devices that enable one to identify more readily a specific organism, much in the manner of the geographical identification of a specific locality. The following may serve to illustrate my meaning (if I may be forgiven the use of my own working area):

Hemisphere Western
Country United States of America
State California
County Santa Barbara
City Santa Barbara
Locality university campus

The largest grouping, hemisphere, contains the greatest number of specific areas, while the county or city contains fewer such areas.

19.2 Scientific Name

As mentioned in the previous section, the smallest two categories to which a plant belongs, the genus and species, constitute its **scientific name** and are usually followed by the name or initial of the person who first described the species fully. All scientific names are **binomials** (i.e., a name consisting of two parts) in which both terms are underlined or italicized and the genus is capitalized, for example, *Sequoia sempervirens* (D. Don) Endl. The abbreviated names following the scientific name of this redwood tree refer to the man who first described the species, David Don, and the man who revised that classification, Stephen Endlicher. We are indebted to the great Swedish naturalist Linnaeus (Karl von Linné) for establishing the binomial system of classification in his monumental work *Species Plantarum,* published in 1753. Until that time, plants were referred to by a short descriptive phrase, a polynomial, which was cumbersome to say the least. Such phrases made no attempt at indicating relationships, which is the basis for all modern classification.

Beginning students frequently want to know why a redwood cannot be called a redwood, and why it must be called *Sequoia sempervirens.* This is certainly a logical question and deserves an answer. A standardized name of international recognition is necessary so that each interested individual knows which plant is being discussed and the relationship of this plant to others. This accord would be difficult to achieve if common (vernacular or trivial) names were used, because the same common name is frequently used for entirely different plants, depending upon the locality. No difficulty arises with the redwoods, but, as an example, the word "corn" originally meant the grain of any cereal plant and has come to refer to "maize" in the United States, "wheat" in England, and "oats" in Scotland. The name *Zea mays* refers to one specific plant in all three countries, that which many of us are fond of as corn on the cob. In this case the "us" refers to people in the United States. Different common names are also often applied to the same species in different areas. For example, *Pseudotsuga menziesii* is called Douglas fir, Douglas spruce, red spruce, yellow spruce, and Oregon pine—it is *not* a fir or a spruce or a pine; the name derivation is *pseudos,* false, and *tsuga,* hemlock. Consider also the confusion arising from "poison sumac," "poison ivy," and "poison oak." These are actually closely related species within the genus *Rhus* (respectively, *R. vernix, R. toxicodendron,* and *R. diversiloba*), although the common names would not indicate this. The confusion engendered by the use of common names is quickly clarified when scientific names are utilized. Also, it might be pointed out that no common names exist for many genera or species: chrysanthemum, fuchsia, aster, and asparagus are examples.

The terms used to name plants are derived mostly from either Latin or Greek and are selected for various reasons. In many cases the term reflects a specific characteristic, such as *Lupinus horizontalis,* in which the lower branches are prostrate, or *Trifolium* spp. (spp. = species), which has leaves divided into three leaflets (palmately trifoliate). In other cases the name reflects a location, such as *Pyrus americana* (American mountain ash), or it may refer to a famous person, as in the genus *Erwinia,* which was named after Erwin F. Smith, a noted authority in the field of bacterial plant pathogens. Once a plant has been named and described fully, that name cannot be used again. The correct name is that which first appears in the literature, although changes must sometimes be made as classifications are revised. Under these conditions, however, the species name is maintained if at all possible, even if the plant is placed in a new genus, family, and so forth. Even if changes must be made, the scientific name has stability because such changes are made according to established rules of nomenclature. The Latin form is used in naming plants because it is no longer a spoken language and does not change, and it was at one time the written language of scholars.

The binomial is used because it is precise, simple, and reflects relationships. It frequently is a description reduced to its simplest form, and a proper scientific name usually reveals a great deal about the plant to anyone who has an understanding of the rudiments of taxonomy. For example, *Quercus alba* is the white oak which has light-colored leaves; *alba* is from the Latin and means white. In addition the genus *Quercus* includes only those plants with certain characteristics, and this genus is a member of the Fagaceae or beech family, which also has certain characteristics. Such basic knowledge enables the student to visualize certain descriptive features of the plant under discussion even if he has never seen the plant. In a like manner *Quercus rubra* refers to the red oak (*rubr* = red) whose leaves turn a beautiful red in the fall, and *Q. bicolor* (*bi* = two) refers to the swamp white oak whose leaves are dark green on top and silvery white on the bottom. The student knows that all three plants are closely related and have certain similarities (i.e., the characteristic features of the genus *Quercus* and the family Fagaceae) as well as obvious differences that place them in separate species. Nomenclature is really an important tool used by biologists in their studies of organisms, and the binomial is a "shorthand" or abbreviated method of conveying a lot of information.

19.3 What Is a Species?

Much controversy still exists concerning the definition of a **species,** but it may be regarded as a group containing all the individuals of a particular kind of plant and excluding all those that are different. One may say that two groups are different species if reasonably consistent differences exist between them and if they maintain these differences under natural conditions. Usually, but not always, species are incapable of interbreeding with other species. In any case, the species is considered to be a biological unit delimited primarily by genetical criteria and containing similar individuals.

No exact definition is possible, because species are not really stable but are byproducts of the dynamic evolutionary process and are constantly changing although it is usually a slow process taking many generations. The exact range of such terms as "similarities" and "differences" is also extremely difficult to determine, and they vary from group to group. There are, of course, differences of opinion and interpretation. However, the species concept is an exceedingly important one in taxonomy and accepted by all biologists in spite of the difficulties involved in defining the term.

Identification of a Species

How does one identify a plant or find out its name? The simplest method is to ask someone who knows. This does *not* mean "ask any botanist." A plant anatomist, plant physiologist, or plant cytologist would undoubtedly know some of the more common plants, but his specialty is not taxonomy; the taxonomist should be

351

Sec. 19.4 /
Natural and
Artificial
Systems of
Classification

consulted.[1] If no taxonomist is available, a local manual should be utilized. This is a book that contains descriptive "keys" as an aid to identification. In most instances one finds a **dichotomous key,** one in which two choices are available at each step, such as the following:

1a. Leaves parallel-veined;
   flower parts in 3s 2
1b. Leaves net-veined;
   flower parts in 4s or 5s 10
   2a. Flowers without sepals or petals 3
   2b. Flowers with sepals and petals 8

Each choice leads to another choice of two until the final choice indicates the species. Unfortunately, some groups are not easily "keyed out." A further possibility is to compare the plant with specimens in an **herbarium,** which is an orderly collection of dried, pressed, and labeled plant specimens. One might also visit a botanic garden, or arboretum, which is the botanical equivalent of a zoo and contains many labeled living specimens.

Actually, identifying plants can be a fascinating hobby as well as a profession. Many people find that collecting, preserving, and identifying plants is an excellent outlet for their energy; it is instructive and rewarding. Taxonomy is often that phase of botany which first interests the beginner.

19.4 Natural and Artificial Systems of Classification

The early systems of classification tended to be **artificial,** in that they were based upon arbitrary criteria without regard for evolutionary relationships and emphasized differences such as general growth form, flower color, leaf shape, or type of bark. The natural systems of today have almost completely replaced those of an artificial nature. A **natural system** is one that attempts to classify organisms according to their genetic relationship and reflects the results of evolution. Such a system is based upon both similarities and differences, with the most similar organisms grouped together.

Because natural systems are based upon man's knowledge of plants, they must constantly be revised to keep up with new knowledge. Even when knowledge appears to be complete, the system will be imperfect, because a vast number of

---

[1] The layman usually assumes that all botanists are taxonomists and/or horticulturists. The two most frequent questions asked of a botanist are (1) "What plant is this?" (2) "What is the matter with my ——? It has spots all over it." And yet the same individuals would not think of asking a zoologist to diagnose illness in their children, pets, or livestock. They might, of course, ask a zoologist to identify such diverse items as insects, spiders, snakes, and birds—with about the same results as obtained with botanists identifying plants.

Those fields covered by the term *botany* are so diverse, and include such a tremendous amount of data, that no individual is competent in all phases of the subject. See Chapter 1 for a discussion of the various aspects of botany.

changing, evolving organisms cannot conform to a man-made system of a few categories. And if enough categories could be created, the main purpose of classification would be defeated.

An excellent example of an organism that does not quite fit is *Euglena*. This is a
one-celled organism, motile by means of a single hairlike appendage (a flagellum),
flexible, capable of ingesting food through a gullet, and containing chloroplasts.
All except the last characteristic would indicate that it is an animal; the presence
of chloroplasts indicates that it is a plant. Then what *is* it? The difficulty arises in
the question that is asked. The question *should* be, "What shall we *call* this organism?" *Euglena* can fit into either kingdom, or in a third kingdom (Protista) as suggested by many biologists. In fact, zoologists usually study *Euglena* as a member
of the Class Mastigophora in the Phylum Protozoa, whereas botanists place this
organism in the Division Euglenophyta. This is merely indicative of the fact that
manmade rules do not regulate the activity of living organisms. How we classify
such an organism is not important so long as it is given a name so that it can be
identified, studied, and discussed by interested biologists.

In even our modern natural systems of plant classification we are forced to accept a least some artificiality. For example, the Class Deuteromycetes (or Fungi
Imperfecti) in the Division Eumycota consists of an artificial group of fungi in
which the existence of sexual reproduction has not been determined. This is
strictly a grouping of convenience, and the only relationship involved is to indicate the similarity of these organisms to other true fungi of this division. If a closer
relationship is determined by investigation, a member of this class is transferred to
its more logical position in one of the other classes in the Division Eumycota.
Some fungi without known sexual stages are placed in classes other than Deuteromycetes because of the obvious relationships indicated by various structural characteristics.

There is considerable diversity of opinion as to which organisms should be included in a study of plants. Traditionally, all organisms were classified as either
plants or animals (i.e., two kingdoms), but the more recent trend is to utilize three
to five kingdoms. Such classifications also involve difficulties, and there is as yet
no real agreement with which even the majority of botanists can agree. The propo-
sal that the bacteria and blue-green algae be placed in a separate kingdom (Monera) has received considerable acceptance, however, because these organisms are
prokaryotic while all other organisms are eukaryotic. Some biologists would utilize the kingdom Protista to include protozoa, slime molds, diatoms, and dinoflagellates; while others would also add the rest of the algae, except the blue-greens,
to this kingdom. The Fungi would constitute a separate kingdom, according to
some biologists, but with the slime molds placed in the Protista. On a five kingdom
basis, this would leave multicellular animals in the Animalia and multicellular
plants in the Plantae. The latter kingdom would include multicellular algae in some
systems, but these groups would be in the Protista of other systems. If the algae
are not included, the kingdom Plantae would consist of bryophytes (mosses, liverworts, and so forth) and vascular plants.

There are advantages as well as disadvantages to any scheme of classification,
and it is difficult to decide which groupings to use. It is essential, however, that
some system is utilized. This text will treat classification in the more traditional

manner: the kingdom Monera will be used, but Plantae will include all those other organisms that have been traditionally grouped as plants (i.e., algae, fungi, bryophytes, and vascular plants).

19.5 Types of Reproduction in Plants

Modern systems of classification are based mainly upon reproductive structures, because these are the most stable morphological characteristics. The environment has considerable influence on the general size and shape of the plant; a lack of water or certain minerals, for example, will result in stunted and abnormal development, but the reproductive parts will remain unchanged, provided the plant is capable of reproducing at all. Although plants manifest various kinds of reproduction, it may be thought of as being comprised of two basic types, asexual and sexual, with certain modifications depending upon the plant.

Asexual Reproduction

This type of reproduction is accomplished by cell division and growth, and no fusion of protoplasts (or gametes) occurs.

Vegetative reproduction involves only the ordinary processes of cell division and growth, no special forms of cells being produced. This type of reproduction is accomplished by a variety of methods: the production of runners by strawberry plants, the sprouting of tubers and corms, and the fragmentation of a multicellular plant body into segments each of which is capable of developing into a new plant, as in many algae, are common examples. The rooting of woody cuttings and the grafting of plant parts are methods of vegetative propagation utilized by many individuals, but these do not usually occur under natural conditions. Two methods of vegetative reproduction are sufficiently important to be named and defined separately: fission and budding. **Fission** is the division of a unicellular organism into two new unicellular organisms of approximately equal size and is the method by which bacteria and various other primitive plants reproduce. **Budding,** as in the unicellular yeast plant, is the production of a localized outward bulging (or protuberance) of the cell wall, which receives cytoplasm from the parent cell and becomes constricted at the base; the nucleus of the parent cell divides (mitosis), and one of the daughter nuclei moves into the "bud." A wall forms between the "bud" and the parent cell, and separation of the two structures results in two cells or plants.

Spore formation is that type of asexual reproduction in which are produced specialized reproductive cells or groups of cells that are capable by themselves of producing new plants by cell division and growth. The many different kinds of spores are designated by a variety of names. If the spores are motile, by means of hairlike appendages, they are called **zoospores.** The structure in which spores are produced is a **sporangium** (*-angium* means case or container), although such a structure is sometimes given a more specific name, such as the sporangium in the mosses, which is called a capsule.

Sexual Reproduction

This type of reproduction involves three principal features:

1. The fusion of two protoplasts (or gametes) followed by the fusion of their nuclei, a process termed **syngamy.** The chromosomes, now twice the original number in the gametes, mingle but do not fuse.
2. The subsequent separation of the two sets of chromosomes, each with the original number (haploid), into separate cells called **meiospores.** This process is meiosis (see Section 18.1).
3. Cell division and growth, which may occur entirely between syngamy and meiosis, or entirely between meiosis and syngamy, or at both places in the cycle.

In some of the primitive organisms, the motile gametes are structurally alike, **isogametes,** and their fusion is termed **isogamy.** The term **heterogamy** is used when referring to the fusion of motile gametes that are structurally somewhat dissimilar, **heterogametes.** In the more advanced organisms, the nonmotile female gametes, **eggs,** are usually larger than the male gametes, **sperm,** which may be motile. The fusion of egg and sperm, as in such diverse plants as *Oedogonium* (one of the Chlorophyta or green algae) and *Lilium* (one of the Anthophyta[2] or flowering plants) is termed **oögamy.** The cell resulting from the fusion of two gametes is termed a **zygote,** and it is diploid. The structure in which gametes are produced is a **gametangium** (plural, **gametangia**). Where male and female gametes are differentiated, sperm are produced in a male gametangium or **antheridium** (plural, **antheridia**) while eggs are produced in a female gametangium or **oögonium** (plural, **oögonia**).

19.6 Alternation of Phases

As mentioned in the previous section, cell division and growth may occur between syngamy and meiosis and also between meiosis and syngamy. This is the common situation in most plants, except for some of the lower forms, and is referred to as **alteration of phases** or **alternation of generations.**[3] In the complete life cycle of a higher (or more complex) plant, the zygote develops into a multicellular, diploid body, the **sporophyte,** which produces meiospores following meiosis. These spores develop into a multicellular, haploid phase, the **gametophyte,** which produces gametes. Thus, two distinct structures are involved in the generalized plant life cycle as diagrammed in Figure 19-1. This life cycle with various modifications and embellishments will be used throughout the discussion of representative members of the plant kingdom.

As various plant groups are discussed, the similarities as well as the differences will be emphasized, and the fact will become obvious that the life cycle represent-

---

[2] The Anthophyta and the Chlorophyta are discussed in subsequent chapters.
[3] The student should be able to determine why the first phrase is preferable to the second.

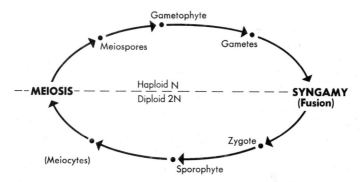

Figure 19.1 Diagram showing generalized plant life cycle.

ative of a particular group can be superimposed upon the word diagram of Figure
19-1. In some the gametophyte is the dominant portion consisting of a relatively
large multicellular plant body, while the sporophyte is relegated to an almost insignificant portion of the total life cycle. A few of the algae, for example, produce no
multicellular sporophyte, and the zygote is the only diploid structure. The other
extreme, where the gametophyte is tiny and inconspicuous, occurs in other algae

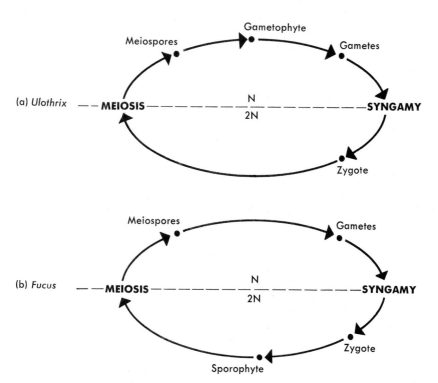

Figure 19-2. **(a)** Simplified diagram of life cycle of *Ulothrix;* no multicellular sporophyte, the zygote functioning as a meiocyte. **(b)** Similar diagram of *Fucus;* no multicellular gametophyte, the
meiospores producing gametes directly.

(e.g., *Laminaria*) and in the flowering plants. Different as these two extremes may appear to be, one is quickly impressed with many basic similarities upon close examination of the structures involved. The basic sequence of events and the basic functions remain the same, even though the structures may be of various sorts:

Zygote → Sporophyte → Meiospores by meiosis →
Gametophyte → Gametes → Zygote by syngamy

The various structural modifications are examples of evolutionary adaptations, and in some cases one or more portions of the generalized life cycle may be by-passed or lost (Figure 19-2).

19.7 Classification of the Plant Kingdom Followed in This Text

In subsequent chapters representative individuals from plant groups will be considered in more detail. No attempt will be made to include all the major groups; rather, selected groups will be utilized to emphasize evolutionary advances and the importance of plants to all life. The following is a brief descriptive account of most of the major plant groups, some of which will be discussed later (a few in more detail than others):

KINGDOM MONERA

Blue-green algae and bacteria. The plant body is unicellular or colonial, with the cells lacking a membrane-bound nucleus and having neither plastids nor mitochondria; the structures within the cell do not have double membranes, and some are not bounded by membranes at all. DNA (not associated with protein) is present, and possibly chromosomes in some. Pigmentation is variable. Reproduction is by fission; sexual reproduction is very rare and does not include fusion of gametes. The outer wall layer is frequently mucilaginous.

Division Cyanophyta

Blue-green algae. Chlorophyll *a* is usually masked by a blue pigment, phycocyanin, or a red pigment, phycoerythrin; there is no chlorophyll *b*. Food reserve is cyanophyte starch, which is similar to glycogen, a type of starch. No sexual reproduction is known. Genera: *Oscillatoria, Nostoc, Gleocapsa,* and so forth.

Division Schizophyta

Bacteria. Chlorophylls are lacking in almost all. They are largely heterotrophic, although a few are chemosynthetic autotrophs. The small group of photosynthetic bacteria contains bacteriochlorophyll, which resembles chlorophyll *a* but has its maximum absorption spectrum in the infrared. Genera: *Nitrosomonas*, *Clostridium*, *Erwinia*, *Vibrio*, and so forth.

KINGDOM PLANTAE

Division Myxomycota

Slime molds. The vegetative body is a naked, uninucleate or multinucleate, nonseptate, amorphous mass of protoplasm. They are heterotrophic, ingesting solid food particles. Asexual reproduction is by nonmotile spores; sexual reproduction is by fusion of amoeboid gametes. Genera: *Plasmodiophora*, *Spongospora*, *Physarium*, and so forth.

Division Eumycota

True fungi. The plant body is unicellular, unbranched or branched filamentous, or pseudoparenchymatous. The cell wall is composed primarily of cellulose or chitin, or a mixture of both. They are heterotrophic. The reserve food is glycogen. Asexual reproduction is by many kinds of spores, some of which are produced in definite sporangia; sexual reproduction is varied. Certain spores are produced only after sexual reproduction and form immediately following meiosis.

Class Phycomycetes (algal-like fungi). The plant consists of nonseptate (coenocytic), branching filaments not organized into compact bodies of definite form. Asexual reproduction is by means of spores (sporangiospores) produced in a sporangium; sexual reproduction is similar to Chlorophyta. Genera: *Rhizopus*, *Saprolegnia*, *Phytophthora*, *Pythium*, and so forth.

Class Ascomycetes (sac fungi). The members of this group are unicellular to multicellular. In most of the latter the septate filaments combine to form a definite macroscopic reproductive body, the ascocarp. Asexual reproduction is usually by chains of spores (conidia) produced at the tips of filaments, although sometimes by budding; sexual reproduction results in the development of a sac, or ascus, containing four or eight ascospores. Genera: *Peziza*, *Aspergillus*, *Penicillium*, *Sacchoromyces*, *Claviceps*, *Venturia*, and so forth.

Class Basidiomycetes (club fungi). Asexual reproduction is usually by chains of spores (conidia) produced at the tips of filaments. The septate filaments usually form a definite macroscopic reproductive body, the basidiocarp. Following nuclear fusion and meiosis, a club-shaped basidium is formed, bearing four basidiospores on small stalks. Genera: *Agaricus*, *Amanita*, *Puccinia*, *Fomes*, *Lycoperdon*, *Ustilago*, and so forth.

Class Deuteromycetes (imperfect fungi). A large, miscellaneous, artificial group of fungi in which sexual reproduction does not occur or has not been found. Genera: *Candida, Alternaria, Colletotrichum, Fusarium, Botrytis, Cercospora,* and so forth.

Class Lichens. These plants are associations of an alga (Chlorophyta or Cyanophyceae) and a fungus (Ascomycetes or Basidiomycetes). They are placed in this division because lichen taxonomy is essentially fungal taxonomy; gross morphology is determined by the fungal partner, and the sexual reproductive structures are strictly fungal.

Division Chlorophyta

Green algae. The plant body is unicellular, colonial, or multicellular, with the cells containing organized and membrane-bound nuclei and plastids. Chlorophylls *a* and *b* and carotenoids are present, as in the higher green plants. Food reserves are mostly starch and fats or oils. Asexual reproduction occurs by cell division and zoospores; sexual reproduction is isogamous, heterogamous, or oögamous. Motile cells usually bear two or four equal flagella anteriorly placed. Genera: *Chlamydomonas, Ulothrix, Oedogonium, Ulva,* and so forth.

Division Euglenophyta

Euglenoids. These are unwalled, unicellular, and typically uniflagellate, with definite nuclei and plastids. Chlorophylls *a* and *b* and carotenoids are present as in Chlorophyta. Food reserves are paramylum (a polysaccharide similar to starch) and fats or oils. Asexual reproduction is by longitudinal cell division. A gullet, which functions to ingest food, is present in some genera. Genera: *Euglena, Colacium,* and so forth.

Division Chrysophyta

Golden-brown algae and the diatoms. The plant body is unicellular, colonial, or multicellular, with organized nuclei and plastids. Chlorophylls *a* and *c* are present but may be masked by various yellow or brown pigments. Food reserves are oils and chrysolaminarin, a complex carbohydrate. Asexual reproduction is by cell division, motile and nonmotile spores; sexual reproduction is isogamous or heterogamous. The wall of each cell is generally composed of two overlapping halves, in either the metabolic or cyst stages; silica frequently composes part of the wall. Flagella, when present, are of unequal size and anteriorly placed. Genera: *Vaucheria, Pinnularia, Navicula,* and so forth.

Division Pyrrophyta

Dinoflagellates. These are typically unicellular, with a wall composed of interlocking plates and containing two grooves, one encircling the cell transversely while the other runs longitudinally along one side. They have two flagella, one lying in the transverse groove, the other extending out from the longitudinal groove. Chlorophylls *a* and *c* are somewhat masked by a brown pigment, peridinin. Food reserves are starch and fat or oil. Reproduction is primarily by longitudinal cell division. Genera: *Ceratium, Goniaulax, Gymnodinium,* and so forth.

Division Phaeophyta

Brown algae. The plant body is a filamentous to complex parenchymatous structure with considerable differentiation. Chlorophylls *a* and *c* are usually masked by a brown pigment, fucoxanthin. Food reserves are sugars, laminarin (a complex carbohydrate), oils, and complex alcohols (such as, mannitol). Asexual reproduction is usually by motile spores; sexual reproduction is isogamous or oögamous. Motile reproductive cells are pear-shaped and have two lateral, unequal flagella. Genera: *Ectocarpus, Laminaria, Fucus,* and so forth.

Division Rhodophyta

Red algae. These plants are essentially branching filamentous structures, which may build up a relatively complex body. Chlorophylls *a* and *d* are usually masked by a red pigment, phycoerythrin; a blue pigment, phycocyanin, may also be present. Food reserve is "floridean" starch. Asexual reproduction is by non-motile spores; sexual reproduction is oögamous. No flagellated cells are produced. Genera: *Nemalion, Polysiphonia, Porphyra, Gelidium,* and so forth.

Division Bryophyta

Mosses and allies. These are small plants that lack true roots, stems, and leaves; they are parenchymatous; they lack vascular tissue. Water is necessary for the transfer of sperm prior to fertilization. The gametophyte is dominant and nutritionally independent, whereas the sporophyte is permanently attached to the gametophyte and dependent upon it. The sex organs and sporangia are multicellular and have an outer layer of sterile cells. They are oögamous. Chlorophylls *a* and *b* and carotenoids are present in plastids, as in Chlorophyta. They are land plants; a cuticle is present. A definite alternation of phases occurs.

Class Musci (mosses). The gametophyte has an early transitory protonemal stage, which gives rise to "leafy" shoots; the mature gametophyte is usually erect and "blades" are radially arranged on a stalk. The meiospores are released from the sporophyte by a transverse splitting of the sporangium or capsule; no elaters are present. Genera: *Mnium, Polytrichum, Sphagnum,* and so forth.

Class Hepaticae (liverworts). The mature gametophyte is prostrate and dorsiventrally differentiated; there is no protonemal stage or only a brief, indistinct one. The meiospores are released by a longitudinal splitting of the sporangium (capsule); elaters are present. Genera: *Marchantia, Pellia, Riccia, Porella,* and so forth.

Class Anthocerotae (hornworts). The mature gametophyte is prostrate with dorsiventral symmetry. The sporophyte is an upright, elongated, cylindrical structure with a basal meristem. The capsule (sporangium) splits longitudinally along two vertical lines of weakness; elaters are present. Genus: *Anthoceros.*

Vascular Plants

The plants of the following groups have a well-developed vascular tissue, xylem and phloem, in the sporophyte, which is dominant and independent at maturity. They are oögamous. Chlorophylls *a* and *b* and carotenoids are present in plastids, as in Chlorophyta. They are land plants; a cuticle is present. A definite alternation of phases occurs.

Division Psilophyta

The sporophytes have scalelike leaves or no leaves, no roots; no leaf gaps (see Section 25.1) are present in the vascular cylinder. They are dichotomously branched, with terminal sporangia containing only one type of meiospore (homosporous). Water is necessary for the transfer of sperm previous to fertilization. Sex organs and sporangia are multicellular and have an outer layer of sterile cells. Both sporophyte and gametophyte are nutritionally independent. Genera: *Psilotum* and *Tmesipteris.*

Division Lycophyta

Club mosses. The sporophytes may be homosporous or heterosporous (two kinds of meiospores); and have roots, stems, and small leaves, but no leaf gaps. Single sporangia are borne on the upper surface of leaves (sporophylls), which are usually arranged in the form of a cone or strobilus. The sporophyte is dominant, and the gametophyte is small; both are nutritionally independent. Water is necessary for sperm transfer. Sex organs and sporangia are multicellular and have an outer layer of sterile cells. Genera: *Lycopodium, Selaginella,* and so forth.

Division Sphenophyta

Horsetails. The sporophytes have roots, stems, and small leaves, but no leaf gaps. The leaves and branches are whorled; the stem is hollow, jointed, and contains silica. Groups of sporangia are borne on stalked, umbrellalike structures, which are grouped to form strobili; elaters present on meiospores. The sporophyte is dominant, the gametophyte is small, and both are nutritionally independent. Water is necessary for sperm transfer. Sex organs and sporangia are multicellular and have an outer layer of sterile cells. Genus: *Equisetum.*

Division Pterophyta

Ferns. The dominant sporophytes usually have roots, stems, and large leaves; leaf gaps are generally present. Leaves are typically compound and uncoil as they develop (circinate vernation). Roots are usually adventitious from a horizontal rhizome, except in tree ferns. Most are homosporous. The gametophyte is nutritionally independent, as is the mature, much larger sporophyte. Water is necessary for fertilization by swimming sperm. Sex organs and sporangia are multicellular and have an outer layer of sterile cells. Sporangia are borne on the lower surface of leaves or sporophylls. Genera: *Polypodium, Pteris, Ophioglossum, Marattia,* and so forth.

Division Cycadophyta

Cycads. The dominant sporophytes have roots, stems, and large leaves. Naked ovules, which develop into seeds, are borne on megasporophylls, which are usually parts of cones. Leaves are large and fernlike. These plants are heterosporous; the very small gametophytes are nutritionally dependent upon the sporophyte, with the female gametophyte retained within the megasporangium. They are wind-pollinated, pollen tubes are formed, but ciliated swimming sperm are produced. Genera: *Zamia, Dioon, Cycas,* and so forth.

Division Coniferophyta

The dominant sporophytes have roots, stems, and leaves. Naked ovules, which develop into seeds, are produced on the upper sides of scales, which are usually parts of cones. Leaves are usually evergreen needles or scales. These plants are heterosporous (i.e., produce two kinds of meiospores); the very small gametophytes are nutritionally dependent upon the sporophyte, with the female gametophyte retained within the megasporangium. They are wind-pollinated; pollen tubes are formed. Genera: *Pinus, Picea, Abies, Sequoia, Thuja,* and so forth.

Division Anthophyta

Flowering plants. The dominant sporophytes have roots, stems, and large leaves; leaf gaps are present. Sporangia are borne on modified branch systems, the stamens and carpels, which are parts of the flower. Seeds develop from ovules, which are enclosed in the carpels. These plants are heterosporous; the gametophytes are very reduced and dependent upon the sporophyte, with the female gametophyte retained within the sporangium. They are wind- or insect-pollinated; pollen tubes are formed. "Double fertilization" occurs.

Class Dicotyledoneae (dicots). These plants have two cotyledons (seed leaves); the flower parts are mostly in multiples of fours and fives; the leaves are net-veined; a cambium is usually present. Genera: *Ranunculus, Magnolia, Ulmus, Brassica, Rosa, Pyrus, Phaseolus, Acer, Quercus, Aster, Zinnea,* and so forth.

Table 19-1 Comparison of three classifications. The one adopted for this text is the one at the left

Kingdom Monera	Kingdom Monera	Kingdom Plantae
		Subkingdom Thallophyta
		Division Schizophyta
Division Schizophyta	Division Schizophyta	Class Schizomycetes
Division Cyanophyta	Division Cyanophyta	Class Cyanophyceae
Kingdom Plantae	Kingdom Protista	
Division Euglenophyta	Division Euglenophyta	Division Euglenophyta
Division Chrysophyta	Division Chrysophyta	Division Chrysophyta
Division Pyrrophyta	Division Pyrrophyta	Division Pyrrophyta
	Kingdom Fungi	
Division Myxomycota	Division Myxomycota	Division Myxomycophyta
Division Eumycota		Division Eumycophyta
Class Phycomycetes	Division Phycomycota	Class Phycomycetes
Class Ascomycetes	Division Ascomycota	Class Ascomycetes
Class Basidiomycetes	Division Basidiomycota	Class Basidiomycetes
Class Deuteromycetes	Division Deuteromycota	Class Deuteromycetes
Class Lichens	Division Lichens	Class Lichens
	Kingdom Plantae	
Division Chlorophyta	Division Chlorophyta	Division Chlorophyta
Division Phaeophyta	Division Phaeophyta	Division Phaeophyta
Division Rhodophyta	Division Rhodophyta	Division Rhodophyta
		Subkingdom Embryophyta
Division Bryophyta	Division Bryophyta	Division Bryophyta
Class Musci	Class Musci	Class Musci
Class Hepaticae	Class Hepaticae	Class Hepaticae
Class Anthocerotae	Class Anthocerotae	Class Anthocerotae
		Division Tracheophyta
Division Psilophyta	Division Psilophyta	Subdivision Psilopsida
Division Lycophyta	Division Lycophyta	Subdivision Lycopsida
Division Sphenophyta	Division Sphenophyta	Subdivision Sphenopsida
		Subdivision Pteropsida
Division Pterophyta	Division Pterophyta	Class Filicineae
Division Cycadophyta	Division Cycadophyta	Class Gymnospermae
Division Coniferophyta	Division Coniferophyta	Class Gymnospermae
Division Anthophyta	Division Anthophyta	Class Angiospermae
Class Dicotyledoneae	Class Dicotyledoneae	Subclass Dicotyledoneae
Class Monocotyledoneae	Class Monocotyledoneae	Subclass Monocotyledoneae

Class Monocotyledoneae (monocots). These plants have one cotyledon; flower parts are mostly in threes; the leaves are usually narrow, elongated, and with parallel veins; a cambium is usually lacking. Genera: *Typha, Lilium, Allium, Carex, Bromus, Triticum, Hordeum, Festuca, Oryza, Avena, Zea,* and so forth.

To repeat, the preceding classification of the plant kingdom is based upon apparent relationships of plants; gaps exist in our knowledge of exact relationships between some of the groups; changes quite likely will have to be made in this system as further investigations add to our knowledge. Table 19-1 compares the classification used in this text with other systems that are used. No one system is accepted by all botanists; in each case the author must decide which he believes best demonstrates natural relationships.

Summary

1. Taxonomists, in studying plant classification, attempt to arrange plants into categories or groups that reflect the relationships among the various plants. In some instances artificial categories must be used until relationships are clarified by further investigations. In order to prevent confusion, a plant should always be referred to by its scientific name; this consists of the genus and species categories.

2. A species is rather difficult to define, but it may be considered as a group containing all the individuals of a particular kind of plant and excluding all those that are different. Usually species are not capable of interbreeding.

3. Plants may reproduce asexually or sexually. In the former type no fusion of gametes occurs, and such reproduction is usually accomplished by fission, budding, spore formation, the formation of tubers or corms, runners, and other methods. Sexual reproduction involves fusion of gametes, cell division and growth, and meiosis.

4. In plants the zygote, as a result of mitotic divisions, develops into a multicellular diploid phase (the sporophyte), which produces meiospores as a result of meiosis. These spores develop into a multicellular, haploid phase (the gametophyte), which produces gametes. This rotation of a diploid and a haploid structure is termed *alternation of phases*. In some plants the gametophyte is dominant, in some the sporophyte, and in others both phases are of equal importance.

5. The classification of the plant kingdom presented in this text is based upon apparent relationships of plants. Systems of classification are being revised continually as new information is obtained, and various changes probably will have to be made in the present system.

Review Topics and Questions

1. Discuss the difference between a natural and an artificial system of classification. Why are all systems at least partially artificial?

2. What is meant when one speaks of the "scientific name" of an organism? Why are such names necessary?
3. Why is a "species" difficult to define?
4. Explain in detail what you should do in order to identify a plant you have found. If you found only the roots, stems, and leaves of the plant, why would this make your task of identification more difficult?
5. Explain what is meant by "alternation of phases." Why is this term preferable to "alternation of generations"?

Suggested Readings

BOLD, H. C., *Morphology of Plants,* 3rd ed. (New York, Harper and Row, 1973).
BURNS, G. W., *The Plant Kingdom* (New York, Macmillan, Inc., 1974).
TORTORA, G. J., et al., *Plant Form and Function* (New York, Macmillan, Inc., 1970).

Kingdom Monera

20.1 General Characteristics of the Kingdom Monera

The Kingdom Monera includes the blue-green algae (Division Cyanophyta) and the bacteria (Division Schizophyta), of which the latter are by far the more important. The plant body is unicellular or colonial with the cells containing nuclear material not separated from the cytoplasm by a membrane and no plastids. The double helix of DNA of the cell occurs as a circular structure tightly packed within the bacteria but may extend as a three-dimensional network throughout the cell in blue-green algae. It is not associated with protein as is characteristic of chromosomes. The DNA is, however, the material involved in transmitting hereditary characteristics to the new cells that form. Ribosomes are abundant, but there are no mitochondria. Enzyme assemblies involved in the oxidation of food materials (in respiration or fermentation) are apparently located on the cell membrane, which may have various infoldings. The outer wall layer is frequently mucilaginous, resulting in a slimy mass when large numbers of organisms are present. Reproduction is typically asexual by fission, although transfer of genetic material has been demonstrated in one or two bacterial forms. Such transfer occurs but rarely and probably is not of any great significance in the usual development of bacterial populations. No type of sexual reproduction has been demonstrated in the blue-green algae.

20.2 Characteristics of the Division Cyanophyta (Blue-green algae)

Included in the Cyanophyta are the simplest green plants, consisting of but one cell or colonies of individual cells; the pigments are in the peripheral region of the cell and not in plastids, and the genetic material (DNA) is in a central region, which has no membrane, or extends throughout the cell. Chlorophyll *a* is usually masked somewhat by a blue pigment, **phycocyanin;** a red pigment, **phycoerythrin,** is present in the cells of some species. The chlorophyll is bound to lamellae, which form a complex network usually at the periphery of the cell. The stored food material is mainly **cyanophyte starch,** similar to the glycogen found in many animals. Reproduction is by simple cell division, and the resultant cells frequently remain closely associated in the form of filamentous colonies. The cells and colonies are often enclosed in a slimy gelatinous sheath. A few species form resistant spores, called **akinetes,** but this is not a method of reproduction. A single cell enlarges and develops a thick wall. After a period of rest, during which the cell is very resistant to adverse environmental conditions, the wall ruptures, and the contents then function as a normal cell, usually dividing within a short period and forming a new colony. No increase in numbers occurs as a result of such spore formation.

This group is of no real importance economically. It is a primitive group that never evolved further. The blue-green algae are widely distributed and are either aquatic or found on various damp surfaces. They are known to occur in hot springs at a temperature of up to at least 80°C. Some blue-greens may be food for various organisms, and they contribute to the organic matter in soil. Several members of the Cyanophyta are capable of utilizing nitrogen gas and may be beneficial by increasing the nitrogen supply in the soil (see Section 20.6 for a discussion of nitrogen fixing). They are believed to play a role in maintaining the fertility of rice paddies as a result of this ability to fix nitrogen. Members of this group are sometimes a nuisance in reservoirs and other bodies of water by imparting a fishy taste and odor, probably from waste products, the slimy sheath, or decay products. *Oscillatoria* is of some interest because of its motility. The filament glides back and forth while the tips swing from side to side. As in all members of this group, no organs of locomotion are present, and the mechanism of these movements is not known.

20.3 Characteristics of the Division Schizophyta (Bacteria)

The bacteria are an extremely important, although structurally simple, group of plants. These unicellular plants generally lack chlorophyll and thus are largely heterotrophic; a few species contain pigments similar to chlorophyll and are photosynthetic. Bacteriochlorophyll is similar to chlorophyll *a* but its maximum absorption spectrum is in the infrared. Some of the bacteria are chemosynthetic autotrophs, and their importance will be emphasized in a later section (20.6). Bacterial metabolism is extremely variable and sometimes quite complicated, but it is useful in the classification of this group. Because of the bacteria's small size averaging about 0.5 to 3 μm in diameter or length[1] and general lack of structural features, bacteriologists must supplement morphological criteria with various physiological characteristics, including staining, for identification. They determine whether the bacterium is aerobic or anaerobic, which sugars and alcohols it utilizes, what acids and gases it produces, whether or not it can digest protein (i.e., liquefy gelatin), what pigments are produced, how it stains with various dyes, and many other effects of bacterial action. Texture, color, and shape of bac-

---

[1] The high-power objective of the microscope must be used. In fact, most bacteria are viewed through an oil-immersion objective at a magnification of about 950. One μm = about $\frac{1}{25,000}$ of an inch.

367

Sec. 20.3 /
Characteristics
of the Division
Schizophyta
(Bacteria)

terial colonies may also be used. The identification of bacteria is not an easy task, and it is beyond the scope of this text. We will discuss bacteria in terms of their general importance to other forms of life.

Bacteria have three basic body forms (Figure 20-1), which are (1) **coccus,** or spheres, (2) **bacillus,** or rods, and (3) **spirillum,** or spirals (one or more curves in a rod). The bacilli and spirilli may be motile by means of **flagella** that extrude from all surfaces, in localized tufts at the ends, or singly at the ends. In many species the spheres or rods may cohere to form chains or colonies of cells. Some of the rodforms produce resistant spores, but these are not reproductive structures; one cell produces one spore, which germinates under favorable conditions to produce one cell.

Under favorable conditions, some bacteria may divide about once every 30 minutes, which in 30 days would result in about 10^{20} cells, weighing about 1500 tons, if they all survived. Obviously, this does not happen, and death is usually caused by a lack of food or the accumulation of toxic waste products.

The substrates utilized by various bacteria include almost all forms of organic matter and even some inorganic materials, such as ammonia, nitrate salts, carbon monoxide, sulfur, hydrogen, and nitrogen. The products resulting from bacterial metabolism are just as varied in their way as the substrates used. Some bacteria require oxygen, some can grow either with or without oxygen, and some can exist only in the absence of oxygen or at extremely low concentrations of oxygen. Because of the enormously great variations in bacterial metabolism, these organisms are of almost universal occurrence. Many can withstand rather high temperatures, others are not adversely influenced by low temperatures, and still others can exist deep in the oceans under high pressures. Only a very few places exist where bacteria cannot be found. The many thousands of bacteria make impossible the discussion here of more than a few of these. We will discuss the general importance of bacteria to other plants and to humans and indicate the activities of some of the more important bacteria.

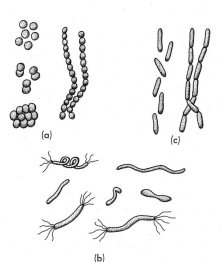

Figure 20-1. Bacterial forms. **(a)** Coccus (spheres); **(b)** spirillum (one or more curves in a rod); **(c)** bacillus (rods).

Recombination in Bacteria

Typical sexual reproduction, involving fusion of haploid gametes (and their nuclei) and followed eventually by a reduction of chromosome numbers, has not been demonstrated in bacteria. In fact as there are no chromosomes, only DNA, bacteria do not undergo either mitosis or meiosis. However, in some bacteria genetic material is transferred from a *donor* cell to a *recipient* cell, but usually only a portion of the DNA is transferred. The resultant recipient cell is not really diploid or is incompletely diploid. Thus, there is transfer of genetic material in some bacteria but it is questionable whether this should be termed sexual reproduction. Such recombination, or rearrangement, of genetic material has been shown to occur in several ways: (1) transformation, (2) conjugation, and (3) transduction.

Transformation. In 1928 F. Griffith showed that inoculating mice with a noncapsulated (avirulent) type of *Streptococcus pneumoniae* brought about no infection. If the capsulated (virulent) type of *S. pneumoniae* was injected, the mice died of pneumonia. Heat-killed virulent type had no effect upon the mice. When he injected the avirulent type along with the heat-killed virulent type, he found that the mice contracted pneumonia, and he was able to recover living capsulated (virulent) bacteria. A **transformation** of the avirulent type to the virulent type must have occurred. In 1944 other investigators were able to show that the transforming principle was DNA. They extracted DNA from heat-killed virulent bacteria, mixed this with the nonvirulent type of bacteria, and obtained living virulent bacteria. Later studies showed that the donor DNA was broken into fragments about 1/200th of the total donor DNA. These fragments, containing one or more genes, enter the recipient cells and replace similar regions of the recipient DNA, thus forming recombinants. Note that the donor and recipient cells need not be in contact.

Conjugation. In 1946 J. Lederberg and E. L. Tatum found a transfer of genetic material between two types of *Escherichia coli,* a common bacterial inhabitant of man's intestinal tract. Using strains that differed in their ability to synthesize certain growth factors, they found that growing large numbers of these two strains together gave rise to some bacteria having characteristics of both original types. Other investigators showed that genetic material is transferred from a donor cell to a recipient cell through a cytoplasmic bridge, which forms after chance collision and surface contact. This is termed **conjugation.** Usually only a part of the donor DNA is transferred and becomes integrated into the recipient DNA.

Transduction. In 1952 N. D. Zinder and J. Lederberg discovered that bacterial viruses, or phages, may act as a carrier for genes from one bacterial cell to another. Contact between bacterial cells is not necessary. Viruses are composed of a protein shell and an inner core of nucleic acid, either DNA or RNA; those attacking bacteria contain DNA. Once inside the bacterial cell, the viral DNA assumes genetic control over the chemical machinery of the host and new virus particles are synthesized. The host cell bursts and the new viruses are free to infect other bacterial cells. During the synthesis of new virus particles within the bacterial cell, small portions of the host DNA may become incorporated into the virus. When a new bacterial cell is infected, these host DNA fragments may become incorporated into the DNA of the recipient cell, thus bringing about recombinations. This process has been termed **transduction** and has been demonstrated in various bacteria.

20.4 Importance of the Monera

The bacteria and the blue-green algae are the oldest of living organisms. The oldest fossils (possibly over three billion years old) are actually examples of Cyanophyta. However, the immense iron deposits are believed to be products of iron-oxidizing bacteria and would indicate that bacteria were one of the first forms of life on earth. Botanists generally consider the Monera as not having evolved much beyond the primitive organism. They are probably holdovers on a side branch of the evolutionary family tree and have not given rise to any of the more complex plants or animals. They are, however, an extremely successful group.

Bacteria are exceedingly important to the total ecological system of the world. The recycling of materials depends to a great extent upon the activities of bacteria in bringing about decomposition of complex molecules and various interconversions that produce materials that other organisms can utilize. In this way atoms, such as carbon and nitrogen, are recycled and not lost from use by remaining tied up in complex molecules that most organisms cannot utilize. These aspects, as well as the harmful aspects of bacteria (disease), will be discussed in some detail in the next section.

20.5 Harmful Aspects of the Bacteria

Probably the first point to emphasize about harmful bacteria is that they are in the minority; not even 1 per cent of the known bacteria are harmful, and of these only a few are really of any great importance. We will discuss the harmful aspects first, but the student should bear in mind that most bacteria are beneficial and that the great importance of the Schizophyta relates mainly to their beneficial activities. This is not to imply that harmful bacteria are of no consequence. As will be discussed shortly, disease-producing bacteria have had and still do have a profound effect upon human populations. However, the beneficial aspects of the bacterial plants far overshadow the harmful aspects of a small minority of these organisms.

Human Diseases

In Table 20-1 are listed some of the bacterial diseases of man, the bacterium involved, and a brief comment about possible control. Of course, a variety of other human diseases are caused by bacteria, as well as diseases of other animals,

Table 20-1 Some bacterial diseases of man

Disease	Pathogen	Control
Plague	*Yersinia pestis*	Rat extermination, sanitation; vaccine; antibiotics
Asiatic cholera	*Vibrio cholerae*	Isolation, sanitation, disinfection[a]; vaccine; antibiotics
Typhoid fever	*Salmonella typhi*	Isolation, sanitation; vaccine; antibiotics
Pneumonia (lobar)	*Streptococcus pneumoniae*	Isolation; penicillin, sulfa drugs
Boils, abscesses, pus-forming inflammations, etc.	*Staphylococcus aureus*	Disinfection,[a] cleanliness; penicillin
Scarlet fever, sore throat, tonsillitis, etc.	*Streptococcus pyogenes*	Disinfection[a]; antitoxin for scarlet fever, sulfa drugs, penicillin
Gonorrhea	*Neisseria meningitidis*	Penicillin, sulfa drugs
Epidemic meningitis	*Neissera intracellularia*	Antiserum, sulfa drugs, penicillin
Diphtheria	*Corynebacterium diphtheriae*	Antitoxin; antibiotics
Tuberculosis	*Mycobacterium tuberculosis*	Isolation, rest, disinfection[a]; chemotherapy
Tetanus (lockjaw)	*Clostridium tetani*	Antitoxin; vaccine
Botulism	*Clostridium botulinum*	Careful processing of canned foods; antitoxin

[a] Disinfection of dressings, discharges, clothing, and so forth.

such as livestock, cats, and dogs. A few of these diseases are somewhat more drastic in their influence on human populations; we will indicate the effect of some of them.

Plague. *Yersinia pestis* causes bubonic and pneumonic plague in man, a disease that was known as the Black Death in the fourteenth century. Historians estimate that several cycles of bubonic plague during that century killed some 25 million people in Europe, about one fourth of the total population at the time. The Great Plague of London in 1664–1665 killed about 68,000 people out of 460,000, almost 15 per cent of the population. Plague epidemics in other cities and coutries have had similar results, and as late as 1905 over 900,000 persons died in India as a result of this disease. It is still responsible for thousands of deaths now, although in most areas it has almost been eliminated.

At the present time a plague vaccine is available, which can be utilized to immunize people, but an easier control method is through the establishment of sanitary facilities, cleanliness, and rat extermination. This last point is of utmost importance. Plague is transmitted by the bite of infected fleas, which obtain the bacterium mainly from the blood of rats. Infected rats act as a reservoir for the organism. This disease is almost never found in any locality in which modern sanitary facilities are available. It is a disease of crowded, poverty-ridden areas. In the United States only a few isolated cases occur, but the disease is still present in some Asiatic countries. *Y. pestis* is susceptible to the action of streptomycin, chloramphenicol, and tetracyclines; and these antibiotics may be used singly or in combination. The latter use is preferable because resistance to streptomycin, at least, may develop rather rapidly.

370

Asiatic cholera. *Vibrio cholerae* can exist in the human intestines and is spread by polluted water and food, found in areas where sanitary facilities break down or are nonexistent. The intense diarrhea, prostration, and dehydration result in high rates of mortality. During the Crimean War in 1855–1856, at least half of the French troops were hospitalized by an outbreak of cholera. A similar outbreak in the ranks of the Chinese and North Korean armies during the Korean War in 1952 led to charges of bacteriological warfare.

Immunizing vaccines are available, but the establishment of sanitary facilities is a simpler and less expensive method of control. Asiatic cholera is an ever-present danger in the Orient and will remain so until the crowded conditions and poverty are alleviated. The mortality rate can be decreased drastically by the replacement of lost fluids and salts (i.e., by the use of solutions containing sodium chloride, potassium chloride, and sodium bicarbonate). Tetracycline therapy reduces fluid loss and eliminates *V. cholerae* from feces. The latter effect is a great aid in decreasing the spread of the disease.

Typhoid fever. *Salmonella typhi* is also an intestinal pathogen, and the disease is spread by polluted water, by flies, and by food handled by individuals harboring the organism. Unfortunately, the bacterium may be present in some individuals without causing disease symptoms. Such persons, called **carriers,** are exceedingly dangerous to others if they are allowed to come in contact with food or the preparation of food.

Some historians believe that more battles have been lost because of typhoid fever epidemics and associated diseases, such as dysentery and cholera, than because of military opposition. Napoleon's campaign into Russia (1812) may be cited as an example. Although the extremely cold winter was one factor in the famous retreat, most of Napoleon's army was unfit for duty because of various epidemics, particularly typhoid fever and dysentery resulting from the crowded and unsanitary conditions, which made possible the rapid spread of disease.

Immunizing vaccines and antibiotics are now available, but sanitation is once again of prime consideration. Occasional outbreaks of typhoid fever occur in almost all countries and cause the public health authorities to function as detectives. They must locate the source of infection, and this may involve considerable difficulty if a carrier is involved.

Sterilization and Pasteurization of Food Materials

Foods consumed by humans are also excellent media for the growth of various bacteria or are means whereby pathogens are transmitted from one person to another or from food animal to human. The decomposition of foods rich in carbohydrates generally results in various types of fermentation, which are usually harmless although somewhat unappetizing. Bacteria acting upon high-protein foods bring about putrefactions, which are objectionable and frequently dangerous. To prevent spoilage, food must be kept free of organisms, or these organisms must be prevented from multiplying, because such decomposition of food results from the activity of microorganisms. The method usually used to achieve this end is heat treatment, a process basic to the canning and milk industries.

In **sterilizing** materials all organisms are destroyed by the high temperatures that are used. The amount of heat and the time of treatment vary with the kind and amount of food that is being processed; in general 60°C for one hour in a liquid medium is sufficient to kill nonspore-forming organisms, but 121°C for 15 minutes is utilized to destroy all forms of life. This heating is usually done under pressure in an autoclave (or pressure cooker), because steam under pressure is hotter than free-flowing steam. Free-flowing steam has a temperature of 100°C, compared to 121°C for steam under 15 lb of pressure. If the food is sealed in a container, reinoculation of the contents by the universally present airborne microorganisms is not possible. Food carefully processed in this fashion may be preserved indefinitely.

Pasteurization refers to the process of heating milk or milk products (or other materials) to 62°C for 30 minutes or to 71°C for 15 seconds. This treatment is sufficient to destroy disease organisms without ruining the flavor of the milk. Pasteurization is not the same as sterilization. In the former treatment many nonpathogenic bacteria survive, and pasteurized milk is safe but not sterile.

Foods may also be preserved by drying, by adding salt, or by adding sugar. All of these treatments are designed to decrease the possibility of microorganisms being able to multiply in the foods. Materials having a high osmotic concentration result in the plasmolysis of spores or cells that may have gained access to such foods (see Section 5.5). A deficiency of water would also prevent or greatly decrease bacterial multiplication. Rapid freezing of food also prevents growth of bacteria, but does not kill the bacteria, which can start to grow and multiply as soon as the food is thawed.

Plant Diseases

Although everyone is aware that bacteria cause animal diseases, very few people realize that a great many diseases of plants also are caused by bacteria. A detailed discussion of such diseases would fill several volumes; the 1953 *Yearbook of Agriculture,* published by the United States Department of Agriculture, is devoted to plant diseases and includes some of the more important bacterial diseases. Table 20-2 lists some of the more important plant diseases and the pathogenic organisms involved. A bacterial disease of plants is rather difficult to control once it has become established, because the pathogen is usually in the vascular tissue or other tissues of the susceptible plant, and sprays or dusts do not penetrate readily to these areas. In general, the use of resistant plants, the use of disease-free seeds and stocks, and the surface disinfecting of seeds are utilized in attempts to control bacterial diseases of plants. In those diseases where the bacteria are carried from plant to plant by insects, the use of insecticides has greatly decreased the severity of the disease. However, the use of disease-free materials, resistant plants, and crop rotation is the only effective means of controlling bacterial infections, although soil sterilization by heat or chemicals can be utilized in greenhouses. Fortunately, plant pathogenic bacteria are poor competitors in the soil

Table 20-2 Some bacterial diseases of plants

Disease	Pathogen	Susceptible plants
Fire blight	*Erwinia amylovora*	Pear, apple
Bacterial blight of stone fruits	*Xanthomonas pruni*	Plum, cherry, peach
Common blight of bean	*Xanthomonas phaseoli*	Bean
Halo blight of bean	*Pseudomonas phaseolicola*	Bean
Soft rot	*Erwinia carotovora*	Storage disease of carrots, potatoes, onions, bulbs, etc.
Bacterial blight of cotton	*Xanthomonas malvacearum*	Cotton
Bacterial wilt of corn	*Bacterium stewarti*	Corn, gama grass
Alfalfa wilt	*Corynebacterium insidiosa*	Alfalfa and white sweet clover
Bacterial wilt of cucurbits	*Erwinia tracheiphila*	Cucumber, cantaloupe, squash, watermelon, etc.
Crown gall	*Agrobacterium tumefaciens*	Apple, cherry, prune, grape, alfalfa, cotton, etc.
Brown rot of potato	*Xanthomonas solanacearum*	Potato, tomato, tobacco, etc.

and are crowded out or destroyed by the true soil saprophytes. Thus, the pathogenic types do not remain alive for long periods in the soil.

In the United States and Canada, great losses from bacterial diseases occur in potatoes, fruit trees, beans, tomatoes, tobacco, and alfalfa. In California fire blight has caused as much as a $2,000,000 loss in one year, including a loss of 100,000 trees, while the annual loss in the United States has gone as high as 5,000,000 bushels of apples and 500,000 bushels of pears. Fire blight, caused by *Erwinia amylovora,* is of particular interest because it was the first plant disease proven to be of bacterial origin. T. J. Burrill of the University of Illinois is considered the first to relate a bacterium to the cause of a plant disease; his reports were issued between 1878 and 1884. Bacteria wilt of alfalfa has become the most important disease of this crop and is a cause of much concern because of its widespread occurrence in most alfalfa-growing areas of the United States. In many fields of the southern United States, tobacco, tomatoes, and potatoes cannot be grown successfully because of a bacterial wilt disease known as brown rot. Bacterial blight annually may cause up to 50 per cent loss in beans in certain locations. In addition to these various diseases of plants in the field, serious losses of up to 100 per cent occur in stored crops, such as carrots, turnips, and potatoes, as a result of soft rot, which is of universal occurrence.

Cause of Symptoms

Diseases are generally recognized by characteristic symptoms, which result from the activity of a pathogenic organism.[2] In some cases **toxins,** poisonous waste products of a protein nature, are excreted by the bacteria, and these poisons may be general or selective in their action on host tissues. In animals various cells

---

[2] Some diseases, of course, are not caused by pathogenic organisms, but we are not concerned with these at this time. Vitamin deficiency diseases in animals and mineral deficiency diseases of plants are excellent examples of diseases that are not the result of a pathogen.

of the body may produce substances, **antibodies,** that neutralize the toxins and render the animal **immune** to the disease. This **naturally acquired immunity,** a result of the animal actually having the disease and recovering from it, is quite different from **artificial immunity.** The latter type of immunity may be either active or passive. In **active artificial immunity** the patient's body is stimulated to produce antibodies by being injected with a **vaccine,** preparations of dead or weakened bacteria or nonpoisonous products[3] of these bacteria. Vaccines are used against smallpox (a virus disease), typhoid fever, tetanus, cholera, diphtheria, and others; the period of immunity is from several years to a lifetime. **Passive artificial immunity** results from injecting the patient with antibody-containing serum; he is immune immediately, with no latent period during which the patient's body produces antibodies; he has them injected "ready-made." Such antibody-containing, or antitoxin-containing, serum is usually obtained from horses that have had long series of injections of the proper vaccine; the horse has developed active artificial immunity. This type of serum is especially useful in treating tetanus and diphtheria. One should remember that such immunity is transitory and not lasting. Unfortunately, antibodies are not produced by plants. Some plants are thought to develop immunity to certain virus diseases, but antibody production has not been demonstrated.

Symptoms in plants may result from a removal of food from host tissues, which results in various abnormal growth responses and reduced yields of the latter, and also from mechanical plugging of xylem cells by large masses of the pathogen or by slimy and gummy secretions of the pathogen, which results in wilting of the susceptible plant. In some plant diseases, such as crown gall, the pathogen stimulates the host cells to excessive cellular divisions and growth; such abnormal enlargements are termed **hypertrophies.** In those diseases wherein the host cells are actually destroyed, the symptoms may be localized (e.g., leaf spots), or of a more general nature, in which individual branches or entire plants are severely injured or killed.

Many symptoms in plants, as in humans, are quite characteristic of specific diseases. However, proof that a disease is caused by a certain bacterium depends upon the application of Koch's[4] postulates. Basically, these postulates define the evidence that is necessary to prove an organism to be the cause of a disease:

1. The organism must be associated with all cases of a given disease.
2. The organism must be isolated from the diseased individual.
3. When this organism is subsequently inoculated into susceptible plants or animals, it must reproduce the disease.
4. The organism must be reisolated from such experimental infections.

All these postulates cannot be fulfilled in every case. Some diseases are caused by ultramicroscopic viruses; the pathogens for other diseases have never been

---

[3] Toxins are rendered nonpoisonous in a variety of ways and may then be used as vaccines.
[4] Robert Koch (1843–1910) was an eminent German bacteriologist who developed many techniques of staining and culturing bacteria.

grown in artificial media. Some diseases of man are not communicable to lower animals, and thus the third postulate might not be satisfied. However, in most such cases sufficient circumstantial evidence is available to indicate relationships between diseases and pathogen.

Control of Plant Diseases

Many techniques are utilized in the control of plant diseases, but they may be discussed as four general types. In all cases **control** refers to a reduction in the amount of injury caused by the disease-producing agent. This may mean only a slight diminution of injury or a complete eradication of the disease in question.

1. Exclusion. This term refers to any method that prevents the entrance and establishment of the pathogen into uninfested areas. Plant quarantine laws are an attempt to prevent the entrance of certain materials into the United States or, in some cases, from one state to another. Some materials can be disinfected and then brought into the area. Unfortunately, this cannot prevent the spread of pathogens over short distances by wind, insects, or water.

2. Eradication. The methods included under this category are those that seek the removal, elimination, or destruction of the pathogen from a given area. Various chemical sprays are useful to destroy the pathogen, or crop rotation may be used to decrease the amount of food available to the pathogen and thus eliminate it. It is also possible, of course, to remove and destroy (burn) the diseased plant along with the pathogen.

3. Protection. Protection refers to the interposition of an effective barrier between the susceptible plant and the pathogen. Many spray and dust applications, at least those applied early in the growth season, are protective, rather than eradicant, in nature. Once the pathogen becomes established, the same sprays and dusts are frequently used to destroy the organism (eradication).

4. Immunization. This term refers to methods utilized to render the plant immune, resistant, or tolerant to the pathogen. This is where plant breeding is important. Naturally resistant varieties are sometimes available, but the production of resistant hybrids by breeding programs has made the most significant advances. In some diseases the use of resistant varieties is the only economically feasible means of control.

Three basic difficulties are involved in the breeding of plants for resistance to a specific disease: (1) more than one genetic factor is usually involved, (2) quality and quantity must not be sacrificed, and (3) the pathogen frequently produces new varieties, which may attack previously resistant crop plants. This last point is especially true of plant diseases caused by members of the Division Eumycota (see Chapter 21). However, if resistant plants are available or resistant hybrids can be produced, the use of such plants immediately results in monetary advantages to the farmer: one of the expenses involved in any type of farming is that incurred through spraying and dusting programs for disease control; resistant plants need little or no such treatment.

20.6 Beneficial Aspects of the Bacteria

As has been mentioned before, and should be emphasized again, most bacteria (at least 99 per cent) are not harmful, and almost all of these are beneficial. Any discussion of bacteria should really stress this point, especially since most humans are prone to investigate harmful factors in attempts to negate such activity and then almost completely ignore the good derived from bacterial activity.

Decay

Undoubtedly the most important activity of bacteria is the decay of complex compounds. **Decay** specifically refers to the decomposition or breakdown of compounds, eventually resulting in carbon dioxide, water, and minerals if decay is complete. As we have already seen, green plants use these simple materials in the presence of light and chlorophyll to produce the complex components of their cells; animals and nongreen plants use the complex products of green plants to produce their own component parts. If this were the end of the matter, an enormous mass of complex organic compounds in the form of dead plants and animals would gradually accumulate upon the earth; life, of course, would then cease. Obviously this does not occur, and it does not occur because of the activity of untold millions of bacteria and fungi.[5] These are the organisms that make it possible for elements to be used over and over again. Complex materials are made by green plants and utilized by animals and nongreen plants, they are converted to simple materials by microorganisms, and then green plants make additional complex compounds from these simple products of decay.

In the book of Genesis of the Bible may be found the statement, " . . . for dust thou art and unto dust shalt thou return." In a way this is a fairly accurate, although oversimplified, statement. The addition of carbon dioxide and water to the "dust" is all that is needed. But after death, all bodies deteriorate and eventually disappear (dust?). Food comes from plants growing out of the earth (dust?) or from animals that eat plants. Thus, a cycle apparently exists, and we shall now examine this cycle in some detail.

Every bacterium does not decompose all organic material, and various bacteria specialize in the substrates they utilize. However, no organic matter is produced by living organisms that is not decomposed by some bacteria or fungi. The decomposition of dead plant or animal material or their excretory products is not a sim-

---

[5] In this context "fungi" refers to the true fungi, members of the Division Eumycota. These will be discussed in Chapter 21; therefore, this chapter will be limited to a discussion of the bacteria, members of the Division Schizophyta.

ple process but represents the combined activities of many different kinds of bacteria and fungi. The final result, the production of carbon dioxide, water, and minerals, may take a long time, but it is always the same. Some organic materials merely decompose more slowly than others, and some may escape decomposition for a time because of certain environmental conditions, as in coal formation.

The decay of carbohydrates[6] and fats results in the production of carbon dioxide, water, and energy. The importance of such decompositions was discussed in Section 14.4, and we shall now deal more with nitrogen-containing organic compounds. Many of the minerals present in cells are there as salts or in loose combination with various organic materials. These minerals are readily released and restored to their mineral state upon the death of the cell and decomposition of the organic compounds. Three important elements, however, do form intimate parts of certain organic molecules of the living cell; they are phosphorus, sulfur, and nitrogen. Organic phosphorus, as in sugar phosphates, DNA, RNA, and in adenosine triphosphates (ATP), is readily converted to inorganic phosphates by bacteria, and such phosphates are readily available to green plants. The situation with sulfur and nitrogen is more complex.

Sulfur conversions.[7] The element sulfur is found in the cell as sulfate salts, as part of various protein molecules, or as part of other complex organic molecules. The inorganic sulfates are released when the cell ruptures, and they are then available for use by plants. The sulfur in organic compounds, predominantly proteins, is liberated as sulfates during the aerobic decomposition of these compounds by various bacteria:

$$CHONS + O_2 \rightarrow CO_2 + H_2O + NH_4^+ + SO_4^= + Energy$$

In the reaction as shown, the complex protein molecule is represented by a general formula indicating the kind of atoms present (CHONS). If oxygen is not present or is used up, anaerobic decomposition of the protein takes place and hydrogen sulfide gas (H_2S) is produced (rather than sulfates) from the sulfur component of the molecule.

$$CHONS + H_2O \rightarrow CO_2 + CH_4 + NH_4^+ + H_2S + Energy$$

This gas is one of the evil-smelling odors that result from putrefaction and is poisonous. Putrefaction generally refers to anaerobic decomposition of proteins and is usually accompanied by the production of various obnoxious-smelling materials. Complete aerobic decomposition does not produce such odors. Note that

---

[6] An interesting type of decay is the cellulose decomposition occurring in ruminants (cattle, sheep, goats, camel, deer, and so forth). These animals, which feed mainly upon grass, have a preliminary digestive system of several compartments between the mouth and stomach. Food first enters this rumen and is thoroughly mixed with the semiliquid mass contained therein, which includes large numbers of cellulose-fermenting bacteria. The food is regurgitated and thoroughly chewed, during which time bacterial fermentations decompose the cellulose to various materials readily absorbed by the animal. Ruminants are not capable of using cellulose and can exist on grass only because of the activity of the bacteria they harbor in the rumen.

[7] The decomposition, or breakdown, portions of these conversions are part of the general processes included in the term *decay;* they are considered separately because of the somewhat greater complexity involved.

methane gas (CH_4) is also produced anaerobically. Fortunately, certain bacteria can oxidize hydrogen sulfide to elemental sulfur and then further oxidize the sulfur to sulfates, probably in the following manner:

$$2H_2S + O_2 \rightarrow 2S + 2H_2O + Energy$$
$$2S + 3O_2 + 2H_2O \rightarrow 2H_2SO_4 + Energy$$

Sulfide minerals of the soil, such as ferrous sulfide (FeS), are also oxidized by soil microorganisms. Where there is good aeration, moisture, and a suitable temperature, FeS is oxidized by certain bacteria to elemental sulfur:

$$2FeS + O_2 + 2H_2O \rightarrow 2Fe(OH)_2 + 2S + Energy$$

Note that the sulfur can be oxidized further as shown previously. These oxidations supply energy that the bacteria use to produce organic compounds from carbon dioxide gas or dissolved carbonates; these bacteria are chemosynthetic autotrophs. The bacteria, such as members of the genus *Thiobacillus,* actually produce sulfuric acid (H_2SO_4) as a waste product, and this reacts readily with soil particles to form sulfate salts. Such acid formation quite likely is of considerable importance in soil production as a factor in the chemical disintegration of parent rock material. Members of the genus *Beggiatoa* actually store sulfur granules as we might store fat from a diet of excessive carbohydrates. When the supply of hydrogen sulfide is exhausted, these bacteria oxidize the reserve sulfur granules to sulfates. Sulfate is the mineral form in which most plants, especially the green plants, obtain their sulfur supply.

A few species of photosynthetic bacteria, the purple sulfur bacteria and the green sulfur bacteria, utilize hydrogen sulfide (H_2S) instead of water (H_2O) as a hydrogen source. In these organisms sulfur is liberated rather than oxygen. Both these organisms contain bacteriochlorophyll, which is similar to chlorophyll *a* but has its maximum absorption spectrum in the infrared. They are capable of making carbohydrates from carbon dioxide in the presence of light (and using hydrogens from H_2S). In fact, they can carry on this process of photosynthesis at the invisible wavelengths of infrared. The sulfur that has been produced may be utilized by other nonphotosynthetic sulfur bacteria as just discussed.

Figure 20-2 represents the various interconversions of sulfur and sulfur-containing compounds as they tend to occur in nature. This is a somewhat simplified version; not all the various sulfur compounds are indicated, nor are all types of sulfur bacteria included.

Nitrogen conversions.[8]   All organisms require nitrogen. Green plants obtain nitrogen in the form of nitrate (NO_3^-) salts or, less readily, as ammonium (NH_4^+) salts, whereas animals and nongreen plants are almost exclusively dependent upon green plants for their nitrogen in the form of proteins or amino acids. The exceptions to the latter part of the preceding statement are certain groups of bacteria, chemosynthetic autotrophs, which can use inorganic nitrogen salts. Obviously, then, a supply of nitrogen salts in the soil is essential for all life. However,

---

[8] See footnote number 7.

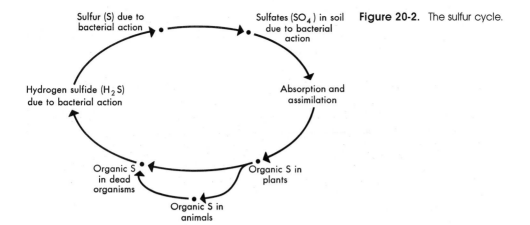

Sulfur (S) due to bacterial action

Sulfates (SO_4) in soil due to bacterial action

Figure 20-2. The sulfur cycle.

Hydrogen sulfide (H_2S) due to bacterial action

Absorption and assimilation

Organic S in dead organisms

Organic S in plants

Organic S in animals

such salts are not found as minerals in the soil, the way potassium salts and phosphate salts are found. Nitrogen salts come from the decomposition of dead plants and animals or their excretory products, and this decomposition depends upon several groups of bacteria.

Many kinds of bacteria use as a substrate or food supply various organic nitrogen-containing compounds, which are mostly proteins, and excrete ammonia as a waste product. (See previous reactions in discussion of sulfur conversions.) This conversion of organic nitrogen compounds to ammonia is frequently termed **ammonification.** Members of the genus *Nitrosomonas* oxidize ammonia to nitrite (NO_2^-). This is an oxidative process that yields the energy for growth; carbon dioxide and water are used in the production of organic compounds as in the green plants. Ammonia (NH_3) dissolved in soil water will form ammonium (NH_4^+) ions. These forms of nitrogen are interconvertible:

$$NH_3 + H_2O \rightleftharpoons NH_4^+ + OH^-$$

The reactions may be summarized:

(a) $NH_4^+ + 2O_2 \longrightarrow NO_2^- + 2H_2O + Energy$

(b) $CO_2 + H_2O + Energy \longrightarrow Organic\ compounds$

Nitrite is toxic to most plants, but it does not accumulate in the soil because of the presence of another group of bacteria. Just as the preceding group oxidized ammonia, *Nitrobacter* species oxidize nitrites to nitrates (NO_3^-) and thereby obtain energy for growth and maintenance:

(a) $2NO_2^- + O_2 \longrightarrow 2NO_3^- + Energy$

(b) $CO_2 + H_2O + Energy \longrightarrow Organic\ compounds$

The conversion of ammonia to nitrite and then to nitrate is frequently termed **nitrification,** but one should be aware of the fact that two steps are involved, as well as separate groups of bacteria. Nitrification by soil bacteria is quite rapid, and ammonia seldom accumulates to any extent. Actually, there are several nitrogen salts found in soils (e.g., NH_4SO_4, KNO_3, $NaNO_3$), but ionization takes place in solution and so the various reactions have been shown with the appropriate nitrogen-containing ion (i.e., NH_4^+, NO_2^-, NO_3^-).

The end result of this stepwise decomposition of organic nitrogen (proteins) to ammonia to nitrite to nitrate is that nitrogen, as nitrates, is made available to green plants. Organic nitrogen compounds are not generally available for use by green plants. They require nitrogen in the form of salts, which are absorbed by the roots from the soil. Without the bacterial decomposition taking place in the soil, nitrogen would soon be lost from use, and life would cease. So, once again, we are saved by the lowly microbe.

We have just discussed how organic nitrogen is converted to inorganic nitrate salts, which are then absorbed by plants. But where did that organic nitrogen come from? We have already seen that green plants make their own organic nitrogen (review Sections 12.1 and 13.10 if necessary) and that animals obtain their supply of nitrogen by eating plants. Now we seem to be going around in circles. Thus, it is necessary to consider yet another step in the various nitrogen conversions that take place.

The atmosphere has a tremendous amount of nitrogen, which is not available except to a few groups of bacteria. Nitrogen gas comprises a little over 79 per cent of the air we breathe, but this material cannot be used by any animals or by most plants. Some blue-green algae (Cyanophyta) have been shown to utilize gaseous nitrogen in the synthesis of amino acids and proteins, but these are relatively unimportant compared to three genera of bacteria: *Azotobacter, Clostridium,* and *Rhizobium.* There is evidence, however, that blue-green algae may be important in maintaining the nitrogen fertility of rice paddies, where they grow vigorously in the flooded fields.

The **nitrogen-fixing** bacteria utilize organic compounds as their source of energy (respiration) and can use some of this energy to convert nitrogen gas (N_2) to organic nitrogen compounds (mainly amino acids and then proteins):

$$\text{(a) Organic compounds} \longrightarrow CO_2 + H_2O + \text{Energy}$$

$$\downarrow$$

$$\text{(b) } N_2 + \text{Energy} \longrightarrow \text{Organic nitrogen compounds}$$

Organic acids and hydrogens are also available in the bacterial cell as a result of respiration. The hydrogens along with the nitrogens from nitrogen gas are used to form amino groups ($-NH_2$), which are then added to the organic acids to form amino acids. The amino acids are in turn converted to proteins. These organic nitrogen compounds are basic constituents of the bacterial cell and are really not available to green plants. As the nitrogen-fixing bacteria die, however, ammonification and nitrification processes convert such nitrogen compounds to nitrogen salts, which are then readily absorbed by plant roots. *Azotobacter* is a free-living organism that carries on aerobic metabolism, while *Clostridium* is also a free-living soil organism but functions anaerobically.

Although *Rhizobium* is also a soil organism, it is not capable of fixing nitrogen unless it first invades the roots of various plants, especially legumes.[9] The bacteria penetrate through the root hairs and into the inner tissues where the cortical cells are stimulated to divide and grow. The clearly visible, localized swelling or enlargement of the root is termed a **nodule,** and the cells are filled with masses of bacteria; *Rhizobium* is frequently termed the **root nodule bacterium.** Once the relationship between the root cells and *Rhizobium* has been established, nitrogen fixation readily occurs. Neither the legume nor the bacterium alone is capable of fixing nitrogen, but when together nitrogen gas is converted to organic nitrogen by the bacterial cells. This is a relationship of mutual benefit. The green plant obtains nitrogen compounds from the bacteria that are within the root cells or from the dead bacteria by way of ammonification and nitrification. The bacteria (*Rhizobium*) obtain organic food materials from the root cells of the green plant—such food having been produced by photosynthetic activity of the higher plant, of course. Actually, the legume probably supplies more than just food for the bacteria, since *Rhizobium* can live freely on soil organic matter but without fixing nitrogen. The exact relationship involved in nodule formation and nitrogen fixation is not known. It is common practice to inoculate legume seeds with *Rhizobium* before planting to ensure the formation of nodules and the resultant fixing of nitrogen. This can be done quite readily by mixing seed with dried bacterial cells or by dipping seeds in a bacterial suspension.

The amount of nitrogen fixation that occurs in a field varies considerably with environmental conditions and soil types. In general, nonsymbiotic nitrogen fixation varies from 0 to 25 kg per acre per year (1 kg = about 2.2 lb). Data obtained for symbiotic nitrogen fixation vary from 22 to 130 kg per acre per year with one report indicating 270 kg per acre by a single legume crop in a single season. These amounts are significant in agriculture. The greatest benefit to succeeding crops is obtained by plowing under the entire legume crop because of the large amount of nitrogen in the aboveground portion. The rotation of crops with legumes as one of the plants, or the planting of legumes with other plants (such as planting clover seed along with grass seed for a lawn), is beneficial because of the addition of nitrogen to the soil that is obtained with legumes. The succeeding crop, or the associated plants, thus have a supply of nitrogen made available.

It should be pointed out at this time that denitrification occurs occasionally in some soils under anaerobic conditions (e.g., waterlogged) and in the presence of organic matter (substrate). Several species of *Pseudomonas* and *Bacillus* are capable of reducing nitrates to nitrous oxide gas or to nitrogen gas. This is harmful to green plants, because it converts nitrogen from an available form (NO_3^-) to a nonavailable gaseous form (N_2O or N_2).

Figure 20-3 summarizes various important interconversions of nitrogen-containing compounds as discussed in the preceding paragraphs.

[9] The term *legume* refers to members of the bean family, *Leguminosae,* and includes beans (*Phaseolus*), peas (*Pisum*), alfalfa (*Medicago*), clover (*Trifolium*), vetch (*Vicia*), sweet clover (*Melilotus*), and so forth.

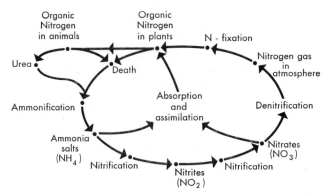

Figure 20-3. Nitrogen transformations.

Sewage Disposal

One of the difficult problems confronting city dwellers is sewage disposal. The liquid that runs through sewer pipes contains human wastes, factory wastes, and water that is used in many daily operations of the home and at work (e.g., washing, bathing, and so on). The actual amount of material usually varies from 15 to 25 gal per person. In small towns and isolated dwellings, this sewage can be disposed of in streams or applied to the land, where bacterial activity will convert the organic materials to carbon dioxide, water, and minerals—none of which are evil-smelling. In large cities such treatment would endanger the health of the population, as well as offend its aesthetic senses, because of the large amounts involved and the anaerobic conditions that soon develop as a result of rapid bacterial activity; the unwanted odors that accompany sewage are actually a result of fermentations or anaerobic bacterial activity. The problem thus revolves about producing an environment that is suitable for rapid, aerobic bacterial decay. This is the basis upon which all sewage disposal plants operate. One must remember that sewage can be disposed of only by bacterial activity. One may discard sewage by dumping it into a stream and thus eliminate a nuisance. The people downstream, however, would not maintain friendly relations with the people upstream under those conditions. Imagine for a moment the condition of the Mississippi River in the vicinity of New Orleans if all cities and towns on the river and its tributaries merely emptied sewer pipes into the streams and rivers. Not a very pleasant thought, but much of this is still being done.

Many methods of sewage disposal are available, from small septic tanks to enormous plants capable of handling millions of gallons of sewage daily. We shall discuss one method of sewage disposal. The material is screened to remove sizable matter (e.g., pieces of wood, bundles of rags, wire); then it enters a large tank in which most of the solids settle to the bottom, and the turbid liquid passes over to a second settling tank and finally to filter beds. The latter usually consist of deep layers of porous rock, such as coke, upon which the liquid is sprayed and then allowed to percolate down. The use of porous rock, the sprinkling, and the percolation are all designed to increase the amount of air (actually oxygen) to which the turbid liquid is exposed. Bacterial growth and development under these conditions are phenomenal, and the rock becomes covered with a slimy mass of bacteria. These microorganisms rapidly decompose the suspended organic material, and the effluent is clear, pure water, which is usually chlorinated to ensure the death of any harmful organisms. This liquid could be returned to the water supply of the city, but in most instances the purified water from a sewage disposal plant is foolishly discarded into the ocean or streams nearby.

The solid material, or sludge, is inoculated with sludge from a previous treatment. The latter contains enormous masses of bacteria and protozoa, which ensure rapid decomposition of the second batch of sludge, especially since this mass is aerated under pressure or stirred. Aerobic bacteria bring about decomposition of the organic matter and without obnoxious odors if aeration is adequate. The treated sludge contains masses of bacteria, soil and debris, and some nondecomposed material. It is dried and can be used as an organic fertilizer or as fill material; a portion of it is saved to inoculate the next batch of sludge. In the process of digestion the microorganisms produce methane gas, which may be collected and used to heat the digestion tanks and buildings and used in gas engines that drive generators and pumps.

Sewage disposal is really a matter of encouraging the proper bacteria to multiply rapidly and to metabolize actively. These are not special bacteria treated in a special way. The same results are obtained when small amounts of sewage are emptied into streams and rivers, or onto soil, or into septic tanks. A sewage disposal plant is necessary only in case of relatively large amounts of sewage. If this material were emptied indiscriminately into rivers, bacterial activity would rapidly deplete the waters of their oxygen supply, and at this point fermentations and putrefactions would give rise to noxious odors. To prevent these odors and to dispose of sewage rapidly, the sewage disposal plants of the city have been developed. Since less than 5 per cent of the water used in a city is actually consumed, to purify sewage water and then discard it is very wasteful. In some cities the water from sewage disposal plants is at least used for irrigation. All cities should build complete sewage disposal plants that can empty purified water into the reservoir systems of the cities. This would solve the water problem of many cities for a good number of years.

When waters become contaminated with sewage, this creates health problems (i.e., the possibility of pathogenic organisms being present) and aesthetic problems in terms of odors and appearance, and frequently causes death of fish and other aquatic organisms. Death occurs primarily from a lack of oxygen. The tremendous bacterial activity in utilizing the organic matter in sewage also involves the use of oxygen by aerobic types of bacteria. The waters are rather rapidly depleted of oxygen, and fish cannot survive. It is obvious that such a situation should not be allowed to occur.

Commercial Uses of Bacteria

The use of bacteria in sewage disposal plants is not considered to be a commercial venture, because such plants are almost always managed under the auspices of a municipal or county government. However, many commercial ventures do depend upon harnessing bacterial activity.

Cotton fibers occur separately and are readily used in manufacturing textiles. Other fibers, however, are embedded in stem or leaf tissues and must be freed before they can be used. **Retting** is the process whereby pectin-digesting bacteria bring about the separation of cells from one another. This is important in obtaining fibers from plants such as hemp (*Cannabis sativa*), jute (*Corchorus capsularis*), and flax *(Linum usitatissimum)*. The plant material is placed in ponds or tanks where various anaerobic pectin-destroying bacteria, mainly species of *Clostridium,* rapidly separate fibers from other material. The plant material must be removed before too long, or cellulose-digesting bacteria will ruin the fibers. After being washed in running water, the fibers are removed mechanically and used.

Coffee "beans" and cocoa "beans" are subjected to preliminary bacterial activity before further processing. These are not "beans" in the strict sense. They are fruits of *Coffea arabica* and *Theobrama cacao* plants, although it is the seed that is eventually utilized. The remains of the pulpy material of the coffee fruits are removed from the seeds by fermentations, and fermentation of the cacao seeds produce "precursors" of the chocolate flavor that develops during subsequent roasting. Certain choice tastes and aromas of tobacco develop during the curing of green leaves and are probably caused by bacterial fermentations. Vinegar and acetic acid are produced when *Acetobacter* oxidizes alcohol; acetic acid is the main component of vinegar and is obtained by distillation. Various other products are obtained as a result of bacterial activity: propionic acid, lactic acid, acetone, butanol (butyl alcohol), and so forth.

Various kinds of food materials are made available to us through the generous activity of various bacteria. Sauerkraut, pickles, butter, buttermilk, and certain cheeses obtain their flavors and aromas from bacterial metabolism. Soft cheeses, such as Limburger and Liederkranz, and hard cheeses, such as Swiss and Cheddar, are ripened by bacteria; the holes in Swiss cheeses result from gas production as the cheese solidifies. The different types of cheeses are produced by different treatments, which favor certain groups of bacteria over others. In most cases, pure cultures of bacteria are added to the milk, the type of culture depending upon the desired cheese.

After a bit of reflection, one should see clearly that bacteria are not only essential for our very existence but also supply us with a few added niceties that serve to make life somewhat more enjoyable.

Summary

1. The Kingdom Monera consists of unicellular or colonial forms, which contain nuclear material not separated from the cytoplasm by a membrane, and no plastids or mitochondria; reproduction is typically asexual by fission. In the Division Cyanophyta (blue-green algae), chlorophyll is usually somewhat masked by other pigments, and the stored food material is mainly cyanophyte starch; this group is of minor importance.

2. The Division Schizophyta (bacteria) consists of very small plants, which are

mainly heterotrophic. Metabolic activity of the bacteria is exceedingly diverse and is a useful tool for classifying organisms that have so few structural features.

3. Although bacteria cause a variety of plant and animal diseases, the great majority of these organisms are very beneficial to other living individuals. Microbial activity, in bringing about the decay of complex organic compounds, allows green plants to utilize the component parts of such complex molecules. These plants in turn provide other organisms with food and minerals.

4. Nitrogen salts are not found as minerals in the soil. The source of such salts is the nitrogen found in plants and animals and the nitrogen gas of the atmosphere. However, the bacteria are essential for the conversion of nitrogen gas to organic nitrogen and eventually to nitrogen salts, which may then be absorbed and utilized by green plants. Other organisms receive their essential nitrogen supplies from these plants or from animals that have eaten the plants. The processes of ammonification, nitrification, and nitrogen fixation are basic to the continuation of life on earth.

5. Man has made use of many of the activities of various bacteria: retting; organic acid production; production of sauerkraut, pickles, cheese, vinegar; production of various organic solvents, and so forth.

6. Sewage disposal utilizes the general decay activities of microbes to destroy organic wastes rapidly. Sewage disposal plants are basically designed to produce an environment that is suitable for rapid aerobic bacterial activity. The obnoxious odors associated with sewage are a result of anaerobic decomposition.

Review Topics and Questions

1. What are the diagnostic characteristics of the Kingdom Monera?
2. Most systems of classification utilize morphological features. Since this is true, why are the Schizophyta classified primarily on the basis of physiological characteristics?
3. List at least three different ways in which bacteria are harmful.
4. Compare sterilization and pasteurization. Why are both methods used for preserving foods? Wouldn't either one alone be enough?
5. Define the following: toxin, antibody, acquired immunity, passive immunity, and vaccine.
6. Discuss the techniques utilized in the control of plant diseases. Explain which of these techniques are useful in controlling human diseases.
7. What is meant by "decay" and why is it important?
8. List at least ten ways in which bacteria are used by man.
9. A labeled nitrogen atom in a nitrogen gas molecule over a bean field collides with a labeled carbon atom in a carbon dioxide molecule. The nitrogen atom is next discovered as part of a nitrate salt in the soil, while the carbon atom exists as part of a sugar molecule in a bean stem. Later, these two atoms are found in the same protein molecule in a bean seed.
   Explain in detail and in logical sequence the processes involved and where

they take place, the cells and tissues and structures concerned, and the forces and essential factors necessary to bring about such transformations and movements of the two atoms. Where more than one pathway is available, discuss all possibilities.

10. A labeled sulfur atom is found in a protein molecule of a dead earthworm in the soil under a pea plant. Later this atom is absorbed by the pea plant and eventually is found in a protein molecule in a pea seed.

    Explain in detail and in logical sequence all the processes and transformations involved in the movement of the sulfur atom, including means of transport, chemical changes, energy changes, essential factors, and the cells and tissues in which each occurs.

Suggested Readings

DAVIS, B. D., et al., *Microbiology,* 2nd Ed. (New York, Harper and Row, 1973).

SAFRANY, D. R., "Nitrogen Fixation," *Scientific American* **231,** 64–80 (April 1974).

SALLE, A. J., *Fundamental Principles of Bacteriology,* 7th Ed. (New York, McGraw-Hill, 1973).

Division Eumycota

21.1 General Characteristics of the Algal and Fungal Groups

There are several distinct divisions of the plant kingdom, which, although they are not really closely related, are conveniently discussed together because they do have certain features in common. Even though some of these basically simple plants appear to have bodies that are differentiated, they are considered not to have true roots, stems, or leaves because they lack the vascular tissue and other characteristics of such structures in higher plants. The sex organs and sporangia are usually one-celled, or, if multicellular, the gametes and spores are not enclosed within a wall formed by a layer of sterile (nonreproductive) cells.

The members of these groups range in size up to extremely large multicellular forms often more than 70 m (200 ft) in length (brown algae). They are the most primitive members of the plant kingdom. Those that contain chlorophyll are autotrophic, while the rest are mainly heterotrophic. The methods of reproduction vary greatly and will be discussed individually for certain typical examples from the different divisions.

Some members of these divisions are extremely important to life in general as food organisms, as oxygen producers, or as decay organisms. Others are important in demonstrating evolutionary advances. Specific individuals will be examined in some detail to emphasize these points.

In many older systems of classification, the simple plants just mentioned were subdivided into Algae and Fungi. An **alga** was considered to be any member of these groups that contained chlorophyll, while a **fungus** was any member that did not contain chlorophyll. This separation is very artificial and not particularly useful in classifying plants. However, to speak of "algae" and "fungi" is sometimes convenient, so these words have been retained, although they have no real taxonomic significance. In any of the sciences, care must be taken that words are well defined and used accurately.

In classifying organisms, a characteristic such as the presence or absence of a single pigment is usually not considered to be sufficiently important for use as a diagnostic feature for a major group. These plants consist of several divisions of equal rank that probably evolved independently in a parallel series from a common ancestral form. In some systems of classification, however, a separate King-dom Fungi is utilized (see Section 19.4).

21.2 Characteristics of the Division Eumycota (True Fungi)

The true fungi constitute a large and diverse group of nonchlorophyll-containing simple plants; they are usually filamentous, although some unicellular forms are present. The major component of the cell wall is cellulose or chitin; both are poly-saccharides. **Chitin** is a complex molecule (also found in the exoskeleton of in-sects and crustaceans) made of glucoselike units that contain some nitrogen. Vari-ous types of spores are formed, which produce elongated, cylindrical threads upon germination. A single thread (filament), or a branch, is termed a **hypha,** and the entangled mass of **hyphae** that constitutes the thallus or vegetative plant body is termed a **mycelium;** unicellular forms obviously lack a mycelium. All these plants are heterotrophic, mostly saprophytic and aerobic, and are usually color-less, although a few do produce nonphotosynthetic pigments. In most of the latter, the asexual spore contains the pigment. Starch is not found in the Eumycota, the soluble food being transformed into **glycogen,** a storage product which is similar to starch and commonly found in animal tissues.

The majority of fungi form two or more kinds of spores in the life cycle. The asexual spores are sometimes produced in a sporangium and sometimes individu-ally or in chains at the end of specialized hyphae, the **conidiophores.** Spores of the latter type are termed **conidia** (singular, **conidium**). Sexual reproduction is varied, as in the algae. In the Ascomycetes and Basidiomycetes, at least, the sexual spores are produced only after nuclear fusion and form immediately following meiosis. The separation of the true fungi into four classes is based upon differ-ences in the sexual stage, with the Class Deuteromycetes being reserved for those having asexual stages that have not been identified with any sexual stage. The lat-ter group is an arbitrary one of convenience, and the organisms placed therein properly belong in one of the other classes, probably the Class Ascomycetes, for the most part. When one of the Deuteromycetes is proved to be an asexual stage of a known member of one of the other classes, it passes automatically to that group; many such transfers are made. In a number of cases, fungi with only asex-ual stages are placed within the natural groupings, and not in the Deuteromycetes, because of other characteristics that clearly indicate the proper relationships.

The outstanding characteristics of the Phycomycetes, Ascomycetes, and Basid-iomycetes will be treated briefly, and an example or two of each will be discussed. For an extensive discussion of the Eumycota, the student is referred to a mycol-ogy text, a few of which are listed at the end of this chapter.

21.3 Characteristics of the Class Phycomycetes (Algal-like Fungi)

The plants of this heterogeneous group are frequently referred to as the algal-like fungi, and the methods of sexual reproduction are quite similar to those of the algae, in that isogamy, heterogamy, and oögamy are all represented (see Section 19.5). The nonseptate **(coenocytic),** branching filaments do not form a compressed body but remain a cottony mass; a few of the simpler forms are unicellular. Cross walls, or septations, usually form only when reproductive structures are produced. Asexual reproduction is typically by means of spores **(sporangiospores)** borne in a sporangium, and they may be motile or nonmotile; the hypha on which a sporangium is borne is termed a **sporangiophore.** The sporangiospores are formed by the cleavage of multinucleate sporangial protoplasts, and the motile ones develop hairlike **flagella** (singular, **flagellum**) that project from the cytoplasm. Some Phycomycetes are parasitic on other plants or on animals (usually fish or insects).

21.4 Representative Members of the Phycomycetes

The Phycomycetes are a heterogeneous, and probably artificial, grouping of primitive fungi whose taxonomic treatment varies considerably. Some botanists subdivide them into eight orders and their component families and genera, while other botanists suggest that the group should be divided into a number of Classes of equal taxonomic rank. It is beyond the scope of this text to consider the technicalities involved in such groupings, and only two genera will be discussed in any detail.

Saprolegnia

The members of this genus (Figure 21-1) are mainly saprophytic water molds, but some species are serious parasites of fish. The body consists of slender, branched, loosely tangled, coenocytic hyphae anchored to the substratum by profusely branched hyphae termed **rhizoids.** Asexual reproduction is by means of biflagellated zoospores produced in a **zoosporangium,** which develops as a swollen hyphal tip delimited from the rest of the hypha by a cross wall. These **primary zoospores** are liberated through a pore at the end of the zoosporangium, swim about, become quiescent, and develop a resistant wall. Later this wall ruptures, liberating a biflagellated **secondary zoospore,** which develops into a new mycelium. Sexual reproduction is oögamous, with an oögonium containing several eggs usually developing next to an antheridium containing many sperm. **Fertilization tubes** from one or more antheridia penetrate the oögonium and reach the eggs; eventually an antheridial nucleus fuses with each egg to produce several zygotes in the oögonium. The fertilized egg develops a thick wall and usually does not germinate for some time. This relatively resistant structure is termed an **oöspore.** Under favorable conditions, the oöspore undergoes meiosis and sends out new hyphae. There is disagreement concerning the occurrence of meiosis. Recent research indicates that this process may occur in the antheridia and oögonia preceding gamete formation. However, the more traditional view has been retained until the situation is clarified.

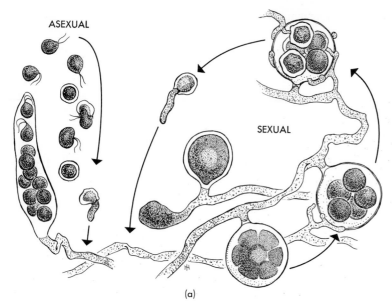

ASEXUAL

SEXUAL

Figure 21-1. *Saprolegnia.* **(a)** Life cycle. At the left, primary zoospores are liberated from the zoosporangium. Later, this zoospore develops into a secondary zoospore, which then forms a new mycelium. At the right, fertilization tubes from the antheridia penetrate to the eggs in the oögonium and zygotes are formed. After meiosis, these zygotes develop into new haploid mycelia. **(b)** Diagram of life cycle.

(a)

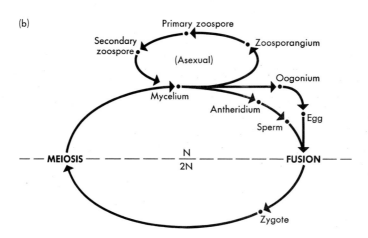

(b)

Primary zoospore

Secondary zoospore

(Asexual)

Zoosporangium

Oogonium

Mycelium

Antheridium

Sperm

Egg

MEIOSIS — — — — — — — $\dfrac{N}{2N}$ — — — — — FUSION — —

Zygote

Rhizopus

The members of this genus (Figure 21-2) are commonly found growing on damp organic material, such as bread and ripe fruit. The mycelium is usually a dense mat of intertwined hyphae, the main branches of which grow horizontally over the surface of the substratum. Such hyphae are termed **stolons,** and, where they touch the surface, slender much-branched filaments called rhizoids grow downward, penetrating into the substratum. The rhizoids serve in a manner analogous to roots; they anchor the plant and absorb nutrients. The fungi are heterotrophic, and the material absorbed is mainly water and dissolved organic matter. Enzymes secreted by the hyphae bring about this digestion and dissolution. Usually at the

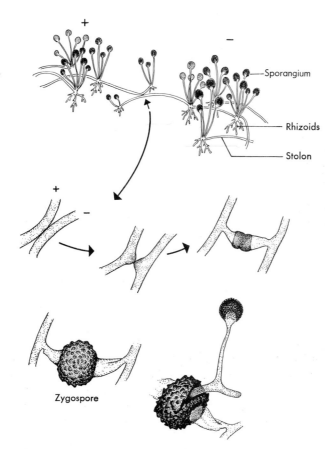

+

−

— Sporangium

— Rhizoids

— Stolon

+

−

Zygospore

Figure 21-2. *Rhizopus* life cycle. The sporangia formed during asexual reproduction are shown in the uppermost figure. The arrows indicate sexual reproduction, which culminates in the formation of a thick-walled zygote, shown at the lower left. Meiosis occurs when the zygote germinates (lower right) and the resultant sporangium contains haploid sporangiospores.

point of contact, upright hyphae are produced, which bear single sporangia at the tip. A sporangium is formed by the enlargement of the tip of a sporangiophore, and the outer portion of the protoplast in this enlargement is subdivided into large numbers of asexual spores. The rest of the protoplast in the tip is delimited from the developing sporangiospores by the formation of a wall, projects into the sporangium as a sterile portion known as the **columella,** and is continuous with the rest of the sporangiophore (Figure 21-3). When the sporangial wall ruptures, the spores are blown about by the wind and develop into new mycelia if they happen to land in a suitable environment.

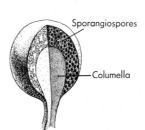

Sporangiospores

Columella

Figure 21-3. *Rhizopus*, development of sporangium and sporangiospores.

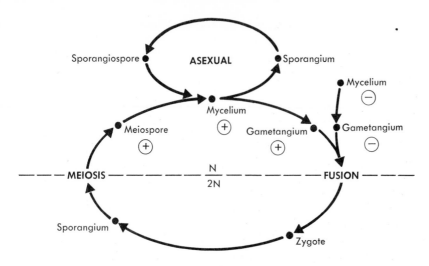

Figure 21-4. Diagram showing the life cycle of *Rhizopus*.

Sporangiospore ●   **ASEXUAL**   ●Sporangium

●Mycelium
⊖

Meiospore ●   Mycelium

⊕   Gametangium ●   ●Gametangium

⊕   ⊕   ⊖

— — — —MEIOSIS — — — — — — $\frac{N}{2N}$ — — — — — FUSION — — —

Sporangium ●   ●Zygote

Rhizopus is **heterothallic,** and zygotes develop only when hyphae from different mycelia come into contact (Figure 21-2). At the point of contact of the two plants, small side branches with enlarged tips are produced, and each tip is cut off from the rest of the hypha by a cross wall. These two terminal cells are touching, the wall between them disintegrates, and the cells fuse, forming a single-celled zygote. The cells that fuse to form the zygote each have many nuclei. They should probably be thought of as gametangia that function as gametes. The zygote is multinucleate, but nuclei from the different hyphae pair and fuse, producing diploid nuclei. The zygote enlarges, develops a thick wall, and becomes a resistant **zygospore.** On germination, the thick wall ruptures, and a sporangiophore emerges. Meiosis occurs at this time, and the spores that are produced are haploid. Germination of these meiospores results in the formation of two kinds of mycelia, only one mycelium per spore, of course. Because the two kinds of mycelia look alike, they are referred to as plus (+) and minus(−) strains; no visible sexual difference allows the use of the terms male and female. Certainly a difference does exist, even though it is not obvious, and zygotes are produced only when + and − strains touch, never between + and + or − and −. A diagram of the life cycle is shown in Figure 21-4.

21.5 Importance of the Phycomycetes

The origin of the Phycomycetes is debatable; two main points of view are current. The first suggestion is that these fungi arose from various members of the Chlorophyta. This is based upon the similarity in reproduction between certain Phycomycetes and certain green algae, as in *Rhizopus* and *Spirogyra*. The different types of reproduction in the Phycomycetes result from the origin of these fungi from Chlorophyta with similar variations in their type of sexual reproduction. The other viewpoint holds that the Phycomycetes and the Chlorophyta were derived from some common ancestor and that the similarities in reproductive structures are an example of **parallel evolution,** in which genetically different plants followed similar pathways. In the Phycomycetes the trend is for sporangia to be transformed into **sporangioles** (small sporangia with few spores) and these into conidia; a small sporangium containing a single spore is quite similar to a conidium. We shall not consider this trend in detail except to indicate the possibility that such a tendency could lead toward the Ascomycetes and Basidiomycetes, where conidia are quite typical. The Phycometes are diverse organisms, and it is more a matter of convenience rather than relationships that groups them together. Until the matter is clarified, the more traditional classification will be retained in this text.

The Phycomycetes are very important in bringing about the decay of dead plants and animals and their wastes, especially cellulose materials. A few members of this class of fungi are useful commercially in the production of fumaric acid, oxalic acid, and various enzymes. Some of these fungi are also important in the spoilage of stored fruits and vegetables, and others are parasitic on various crop plants (Table 21-1). The late blight of potato, caused by *Phytophthora infestans,* is one of the most famous of plant diseases and was the basic cause of the migration of 1,500,000 Irish people to the United States in the late 1840s. The staple crop of Ireland was the potato, and the economy of that land was built upon that crop; no other crops were of any significance. The late blight disease struck with disastrous suddenness in the 1840s and resulted in the tremendous famine of 1845, when the potato crop was almost completely destroyed. Almost 1 million people starved to death. The total reduction of population, death plus migrations, caused by this one plant disease was somewhat more than two-and-one-half million people, or almost one third of the total population. Although the discovery of Bordeaux mixture and organic fungicides, such as dithiocarbamate, as control sprays has greatly decreased the severity of this disease, losses in the United States still range from 1 million to 50 million bushels per year. The higher figures result from failure to spray crops, usually as a result of some "good years" with little loss causing growers to become indifferent. The United States lost over 54 million bushels, nearly one eighth of its potato crop, in 1938; up to 45 per cent of the crop was lost in some areas.

Table 21-1 Some plant diseases caused by Phycomycetes

Disease	Pathogen	Control
Late blight of potato	*Phytophthora infestans*	Bordeaux mixture[a] as a spray
White rust of crucifers (cabbage, cauliflower, radish, turnip, etc.)	*Albugo candida*	Crop rotation, destruction of infected crops
Downy mildew of grape	*Plasmopara viticola*	Bordeaux mixture[a]
Damping-off of seedlings	*Pythium debaryanum*	Seed treatment with Ceresan[b] or Semesan,[b] soil disinfestation
Club root of crucifers	*Plasmodiophora brassicae*	Crop rotation, soil disinfestation

[a] Bordeaux mixture: 4 lb copper sulfate, 4 lb lime, and 50 gal water.
[b] Ceresan and Semesan are organic mercury compounds.

A few of the Ascomycetes are unicellular, but in most the hyphae are septate, and in the majority a fruiting body, or **ascocarp,** develops as a result of sexual reproduction. This group is primarily characterized by the production of **asci** (singular, **ascus**), each containing typically eight **ascospores,** within the ascocarp. These fungi are usually heterothallic, and the mycelium is really a mixture of male and female strains. The female gametangium is termed an **ascogonium** and consists of a swollen, multinucleate, basal cell and an elongated cell, or **trichogyne.** The male gametangium, or **antheridium,** is long and slender and usually curves around the ascogonium, eventually fusing with the trichogyne. The nuclei of the trichogyne and the wall at its base disintegrate as the nuclei of the antheridium move to and pair with the nuclei of the ascogonium. These nuclei do not fuse; they divide synchronously. **Ascogenous hyphae** grow out from the ascogonium and intertwine with vegetative hyphae to form the ascocarp. The cells of the ascogenous hyphae differ from the vegetative hyphae in that the former have paired nuclei, a **dicaryon** (N + N) condition. Haploid is designated as N, diploid as 2N. In the ascocarp, which is formed primarily of haploid vegetative hyphae, the end cells of various ascogenous hyphae become somewhat swollen and enlarged, or sac-shaped. The paired nuclei in these cells fuse, the condition now being diploid (2N), and then undergo meiosis and mitosis, resulting in eight haploid (N) nuclei. Walls form around the eight nuclei and a small amount of cytoplasm for each, and the resultant structures are ascospores contained within an ascus (see Figures 21-5 and 21-6).

Ascocarps are of three general types, depending upon the particular fungus involved. In all these the asci are produced in a layer interspersed with sterile hyphal tips; such a layer is termed the **hymenium** (Figure 21-7). The rest of the ascocarp (or fruiting body) is composed of densely interwoven hyphae, the cells of

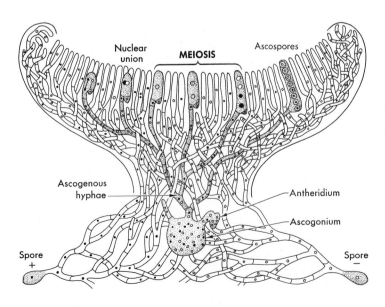

Figure 21-5. Diagram of the sexual life cycle of a heterothallic ascomycete. The early stages are in the lower portion of the diagram and later stages in the upper portion. Long before the apothecium and the spores are mature, the sex organs have disappeared. (Redrawn from *Fundamentals of Cytology,* by L. C. Sharp, copyright 1935 by McGraw-Hill, used with permission.)

395

Sec. 21.6 /
Characteristics
of the Class
Ascomycetes
(Sac Fungi)

(a)

Figure 21-6. Ascomycete. **(a)** Ascus with ascospores; from *Morchella.* **(b)** Diagram of sexual cycle.

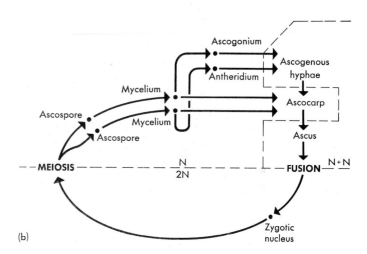

(b)

which may be so swollen as to give a parenchymatous aspect. An **apothecium,** as in *Peziza,* is an open cup-shaped structure with the hymenium lining the concave surface of the cup (Figures 21-7 and 21-11). A **cleistothecium,** as in *Erysiphe,* is a spherical structure in which the asci are completely enclosed (Figure 21-8); the cleistothecium splits and the asci protrude when they are mature. A **perithecium,** as in *Venturia,* is similar to a cleistothecium except that the former has a pore through which the hyphae may protrude. (Figure 21-9).

Asexual reproduction in the Ascomycetes is typically by the formation of conidia, although budding occurs in the yeasts. The shape and branching of the conidiophores and the color of the conidia are useful as diagnostic characteristics in classifying these organisms.

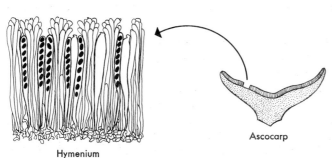

Hymenium

Ascocarp

Figure 21-7. An ascocarp (apothecium) showing the location of the hymenium. The enlargement at the left shows several asci with ascospores and numerous sterile hyphal tips.

Figure 21-8.

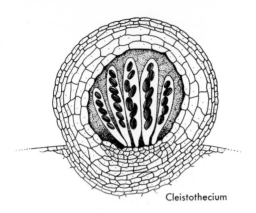

Cleistothecium

21.7 Representative Members of the Ascomycetes

The classification of this group is in a state of flux and is variously subdivided into numerous orders, families, and genera, but only a few of the more common forms will be discussed.

Saccharomyces

The yeasts (Figure 21-10) are one of the few members of the Ascomycetes that are unicellular. Reproduction in these plants is almost always asexually by **budding,** in which a small protuberance develops from the parent cell, the nucleus divides by mitosis, and one of the daughter nuclei migrates into the bud, after which a wall forms delimiting the bud from the original cell. Several buds may be produced before they actually separate from the parent plant. In a few species ascospores are produced after two yeast cells fuse, the entire cell functioning as a single ascus; or the diploid cell may undergo repeated budding. In the latter case, some of the diploid nuclei of the resultant cells eventually undergo meiosis and produce four ascospores with the cell thus becoming an ascus.

Figure 21-9.

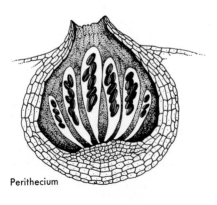

Perithecium

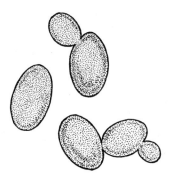

Figure 21-10. Yeast (*Saccharomyces cerevisiae*). The uppermost plant on the right has one bud, whereas the one beneath has a second bud projecting from the first bud.

Yeasts, particularly *Saccharomyces cerevisiae,* are extremely important commercially in the production of alcoholic beverages and industrial alcohol. Sugars contained in grapes and other fruits are metabolized anaerobically (fermentation), producing alcohol, carbon dioxide, and various by-products that serve to flavor the resultant liquid or wine. In the production of beer, germinated barley grains are ground and mixed with water plus various adjuncts, such as carbohydrates from corn or rice. Enzymes from the barley, especially diastase, digest these carbohydrates, and the soluble materials are used as food by the yeasts (which are added later). The germinated barley grains mainly digest starch to sugar, a necessary preliminary step caused by the inability of yeast cells to utilize starch. The grains are finally removed, hops (the ripe, dried female flower clusters of a twining vine, *Humulus lupulus*) are added for flavoring, and then special strains of yeast are added. Fermentation by these yeast plants results in the production of beer. The actual flavor of the beer depends upon all the various factors: barley, kind of adjuncts, strain of yeast, hops, length of fermentation, temperature, and water. Industrial alcohol is made by adding yeast to any inexpensive supply of sugar, such as molasses.

The production of carbon dioxide gas by yeast plants as they respire causes dough to "rise" when making bread. The heat of the oven eventually kills the yeast, evaporates any alcohol that may have been produced, and causes the dough to harden in the raised condition. The production of alcohol (over 350 million gal per year in the United States alone) and the making of bread products constitute the most important uses of yeast. An interesting note is that these two important aspects of yeast culture utilize two quite different physiological processes of the yeast plant. In the production of alcohol, fermentations are utilized; to produce the large quantities of yeast used in the bread industry, the yeast cultures are aerated to ensure respiratory activity rather than fermentation. (Why?)

Yeast plants contain a high percentage of proteins, some fats, and certain vitamins (thiamin, riboflavin, biotin, pyridoxine, pantothenic acid, and calciferol), and their utilization as food or a food supplement may become exceedingly important in the future. A few yeasts are pathogenic to man, causing skin infections or infections of the lung and central nervous system, but these are not really of any great importance.

Figure 21-11. *Peziza*, habit sketch of the apothecium.

Peziza

The members of this genus (Figure 21-11) are saprophytes usually found growing on fallen trees or in soil that contains much organic matter. The apothecium is the conspicuous portion of the plant, but an extensive vegetative mycelium penetrates the substratum (e.g., decaying wood) and obtains food materials. Asci are produced along the inner surface of the cup-shaped fruiting body, and the spores are discharged when a lid at the tip of each ascus opens. Ascospores are disseminated by the wind and give rise to vegetative mycelia when they germinate. Apothecia, asci, and ascospores are produced only as a result of sexual reproduction.

Morchella

In spite of its somewhat different appearance, *Morchella* (Figure 21-12) is quite similar to *Peziza* in basic structure. The former, sometimes called a morel or sponge fungus, has a stalked spongelike ascocarp in which each depression is comparable to the cup-shaped ascocarp of *Peziza*. There is an extensive vegetative mycelium penetrating the substratum. The morel is one of the most delicious of the edible fungi.

Figure 21-12. *Morchella,* habit sketch of the ascocarp.

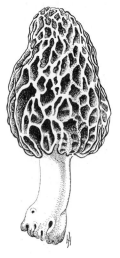

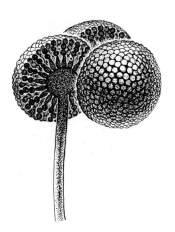

Figure 21-13. *Asperigillus niger,* conidia borne on a conidiophore.

Aspergillus

The members of this genus (Figure 21-13) are usually encountered as conidial fungi growing on all sorts of organic material in damp places. The plant body is a loosely interwoven mass of hyphae of no definite form, and the sexual stage is found in only a few species. However, the asexual structures (conidia and conidiophores) are so characteristic that even those reproducing only asexually are placed in this group rather than among the Deuteromycetes (Fungi Imperfecti). *Aspergillus* species cause the decay of many stored fruits and vegetables, bread, leather goods, and other fabrics; some cause ear and lung infections in domesticated mammals and in man. *Aspergillus niger* is used in the production of citric and oxalic acids.

Penicillium

The blue-green molds are also found mainly as conidial stages, the color being caused by masses of pigmented conidia that are produced. As in *Aspergillus,* some species of *Penicillium* (Figure 21-14) reproduce sexually, forming a cleis-

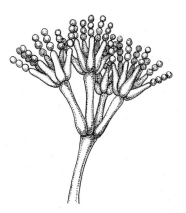

Figure 21-14. *Penicillium notatum,* conidia borne on a branched conidiophore.

tothecium containing asci. Various species are responsible for the spoilage of fruits (especially citrus), paper, books, leather, lumber, and bread. *Penicillium camemberti* produces the flavoring in Camembert cheese by means of the products of casein hydrolysis; similar hydrolysis of butter fats by *P. roqueforti* imparts flavors to Roquefort cheese. The conidia, which are produced in enormous numbers, are visible as bluish-green streaks in these cheeses. One of the most important members of this genus is *Penicillium chrysogenum*, which is famous as the producer of the antibiotic penicillin. Antibiotics are substances that are produced by one organism and inhibit or destroy other organisms. Penicillin has proved effective against many bacterial diseases of man: gonorrhea, syphilis, pneumonia, *Streptococcus* infections, *Staphylococcus* infections, and so forth.

Many antibiotics have been isolated from various molds, bacteria, and actinomycetes. The actinomycetes are members of the Division Schizophyta, as are the bacteria, and one species produces streptomycin, an antibiotic second only to penicillin in usefulness. Unfortunately, most antibiotics are too toxic to man to be used, but at least a dozen or so can be utilized to control various diseases. Other antibiotics very likely will be isolated in the future.

21.8 Importance of the Ascomycetes

The origin of the Ascomycetes is as debatable as that of the Phycomycetes. The viewpoint held by most mycologists is that the Ascomycetes are probably descended from Phycomycetes. In addition to the sporangium-conidium relationship already discussed (Section 21.5), the similarity between the ascogonium and the oögonium (as in *Saprolegnia*) is presented as evidence in favor of this view. Further discussion of such relationships is beyond the scope of this text.

The Ascomycetes are important decay organisms, which bring about spoilage of various materials; some are parasitic on various crop plants (Table 21-2). A number of commercial applications of various members of the Class Ascomycetes have already been mentioned. We might emphasize once more that the beneficial aspects of these fungi as decay organisms greatly outweigh any possible harm they

Table 21-2 Some plant diseases caused by Ascomycetes

Disease	Pathogen	Control
Scab of cereals	*Gibberella zeae*	Sanitation[a] and crop rotation, seed treatment with Semesan[b]
Apple scab	*Venturia inaequalis*	Bordeaux mixture[c] and lime sulfur
Chestnut blight	*Endothia parasitica*	None
Ergot of grains and grasses	*Claviceps purpurea*	Crop rotation, use of disease-free seed, sanitation[a]
Powdery mildew of cereals	*Erysiphe graminis*	Resistant varieties
Peach leaf curl	*Taphrina deformans*	Bordeaux mixture[c]

[a] Sanitation refers to the destruction of crop residues.
[b] Semesan is an organic mercury compound.
[c] Bordeaux mixture: 4 lb copper sulfate, 4 lb lime, and 50 gal water.

may cause as parasitic agents. This is not to imply that plant diseases caused by Ascomycetes are unimportant; some of the diseases result in enormous losses. Remember, however, that "harmful" and "beneficial" are relative terms, and an overall view indicates that the beneficial aspects predominate.

The ergot disease of grains, especially of rye, is interesting in that this disease not only decreases the yield of crop plants but also results in the poisoning of animals that feed upon diseased grains. The "Holy Fire" of the Middle Ages was actually ergotism, resulting from the use of diseased rye grains in baking bread. This fungus, *Claviceps purpurea,* transforms rye grains into enlarged purplish bodies filled with hyphae. Ergot poisoning is caused by an alkaloid, ergotinin, which is contained within these purplish structures. In humans blindness, convulsions, hallucinations, and paralysis may occur; blood vessels are constricted and the limited supply of blood to the extremities may result in gangrene. Since the discovery of the cause of ergotism around 1845, this disease has been relatively rare in humans, although grazing animals are frequently affected. However, as recently as 1951, ergotism flared up in France as a result of using infected rye grain, and over 200 people required medical attention. Surprisingly, a drug is prepared from ergot that is useful as a constrictor of smooth muscle in cases of excessive bleeding (hemorrhage) and as an aid in childbirth. Ergot was also the initial source for the psychedelic drug lysergic acid diethylamide (LSD).

21.9 Characteristics of the Class Basidiomycetes

In most of the members of this class, the septate hyphae form a fruiting body, the **basidiocarp,** after sexual reproduction has occurred. The primary characteristic of the Basidiomycetes is the production of **basidia** (singular, **basidium**), each bearing four **basidiospores** externally on small stalks (**sterigmata;** singular, **sterigma**). The plant may be homothallic or heterothallic, and no sexual organs are formed; sexual reproduction results from the fusion of hyphae, or a hypha and a single cell (**spermatium**). Nuclear fusion is delayed, as in the Ascomycetes, and the dicaryon

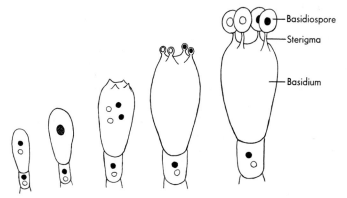

Basidiospore

Sterigma

Basidium

Figure 21-15. Development of a basidium and basidiospores. *Second from the left:* meiosis occurs immediately after fusion, and the four resultant nuclei are haploid. *Far right:* four mature basidiospores borne externally on a basidium.

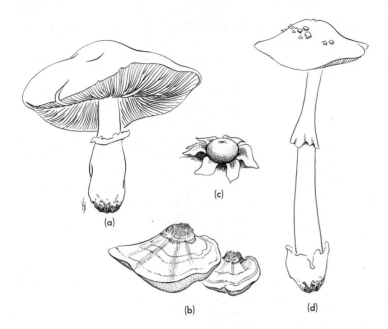

Figure 21-16. The basidiocarp (fruiting body) of representative Basidiomycetes. **(a)** *Agaricus;* **(b)** *Fomes;* **(c)** *Geaster;* **(d)** *Amanita.*

(c)

(a)

(b)   (d)

hyphae (N + N) form the basidiocarp where the hymenium is located. The end cells of various hyphae in the basidiocarp become swollen and club-shaped, the paired nuclei fuse (2N), and then they undergo meiosis, forming four haploid (N) nuclei. The club-shaped basidium produces small protuberances into which the nuclei migrate and are cut off by the formation of walls at the base of each projection. Enlargement of the structures in which the nuclei are now located completes the formation of basidiospores, borne externally on the basidium (Figure 21-15); in some Basidiomycetes the basidium is septate. The fruiting body is of various sizes and shapes, some of which are shown in Figure 21-16. Asexual reproduction is typically by the formation of conidia, but such spores are not nearly as prevalent in the Basidiomycetes as they are in the Ascomycetes; some of the former have never been found in the conidial stage. A few of the parasitic Basidiomycetes produce more than two kinds of spores and may have two hosts in their life cycle, but all of them produce basidia and basidiospores at some time.

21.10 Representative Members of the Basidiomycetes

The classification of this group of fungi has engendered considerable controversy, but we will limit our discussion to a few of the more common or interesting genera.

Agaricus

Agaricus (Figure 21-16[a]) is the common edible mushroom that may be purchased in grocery stores, either fresh or in cans. The terms *mushroom* and *toad-*

stool have no botanical significance except as convenient descriptions. The latter term usually means an ''inedible mushroom'' and will not be used at all. It is a saprophytic plant deriving its food supply from various organic materials in the soil, the vegetative mycelium penetrating extensively through the substratum. The basidiocarp grows upward above ground and consists of a densely woven mass of dicaryon hyphae that form a fleshy structure; the **stipe** (stalk) is capped by an umbrella-shaped **pileus** (cap). On the underside of the pileus are thin, sheetlike **gills** extending from the stipe to the edge of the cap; basidia cover the surfaces of the gills (Figure 21-17). The basidiospores are blown about by the wind and germinate to form new mycelia.

Amanita

Amanita (Figure 21-16[d]) is very similar in appearance to *Agaricus*, but the former is quite poisonous. (The student should not collect and eat wild species. The cultivated mushroom purchased in a store is much safer.)

Fomes

The genus *Fomes* (Figure 21-16[b]) includes perennial woody bracket or pore fungi, which are most commonly found as shelflike growths on the sides of trees or logs. The undersurface of the basidiocarp has many pores, which are actually the openings of long tubes; the hymenium lines these tubes, and the basidiospores are blown about by the wind after they fall to the outside. The vegetative mycelium develops rather extensively through the tree trunk before the fruiting body de-

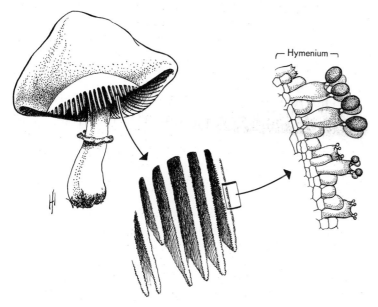

Figure 21-17. *Agaricus. Left:* The basidiocarp with a small portion cut off so that the gills are visible. *Center:* Gills enlarged. *Right:* A portion of a gill enlarged showing basidia and basidiospores with some of the latter not yet mature.

Hymenium

velops. *Polyporus* is similar to *Fomes* but produces only one layer of pores, whereas the latter may produce a new layer each year.

Lycoperdon

The puffballs have a spherical fruiting body, which is homogeneous at first but later separates into a thin outer layer and a central region. Cavities lined with hymenia develop in the central region. At maturity, most of the inner hyphae and basidia disintegrate, and the puffball consists of an outer layer with an apical pore enclosing a powdery mass of basidiospores. Mechanical disturbances result in a puff of "powder," and the spores are then distributed by the wind. Puffballs are exceedingly tasty before they are mature. Once the spores have formed, eating a puffball would be like eating talcum powder. Some puffballs are extremely large, the fruiting body approaching 100 cm in diameter.

21.11 Importance of the Basidiomycetes

The Basidiomycetes are generally regarded as being descended from the Ascomycetes. The dicaryon condition of certain hyphae and the formation of conidia in both groups are characteristics that indicate such a relationship. The general development of the basidium is essentially like that of the ascus; fusion of plus and minus nuclei, followed by meiosis, occurs immediately preceding the formation of both basidiospores and ascospores.

The Basidiomycetes are especially active in bringing about the decay of woody materials but, as with most fungi, they also attack a great variety of organic matter. A few members of this group are edible, especially *Agaricus campestris,* but others are distinctly poisonous if eaten, such as *Amanita* species. Although most of these fungi are saprophytic, some are parasitic and cause enormous economic losses. Table 21-3 lists a few of the more common plant diseases caused by Basidiomycetes.

The stem rust disease of grains and grasses, caused by *Puccinia graminis,* deserves attention because of the severity of losses, the great variety of host plants involved, and the excellent example this disease provides of competition between artificial evolution (plant-breeding programs) and natural evolution. As is true of most plant diseases, stem rust is very dependent upon weather conditions, and crops will suffer negligible losses some years, while other years may result in a loss of 60 per cent of the wheat crop in some areas. Severe rust years result in millions of bushels of wheat lost—for example, 160 million bushels in 1935—at a total cost of millions of dollars. Although stem rust of wheat is the most economically important rust disease, *Puccinia graminis* is subdivided into a number of varieties: *P. graminis tritici* on wheat, *P. graminis avenae* on oats, *P. graminis secalis* on rye, and others, mainly on grasses. Each variety in turn is subdivided into various numbered physiological races. Each of these races is able to attack

Table 21-3 Some plant diseases caused by Basidiomycetes

Disease	Pathogen	Control
Stem rust of grains and grasses	*Puccinia graminis*	Resistant varieties; sulfur dusting; eliminate barberry
Cedar apple rust	*Gymnosporangium juniperivirginianae*	Eliminate one host; sulfur dusting
Blister rust of pine	*Cronartium ribicola*	Eliminate *Ribes* host
Bunt of wheat	*Tilletia tritici* and *T. laevis*	Seed treatment with Ceresan,[a] etc.; resistant varieties
Loose smut of wheat	*Ustilago tritici*	Resistant varieties, hot-water treatment of seeds[b]
Common smut of corn	*Ustilago zeae*	Crop rotation, sanitation[c]
Rhizoctonia disease (black scurf) of potatoes	*Corticium vagum*	Crop rotation, use of clean "seed pieces"

[a] Ceresan is an organic mercury compound.
[b] Enough heat is involved to kill the fungus mycelium but not the plant embryo (in the seed).
[c] Sanitation refers to destruction of crop residues.

certain varieties of wheat, for example, and unable to attack other wheat varieties. Well over 150 physiological forms of *P. graminis tritici* are known.

The wheat rust fungus produces five types of spores and utilizes two hosts to complete its life cycle. Control of this disease by eradicating common barberry (*Berberis vulgaris,* one of the hosts) is important in isolated areas having cold winters and in any area to prevent sexual reproduction of the fungus, since the sexual stage develops on barberry. Eliminating barberry cannot prevent stem rust in areas with mild winters, because some of the spores that reinfect wheat without barberry as an intermediate may survive. The southerly winds in the Plains area of the United States serve to blow these spores from the warmer Southern states, where they overwinter, to the Northern states where they cannot survive. Eradicating barberry does not prevent infection. It does, however, prevent sexual reproduction of the fungus. Control of stem rust is accomplished mainly by the use of resistant wheat varieties; sulfur dusting is effective but expensive. Such wheat varieties do not retain their resistance; they become attacked more and more severely and are soon listed as susceptible varieties. This loss of resistance is not caused by any change in the wheat but by the appearance of new physiological races of the stem rust fungus. New races are a normal result of variations because of new genetic combinations that occur during sexual reproduction or as a result of possible mutations in some of the numerous spores that are produced. The plant breeder is in a continual struggle to produce resistant varieties more rapidly than the fungus develops new strains of rust. This is what was meant by the comment concerning the competition between artificial and natural evolution.

21.12 Characteristics of the Class Lichens

The members of this group are actually composed of close associations between a fungus, usually an Ascomycete (but sometimes a Basidiomycete), and an alga, either Chlorophyta or Cyanophyta. Their exact position in the plant kingdom is not clear, but they are usually placed with the fungi because the greater proportion of the plant body consists of the fungal component, which determines the general form and structure. There are three general types of lichens (Figure 21-18), which are (1) **crustose,** or crustlike, (2) **fruticose,** or shrublike, and (3) **foliose,** or leaflike. Closely interwoven hyphae form the structure, including rhizoids, and enclose algal cells near the upper surface; in some cases **haustoria** (short hyphal branches) penetrate the algal cells. Reproduction is by means of **soredia,** small masses of hyphae enclosing a few algal cells, which form as powdery masses on the surface of the lichen. The soredia blow about and are capable of developing into a lichen in a suitable environment. In many of these plants, apothecia are formed, and the ascospores may germinate upon liberation. If the proper species of alga is encountered, the hyphae will continue to form a lichen; the fungus alone cannot survive although the alga can under optimum conditions.

The relationship between the fungus and the alga in a lichen is rather peculiar. The latter obtains water and minerals from the former, while the fungus obtains food from the photosynthetic process of the alga. The penetrating rhizoids and the organic acids produced during the metabolic activity of the fungus undoubtedly aid in dissolving and obtaining essential minerals from the substratum. The lichen thus can exist where neither component alone could survive. Although the alga can ''go it alone'' in a suitable environment, the fungus cannot. Some botanists maintain that a lichen is a good example of **symbiosis,** the living together of two organisms in an association that is mutually beneficial. Other botanists claim that the fungus is parasitic on the alga, because the former cannot live alone. The true relationship is probably somewhere between these two points of view; the fungus is surely parasitic, but the alga derives considerable benefit from the association.

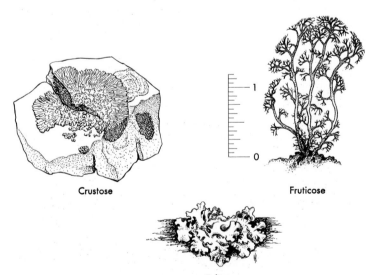

Figure 21-18. The three general forms of lichens.

Crustose

Fruticose

Foliose

The lichens are important as pioneers in plant succession, especially upon bare rock where other organisms cannot survive. The surface of the rock is gradually broken as a result of mechanical forces and chemical disintegration, thus setting the stage for gradual soil formation. Debris blown by the wind accumulates in the rough surfaces of the lichen structure, and this, plus the plants themselves when they die, adds to soil development. Eventually, when enough soil is accumulated, other plants become established. In this fashion more and more debris, including organic matter from dead plant structures, accumulates and converts a barren area into one that can support vegetation.

Some lichens are sources of food for grazing animals in the Arctic regions, the so-called "reindeer moss." Other lichens are used in the perfume industry, in dyeing processes, and as a source of litmus. The latter is used in chemistry laboratories as an indicator for acids and alkalies. The natural purple color of litmus is changed to red by acids and to blue by alkalies.

21.13 Mycorrhiza

The hyphae of numerous fungi, especially members of the Ascomycetes and Basidiomycetes, form intimate associations with plant roots either as a surface mantle with some hyphae penetrating the root or by actually invading the root and living within the cells. This complex association of root and fungus is termed a **mycorrhiza.** Such mycorrhizae are very common, and most species of woody plants appear to have them.

This relationship seems to be beneficial to the higher plant as well as to the fungus. In some species of pine the seedlings grow very slowly and poorly if the suitable mycorrhizal fungus is not available, as in sterilized soil. If the soil is inoculated with the appropriate fungus, mycorrhizae form and the growth of the seedlings improves markedly. Orchids never occur free of mycorrhizae. They can be grown without fungi in the laboratory, however, if they are supplied with nutrients, especially sugars.

It is likely that the fungi digest some of the organic matter in soils and thus make nitrogen, sugars, and minerals available to the roots. The extensive development of the fungus mycelium into the soil is believed to increase substantially the absorbing surface through which minerals and possibly water may enter and eventually become available to the host plant. Investigations using radioactive isotopes of calcium, phosphorus, and potassium have shown that these minerals are absorbed in greater amounts by plants with mycorrhizae than by plants without them. There is also evidence that nitrogen fixation may occur in some mycorrhizae, again making nitrogen available to the root cells. In any event the presence of the fungi appears to enable plants to grow under unfavorable soil conditions. The fungi, on the other hand, apparently obtain growth-promoting substances and food from the root cells. If radioactive carbon (in carbon dioxide) is fixed by the host plant (photosynthesis), it can later be detected in the fungus.

Summary

1. The algal and fungal groups consist of several quite diverse divisions. However, these divisions all consist of plants that lack vascular tissue and in which the sex organs and sporangia are either one-celled or at least have no outer layer of sterile cells.

2. The division Eumycota consists of nonchlorophyllous simple plants, which are usually filamentous. A single filament is termed a *hypha,* and a mass of hyphae is termed a *mycelium.* Two kinds of spores are produced; one of these types develops only after sexual reproduction. If a sexual stage is not found and if relationships are not obvious, the fungus is placed in the Class Deuteromycetes. The other classes in this division are Phycomycetes, Ascomycetes, and Basidiomycetes.

3. Members of the Class Phycomycetes are readily distinguishable because of their coenocytic filaments. Septations form only when reproductive structures are produced. The Phycomycetes and the Chlorophyta probably are descended from some common ancestor.

4. Phycomycetes are important decay organisms. Some of them are utilized commercially in the production of various organic acids and enzymes. A number of serious plant diseases are caused by members of this group.

5. The Ascomycetes produce an ascocarp, containing asci and ascospores, as a result of sexual reproduction. This fruiting body may be variously shaped. This group is probably descended from the Phycomycetes.

6. Many of the Ascomycetes are of great economic importance. They are used in the production of alcohol, bread, cheese, antibiotics, and a variety of other products. They are also useful decay organisms. Some members of this group cause plant diseases.

7. The Basidiomycetes produce a basidiocarp, containing basidia and basidiospores, as a result of sexual reproduction. This fruiting body may be of various shapes. This group is probably descended from the Ascomycetes.

8. The Basidiomycetes are active decay organisms. Some of them are edible, but others are distinctly poisonous. A number of plant diseases are caused by members of this class.

9. The Lichens are actually close associations between a fungus and an alga; the greater portion of the plant body consists of fungal hyphae. This association enables the lichen to grow in areas that would not support the growth of either component alone. These plants are important in soil development and plant succession.

10. A mycorrhiza is an intimate association of a fungus with a plant root. This relationship is probably beneficial to both organisms.

Review Topics and Questions

1. Define the following: conidium, conidiophore, ascus, ascospore, basidium, ascocarp, basidiocarp, hypha, and mycelium.

2. If you had collected a fungus specimen, describe in detail what you would have to do to determine the identity of the plant(differentiate as to which class of the Eumycota it belongs).

3. If a fungus specimen had no sexual stages, what other characteristics could be used for identification purposes? On what bases could you restrict this specimen to one or more of the classes of Eumycota?

4. A piece of moist bread left on a table gradually becomes covered with a cottony mold growth, which eventually contains many black specks and a fewer number of small, dark, globose structures. Examination with the microscope enables you to identify *Rhizopus nigricans*. Describe in detail the development of this organism, including the reproductive structures, the food source, and the kinds of hyphae produced.

5. In what ways are the Phycomycetes important to humans?

6. Wine production by individuals usually consists of merely crushing grapes and allowing them to ferment. What is the source of the yeast plants? Describe the growth of the yeast plant. Discuss the activities of the yeast plants that result in the formation of wine.

7. List at least ten ways in which the Ascomycetes are important to humans.

8. Describe the life cycle of *Agaricus*.

9. Explain why it is unlikely that a resistant wheat variety will remain resistant to stem rust (caused by *Puccinia graminis*).

10. Discuss how lichens aid in the establishment of plants on rock surfaces.

11. Decay by fungi frequently creates serious problems in the tropics. Explain.

Suggested Readings

ALEXOPOULOS, C. J., and H. C. BOLD, *Algae and Fungi* (New York, Macmillan, Inc., 1967).

BONNER, J. T., "The Growth of Mushrooms," *Scientific American* **194,** 97–106 (May 1956).

CHRISTENSEN, C. M., *The Molds and Man,* 3rd ed. (New York, McGraw-Hill, 1965).

EMERSON, R., "Molds and Man," *Scientific American* **186,** 28–32 (January 1952).

GRAY, W. D., *The Relation of Fungi to Human Affairs* (New York, Holt, 1959).

LAMB, I. M., "Lichens," *Scientific American* **201,** 144–156 (October 1959).

NIEDERHAUSER, J. S., and W. C. COBB, "The Late Blight," *Scientific American* **200,** 100–102 (May 1959).

Algal Groups

Those plants that are usually referred to as algae comprise a really diverse ensemble of organisms from very simple unicellular forms to rather complex and large multicellular plants with a variety of reproductive patterns. It is beyond the scope of this text to discuss all of the Divisions that are included in the term "algae," and we will be selective in order to emphasize certain relationships and the importance of some groups to other organisms, especially humans. The reader should refer to more specialized texts, such as some of those listed at the end of this chapter, for further details and coverage of other groups. We will study representative individuals of the following divisions: Chlorophyta, Chrysophyta, and Phaeophyta.

22.1 Characteristics of the Division Chlorophyta (Green Algae)

The green algae include organisms that vary greatly as to structure and methods of reproduction. They are unicellular, colonial, or multicellular, the cells containing organized nuclei and plastids. A **colony** merely consists of unicellular individuals that cohere to form a clump or group of various sizes and shapes, whereas a **multicellular organism** implies a differentiation of cells or tissues. In the former the cells are all alike and are individual plants themselves; in the latter the entire mass of cells comprises one plant, some cells of which are specialized or modified structurally and functionally and thus are not all alike. The chloroplasts contain chlorophylls *a* and *b* and carotenoid pigments in the same ratio as is found in the higher green plants; filamentous types of green algae may have been the ancestors of the land plants. Food reserves are mostly starch and some fats or oils, which is also similar to the higher green plants.

Most of the species (about 90 per cent) are freshwater forms or marine forms, but some can exist on various damp surfaces, such as soil, bark, shingles, and bricks. Some unicellular, nonmotile green algae are found growing in an intimate relationship with certain fungi, forming a lichen (see Section 21.12), or sometimes enmeshed in the tissues of an animal, such as *Chlorohydra* (a coelenterate). These are probably associations of mutual benefit.

Asexual reproduction in the green algae may be by fission, fragmentation, or zoospores. Sexual reproduction involves isogamy, heterogamy, or oögamy (see Section 19.5). Thus, an alternation of haploid and diploid phases occurs in most species, but many form no multicellular sporophyte. In these, meiosis occurs when the zygote germinates, and the only diploid structure is the one-celled zygote (which functions as a meiocyte). However, in *Ulva* (sea lettuce) a definite alternation of a multicellular gametophyte and a multicellular sporophyte occurs.

There are about 7000 species of green algae, but only a few representative members of this diverse group will be discussed.

22.2 Representative Members of the Chlorophyta

The various kinds of green algae probably evolved from a primitive motile unicellular type somewhat similar to, though more primitive than, *Chlamydomonas*. This organism will be discussed first, and then other individuals will be selected to indicate more advanced structures.

Chlamydomonas

Chlamydomonas is a large genus of unicellular flagellates occurring in stagnant pools of fresh water and on damp soil. The wall of this unicellular plant (see Figure 22-1[a]) is mainly composed of cellulose, and two equal hairlike **flagella,** which project from the cytoplasm at the anterior end, pull the organism through the water. The nucleus is partially enclosed by a single, cup-shaped chloroplast, upon which is located a **pyrenoid.** This latter is a rounded proteinaceous body, which seems to be the center of starch formation. Two small contractile vacuoles and a red pigment spot (frequently called an eyespot) are located near the anterior end of the cell. These vacuoles are possibly excretory in function, while the pigment spot is light-sensitive and results in the plant responding positively to a light stimulus.

Asexual reproduction is accomplished by the protoplast undergoing mitosis and forming into four to eight zoospores, similar in structure to the parent, which swim out and become new individuals when the parent cell wall ruptures. In sexual reproduction the protoplast produces 16 to 32 isogametes, which are also like the parent but much smaller. These fuse in pairs, forming zygotes, which develop a thick wall and are resistant to adverse environmental conditions. The diploid zygote undergoes meiotic division, forming four haploid meiospores, which are very similar to zoospores and which become mature plants when liberated. The life cycle of this plant is shown in Figures 22-1(b) and 22-2.

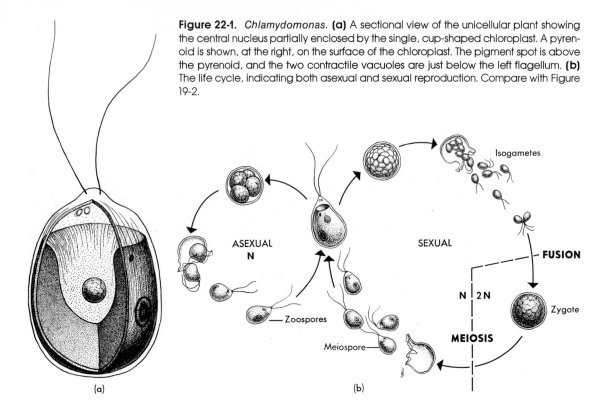

Figure 22-1. *Chlamydomonas.* **(a)** A sectional view of the unicellular plant showing the central nucleus partially enclosed by the single, cup-shaped chloroplast. A pyrenoid is shown, at the right, on the surface of the chloroplast. The pigment spot is above the pyrenoid, and the two contractile vacuoles are just below the left flagellum. **(b)** The life cycle, indicating both asexual and sexual reproduction. Compare with Figure 19-2.

ASEXUAL
N

SEXUAL

Isogametes

FUSION

Zygote

N ┆ 2N

MEIOSIS

Zoospores

Meiospore

(a)

(b)

Ulothrix

This plant (Figure 22-3) occurs in fresh water, often in flowing streams, and is a truly multicellular organism; many cells form one individual having some differentiation of structure and function. This is a considerable advantage in efficiency over a structure such as the unicellular *Chlamydomonas* or various colonial forms. *Ulothrix* consists of a single unbranched filament of indefinite length in

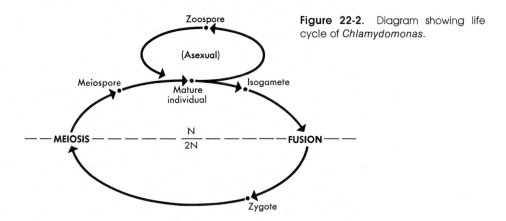

Figure 22-2. Diagram showing life cycle of *Chlamydomonas.*

Zoospore

(Asexual)

Meiospore

Mature
individual

Isogamete

— — —**MEIOSIS**— — — — — $\dfrac{N}{2N}$ — — — — **FUSION**— — —

Zygote

which the basal cell is a **holdfast** (fingerlike projections anchor the filament to solid surfaces). Each cylindrical cell has a single collar-shaped chloroplast, with pyrenoids; the nucleus is usually partially enclosed by the chloroplast. The plant increases in length by mitotic divisions of the component cells.

Asexual reproduction is by fragmentation or by the production of zoospores. These latter are produced by certain cells, called **zoosporangia,** when the contents divide mitotically into several (4 to 16) protoplasts. Each zoospore has four flagella, a chloroplast, a nucleus, and a pigment spot; it is reminiscent of a *Chlamydomonas* cell. After escaping when the wall of the zoosporangium ruptures, the zoospore swims about, attaches to a surface, and produces a new filament. In sexual reproduction some of the cells of a filament become **gametangia,** in which the contents divide mitotically to form 32 to 64 isogametes. These gametes are liberated into the water, but those from the same plant do not fuse. For isogamy to occur, isogametes from different filaments must be present, a condition termed **heterothallic.** These gametes are like small editions of zoospores but have only two flagella. The resistant, thick-walled diploid zygote eventually undergoes meiosis, producing four haploid meiospores, each of which is capable of developing into a new filament. There is no multicellular sporophyte phase; the zygote functions as a meiocyte.

The general size and shape of the isogametes and zoospores of *Chlamydomonas* and *Ulothrix* have led to the suggestion that sexual reproduction probably first arose in a *Chlamydomonas*-type unicell: when small zoospores that did not contain sufficient stored foods to develop into mature new individuals were produced, the fusion of two such small structures would be a great advantage. The fact that isogametes of *Ulothrix* have two flagella while the zoospores have four is similarly suggestive, especially in view of the existence of a unicellular alga, *Carteria,* which differs from *Chlamydomonas* only in having four flagella instead of two and in being larger.

The advantage of even a very simple multicellular organism, such as *Ulothrix,* over a unicellular one arises from the increased number of zoospores and gametes that are produced by the former, as well as the differentiation that leads to greater efficiency. In the multicellular form vegetative cells are always carrying on photosynthesis at the same time that reproductive cells are being produced. This is, of course, not possible with unicells.

Ulva

This plant (Figure 22-4) is an edible marine alga commonly known as sea lettuce. The cells divide mainly in two planes, resulting in a flat, sheetlike plant body two cells in thickness. A plant body such as this, which is not differentiated into tissues, is termed a **thallus** (plural, **thalli**). *Ulothrix,* of course, divides in one plane, resulting in its typical filament. The basal portion of *Ulva* forms a holdfast, which consists of an entangled mass of fingerlike outgrowths of many cells located in that region. This is quite different from the single-cell holdfast of *Ulothrix.* Each cell contains one nucleus and a cup-shaped chloroplast with a single pyrenoid.

Ulva undergoes a very definite alternation of multicellular phases, a condition not found in *Ulothrix,* where the diploid condition consists of the one-celled zygote. In *Ulva* biflagellate isogametes are formed by the protoplast of various cells of the haploid thallus dividing mitotically; such cells are gametangia. Gametes from different thalli fuse to form a zygote, which gives rise to a multicellular diploid thallus by a series of mitotic divisions. Various cells near the margin of this thallus become sporangia, and meiosis results in the production of haploid meiospores within these cells. The meiospores (frequently termed zoospores) have four flagella, swim about, settle, and grow by a series of mitotic divisions, giving rise to haploid thalli that produce only gametes. Although the chromosome num-

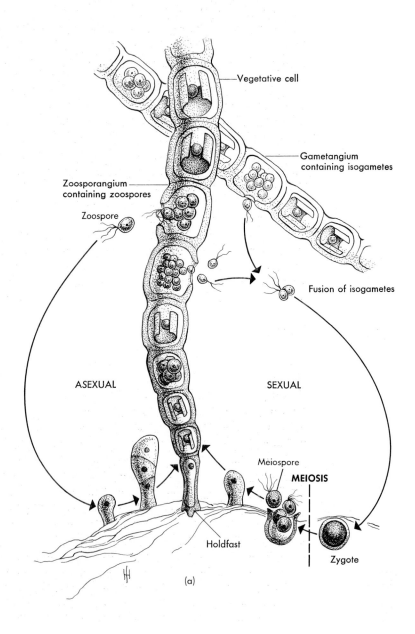

(a)

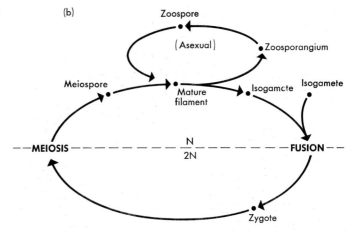

(b)

Zoospore

(Asexual)

Zoosporangium

Meiospore

Mature filament

Isogamcte Isogamete

MEIOSIS — — — — — $\dfrac{N}{2N}$ — — — — — FUSION — —

Zygote

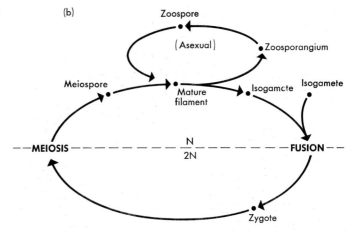

Figure 22-4. *Ulva.* **(a)** Habit sketch; (b) zoospores; **(c)** isogametes; **(d)** diagram of life cycle.

(b)

(c)

(a)

(d)

Gametophyte

Gametangium

Isogamete

Meiospore

Isogamete

MEIOSIS — — — — — $\dfrac{N}{2N}$ — — — — — FUSION —

Zygote

Sporangium

Sporophyte

bers of the cells comprising the two kinds of thalli differ, the thalli look alike and can be distinguished only when reproductive cells are being formed. This alternation of phases is an important feature of plant life cycles; it is characteristic of most Phaeophyta, Rhodophyta, Bryophyta, and the more complex plants. All sexually reproducing plants undergo an alternation of phases, although the relative emphasis on gametophyte and sporophyte differs among the several groups.

Pandorina

This plant is a spherical colonial form that is of interest because of the fusion of unequal gametes that takes place during sexual reproduction; though both are biflagellate, a small active gamete fuses with a large sluggish gamete. This is an example of **heterogamy** and is regarded as a very primitive stage in the differentiation of sex.

Oedogonium

This plant (Figure 22-5) is an unbranched filament consisting of cylindrical cells, each of which contains a nucleus and a single large chloroplast of irregular netlike shape with many pyrenoids. The basal cell is modified as a holdfast, and growth of the filament occurs as various cells divide mitotically. *Oedogonium* occurs in ponds and slowly moving streams, usually floating in masses when mature.

Asexual reproduction is by fragmentation or by the production of a single large zoospore from the protoplast of a cell, which is then termed a **zoosporangium.** The anterior region of the zoospore is clear and is encircled by a crown of **cilia,** which are similar to flagella but much shorter. The zoospore is liberated by a rupture of the wall of the zoosporangium, swims about, attaches to a surface, and develops into a new filament by a series of mitotic divisions. Sexual reproduction is of the oögamous type. Two sperm are produced by a mitotic division of the protoplast of a small cell of the filament. These resemble zoospores but are much smaller. The sperm-producing cells are termed **antheridia** (singular, **antheridium**), and occur in groups (Figure 22-5). The sperm swims to an enlarged globose cell, the **oögonium** (plural, **oögonia**), enters through a pore in the wall of this cell, and fuses with the single egg contained therein. In some species the antheridia and oögonia are produced on the same filament (homothallic); in others they are segregated in different filaments (heterothallic). The diploid zygote develops a thick wall and is very resistant to adverse conditions. When the zygote germinates, it produces four haploid motile meiospores (which resemble zoospores) as a result of meiosis, each of which may develop into a new filament.

Oedogonium, because it has evolved oögamy, is considered to be a more advanced filamentous type than *Ulothrix.* Such a differentiation of gametes has considerable survival value because of the presence of a relatively large amount of food in the egg, which is then available for the subsequent germination of the zygote and the growth of the resultant meiospores. Also, at least one of the gametes remains protected by the walls of the cell in which it is formed; the egg is not lib-

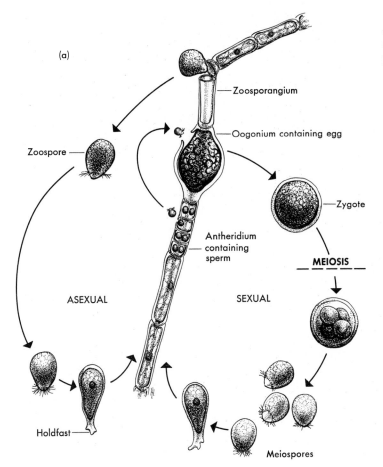

(a)

Zoosporangium

Oogonium containing egg

Zoospore

Zygote

Antheridium
containing
sperm

MEIOSIS

ASEXUAL

SEXUAL

Holdfast

Meiospores

Figure 22-5. *Oedogonium.* **(a)** Life cycle, indicating both asexual and sexual reproduction of this oogamous plant. **(b)** Diagram of life cycle.

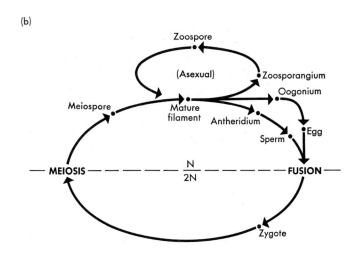

(b)

Zoospore

(Asexual)

Zoosporangium

Oogonium

Meiospore

Mature
filament

Antheridium

Egg

Sperm

MEIOSIS — — — $\dfrac{N}{2N}$ — — — **FUSION**

Zygote

erated into the surrounding water, as is the sperm. In the life history of any organism, the gametes are always the most delicate structures. They never have thick walls; any gametes that might by chance develop thick walls would find fusion with other gametes a rather difficult proposition. An organism that produced such peculiar gametes would have difficulty surviving. Therefore, any protection afforded to the gametes would be very advantageous to the plant. This should be borne in mind as various groups of the plant kingdom are studied.

Spirogyra

A *Spirogyra* plant (Figure 22-6) consists of an unbranched filament composed of cylindrical cells, each of which contains one or more ribbonlike chloroplasts in the form of a spiral band; pyrenoids are conspicuous on a chloroplast. Most of the interior of the cell is occupied by a large vacuole. Cytoplasm lines the cell wall and extends as strands toward the central region, where it surrounds the nucleus. These plants are commonly found as bright green masses on the surfaces of ponds and slowly moving streams; they are often termed "pond scum."

No zoospores are produced, and asexual reproduction is by fragmentation only. During sexual reproduction, when two filaments come in contact, projections grow singly from cells of each filament, and the wall between the projections eventually dissolves. The protoplast of each cell shrinks, as a result of water loss, and becomes a nonflagellated gamete. The protoplast of one cell moves through the connecting **conjugation tube** to the opposite cell and fuses with the protoplast of that cell. Usually all the cells of a filament act alike so that, after fusion, one filament contains zygotes and the opposite filament consists of empty cells. The differentiation of gametes is so slight that sexual reproduction is classified as isogamous. The zygote develops a thick wall and is resistant to adverse conditions. On germination the diploid zygote nucleus undergoes meiosis, forming four haploid nuclei, three of which degenerate. The cell with the remaining nucleus grows into a new haploid filament by mitotic divisions and subsequent cell enlargements.

Spirogyra and similar Chlorophyta are side branches in the evolutionary scheme of things; they are a dead end and have given rise to no other forms. This plant was discussed to indicate another type of sexual reproduction, a type that evolved independently in the Eumycota, for example, *Rhizopus* (Section 21.4).

22.3 Importance of the Chlorophyta

The green algae probably evolved from motile unicells similar to *Chlamydomonas* and produced a variety of forms and types of reproduction, only some of which have been discussed. This group presents a clear example of radiating evolutionary development. The evolution of sexuality has been emphasized because of the great importance of this type of reproduction in the production of variations, which are so significant in the survival of any group.

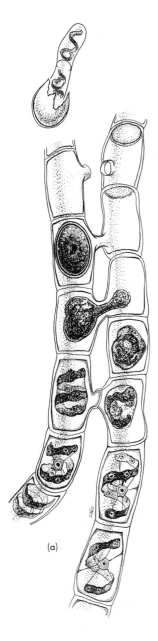

Figure 22-6. *Spirogyra.* **(a)** Two filaments showing sexual reproduction. The bottom two cells in each filament are in the vegetative condition. In the cells above conjugation tubes have formed; the protoplast of one cell is shown moving through the tube and fusing with the protoplast of the cell to the left. A thick-walled zygote is present in one of the cells at the upper left. A similar zygote is shown germinating at the top left of the diagram. **(b)** Diagram of life cycle.

(b)

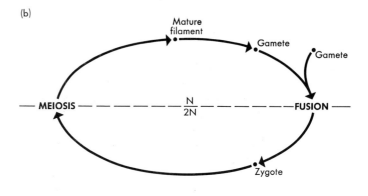

(a)

The green algae are also thought to have given rise to all of the higher green plants.

Many of the Chlorophyta in both fresh and salt water comprise food supplies for animals. Some (*Ulva*, for example) are utilized by man as food. In addition to supplying food, the photosynthetic process supplies the oxygen that is so necessary for life in the waters. Although the green algae themselves require oxygen for their respiratory processes and obtain their supply from photosynthesis, their rate of photosynthesis fortunately is much greater than the rate of respiration, and sufficient oxygen is left over to supply the needs of aquatic animals.

The green algae, along with other algae, occasionally cause problems when they become so abundant in water supplies that they impart obnoxious odors and flavors to the water. Such large numbers may also cause difficulties in the filtration processes of water treatment plants. This will be discussed more fully in Section 32.3.

22.4 Characteristics of the Division Chrysophyta (Diatoms and Golden-brown Algae)

This division includes several groups (golden-brown algae and the diatoms), but only the diatoms will be discussed (Figure 22-7). These plants are unicellular, sometimes forming colonies, each cell containing a nucleus and two or more chloroplasts in which are found chlorophylls *a* and *c* plus various yellow and brown pigments. The food reserves are in the form of chrysolaminarin (a complex carbohydrate) and oils; no starch is formed. The wall of each cell consists of two overlapping halves composed of pectic substances impregnated with silica (one of the chief constituents of glass). The walls do not decay, at least not the silica portion. Exceedingly beautiful and delicate markings are found on the walls; microscopic examination reveals that these ''lines'' are actually rows of tiny dots or pits. (The effective resolution of a microscope is frequently tested by examining these markings.) Many diatoms move about slowly as a result of the streaming of cytoplasm, which extrudes slightly through a long, narrow opening or cleft that

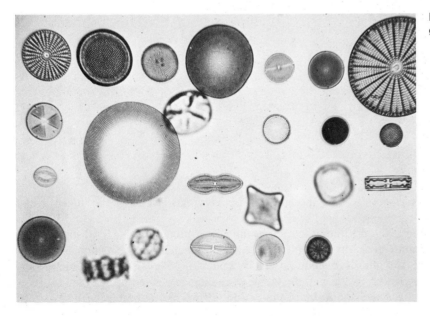

Figure 22-7. Photomicrograph of various diatoms.

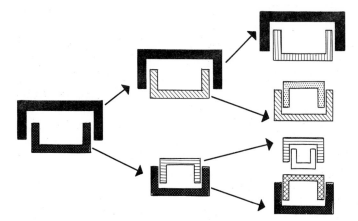

Figure 22-8. Cell division in diatoms. Note that the new half-wall forms inside of the old one. Some of the resultant diatoms, thererfore, are smaller than others. The protoplasts are not shown. See text for further discussion.

runs down the center of the top cell half. Diatoms are common in both fresh and salt waters, where they may be free floating or attached to various objects.

Asexual reproduction is by cell division (mitosis), during which the overlapping walls come apart, each partially enclosing one of the new protoplasts. A new half-wall is secreted inside of the old and is, of course, always smaller than the older half-wall (Figure 22-8); some of the diatoms have a tendency to become smaller and smaller. The original size is restored by the formation of **auxospores.** The protoplasts of two diatoms fuse to form an enlarged auxospore, which is really a zygote in this case and the two protoplasts are gametes. The zygote increases several times in size, to the maximum for the species, and then secretes new siliceous walls. Divisions of the auxospores then occur as described previously for other diatom cells. The vegetative diatom is presumably diploid, meiosis occurring with the formation of gametes.

22.5 Importance of the Chrysophyta

The various groups of this division have many structural and reproductive features that are quite similar to groups in the Chlorophyta. The scope of this text does not allow us to discuss these similarities except to state that they can be accounted for by parallel evolution of the two distinct divisions. Although the two may have been derived from different primitive flagellated unicells, these ancestors themselves were related.

The diatoms, along with the Pyrrophyta (not discussed in this text), are known as the "grass of the seas" and form the basis of most of the food chains in the seas. Small animals eat these plants and are in turn eaten by larger animals, and so on. The animals obtain their basic building materials for synthesizing cellular constituents from the sugars, oils, and amino acids of these tiny plants. Even man, feeding upon fishes, is dependent upon these organisms, which occur in such enormous numbers in the waters of the world.

The red tide that occasionally appears off the Atlantic or Pacific Coasts of North

America consists of billions of tiny dinoflagellates (Pyrrophyta), usually *Gymnodinium* or *Gonyaulax;* up to 60 million of these organisms may be in 1 liter (quart) of water, which appears reddish because of pigments in the cells. These organisms produce toxins that have resulted in the killing of thousands of fish during such red tide outbreaks. The dinoflagellates are also ingested by such organisms as mussels, which may become dangerous for human consumption as a result of the accumulation of toxins. Such a red tide occurred along the New England coast in the late summer of 1972, and over 20 people required hospital treatment as a result of poisoning by contaminated shellfish.

Since the siliceous walls of diatoms do not decay, great numbers of empty walls accumulate at the bottom of any water in which these plants live. These rocklike deposits of **diatomaceous earth** are frequently very thick, and many are worked as open quarries. This material is used in filters, as insulation, in dynamite as an absorbent for liquid nitroglycerine, in cement to increase the workability and strength of concrete, and as a mild abrasive in metal polishes and toothpastes. One of the largest deposits of diatomaceous earth occurs at Lompoc, California, and is approximately 1000 ft deep and 12 mi sq; since 1 cu in. of diatomaceous earth contains about 40 million diatom remains, this deposit originally contained somewhat more than 10^{24} diatoms (10^{24} equals 1 followed by 24 zeroes). Such deposits that occur above the water surface are the result of geological uplift; the area must have been under water at one time because all diatoms are either marine or freshwater forms. Diatoms are also often found with oil deposits and are frequently of use to geologists as one of the diagnostic characteristics of oil-bearing soils. Diatoms quite possibly were involved in the production of the oil deposits in the first place, as is suggested by the typical oil food reserves in this group.

22.6 Characteristics of the Division Phaeophyta (Brown Algae)

The brown algae, nearly all of them marine, vary in size from small filamentous types to the huge complex **kelps** usually found in cold waters; no unicellular or colonial forms have been found. The plastids contain chlorophylls *a* and *c,* which are usually masked by an accompanying brown pigment, **fucoxanthin.** Some kelps have considerable parenchymatous tissue differentiation; in addition to the holdfast is a **stipe** (or stalk), which bears flattened **blades.** The stipe has a definite **epidermis,** a **medulla** in the central region consisting of elongated cells, and an intermediate **cortex,** which consists of loosely arranged spherical cells. One type of cell in the medulla resembles the sieve tube elements in the phloem of higher plants; it is elongated, arranged end to end, and has perforated end walls similar to sieve plates. However, none of this group has any true vascular tissue and certainly no xylem, and the brown algae are not considered to have true roots, stems, or leaves. Many of the larger forms bear **bladders,** gas-filled enlargements, which serve to buoy the upper portions near the surface. The food reserves are typically sugars, laminarin (a complex soluble carbohydrate), oils, and complex alcohols; no insoluble carbohydrates are formed. The inner layer of the wall is composed of cellulose, but the outer layer consists of a slimy, gelatinous material called *algin.*

Brown algae may increase in size as a result of a growing apex or, as in kelps, by the activity of a meristematic region at the base of the stipe or at the juncture of stipe and blade. Asexual reproduction is occasionally by fragmentation, but usually by zoospores. Sexual reproduction is various; both isogamous and oögamous forms are found. An alternation of phases occurs, with a strong tendency toward a reduction of the gametophyte; the large kelps are sporophytes, and in *Fucus* the gametophyte phase is present only as gametes (no multicellular gametophyte). The motile cells, either zoospores or gametes, are pear shaped with two unequal flagella attached at the side.

22.7 Representative Members of the Phaeophyta

As with the Chlorophyta, a few individual brown algae will be discussed in order to emphasize evolutionary changes.

Ectocarpus

The genus *Ectocarpus* (Figure 22-9) consists of small, branched, filamentous forms in which the sporophyte and gametophyte phases are very similar and can be distinguished only by the type of reproductive cells that are produced. The sporophyte bears two kinds of sporangia: a **plurilocular** (multicellular) **zoosporangium** in which each cell produces a diploid zoospore that develops into a new multicellular sporophyte, and a **unilocular** (one-celled) **sporangium** in which meiosis occurs, followed by mitotic divisions, and the resultant numerous haploid meiospores develop into multicellular gametophytes by mitotic divisions. Although the meiospores and zoospores are similar in appearance, they can be distinguished if it is known in which type of sporangium they were produced; the former are haploid while the latter are diploid, but this does not affect the appearance. The gametophytes bear **plurilocular gametangia,** which are very similar in appearance to the plurilocular zoosporangia, and the liberated isogametes from different plants fuse to produce the diploid zygote; germination of the zygote produces a sporophyte. Isogametes are indistinguishable in appearance from the spores, but they are distinguishable on the basis of the fusion that must occur before development of the sporophyte.

Laminaria

In this plant (Figure 22-10) the diploid sporophyte is very large, frequently reaching a length of five meters or more, usually consisting of long expanded blades with a stout stipe held to the substratum by a holdfast, while the haploid gametophyte is extremely tiny in size, consisting of but a few cells. Sexual reproduction is of the oögamous type; sperm are produced in antheridia by a male gametophyte, and eggs in oögonia by the female gametophyte. Syngamy (fusion of

Figure 22-9. *Ectocarpus.* **(a)** Habit sketch; **(b)** diagram of life cycle.

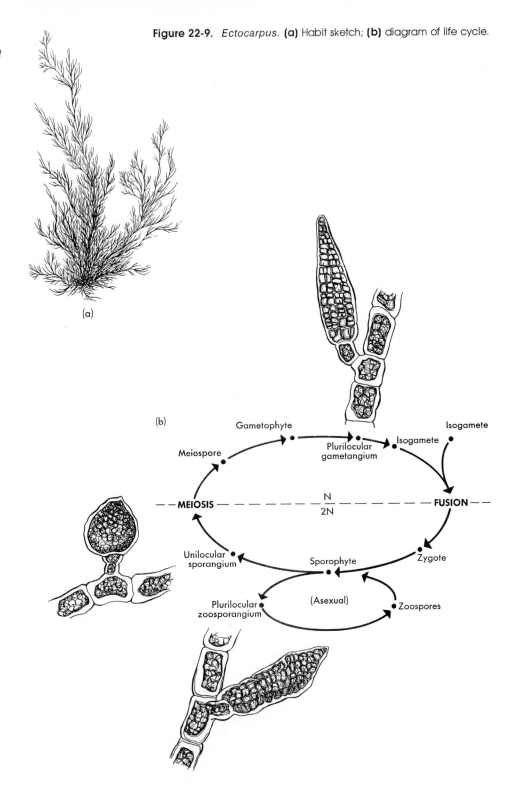

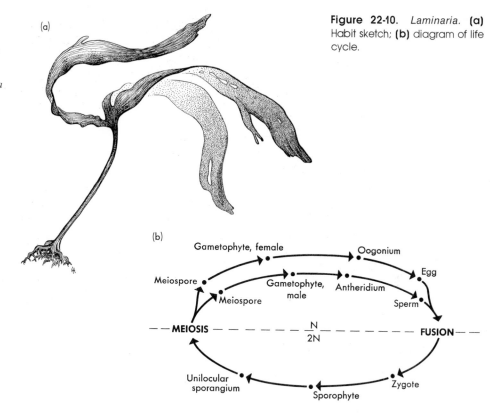

Figure 22-10. *Laminaria.* **(a)**
Habit sketch; **(b)** diagram of life
cycle.

gametes) takes place with the egg partially extruded, and the zygote develops into a sporophyte. The nucleus of a unilocular sporangium of the sporophyte first divides meiotically, and then numerous mitotic divisions finally result in the production of many haploid meiospores, which are liberated and develop into gametophytes (half of which are male and half female). No asexual zoospores are produced.

Fucus

This plant (Figure 22-11) consists of a flattened, dichotomously branched thallus with a stipe and a holdfast; air-filled bladders are frequently present and serve to buoy up the plant. The inflated tips (**receptacles**) of the thallus are covered with small wartlike swellings, which are actually the raised exists from embedded spherical chambers (**conceptacles**). Two kinds of sporangia are produced, either within the same conceptacle or in different conceptacles; in some species only one kind of sporangium is produced on any one plant. The **microsporangia** (male sporangia) are small oval structures borne on branched filaments. As a result of meiosis, haploid **microspores** (male spores) are produced, which undergo mitotic divisions to form numerous biflagellated sperm. The globose **megasporangium** (fe-

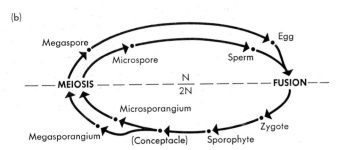

Figure 22.11. *Fucus.* (a) Habit sketch; (b) diagram of life cycle.

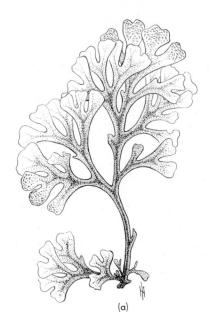

(b)

Megaspore

Microspore

Egg

Sperm

MEIOSIS — — — — — $\dfrac{N}{2N}$ — — — — — FUSION — —

Microsporangium

Megasporangium (Conceptacle) Sporophyte

Zygote

(a)

male sporangium) is produced at the end of a small stalk; meiosis results in the formation of four haploid **megaspores,** which divide once more mitotically to produce eight eggs. There is no multicellular gametophyte; gametes are formed directly from spores. In some texts the sporangia are considered to be gametangia that form the respective gametes with meiosis occurring during this formation. The gametes are released into the water, where fertilization occurs; the eggs float and become surrounded by a swarm of motile sperm. The zygotes sink to the bottom and develop into new sporophyte *Fucus* plants.

22.8 Importance of the Phaeophyta

Because of the lack of any simple brown algae, we cannot trace their ancestry, but they probably arose from flagellated brown unicells in a manner parallel to that of the Chlorophyta. The Phaeophyta have evolved further than the Chlorophyta; the former have considerable differentiation of tissues and some multicellular sex organs. An evolution of sexuality from the isogamous type (*Ectocarpus*) to the oögamous type (*Laminaria* and *Fucus*) has occurred; a general tendency for a reduction in the size of the gametophyte has also been noticed. In *Ectocarpus* the gametophyte and sporophyte are equal; in *Laminaria* the gametophyte is greatly reduced; in *Fucus* the gametophyte is restricted to the meiospores and the gametes (there is no multicellular haploid stage).

Some of the kelps are used as food in China and Japan, and some are fed to cattle in the British Isles. The ash obtained by burning kelps is an excellent source of iodine and potassium, but this is unprofitable because of the availability of these

Figure 22-12. Kelp harvester (Courtesy of Kelco Company San Diego, California.)

materials from other sources; in some areas kelps are still used as fertilizers, however. Kelps are harvested (Figure 22-12) off the Pacific Coast of the United States especially for their algin content. This material is a colloid used in stabilizing many dairy products; it imparts a smooth consistency and prevents the formation of ice crystals.

22.9 Biological Significance of the Algae

As mentioned previously, green plants form the basis of all food chains; they are the food-producing organisms. (The chemosynthetic autotrophic bacteria, which were discussed in Chapter 20, are unimportant as far as food supply is concerned.) Obviously, then, all organisms living in water are dependent upon algae for food. Man may, of course, turn to the algae themselves or to the fish feeding upon algae as a food source. The total food productivity of the seas probably exceeds that of the land by a considerable amount, and future human populations may well turn more and more to the sea for sustenance. Greater efforts will probably be made to "farm" algae, as do the Japanese now. In addition to their food content, algae also produce tremendous quantities of oxygen as a result of photosynthesis. Such oxygen is, of course, also necessary to animal life, as well as to the algae themselves. "Algal blooms," a tremendous increase in numbers of algae resulting when there is an unusual supply of nutrients, may pollute water supplies. Such growth frequently occurs when waters receive drainage from farmlands, industrial wastes, or sewage (see Section 32.3).

Summary

1. The division Chlorophyta is characterized by the presence of chlorophylls *a* and *b* and carotenoids in chloroplasts and by the production of starch and fats or oils as food reserves. Reproduction is of various types, and the plants range from unicellular organisms to multicellular plants of small size. Members of this division represent a clear example of radiating evolution, beginning with simple unicellular forms and progressing to the complex multicellular forms with their specialization of functions. The evolution of sexuality is also apparent, beginning with the simple isogamous types and progressing through the heterogamous types to the more complex oögamous form of sexual reproduction. The higher green plants quite likely arose from the Chlorophyta.

2. The diatoms of the division Chrysophyta contain chlorophylls *a* and *c*, accessory yellow or brown pigments, and food reserves in the form of chrysolaminarin and oils; no starch is formed. The wall of each cell is composed of two overlapping halves, which are impregnated with silica. These plants are enormously important as the basis of food supplies in the seas. In addition to this, diatomaceous earth is used in filters, as insulations, and as a mild abrasive.

3. In the division Phaeophyta chlorophylls *a* and *c* are usually masked by the brown pigment, fucoxanthin, and the food reserves are sugars, laminarin, oils, and alcohols. A well-developed alteration of phases usually occurs, and the motile cells are pear shaped with two unequal flagella attached at the side.

4. The "algae" in general are important as food sources for the animals of the water.

Review Topics and Questions

1. Compare the life cycles of *Ulothrix* and *Oedogonium,* indicating for each structure whether the cycle is haploid or diploid. In what ways is *Oedogonium* considered to be more advanced than *Ulothrix,* and what advantages obtain to these characteristics?
2. Describe the life cycle of *Chlamydomonas.* Why is this plant considered to be more primitive than *Ulothrix?*
3. Discuss the differences among isogamy, heterogamy, and oögamy.
4. What is meant when an organism is said to be "heterothallic"? What might be an advantage of such a situation?
5. Describe the life cycle of *Ulva.* How can the gametophyte phase be distinguished from the sporophyte phase?
6. How could you distinguish between a zoospore and an isogamete of *Ulothrix* if the flagella are not visible?
7. List at least five reasons why diatoms are important to man.
8. Describe the life cycle of *Ectocarpus.* How could you distinguish a plurilocular gametangium from a plurilocular zoosporangium?
9. A friend has collected specimens of an alga, which he sends to you for identification. All the structures are well preserved, but the liquid preservative has

bleached all the pigments. While you are waiting for a letter from your friend describing the color of a living specimen, what should you look for before placing this alga in the Chlorophyta, Chrysophyta, or Phaeophyta?

10. Describe the life cycle of *Laminaria*.
11. Distinguish between microsporangia and megasporangia.

Suggested Readings

ALEXOPOULOS, C. J., and H. C. BOLD, *Algae and Fungi* (New York, Macmillan, Inc., 1967).

BOLD, H. C., *Morphology of Plants,* 3rd ed. (New York, Harper and Row, 1973).

CRONQUIST, A., *Introductory Botany,* 2nd ed. (New York, Harper and Row, 1971).

SMITH, G. M., *Cryptogamic Botany,* Vol. I: *Algae and Fungi* (New York, McGraw-Hill, 1955).

TORTORA, G. J., et al., *Plant Form and Function* (New York, Macmillan, Inc., 1970).

Division Bryophyta

23.1 Characteristics of the Division Bryophyta

The Division Bryophyta consists of the mosses and their allies, subdivided into three classes, of which only two will be discussed. These are all small plants (usually less than 20 cm in height), which are parenchymatous and lack vascular tissues although a central column of elongated cells is present in the gametophyte of some; thus, they have no true roots, stems, or leaves. The absence of vascular tissue with the attendant lack of efficient water conduction is probably the cause of the small size and the general restriction of the bryophytes to moist habitats. The plant body is essentially a thallus with hairlike rhizoids, which penetrate the substratum and absorb water and minerals.

The gametophyte phase is nutritionally independent and structurally more complex than the sporophyte. The Bryophyta are all oögamous, and the sex organs are multicellular, the outer layer of cells forming a sterile jacket enclosing the gametes. This affords additional protection to the delicate gametes, protecting them from drying out. The male sex organ, **antheridium** (Figure 23-1), is usually spherical to oval in shape and generally projects on a stalk from the surface of the gametophyte tissue. The antheridium first consists of densely packed cells, but as it matures the cells in the central region differentiate into sperm, which are enclosed in a cavity that develops as the entire structure enlarges. These sperm are liberated when the antheridium absorbs water and the jacket cells burst apart. The female sex organ, **archegonium** (Figure 23-1), is generally flask-shaped; the enlarged base, or **venter,** containing the single egg is usually exposed on a small stalk. The canal cells of the neck disintegrate, leaving a channel through which the sperm may reach the egg. The antheridium and archegonium likely evolved from plurilocular gametangia (see Section 22.7). If only the outer cells of a plurilocular gametangium became sterile and did not produce gametes, the structure would essentially be an antheridium. The gametophyte may bear both antheridia and archegonia, or male and female gametophytes may be separate. In either case, the sperm are motile and swim to the egg through a film of water although the splashing of raindrops aids in the transfer in many cases. The sperm are attracted to the archegonia by chemical stimuli, and syngamy occurs in the venter. The diploid

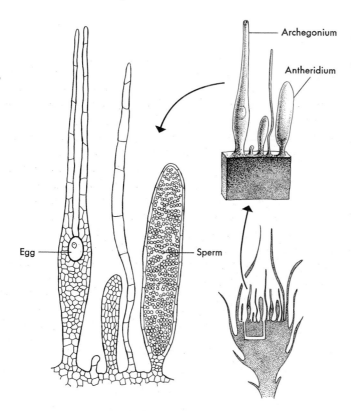

Figure 23-1. The tip of a moss (*Mnium*) gametophyte. The sperms are formed in an antheridium; the egg in an archegonium.

Archegonium

Antheridium

Egg

Sperm

zygote undergoes a series of mitotic divisions, and develops into the multicellular sporophyte, which remains attached to the gametophyte.

The sporophytes produce multicellular sporangia of various sizes and shapes, but all have an outer sterile layer of cells enclosing sporogenous tissue in which the cells undergo meiosis as they develop into haploid meiospores. These meiospores, upon liberation from the sporangium, develop into gametophytes. In some species the sporophyte may contain chloroplasts, as the gametophyte always does, but the former is permanently attached to the latter and derives water and minerals from the tissues of the gametophyte.

The Bryophyta are basically land plants and have a cuticle, but they are rather poorly adapted to a land habitat. They lack conducting and supporting tissues, and they have motile sperm and are dependent upon a film of water to ensure fertilization. A definite alternation of a multicellular sporophyte with a multicellular gametophyte occurs, a feature that is lacking in many algal forms where the only diploid structure is the one-celled zygote. Chlorophylls *a* and *b* and the carotenoid pigments are found in plastids in the same ratio as in the Chlorophyta.

It is quite likely that the Bryophyta arose from the Chlorophyta but independently of the vascular plants. The similarities in chlorophyll pigments and storage of starch indicate such a relationship. However, the presence of antheridia and archegonia suggests the possibility that the Bryophyta may be a degenerate line from some of the simpler vascular plants. Most botanists favor the suggestion of algal ancestry. All agree that the Bryophyta are a dead-end side branch and have given rise to no other group.

In some instances mosses are influential in soil development by their participation in plant succession. They may become established on rock surfaces after lichens have paved the way, or they may be pioneers themselves. The mosses assist in preventing erosion, in collecting debris, and in adding organic matter to the developing soil when they die.

23.2 Characteristics of the Class Musci

The mosses are common plants of the woodlands and various moist areas. The green mat consists of many rather erect and "leafy" gametophytes. These appendages, arranged spirally around the axis, look and function like leaves, but they do not contain conducting tissue (xylem and phloem) and thus are not considered to be true leaves. They are merely lobes of the thallus, and usually one or a few cells in thickness except at the midrib where they may be thicker. At the base of this "leafy" structure are rhizoids, small filaments that function as do the roots of higher plants, and at the apex are found the sex organs. In *Mnium* (Figures 23-1 and 23-2) both antheridia and archegonia are found on the same gametophyte, whereas in *Polytrichum* male gameotphytes produce only antheridia and female gametophytes produce only archegonia. The sperm are flagellated and swim to the archegonium, actually being stimulated to move in that direction by secretions from the archegonium or egg.

The zygote develops into an embryo, which is retained for a short time within the venter of the archegonium. The **embryo** may be considered to be an immature new sporophytic plant, which receives essential nutrients from the parent tissues. Further cellular divisions and differentiations produce the recognizable young sporophyte, the lower end of which (the foot) is buried in the tissues of the gametophyte. The archegonium enlarges considerably, but the elongation of the sporophyte ruptures the archegonium and carries the upper part along as a cap, the **calyptra,** covering the tip of the sporophyte. The upper part of the sporophyte develops into an enlarged **capsule,** or sporangium, separated from the foot by a relatively long stalk, the **seta.** At this time, the calyptra usually shrivels and is easily blown away by the wind. Figure 23-3 represents a moss plant with a young sporophyte projecting from the top of the gametophyte; the foot of the sporophyte is not shown, but the calyptra is still in place. As the sporophyte matures, the capsule enlarges, and sporogenous tissue develops in a cylindrical fashion around a central sterile mass of cells called the **columella.** Figure 23-4(a) is a longitudinal section through such a young capsule, showing also the **operculum,** or lid, and the **peristome,** the ring of toothlike structures just beneath the lid. The sporogenous tissue (Figure 23-4[b]) actually consists of diploid **spore mother cells** (meiocytes), which, after meiosis, give rise to haploid meiospores. The sterile cells break down, the meiospores occur as a powdery mass, and the lid is shed by a transverse splitting of the sporangium. Dissemination of spores is facilitated by the "teeth," which bend in when moist and then bend outward when the air is dry;

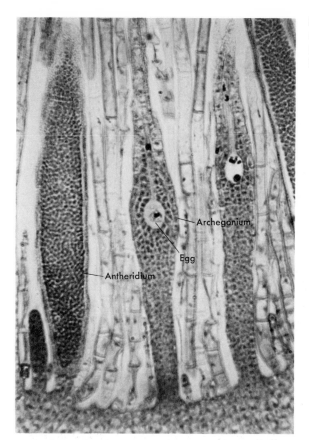

Figure 23-2. Photomicrograph of archegonium and antheridium in a moss, *Mnium* (×450). (Courtesy of Peter Jankay.)

Figure 23-3. Habit sketch of a moss plant with the sporophyte projecting from the top of the "leafy" gametophyte. The calyptra is still in place on the former.

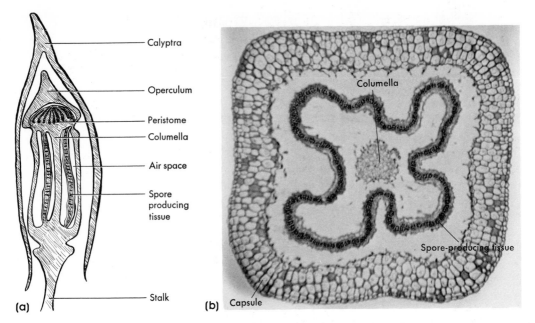

Figure 23-4. **(a)** Moss capsule with the calyptra still in place, longitudinal section. **(b)** Photomicrograph of a moss capsule, cross section (×200). ([**b**] courtesy of Peter Jankay.)

this results in the scooping out of spores or at least the escape of spores, after which they will be blown about more readily by wind currents.

When a moss spore is in a suitable environment, it germinates to produce a **protonema,** which is a long, filamentous, multicellular, branching structure growing along the soil surface. The cells contain chloroplasts; thus the protonema produces its own food supply. At intervals, rhizoids growing from the protonema penetrate the soil; no chloroplasts are present in the cells of the rhizoids. Numerous small projections, or "buds," appear on the protonema and develop into the "leafy" erect structures that will eventually bear the gametangia. Additional rhizoids develop at the base of such erect structures, which become independent individuals upon the death and disappearance of the protonema. Figure 23-5 represents this development of protonema, rhizoids, and **gametophores** (the generally

Figure 23-5. Development of a moss gametophyte. The spore has germinated and produced a horizontal protonema. From the latter, rhizoids project down into the soil and erect gametophores are produced at intervals.

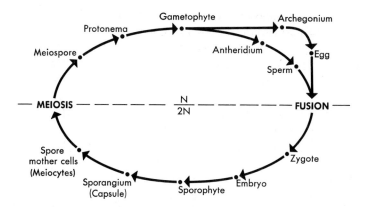

Figure 23-6. Diagram of moss life cycle.

erect structures that will bear antheridia and archegonia). Strictly speaking, all the structures that develop from the spore are parts of the gametophyte phase. Because the "leafy" structure is really only a part of the gametophyte, it is termed the gametophore. The overall life cycle of the moss plant is represented in Figure 23-6.

The *Musci* are a comparatively large group of plants, but little is known of their evolutionary history, although the resemblance of the protonema to a filamentous green alga suggests an algal ancestry. The only genus of any economic importance is *Sphagnum,* the bog or peat moss. Peat actually consists of a mixture of plants, but by far the greatest amount of the material is compressed *Sphagnum.* The water in which these plants grow is very acid. This plus the antiseptic properties of *Sphagnum* result in exceedingly slow decay, and the plant parts gradually accumulate as peat. This material may be dried and used as fuel. It is also useful in horticulture because of its water-holding capacity and is frequently mixed with sandy or humus-poor soils for this reason. As *Sphagnum* develops, large empty cells become dispersed among the smaller living ones; the former absorb and hold water much as does a sponge. *Sphagnum* has even been sterilized and then used as wound dressings because of this liquid-holding characteristic.

23.3 Characteristics of The Class Hepaticae

The liverworts are small, prostrate plants found in moist areas. While the mosses are radially symmetrical, the liverworts grow flat upon the ground and are dorsiventrally differentiated; that is, the upper surface of the gametophyte is considerably different from the lower surface. The sex organs are borne dorsally or at the anterior end; in some species these structures are borne on erect gametophores. Some of these plants are thallus types, while others are "leafy" in that a slender axis bears tiny leaflike structures. The name "liverwort" refers to the supposed liverlike shape of the thallus; "wort" is old English for "plant." The gametophyte of many species may withstand desiccation for several months without injury, growth being initiated again when water is available. Although there are over 8000 species in the Hepaticae, it is beyond the scope of this text to consider more than one example of a thallus type and one ''leafy type.''

Riccia

The simple, thalloid gametophyte branches dichotomously; growth results from cellular divisions at the notches of the lobes (Figure 23-7). Occasionally the older portions die, leaving the lobes separated as individual plants that continue their

Figure 23-7. *Riccia* life cycle. Fragmentation of the thallus (*lower right*) has resulted in two plants. At the stage of maturity shown (*upper right*) the capsule wall has disintegrated and no longer surrounds the mass of spores.

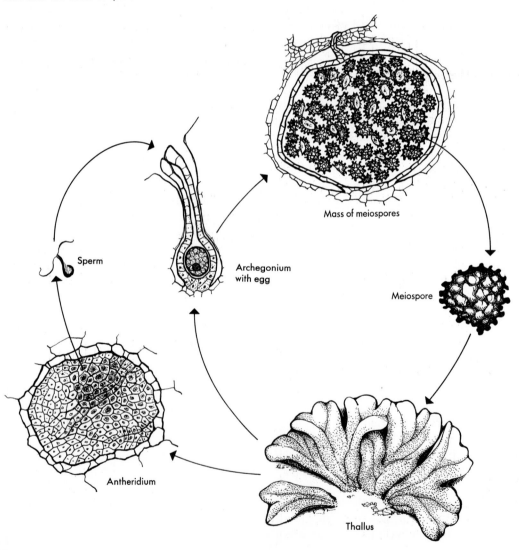

Mass of meiospores

Sperm

Archegonium
with egg

Meiospore

Antheridium

Thallus

growth pattern. The upper epidermis contains pores that lead to **air chambers** between the epidermal layers. The cells in the dorsal region contain chloroplasts, while the lower portion of the thallus consists of colorless parenchymatous cells. The lower epidermis bears rhizoids, which serve to anchor the plant and absorb minerals and water.

Both antheridia and archegonia are usually borne on the same gametophyte and embedded in the thallus along a longitudinal groove on the dorsal side; the neck of the archegonium protudes above the thallus. When the antheridium ruptures, the biflagellated sperm swim to the archegonium where fertilization occurs. The venter of the archegonium enlarges, and cellular divisions of the zygote produce a rather spherical, multicellular sporophyte contained within the archegonium. The cells of the outer layer are sterile and form the wall of the sporophyte capsule; the cells (meiocytes) within undergo meiosis, producing haploid meiospores. The mature sporophyte is a very simple structure, then, consisting simply of a wall surrounding a mass of spores. The latter are liberated upon disintegration of the thallus and the capsule (sporangium). In a suitable environment the meiospores develop into the haploid gametophyte thallus. Figure 23-8 summarizes the life cycle of a liverwort.

Although the wall cells may contain chloroplasts, the sporophyte is dependent upon the gametophyte for minerals, water, and probably most of its food supply. The gametophyte is, of course, independent.

Porella

The "leafy" liverworts, of which *Porella* (Figure 23-9) is an example, are frequently mistaken for mosses. Careful examination of the green gametophyte enables one to see the dorsiventral orientation and the differentiation into a "stem" with three rows of "leaves"; there are two lateral rows and one ventral row of smaller "leaves." The main axis upon which the "leaves" are located also bears rhizoids on its lower surface. Although the gametophyte is branched, there is very

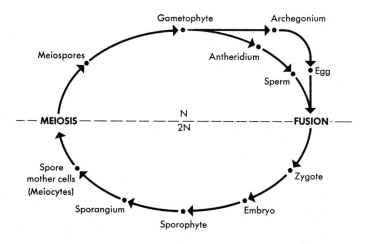

Figure 23-8. Diagram of liverwort life cycle.

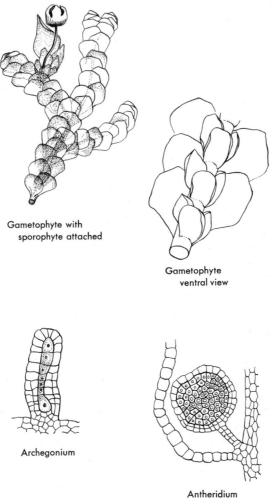

Figure 23-9. **Porella,** a leafy liverwort.

Gametophyte with
sporophyte attached

Gametophyte
ventral view

Archegonium

Antheridium

little internal differentiation; the leaflike structures consist of a single layer of cells.

Antheridia and archegonia are borne on the same gametophyte near the tips of branches. The sperm swim to the archegonium. The mature sporophyte consists of a foot embedded in the gametophyte, a stalk, and the capsule (sporangium) with its haploid meiospores; meiosis occurs in the capsule. The meiospores are liberated by a longitudinal splitting of the capsule into four parts. Upon germination, the meiospore develops into a haploid gametophyte. Figure 23-8 summarizes the life cycle of a liverwort.

As in other members of the Hepaticae, the sporophyte remains attached to the gametophyte and is dependent upon it although there are some chloroplasts in various cells of the former. The gametophyte is an independent structure.

Summary

1. Members of the division Bryophyta are small plants that lack vascular tissues. The gametophyte phase is nutritionally independent and structurally more complex than the sporophyte. The latter is attached to, and dependent upon, the gametophyte. These plants also have a well-developed alternation of phases and are basically land plants.

2. In the mosses (Class Musci) the antheridia and archegonia may be found on the same, or on different, gametophytes. The sperm are flagellated and swim to the archegonium, where fertilization occurs. The zygote develops into an embryo, which is retained within the archegonium for a short time. When the mature sporophyte develops, the basal portion remains embedded in the tissues of the gametophyte. Haploid meiospores are produced in a capsule (sporangium) as a result of meiotic divisions. After the meiospores are liberated, they develop into the filamentous protonema and eventually the mature gametophyte.

3. The general life cycle of a liverwort (Class Hepaticae) is basically similar to that of a moss but without a protonema.

4. The Bryophyta probably arose from the Chlorophyta but have themselves not given rise to any other groups.

Review Topics and Questions

1. Describe the life cycle of a moss, indicating which structures are haploid and which are diploid.
2. Diagram an immature archegonium (before the neck canal cells have disintegrated). Label fully.
3. Diagram a mature capsule in transverse (cross) section. Label fully.
4. Present reasons why the Bryophyta are limited to moist areas.
5. Define the following: protonema, rhizoid, embryo, venter, columella, operculum, and calyptra.
6. A moss gametophyte that had toothed "leaves" with midribs was crossed with a moss gametophyte that had nontoothed "leaves" and no midribs. Assume that these characteristics are determined by the following: T = toothed leaves, T_1 = nontoothed leaves, M = midribs, M_1 = no midribs; T and T_1 are alleles; M and M_1 are alleles.

   The meiospores produced subsequent to this cross developed into the following: (a) 150 plants with toothed leaves and midribs, (b) 160 plants with nontoothed leaves and no midribs, (c) seven plants with toothed leaves and no midribs, and (d) five plants with nontoothed leaves and midribs.

   Diagram the cross described, and give the genotypes of all the plants mentioned (both parents and progeny). Explain the proportions of the various groups in the progeny.
7. Compare the similarities and differences between a moss, such as *Mnium*, and a liverwort, such as *Riccia*.

439

Suggested Readings

BOLD, H. C., *Morphology of Plants,* 3rd ed. (New York, Harper and Row, 1973).

BURNS, G. W., *The Plant Kingdom* (New York, Macmillan, Inc., 1974).

CRONQUIST, A., *Introductory Botany,* 2nd ed. (New York, Harper and Row, 1971).

DOYLE, W. T., *The Biology of Higher Cryptogams* (New York, Macmillan, Inc., 1970).

24
Vascular Plants: Divisions Psilophyta, Lycophyta, and Sphenophyta

24.1 General Characteristics of the Vascular Plants

Representative groups of vascular plants are discussed in this chapter and in the next three chapters. All the plants considered to this point have no distinct vascular tissue, although very primitive conducting strands have been demonstrated in some members of the Phaeophyta (see Section 22.6) and in some members of the Bryophyta. A basic characteristic of the groups to be discussed, then, is the presence of tracheids or their derivatives, which comprise the main portion of the xylem. The conducting elements of the xylem and phloem were considered to indicate close relationship, and, until recently, vascular plants were usually classified in the division Tracheophyta. In the last decade or so, however, other anatomical and morphological differences have been recognized among these groups. The fossil record indicates that many of the modern groups have been distinct lines as far back as there is a clear record of vascular plants. Thus, the division Tracheophyta has been reorganized into several divisions of equal rank, and Tracheophyta has been dropped from this classification. Not all botanists agree with this reorganization, and this is another example emphasizing the temporal nature of any system of classification. New information and new interpretations are appearing at a rapid rate. It is simply not possible to satisfy all (or even the majority of) botanists with a single system.

The plants of these groups have a relatively large, elaborate sporophytic phase, which is dominant and nutritionally independent at maturity; only the embryo or very young sporophyte depends upon the gametophyte. In the primitive forms the gametophyte is nutritionally independent, while in the more advanced forms the gametophyte is dependent upon the sporophyte. All representatives are essentially land plants and have a cuticle. A few of the higher plants have become adapted to a water habitat (e.g., *Lemna*, or duckweed), but they are known to be essentially land plants because the nearest relatives are well-established land plants. They are oögamous, and in the higher forms the sperm do not require free water for transport to the egg. Chlorophylls *a* and *b* and carotenoid pigments are present in plastids, as in the Chlorophyta. Almost all vascular plants have roots, stems, and leaves.

24.2 Characteristics of the Division Psilophyta

Most of the members of this group are known from fossil remains, and only two genera contain living examples. Their importance stems from the fact that they are apparently the most primitive vascular plants, having a central cylindrical core of vascular tissue. No pith or cambium is present, and the phloem consists of thin-walled, vertically elongated, living cells with scattered sieve areas and slanted end walls (considered to be primitive). The plant body itself, utilizing *Psilotum* as the example (Figure 24-1), is a very simple structure consisting of a dichotomously[1] branching axis without roots or leaves, or with exceedingly small scalelike leaves that are mere flaps of tissue; small groups of photosynthetic parenchyma cells are located directly beneath the epidermis. The underground stem, or rhizome, may bear rhizoids. This is the sporophyte, and sporangia are borne terminally on very short branches; only one kind of meiospore is produced, and the plants are termed **homosporous.** In *Tmesipteris* a small unbranched vascular strand penetrates the leaflike structures, which are somewhat larger than those of *Psilotum.*

The meiospore is haploid and develops into a small, bulky, cylindrical, branching gametophyte (a few millimeters long), which lacks chlorophyll and exists saprophytically underground (Figure 24-2) obtaining food through a mycorrhizal (see Section 21.13) association with a fungus. Mature individuals resemble pieces of the sporophytic rhizome, and the nature of the gametophyte was unknown until early in the twentieth century, long after the sporophyte had been described. The antheridia and archegonia are scattered over the surface of the gametophyte; the former project from the surface, while the base of the latter is embedded. Multi-flagellate sperm swim to the egg, and fertilization occurs in the venter of the archegonium. A central strand of tracheids has been observed in at least some gametophytes of *Psilotum,* but this is not usually present. This is the only instance in which vascular tissue has been found in a gametophyte and is of importance with regard to the origin of land plants, as will be discussed in the next section.

The zygote develops into an embryo; the basal cells of the latter produce a conspicuous foot, which penetrates into the adjacent tissues of the gametophyte. For a short period of time the sporophyte absorbs material from the gametophyte. Ultimately the shoot becomes detached from the foot; the rhizome continues to elongate and branch, and eventually some branches emerge above the soil and develop into aerial branches.

Some students of phylogeny (the evolutionary history of a species or larger group) consider *Psilotum* and *Tmesipteris* to be related to certain members of the ferns (Pterophyta; see next chapter), and that they should be transferred to a taxon within that division. In any event, they may still be regarded as the most primitive of vascular plants.

[1] Dichotomous branching is the division into two equal branches, each of which grows for a while and then also forks or divides into two equal branches.

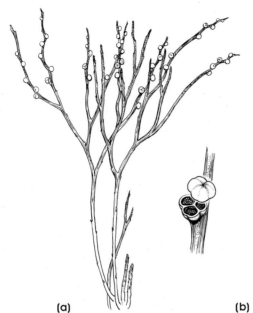

(a)

(b)

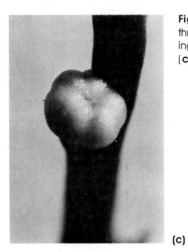

(c)

Figure 24-1. **(a)** *Psilotum.* The enlargement at the right shows three sporangia dissected so that the spores are visible. **(b)** Living *P. nudum* (×1). **(c)** *P. nudum*, sporangia (×20). ([**b**] and [**c**] courtesy of Peter Jankay.)

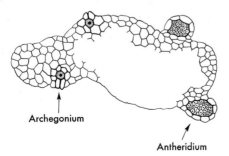

Archegonium

Antheridium

Figure 24-2. *Tmesipteris tannensis* gametophyte, sectioned to show archegonia and antheridia. It is very similar to *Psilotum*, which is described in the text.

443

24.3 The Origin of Land Flora

Botanists are in general agreement that land plants probably arose from a group of algae that had a life cycle similar to that of *Ulva* (see Section 22.2). In such a life cycle, two phases that are essentially alike in structure, size, shape, and even somewhat in reproductive structures alternate regularly with each other. The gametophyte phase is haploid and bears gametangia; the sporophyte is diploid and bears sporangia, which are similar to the gametangia but in which meiosis occurs. The transmigrant, that hypothetical plant that bridged the gap between aquatic and terrestrial existence, thus probably had similar gametophyte and sporophyte phases *both* of which migrated on to the land; both phases are considered to have possessed in a primitive state such necessary land plant characters as vascular tissue and cuticle.

Origin of the Bryophyta

As mentioned in the previous chapter, the Bryophyta probably arose from the Chlorophyta through a transmigrant type but have not themselves given rise to any more advanced group. The earliest known land plants with vascular tissue probably occurred in the Silurian Period about 415 million years ago, whereas Bryophyta remains are not found earlier than about 380 million years ago (in the Devonian Period). The general pattern in fossils is that the remains of structurally simpler organisms are found in layers of the earth that are older than the layers in which the remains of more complex organisms are found. Since the structurally more complex vascular plants are represented by older fossils, it is considered that the bryophytes are not ancestors of the former. In general, the tendency in the bryophytes has been toward a reduction of the sporophyte and an elaboration of the gametophyte, but not much evolutionary change as a whole has occurred. As is generally the case with soft thin-walled forms, lacking the decay-resistant features of the more "woody" plants, a clear chain (or progression) of fossil forms does not exist; and thus the relationships are difficult to establish.

Origin of the Vascular Plants

The vascular plants are also considered to have arisen from some algal group, most likely the Chlorophyta. Actually, no one algal group has all the characters necessary for the land plants, but most such characters do appear in an incipient form somewhere in some algal group. The earliest known land plants, members of the Psilophyta, are not much more than a dichotomizing algal plant with vascular tissue and an advanced type of sporangium.

The type of life cycle found in *Ulva,* a member of the Chlorophyta, and the proportion of specific photosynthetic pigments found in these algae and the vascular plants are evidences of a direct relationship. The close similarity of the sporophyte and gametophyte in some of the Psilophyta, as well as the occurrence of tracheids in the gametophyte, is considered strong evidence that the phases of the life cycle were at one time independent and equal.

Chief Problems of Land Plants

The transition from an aquatic to a terrestrial environment placed the plant in a precarious position with regard to desiccation, absorption, gas exchange, transport of materials, support, and reproduction; problems that do not obtain when the plant is immersed in water. The theory for the transition of plants from an aquatic to a terrestrial habitat assumes that the transmigrant just happened to combine enough of those various characteristics that enable a successful land existence. It is most likely that ancestral types were left behind as shorelines receded because of land uplift or falling water levels, and that land flora evolved gradually over long geologic ages. It seems less likely that there was an active migration or invasion of land. In any case the mechanism of adaptation to a new environment seems clearly to have involved random changes in genetic material, some of which resulted in structural and functional modifications that better fitted the plants for survival under the new environmental conditions.

The aerial parts of land plants could not survive the drying effect of the atmosphere unless there was some means of conserving moisture in the cells. This has been achieved fairly well by those plants in which the secretion of a cuticle happened to evolve as well as the production of cork layers with their suberized cells. Neither cutin nor suberin is completely impervious to water, but they certainly reduce water loss considerably. Such coverings reduce evaporation (transpiration) but they also interfere with the gas exchanges required for photosynthesis and respiration. This presented an advantage to those plants in which functional stomata happened to evolve along with the cuticle and cork; loss of water vapor was decreased while it was still possible for the diffusion of carbon dioxide and oxygen to occur.

Aquatic plants usually absorb water, minerals, and other materials through most of their surfaces from the surrounding water. The water and mineral sources available to land plants require special absorbing structures that penetrate the soil. However, this also presents the problem of distributing such materials to the above-ground portions of the plant, as well as providing the subterranean cells with a food supply from the rather remote photosynthesizing cells. In the most successful land plants there evolved specialized conducting tissues, the xylem and phloem, in which strands of elongated cells transport materials more efficiently than the cell-to-cell diffusion found in simpler forms. The rigid cell walls of much of the xylem tissue, as well as other schlerenchymatous cells, serve as a supportive system replacing the buoyancy of surrounding water and permitting exposure to light.

In aquatic plants and low-growing forms of wet terrestrial habitats, the dependence of syngamy upon liquid water presents no real problem; but establishment in most land habitats would not be possible unless syngamy could be accomplished without reliance upon a film of moisture. The most successful land plants are those in which a mechanism happened to evolve that resulted in the transport

446

Ch. 24 / Vascular
Plants:
Divisions
Psilophyta,
Lycophyta, and
Sphenophyta

of sperm to egg independently of a film of water. Meiospore dissemination is also very different in land plants and aquatic plants. In the former, air currents and insects are the most important agents; and many modifications have evolved that bring about an increased efficiency of such agents, thus presenting advantages to such plants.

In the following discussions of different plant groups emphasis will be placed upon the various evolutionary characteristics that bring about an advantageous exploitation of diverse habitats by members of the specific group.

Characters or adaptations inherited from the algae.   In the algae, especially the Chlorophyta, are found individuals that have the type of life cycle suggested for the transmigrant, the proportion of chlorophylls *a* and *b* found in land plants, and also the multicellular, branching, parenchymatous plant body that is so similar to the primitive land plants. In addition to this, one finds some internal differentiation of tissues, including vascular tissues, and a type of multicellular sex organ in the Phaeophyta that suggests the possibility of such structures in unknown members of the Chlorophyta or the possibility of their evolution in the transmigrant. Oögamous reproduction, so characteristic of land plants, is also found in many groups of algae (e.g., *Oedogonium*). Certain of these characteristics are of benefit with regard to water retention and protection of gametes, while others of these structures are of importance in transporting materials or in support.

Characters or adaptations not found in the algae.   The evolution of a cuticle on the aerial parts of plants and on spores was an extremely important factor in making a land existence possible. The chance occurrence of such a structure resulted in a tremendous decrease in the amount of water lost from such parts and greatly decreased the water requirement of that plant. This is a considerable boon to any existence out of water. The production of an outer layer of sterile cells, forming a protective jacket, as part of a multicellular sex organ enabled gametes to survive more readily in a nonaquatic, as well as an aquatic, environment; a similar change occurred with regard to the sporangium. The retention of the young embryo in the archegonium during its early development also resulted in further protection of the more sensitive portions of the plant. In the simpler land plants sperm swim to the egg, but in the more complex forms sperm are transported to the egg by means other than swimming. This method of transport by means of a pollen tube will be discussed later.

As a result of these adaptive characteristics, plants could exist on land. Additional evolution of vascular tissue, as shown in the more advanced members of the vascular plants, increased the possibility of large plants maintaining themselves on land by enabling them to transport water and minerals from the soil and food from the leaves to all parts of the body. This tissue also provided the support that is essential in exposing leaves to light and a supply of carbon dioxide.

Let us emphasize at this point that none of these adaptive characteristics evolved in order to enable a plant to exist on land, free of its watery environment. As a result of such adaptions arising *by chance,* plants have been enabled to exist on land. Also, these structural adaptations must be inherited for them to be of any

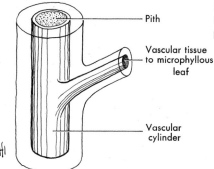

Figure 24-3. Vascular supply to microphyllous leaf; no leaf gap. The central portion of the stem may be pith, as shown, or a solid core of xylem.

significance. (Why?) Further discussion of this idea of chance versus purpose will be found in Section 28.10.

24.4 Characteristics of the Division Lycophyta (Club Mosses)

The dichotomously branching sporophytes have roots, stems, and leaves, and the vascular tissue in most consists of a solid core of xylem surrounded by a cylindrical sheath of phloem. The spirally arranged leaves are very small (**microphyllous).** The small, unbranched vascular bundle projecting into the leaf has no effect on the shape of the vascular tissue, and no leaf gaps are present (Figure 24-3). (See Section 25.1 for a discussion of leaf gaps.) Sporangia are borne singly in the axils of more or less specialized leaves, the **sporophylls,** which are usually arranged in the form of a cone, or **strobilus,** and the plant is either homosporous or **heterosporous** (two kinds of meiospores are produced in the latter). The sporophyte is large and dominant, as compared to the gametophyte, and it is also nutritionally independent at maturity. The strobilus and heterospory found in the Lycophyta are a foreshadowing of the cone and flower structures found in the more advanced plant types (the seed plants).

The haploid meiospore develops into a small independent gametophyte which may be either saprophytic or autotrophic, depending upon the species concerned. In some genera the gametophyte remains enclosed within the spore wall, a very important situation, which will be discussed more fully later. Water is necessary for the transfer of sperm from the antheridium to the archegonium.

24.5 Representative Members of the Lycophyta

Several hundred species are organized into four living genera and several orders, but our discussion will be limited to two genera.

In the genus *Lycopodium* (Figure 24-4) the sporophyte is relatively small and has an underground stem from which the roots extend. In some species the sporo-

448

*Ch. 24 / Vascular
Plants:
Divisions
Psilophyta,
Lycophyta, and
Sphenophyta*

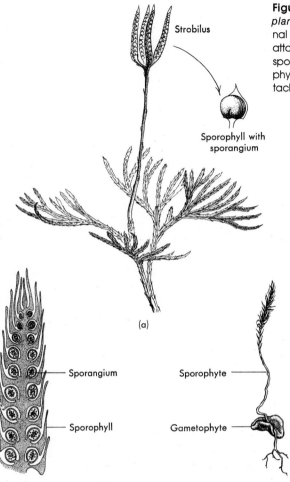

Strobilus

Sporophyll with
sporangium

(a)

Figure 24-4. *Lycopodium.* **(a)** *L. com-
planatum.* **(b)** *L. adpressum,* longitudinal section of a strobilus to show the
attachment of each sporangium to a
sporophyll. **(c)** *L. adpressum,* gametophyte with young sporophyte still attached.

Sporangium

Sporophyll

Sporophyte

Gametophyte

(b)   (c)

phylls are organized in strobili, and all are homosporous. Meiosis occurs in the
sporangium, and the meiospores germinate to produce a gametophyte that bears
both antheridia and archegonia. The sperm swims to the egg, which is retained in
the archegonium, apparently as a result of chemicals secreted by the egg. The
sporophyte remains attached to the gametophyte for a short period, absorbing materials by means of the foot, but soon becomes independent as the roots and
leaves develop. Mature gametophytes may continue to live for some time, supporting one or more young sporophytes in different stages of development.

In *Selaginella* (Figure 24-5) the well-organized strobili typically bear two kinds
of sporangia and spores; the plant is heterosporous. The larger sporangia, containing four large meiospores, are often borne lower down in the strobilus; **megaspores**
are produced by meiosis of a megaspore mother cell (**megasporocyte**) in a **megasporangium** associated with a **megasporophyll.** The smaller sporangia, containing
many small meiospores, are usually borne in the upper portion of the strobilus;

Figure 24-5. *Selaginella.* **(a)** Habit sketch. **(b)** Longitudinal section of a strobilus, the microsporophylls form the upper portion and the megasporophylls from the lower portion of the strobilus. **(c)** A megasporangium with three of the four megaspores. **(d)** A microsporangium with many microspores. **(e)** Strobius (×15). ([e] courtesy of Peter Jankay.)

microspores are produced by meiosis of microspore mother cells (**microsporocytes**) in a **microsporangium** associated with a **microsporophyll.** In prostrate forms, usually the microsporangia are borne in upper ranks and the megasporangia in lower ranks. The size of the meiospores and sporangia is not important except as a means of description and identification. The important factor is the structure that develops when the haploid meiospore germinates.

The megaspore germinates to form a gametophyte, which remains almost completely within the spore wall. This gametophyte, although multicellular and having rhizoids, is very reduced in size and produces small archegonia and no antheridia; it is a female gametophyte, or **megagametophyte,** and contains a considerable amount of stored food in the lower portion (Figure 24-6[a]). The microspore forms a male gametophyte, which is completely retained within the spore wall (Figure 24-6[b]). This **microgametophyte** is greatly reduced in size, consisting only of sterile cells next to the old microspore wall with the sperm differentiating from cells inside this sterile jacket. Both types of meiospores develop into gametophytes while still in their respective sporangia. Syngamy (fertilization) occurs after the gametophytes have fallen to the moist soil, or in some species the megagametophyte is not shed until after fertilization (a situation somewhat akin to that in seed plants). In the latter case microgametophytes sift down to the lower portions of the strobilus.

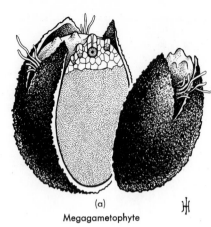

(a)
Megagametophyte

(b)
Microgametophyte

Figure 24-6. *Selaginella.* Both gametophytes are still retained within their respective spores at this stage of development. **(a)** Megagametophyte with one mature and three developing archegonia in the upper area. (Redrawn from H. Bruckmann, *Flora* **104**, 180–224, 1911–1912.) **(b)** Microgametophyte with the enclosed sperms. (Redrawn from R. A. Slagg, *American Journal of Botany* **19**, 106–127, 1932.)

The life cycles of *Lycopodium* and *Selaginella* are presented for comparison in Figures 24-7 and 24-8.

24.6 Importance of the Division Lycophyta

The members of this group are of no real economic significance, but they do indicate certain evolutionary advances over the Psilophyta from which they have probably evolved. The presence of true leaves and roots, the greater development of vascular tissue, and the general arrangement of sporophylls in strobili are all considered to be characteristics of a more advanced nature than those found in the Psilophyta. Even though the Lycophyta do not appear to have given rise to more advanced groups of plants, heterospory, the retention of the microgametophyte within the old microspore wall and the retention of the megagametophyte in the megasporangium (as in some species of *Selaginella*) are reminiscent of important

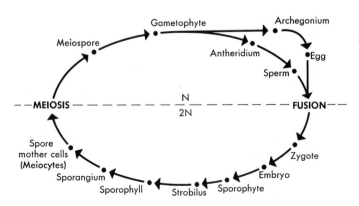

Figure 24-7. Diagram of the cycle of *Lycopodium.*

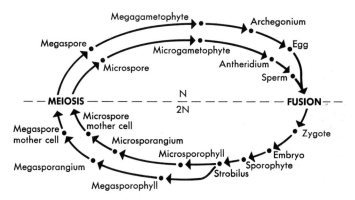

Megagametophyte
Archegonium
Megaspore
Microgametophyte
Egg
Microspore
Antheridium
Sperm
N
2N
— — MEIOSIS — — — — — — — — — — — — — FUSION — —
Microspore
mother cell
Megaspore
mother cell
Microsporangium
Zygote
Megasporangium
Microsporophyll
Embryo
Sporophyte
Strobilus
Megasporophyll

Figure 24-8. Diagram of life cycle of *Selaginella.*

steps toward a seed habit as found in the more advanced groups. This again is an indication that evolutionary trends may be followed in various groups.

24.7 Characteristics of the Division Sphenophyta (Horsetails)

The only living representatives of this group are in the genus *Equisetum* (Figure 24-9) and are generally herbaceous plants that usually do not exceed 1 m (3 ft) in height although a tropical species grows to 5 m. The tiny leaves (microphylls)

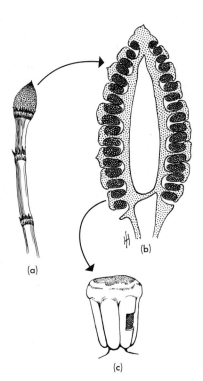

(a)

(b)

(c)

Figure 24-9. *Equisetum.* **(a)** Reproductive stem with a terminal strobilus. **(b)** Longitudinal section of a mature strobilus showing the location of the sporangia and their enclosed spores. **(c)** Enlarged view of a sporangiophore with its sporangia; one sporangium is cut open to show the meiospores.

Figure 24-10. *Equisetum,* Vegetative stems. **(a)** ×1. **(b)** ×5. (Courtesy of Peter Jankay.)

Leaves

occur in whorls at the nodes of hollow stems, which have vertical ridges. The tissues are impregnated with silica, giving such an abrasive consistency to the structures that these plants have been used for scouring kitchen utensils and floors. A perennial underground stem, or rhizome, is present. The members of this group are homosporous, although one extinct genus was heterosporous, and the sporangia are borne on specialized structures called **sporangiophores,** which are arranged in strobili. The sporangiophores have a horizontal stalk and a flattened table-top-shaped upper portion (the sporangiophore is peltate) from which the sporangia hang parallel to the stalk. In *Equisetum* some of the species produce two kinds of stems; vegetative stems are green and branched, while reproductive stems are usually not green, not branched, and bear strobili at the tips (Figure 24-10).

Meiosis occurs in the sporangia, and the mature haploid meiospores are liberated by a longitudinal splitting of the sporangial wall. The wall of the spore is laminated, and four ribbonlike bands, or **elaters,** separate from the wall except for a common point of attachment. The elaters are hygroscopic, uncoiling as they dry out and recoiling with the addition of moisture (Figure 24-11). These movements probably aid in the discharge and dispersal of spores. Tiny, green, ribbonlike gametophytes, usually containing both antheridia and archegonia, develop from the

453

Sec. 24.7 /
Characteristics
of the Division
Sphenophyta
(Horsetails)

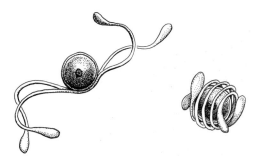

Figure 24-11. *Equisetum* meiospores showing (*left*) elaters uncoiled and (*right*) elaters coiled.

meiospores. After fertilization, the embryonic sporophyte is briefly dependent upon the gametophyte but rapidly develops leaves and an independent existence as the gametophyte degenerates.

These plants are of little economic importance except, as is also true of the Lycophyta, as participants in coal formation during the Carboniferous Period, when they were a significant part of the flora. The Sphenophyta probably arose from the Psilophyta, but they themselves never gave rise to any higher groups.

Summary

1. In vascular plants the sporophyte is usually large, dominant, and nutritionally independent at maturity.

2. The Psilophyta are the most primitive of the vascular plants, consisting basically of a branching axis without roots and usually with small scalelike leaves (or none). Sporangia are borne terminally, and the plants are homosporous. The gametophyte of the Psilophyta exists saprophytically underground.

3. Land plants probably evolved from a group of algae that had a life cycle similar to that of *Ulva* (a member of the Chlorophyta), a life cycle in which the two phases are essentially alike. The tendency in the Bryophyta has been toward a reduction of the sporophyte and an elaboration of the gametophyte; this is just the opposite of the tendency in the more successful land plants. Some of the primitive vascular plants are not much more than a dichotomizing algal plant with vascular tissue.

4. The main problems of land plants are desiccation, transport, and support. Some of the structures found in the more advanced algae tend toward at least the protection of gametes and toward support and transport. The vascular plants, however, have evolved many features that tend to make possible a plant's existence on land: cuticle, extensive vascular tissue, protected gametes, protected embryo, and a relatively large sporophyte phase.

5. The Lycophyta have roots, stems, and small leaves, but no leaf gaps. Sporangia are borne on sporophylls, which are usually arranged as a strobilus; the plants are homosporous or heterosporous. The gametophytes may be saprophytic or autotrophic.

454

Ch. 24 / Vascular
Plants:
Divisions
Psilophyta,
Lycophyta, and
Sphenophyta

6. The Lycophyta probably arose from the Psilophyta but apparently did not themselves give rise to more advanced groups. The retention of gametophytes within the spore walls and the retention of the megagametophyte in the megasporangium (as in some species of *Selaginella*) are reminiscent of the seed habit found in advanced groups, an indication of similar evolutionary trends in various groups.

7. In the Sphenophyta the leaves are borne in whorls, and the sporangia are borne on sporangiophores arranged as strobili. The gametophytes are tiny and autotrophic. This group probably arose from the Psilophyta.

Review Topics and Questions

1. Compare the vascular plants with the Bryophyta. Why were the former more successful in surviving in various terrestrial environments than the latter? If vascular tissue had developed in the gametophyte phase, would the Bryophyta have been as successful as the vascular plants?
2. Describe the life cycle of *Psilotum.*
3. Discuss the ways in which the Lycophyta may be considered as more advanced than the Psilophyta.
4. Describe the life cycle of *Selaginella.*
5. Discuss the theory that attempts to explain the origin of land plants.
6. Define the following: microphyll, homospory, heterospory, sporophyll, megasporophyll, megaspore, microsporophyll, microspore, megagametophyte, microgametophyte, strobilus, and sporangiophore.

Suggested Readings

BIERHORST, D. W., *Morphology of Vascular Plants* (New York, Macmillan, Inc., 1971).

BURNS, G. W., *The Plant Kingdom* (New York, Macmillan, Inc., 1974).

CRONQUIST, A., *Introductory Botany* (New York, Harper and Row, 1971).

FOSTER, A. S., and E. M. GIFFORD, JR., *Comparative Morphology of Vascular Plants,* 2nd ed. (San Francisco, Freeman, 1974).

Division Pterophyta

25.1 General Characteristics of the Division Pterophyta (Ferns)

In addition to the characteristics that are common to all vascular plants, members of the Pterophyta have a dominant sporophyte with roots, stems, and large leaves. Such leaves (**megaphylls**) are believed to have evolved by the modification and fusion of a branch system; they are not merely extensions of the outer tissues, as appears to be the case with microphyllous leaves. The vascular supply to megaphyllous leaves is quite extensive and is associated with a parenchymatous area, known as a **leaf gap** (Figure 25-1), in the vascular tissue of the stem. The **leaf trace** is a vascular bundle connecting the vascular system of the leaf with that of the stem. It extends from the base of the leaf to the area in the stem where it merges with the vascular system of the stem. It appears as though a section of the vascular cylinder of the stem is deflected to one side where the leaf trace diverges into a leaf. Parenchyma instead of vascular tissue differentiates in the vascular region of the stem immediately above the diverging trace. In this region, the parenchyma cells of the pith are continuous with the parenchyma cells of the cortex. Such leaf gaps are not found in the lower vascular plants, which were discussed in the previous chapter, or in a few of the primitive ferns.

Fern leaves (**fronds**) are typically compound and uncoil as they develop, a condition referred to as **circinate vernation** (Figure 25-2). Roots usually arise adventitiously from horizontal rhizomes, although the tree ferns of tropical areas have upright stems appearing much like a tree trunk. In general, the vegetative and fertile (sporangium-bearing) leaves are morphologically similar, most ferns are homosporous, and seeds are not produced. The haploid meiospores develop into small, nutritionally independent gametophytes, and the sperm that are produced require a film of water in their movement toward the egg. The ferns are further subdivided into four orders, of which only one representative member will be discussed.

The fern plant (Figure 25-3), common to many gardens, consists of large conspicuous leaves and the horizontal stem (usually underground) with its tiny adventitious roots. The vascular tissue is quite prominent as the veins of the leaf, and a cross section of the rhizome demonstrates that both xylem and phloem, but no

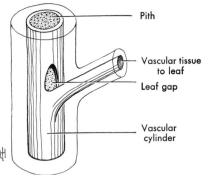

Figure 25-1. Vascular supply to megaphyllous leaf; leaf gap present.

Pith

Vascular tissue
to leaf

Leaf gap

Vascular
cylinder

cambium, are present. On the lower surfaces or at the margins of the leaves are the **sori,** or clusters of sporangia. In some ferns each **sorus** is covered by a small flap of tissue, the **indusium** (Figure 25-4), or by the rolled-in margin of the leaf. Each sporangium in a sorus consists of a stalk and a capsule in which spore mother cells (meiocytes) undergo meiosis in the production of haploid meiospores. Dissemination of the meiospores may be facilitated by the structural development of the capsule. A ring of cells, the **annulus,** extends from the stalk over and around about three fifths of the circumference of the capsule; all except the outer walls of these ring cells are much thicker than the walls of the cells compris-

Figure 25-2. Fern leaf uncoiling. (Courtesy of Peter Jankay.)

457

Sec. 25.1 /
General
Characteristics
of the Division
Pterophyta
(Ferns)

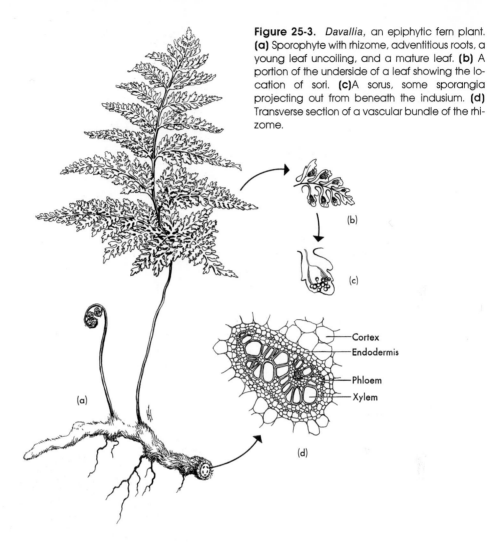

Figure 25-3. *Davallia*, an epiphytic fern plant. **(a)** Sporophyte with rhizome, adventitious roots, a young leaf uncoiling, and a mature leaf. **(b)** A portion of the underside of a leaf showing the location of sori. **(c)** A sorus, some sporangia projecting out from beneath the indusium. **(d)** Transverse section of a vascular bundle of the rhizome.

Cortex
Endodermis

Phloem
Xylem

ing the rest of the capsule (Figure 25-5). As the cells of the sporangium dry, the thick walls of the annulus cells tend to pull together. This places tension on the thin-walled cells of the sporangium, and they rupture. The spores are thus flung out of the capsule and distributed by wind currents.

If a meiospore lands in a suitable environment (shady, moist, cool soil), it germinates to form a filamentous structure that rapidly develops into a green, heart-shaped gametophyte, which is seldom more than 1.5 cm in diameter. Rhizoids project from the lower surface of this **prothallus** and absorb water and minerals from the soil. Also on the lower surface, near the notch of the heart-shaped structure, are found archegonia. Antheridia are located near the basal end among the numerous rhizoids (Figure 25-6). As the antheridium matures, the central cells develop into sperm, the outer cells remaining as a sterile jacket or layer. If mature gametophytes are moistened, the antheridium ruptures and the sperm are lib-

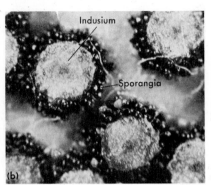

Figure 25-4. Sorus of a fern. **(a)** ×2. **(b)** ×100. (Courtesy of Peter Jankay.)

Sorus

Indusium

Sporangia

(a)

(b)

erated. As the archegonium matures, the canal cells disintegrate, and the cells at the lip of the neck spread apart. Sperm are attracted to the archegonium, possibly by the disintegration products of the canal cells or by secretions from the remaining archegonial cells, and swim down to the egg in the embedded venter, where fertilization (syngamy) takes place. The antheridia tend to mature before the archegonia. This helps to produce cross-fertilization between two adjacent prothalli of slightly different ages.

The fertilized egg (zygote) begins mitotic divisions while in the archegonium, developing into the embryo, which obtains nutrients from the gametophyte through a specialized group of cells called the **foot.** Quite rapidly various groups of cells of the embryo produce a root, stem, and leaf (Figure 25-7); the embryo enlarges, and the young sporophyte is soon independent of the gametophyte, which

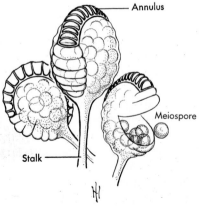

Annulus

Meiospore

Stalk

Figure 25-5. Fern sporangia. Meiospores are being liberated from one.

459

Sec. 25.1 /
*General
Characteristics
of the Division
Pterophyta
(Ferns)*

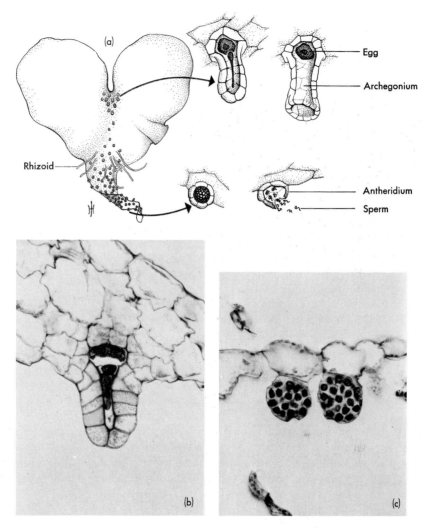

Figure 25-6. **(a)** Fern gametophyte (lower surface) with archegonia and antheridia. **(b)** Photomicrograph of archegonium (×300). **(c)** Photomicrograph of antheridia (×300). (Photographs courtesy of Robert Gill.)

dies. Further growth and development produce the mature fern plant with its sporangia and meiospores.

The many species of ferns have a life cycle similar to that described in Figure 25-8 although they may differ from one another with regard to homospory, heterospory, position of sori, indusium, leaf size and shape, and the presence of specialized reproductive leaves. Most ferns may be propagated vegetatively by cutting the rhizome, each portion of which may then continue growing with its complement of leaves and roots as an individual plant.

Ferns are widely distributed in both tropical and temperate regions. Some grow

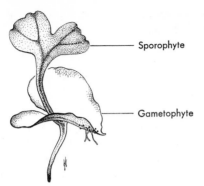

Figure 25-7. Young fern sporophyte still attached to the gametophyte. The latter will die shortly.

Sporophyte

Gametophyte

as floating aquatics and others are epiphytes on tree trunks in moist forests. An **epiphyte** is a plant that grows attached to some larger plant without deriving nourishment from its host.

25.2 Importance of the Division Pterophyta

Except for their contribution to coal formation during the Carboniferous Period, ferns are economically important primarily from an aesthetic viewpoint. They are grown almost universally in gardens, homes, or hothouses (greenhouses), and are often used in bouquets and floral arrangements. Young fronds are used as a green vegetable in many parts of the world. A drug from *Dryopteris filix-mas,* the male fern, has been utilized since the days of Nero in curing tapeworm.

As with all plants with a reduced gametophyte, this situation may be considered to present a distinct advantage in that the motile sperm have a relatively short distance to cover in reaching the egg. Also, both male and female reproductive structures are located on the lower surface of the gametophyte, where the most moisture is likely to be present. The large sporophyte, in addition, ensures the

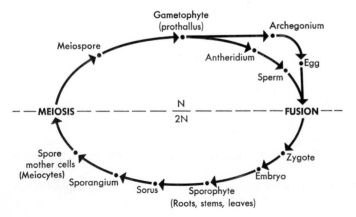

Figure 25-8. Diagram of life cycle of a fern.

widespread dissemination of great numbers of meiospores. Such spores may spread to suitable environments that are at considerable distances from the parent plant.

With the evidence at hand, botanists believe that the Pterophyta probably evolved from the Psilophyta. Some extinct ferns resemble Psilophyta rather closely. Very primitive ferns of the tropics have cylindrical subterranean gametophytes that lack chlorophyll and thus are very similar to the gametophyte of *Psilotum*. In fact some individuals who are studying ferns and their relationships have concluded that living Psilophyta (such as *Psilotum*) belong in the Pterophyta with the ferns. This situation needs a bit more clarification. Many botanists also hold the opinion that certain of the extinct ferns were the ancestors of the seed plants.

Summary

1. The Pterophyta have a dominant sporophyte with roots, stems, and large leaves. The vascular supply to such megaphylls is quite extensive and results in the formation of a leaf gap, a parenchymatous area in the vascular tissue of the stem.

2. In general, the Pterophyta have leaves that uncoil as they develop and a horizontal stem (rhizome) with adventitious roots; they are usually homosporous. The haploid meiospores develop into small, nutritionally independent gametophytes. The sperm swim to the egg, and fertilization takes place in the archegonium.

3. The zygote develops into an embryo and then into a young sporophyte, which is dependent upon the gametophyte for a short period. Sporangia develop in clusters, sori, on the undersides of the leaves. Within these sporangia spore mother cells undergo meiosis, and the resultant haploid meiospores germinate to form gametophytes.

4. The Pterophyta probably evolved from the Psilophyta, and some extinct forms may have been the ancestors of the seed plants. The ferns are considered to be more advanced than groups previously discussed because of the reduced gametophyte and the large sporophyte with their attendant advantages.

Review Topics and Questions

1. What are the diagnostic characteristics of the Pterophyta?
2. Explain in detail what is meant by a "leaf gap." What fills the gap, since it is not actually a hole?
3. What is meant by "circinate vernation"?
4. Describe the life cycle of a fern, indicating which structures are haploid and which are diploid.
5. Define the following: sorus, indusium, annulus, prothallus, rhizome, and meiocytes.

6. In what ways are the ferns more suited to a terrestrial existence than are the Sphenophyta, Lycophyta, and Psilophyta?
7. Why are most ferns limited to fairly moist areas?
8. Why are gametes more sensitive to environmental conditions than are spores? Which factor of the environment is most important as far as the gametes are concerned?

Suggested Readings

BIERHORST, D. W., *Morphology of Vascular Plants* (New York, Macmillan, Inc., 1971).

BURNS, G. W., *The Plant Kingdom* (New York, Macmillan, Inc., 1974).

FOSTER, A. S., AND E. M. GIFFORD, JR., *Comparative Morphology of Vascular Plants,* 2nd ed. (San Francisco, Freeman, 1974).

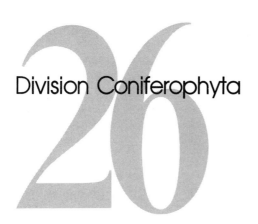

Division Coniferophyta

26.1 Characteristics of the Division Coniferophyta (Conifers)

In addition to the general characteristics of the vascular plants, the plants that comprise the Division Coniferophyta produce unenclosed (naked) seeds located on the upper surfaces of scales, which are usually parts of **cones.** Of the plant groups that have been discussed, this is the first in which the seed habit is found. The conifers are woody plants, chiefly trees, and are sometimes of considerable size with much secondary growth; the leaves are usually evergreen needles or scales. The sporophyte is heterosporous, and the tiny gametophytes are nutritionally dependent upon the sporophyte; the female gametophyte actually develops inside of the megasporangium and never has a free existence. A film of water is not necessary for fertilization; the young male gametophyte, or **pollen grain,** is carried by wind currents to the megasporangium, where the growth of pollen tubes through this tissue brings the sperm to the egg.

Conifers are further subdivided into two groups, but only the genus *Pinus* of the Coniferales will be discussed in detail as a representative of this large group. Any discussion of the various other members is beyond the scope of this text. It might be noted at this point that the plants of the Class Gymnospermae in older classifications are now distributed among several divisions by most taxonomists. One of these divisions is the Coniferophyta. It should be emphasized that this is a judgment made by individual taxonomists based upon the available evidence. As is true of most taxonomic decisions, changes may be made as additional data accumulate. The term "gymnosperm," however, is still commonly used to refer to those plants whose seeds are not enclosed within a fruit (see Section 27.10).

26.2 Life Cycle of Pine

The pine tree usually grows for many years, and much secondary tissue is produced. The general anatomy of pine stems and roots is similar to that of the woody dicotyledonous plant already discussed (Chapters 7 and 9), except that the phloem contains no companion cells and no vessels are found in pine wood, only tracheids. However, large resin ducts are present in pines, and care should be taken not to confuse these with vessels (Figure 26-1). Resin ducts run both radially and longitudinally through the stem. They are lined with parenchyma cells that secrete resin, a material that yields the turpentine (by distillation) and the rosin of commerce.

The photosynthetic needle leaves of this tree occur in clusters on short lateral branches, and the number of leaves in a fascicle depends upon the species involved. The needles generally live from 2–14 years and are shed gradually over the year. Thus, the trees never lose all their leaves at one time the way that **deciduous** plants do. They are shed by an abscission layer forming at the base of the short branch that produced the leaves. The leaf (Figure 26-2) has a thick cuticle, thick-walled epidermal cells, and a layer of sclerenchymatous cells beneath the epidermis. These features, plus the sunken stomata, are considered to be characteristics of xerophytic plants. This type of leaf may be of considerable benefit to such an evergreen plant during winters, when water absorption is difficult. Since leaves are continuously present, transpiration during winter months could be quite detrimental. Pines are also frequently found in drier areas than are neighboring deciduous trees.

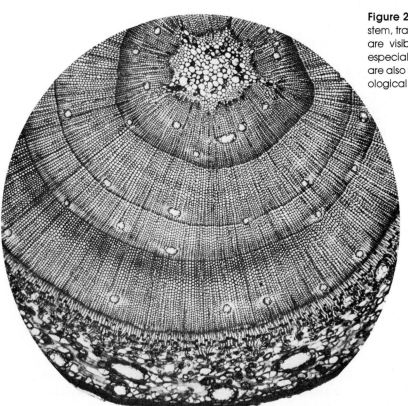

Figure 26-1. Photomicrograph of pine stem, transverse section. The resin ducts are visible as rather large openings, especially in the xylem. Growth rings are also visible. (Courtesy of General Biological Supply House, Chicago.)

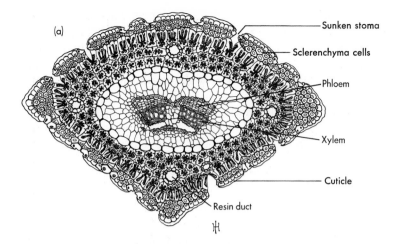

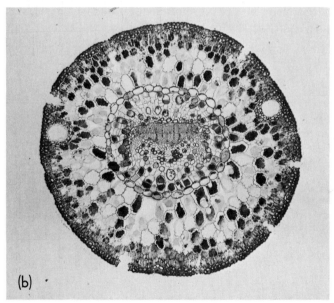

Figure 26-2. **(a)** Leaf of pine, transverse section. **(b)** Photomicrograph of leaf of *Pinus monophylla*, transverse section (×30). (Photograph courtesy of Robert Gill.)

Two distinct types of cones are produced in pines; both the **microsporangiate cones** and the **megasporangiate cones** are borne on the same tree. The former are frequently referred to as pollen cones, while the latter may be called ovulate or seed cones.

Development of Megaspores and Megagametophytes

The ovulate cone is initiated during the late spring or summer, but differentiation is interrupted by the onset of winter, and development continues during the succeeding year. This cone is compound in structure, consisting of a central axis

465

bearing a series of **bracts** (reduced, scalelike leaves), each of which is more or less fused with an ovuliferous scale in the axil of the bract. Two ovules are borne on the upper surface of the scale, which is generally considered to be a reduced simple cone (i.e., equivalent to an axis with fused sterile and fertile appendages). This is of considerable evolutionary significance as evidence relating present-day pines to fossil plants.

The **ovule** (Figure 26-3) consists of a multicellular megasporangium surrounded by and fused with an additional tissue, the **integument,** and contains a single megaspore mother cell **(megasporocyte),** which is readily distinguishable by its larger size. After meiosis, the resultant four haploid megaspores are arranged in a row; three of them disintegrate, while the one farthest from the **micropyle** (a small tubular opening in the integument) remains functional. This megaspore germinates within the megasporangium; enlargement and free nuclear divisions are followed by the formation of cell walls around each nucleus. The result is a multicellular megagametophyte with two to five archegonia differentiated at the micropylar end. The mature archegonium of pine is greatly reduced, consisting of a few neck cells with the large egg cell beneath (Figure 26-4). The various structures that develop when the megaspore germinates are composed of haploid cells, since the megaspore is haploid, and subsequent nuclear divisions are mitotic.

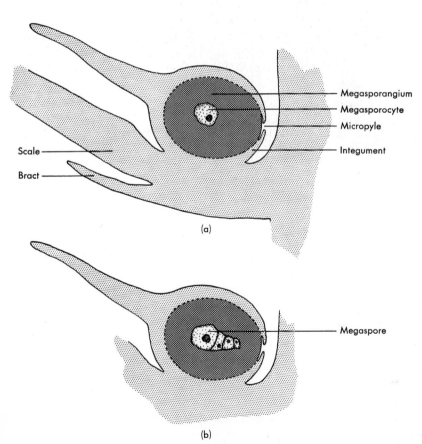

Figure 26-3. Ovule of pine (*Pinus*), longitudinal section, **(a)** Young ovule with a megasporocyte (megaspore mother cell). **(b)** After meiosis, the ovule with four haploid megaspores; three are disintegrating.

Megasporangium

Megasporocyte

Micropyle

Integument

Scale

Bract

(a)

Megaspore

(b)

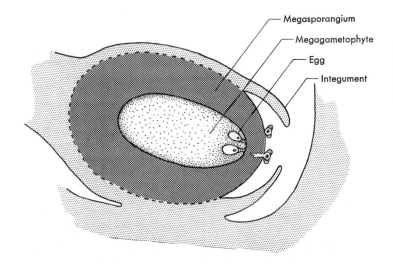

Megasporangium

Megagametophyte

Egg

Integument

Figure 26-4. Ovule of pine (*Pinus*), mature, longitudinal section. Only two archegonia, each with an egg, are shown. Two pollen grains have arrived at the megasporangium; one pollen tube is growing toward an archegonium.

Development of Microspores and Microgametophytes

The pollen cone develops early in the spring and consists of a central axis bearing numerous spirally arranged **microsporophylls** (Figure 26-5). On the lower surface of these modified leaves are borne two microsporangia in which numerous microspore mother cells (**microsporocytes**) undergo meiosis, each producing four haploid microspores. Each microspore undergoes mitotic divisions during which two small **prothallial cells,** a **generative cell,** and a large **tube cell** are formed. The first-named two cells begin to disintegrate rapidly, and the young male gametophyte is set free as a **pollen grain** by the rupturing of the microsporangium (see Figure 26-6). In some species the pollen grains have saclike wings, which are formed by the extension or inflation of the outer covering of the microspore. Large numbers of pollen grains are blown about by the wind, and some sift down

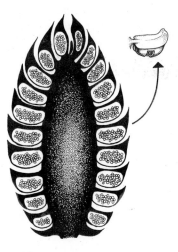

Figure 26-5. Microsporangiate (pollen) cone of pine (*Pinus*), longitudinal section. Upper right figure shows a microsporophyll bearing two microsporangia, one of which is cut open to show the microspores.

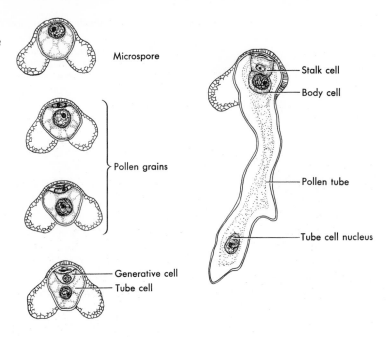

Microspore

Pollen grains

Generative cell
Tube cell

Stalk cell
Body cell

Pollen tube

Tube cell nucleus

Figure 26-6. Development of the microgametophyte of pine (*Pinus*). The young male gameto-phyte is set free as a pollen grain while cellular divisions are continuing. The body cell eventually divides to form two sperms. (From *Gymnosperms, Structure and Evolution*, by C. J. Chamberlain, copyright 1935 by the University of Chicago, used with permission.)

through the young megasporangiate cone, coming to rest in a drop of exudation at the micropyle of the ovule. This transfer of pollen is termed **pollination.** As the drop of liquid dries, the pollen grains are drawn through the micropyle to the mega-sporangium, where a **pollen tube** develops from the tube cell, and the generative cell divides to form a **stalk cell** and a **body cell.**

A considerable period of time elapses between pollination and syngamy (fertil-ization). Meiosis of the megasporocyte occurs about one month after pollination, but growth ceases during the winter and the megagametophyte does not mature until the subsequent year or about 12 months later. After the male gametophyte has reached the stage of development indicated in the preceding paragraph, fur-ther growth during the winter is very slow. During the following spring, develop-ment becomes more rapid, and the body cell divides to form two sperm nuclei, which are carried through the megasporangium to the archegonium by the actively growing pollen tube. Rupture of the tube discharges the sperm nuclei, one of which fuses with the egg nucleus. Eventually the other cells and nuclei of the male gametophyte disintegrate.

Development of the Seed

Although the megagametophyte produces more than one archegonium, only one zygote usually develops into an embryo. During this development, tiers of

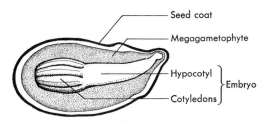

Seed coat

Megagametophyte

Hypocotyl

Cotyledons } Embryo

Figure 26-7. Mature seed of pine (*Pinus*).

cells are produced. The embryo tier is at the apex away from the micropyle and is forced into the gametophyte tissue by great elongation of the **suspensor cells,** which comprise the next tier. Several embryos may begin to develop from the apical tier, but normally only one survives and is pushed farther and farther into the tissues of the female gametophyte. The cells of this latter structure are digested and utilized in the growth of the embryo. The fully developed embryo of *Pinus* consists of a whorl of approximately eight **cotyledons** (seed leaves) surrounding the **epicotyl** (shoot apex) and a bulky **hypocotyl,** that part of the embryo axis below the point at which the cotyledons are attached. The basal tip of this axis is the **radicle,** or root apex.

The various tissues that comprise the ovule enlarge and differentiate further as the embryo develops. The megagametophyte becomes the food storage tissue within which the multicellular, diploid embryo is embedded. The integument enlarges and becomes quite tough and hard, forming the **seed coat.** The remains of the megasporangium can usually be detected as a very thin layer of cells just inside the seed coat. At this point, the seed (Figure 26-7) is mature and consists of the new immature sporophyte embedded within the tissues of the female gametophyte, which in turn is surrounded by modified tissues of the old sporophyte. Growth eventually ceases, and the embryo remains dormant for a varying period

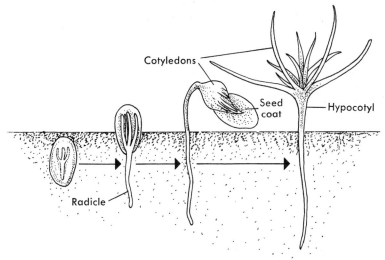

Cotyledons

Seed coat

Hypocotyl

Radicle

Figure 26-8. Germination of pine seed. (From *Plant Form and Function*, by G. J. Tortora et al., New York, Macmillan, Inc., 1970, reprinted by permission.)

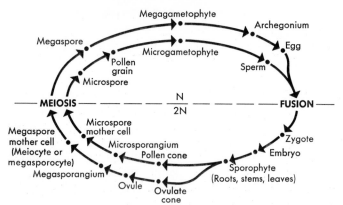

Figure 26-9. Diagram of life cycle of *Pinus*.

Megagametophyte

Megaspore

Archegonium

Microgametophyte

Egg

Pollen grain

Microspore

Sperm

MEIOSIS — — — — — $\dfrac{N}{2N}$ — — — — FUSION —

Megaspore mother cell (Meiocyte or megasporocyte)

Microspore mother cell

Microsporangium
Pollen cone

Zygote

Embryo

Megasporangium

Ovule

Ovulate cone

Sporophyte (Roots, stems, leaves)

of time until conditions are suitable for germination. The seed is really a ripened ovule containing an embryo. In many pines an outgrowth of the ovule serves as a wing, which aids in dispersal of the seed by winds. The seeds are eventually liberated when the female cone opens.

Under suitable environmental conditions, the seed absorbs and imbibes water, swelling in the process. Since proteins and carbohydrates swell to a greater extent than cellulose, the seed coat is ruptured in the early stages of germination; the seed coat is largely cellulose and does not swell proportionately with the embryo. The entire embryo grows rather rapidly, the cotyledons functioning as leaves for a brief period while the epicotyl is developing into the shoot structure (Figure 26-8). The hypocotyl elongates somewhat, which results in carrying the cotyledons up and out of the soil where they are exposed to light, and the radicle develops into the root system with great rapidity and much branching. As indicated in previous chapters, the stem tips and root tips remain meristematic throughout the lifetime of the pine tree, with secondary tissues eventually developing from vascular and cork cambia. As the tree matures, several years of vegetative growth precede any development of cones, but then cones are normally produced each year throughout the lifetime of the pine.

The generalized life cycle of pine is indicated in Figure 26-9.

26.3 Importance of the Division Coniferophyta

The group of plants of which pine was selected as an example constitute an important part of the world's flora, forming great forests in the mountains and in the north. Although they were more abundant in past geological ages, having given way in many areas to the flowering plants, they are still our most important source of timber and timber products. Many of the less expensive grades of paper are made from wood pulp of various conifers, particularly spruce (*Picea*), although Southern yellow pine (*Pinus palustris*) is also much used.

Some of the largest and oldest of living organisms are found among the conifers. The redwood (*Sequoia sempervirens*) frequently grows to a height of 60 m (about

471

Sec. 26.4 /
Evolutionary
Advances
Featured in the
Coniferophyta

200 ft) and a diameter of 6 m and is hundreds of years old. The wood is reddish in color presenting an attractive appearance when used for shingles, siding, and cabinet work; the burls are deeper red, presenting a beautiful color when carved and burnished. A burl is a dome-shaped growth on the root or trunk of a tree. The grain presents a beautiful design because of the tangled growth of xylem and rays. The Big Tree (*Sequoiadendron gigantea*), restricted mainly to California, as is the redwood, is one of the oldest living things; one of these is at least 3500 years old and has a diameter of approximately 12 m. Some of the bristlecone pines (*Pinus aristata*) have been estimated to be at least 4500 years old, but these trees are rather twisted and gnarled, which makes such estimates less exact than those of the Big Tree.

Numerous materials of economic value are obtained from the conifers and their close relatives. Tannins, complex organic compounds that are used in the tanning industry, are obtained mainly from hemlock (*Tsuga canadensis*) bark; this material is also used in the preparation of some kinds of inks. Tanning results from the reaction between the proteins of animal skins and tannins; the soft, pliable, and resistant material that results is leather. Resins are extracted from various conifers, especially *Pinus palustris,* and are used in the manufacture of varnishes, perfumes, linoleum, and various other products. Distillation processes are utilized in obtaining turpentine, certain oils, and methyl alcohol from the wood of various conifers. In addition to these and other products, some pine seeds are edible by man or other animals, especially those seeds from *Pinus edulis, P. torreyana, P. pinea,* and *P. monophylla.*

26.4 Evolutionary Advances Featured in the Coniferophyta

The strobili, or cones, in which are borne the microgametophytes and megagametophytes, are more complex and well developed than in the lower vascular plant groups, especially the megasporangiate cone. The gametophytes are retained within the protective tissues of the cones during a portion of their development or are permanently retained. Even the male gametophyte, as it is transported to the female gametophyte during pollination, is still somewhat protected by the relatively thick wall of the microspore within which it has developed. The gametes are never really exposed; the sperm develop within a pollen tube and are liberated at the archegonium, while the egg is continuously surrounded by gametophyte tissue. The covering (integument) about each megasporangium ensures further protection, not only to the megagametophytes and eggs but also to the developing embryos. Once the megasporangium became surrounded by integumentary tissue, this eliminated liquid water as an agent to effect syngamy. The new means of getting the sperm to the egg was accomplished by the pollen tube of the male gametophyte, of course. The great simplification of both gametophytes and their parasitic existence upon the sporophyte are distinctly advantageous for gamete survival. The seed habit, including dormancy, is considered to be one of the major evolutionary advances within the plant kingdom. *Gymnosperm* (of the older classifications) means "naked seed" and refers to the fact that these seeds are freely exposed and not covered by additional tissues. The latter condition obtains in the flowering plants and will be discussed in the next chapter.

In the conifers one finds protective and nourishing tissues surrounding a zygote and subsequently the developing sporophyte (embryo). The evolution of such tissues provides the new sporophyte with an advantage not afforded the sporophytes of lower groups. It is exceedingly beneficial to have a continued supply of food from the mature sporophyte to the developing gametophytes, the zygote, the embryo, and then storage in the seed where the new sporophyte thus has an immediately available food supply upon germination. Many seeds, in the dry condition, are resistant to adverse environmental conditions and remain viable for years. In fact, in many seed plants, such as the Coniferophyta and the Anthophyta (the flowering plants discussed in the next chapter), the embryo becomes **dormant,** or inactive, before developing into a new mature individual and is rather resistant to adverse environmental conditions while in this state (see Section 29.11). This enables the sporophyte (embryo) to continue further development (germination) when conditions are more favorable. One should remember, however, that seeds will not remain viable indefinitely. In general, the percentage of germination within a group of seeds will decrease during storage; the amount of decrease varies with the kind of seed as well as with the storage conditions.

Many botanists believe that the gymnosperms (which includes the conifers) probably evolved from ancient heterosporous progymnosperms. As additional evidence is gathered, this situation may be clarified.

Summary

1. The Coniferophyta produce naked seeds often on the upper surfaces of bracts, which are usually parts of cones. They are woody plants, usually with evergreen needles or scales. The sporophyte is heterosporous; the tiny gametophytes are nutritionally dependent upon the sporophyte. A film of water is not necessary for fertilization; the pollen grain and the pollen tube are structures that transport the sperm to the egg.

2. *Pinus* species produce microsporangiate (pollen) and megasporangiate (ovulate) cones. The ovules, which develop in the latter, each consist of a megasporangium surrounded by and fused with the integument. An opening, the micropyle, in the integument enables pollen grains to be deposited at the megasporangium. In each ovule, meiosis of the single megasporocyte gives rise to one functional megaspore, which develops into a megagametophyte. Several archegonia, each with an egg, are produced by each female gametophyte.

3. The many microsporocytes in a microsporangium undergo meiosis, producing four microspores each. Several mitotic divisions result in the development of a microgametophyte. The young male gametophyte, or pollen grain, is blown to the ovulate cone, where maturation is completed. As the pollen tube grows toward the archegonium, it carries the two sperm cells to the vicinity of the egg.

4. Although more than one egg may be fertilized and more than one embryo start to develop, only one embryo usually survives. The integument develops into

473

Sec. 26.4 /
Evolutionary
Advances
Featured in the
Coniferophyta

the seed coat, the megagametophyte develops into a food storage tissue, and both surround the embryo; this structure is the seed. On germination of the seed, the basal portion of the embryo, the radicle (which is really the lowest portion of the hypocotyl), forms the root system, and the epicotyl of the embryo produces the shoots.

5. The conifers are important sources of timber and timber products, tannins, resins, turpentine, methanol, and many other materials. Some pine seeds are edible.

6. Various features of the Coniferophyta are distinctly advantageous to a terrestrial existence. The seed habit, including dormancy, is considered to be a major evolutionary advance. Such a structure protects the megagametophyte, the female gametes, the male gametes (after pollination), the zygote, and the embryo. The large sporophyte with its well-developed vascular tissue enables large numbers of spores and seeds to be produced and disseminated.

Review Topics and Questions

1. Discuss those characteristics of the pine that endow it with a greater chance of surviving in a variety of terrestrial environments than is possible in the case of ferns.
2. Describe a pine seed. Of what advantage is the presence of a seed coat?
3. Would the rate of photosynthesis tend to be more rapid in the cells of a pine leaf than in the cells of a broad-leaved plant, such as a bean? Why?
4. Would the rate of transpiration tend to be more rapid from the leaves of a pine than from an equal surface area of a broad-leaved plant, such as a bean? Why?
5. What structural feature results in a pine leaf being rigid or stiff?
6. Diagram the root tip of a pine tree as seen in longitudinal section. Show regions and tissues but not the individual cells.
7. List the tissues and regions from outermost to the center of a 20-year-old pine trunk.
8. Describe a tracheid and a sieve tube element of a pine.
9. Describe the development of a microgametophyte of pine, starting with a microsporocyte (meiocyte). Indicate where the development takes place and whether each structure is haploid or diploid.
10. Repeat problem 9 for a female gametophyte, starting with the megasporocyte.
11. Define the following: pollination, deciduous, integument, micropyle, epicotyl, cotyledons, and radicle.

Suggested Readings

Burns, G. W., *The Plant Kingdom* (New York, Macmillan, Inc., 1974).

Foster, A. S., and E. M. Gifford, Jr., *Comparative Morphology of Vascular Plants,* 2nd ed. (San Francisco, Freeman, 1974).

Division Anthophyta

27.1 Characteristics of the Division Anthophyta (Flowering Plants)

In addition to the general characteristics of the vascular plants, members of the Division Anthophyta produce enclosed seeds in specialized structures, which are collectively termed the flower. Pollen grains are formed, much as in the conifers, but the gametophytes are even more reduced than those of the cone-bearing plants. The flowering plants are the dominant members of the world's flora, consisting of at least 250,000 species, which vary from minute nonwoody types to enormous woody trees. They are of fairly recent origin, fossil remains probably not being older than 130 million years (from the Cretaceous period) to 150 million years (from the Jurassic period), and are considered to be the most advanced of all plant life. The fossil record is not clear and interpretations vary, but the flowering plants were certainly well established (abundant and diverse) by mid-Cretaceous.

The general structure and function of the roots, stems, and leaves have been discussed in previous chapters. The presence of vessel elements in the xylem is characteristic of most Anthophyta. Most of our knowledge concerning the Anthophyta is based upon an intensive study of north temperate floras, modern botany having its origins in Europe. Our knowledge of tropical and arctic floras is quite meager as compared with that of North America and Europe, but future investigations will undoubtedly add to our present concepts and may even cause us to modify some of our suggestions concerning relationships of organisms.

27.2 The Flower

Anthophyta are commonly referred to as flowering plants, and this distinctive structure will be discussed in some detail. Figure 27-1 is a diagrammatic representation of a mature, complete flower, which has developed from a bud. As pointed out in Section 7.1, such buds may be mixed; that is, they develop into both flowers and leaves. The enlarged portion of the flower stalk, to which the various flower parts are attached, is called the **receptacle.** As the flower bud enlarges and unfolds,

474

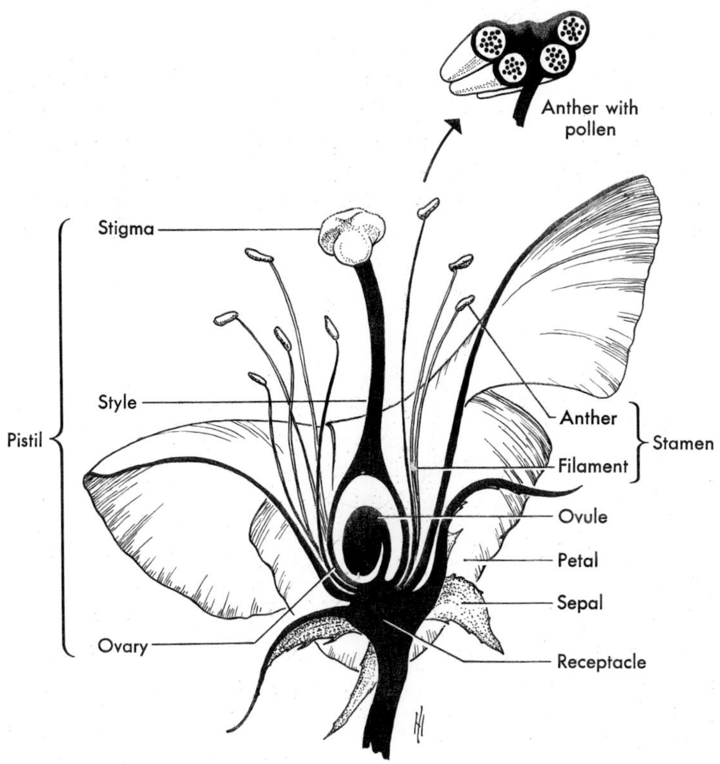

Figure 27-1. Mature, complete flower. Top figure shows one of the anthers enlarged and sectioned to show the pollen grains.

the outermost whorl or parts, the **sepals,** open out to expose the more showy, colored parts, the **petals.** The size, shape, and number of such parts vary with the particular plant involved, and some may even be lacking in certain plants. The sepals are usually green and leaflike and serve to protect the more delicate tissues that are developing within the bud; the sepals collectively are termed the **calyx.** As one would suspect from their appearance, the petals of some flowers serve to attract insects, which then become the unsuspecting agents of pollination in their travels from flower to flower. Such attraction is increased in many plants by the presence of glandular nectaries at the base of the petal. The sweet liquid secreted by these nectaries is utilized by many insects. Essential oils and floral fragrances also ensure the presence of insects at the flowers of different kinds of plants. The petals collectively are termed the **corolla,** while the term **perianth** refers to both sepals and petals. Through anatomical investigations, these structures have been considered to be modified leaves.

The next innermost whorl of parts comprises the **androecium** made up of the **stamens,** each consisting of a slender **filament** (or stalk) and an enlarged portion called the **anther.** Within the anther the pollen grains are formed; the swollen, saclike portions are actually microsporangia. The size, shape, and number of stamens vary considerably from one kind of plant to another but are constant within any one type.

In the central portion of the flower is the **pistil,** consisting of the **stigma** at the

475

tip, an elongated **style,** and the bulbous **ovary** at the base. Pollen lands on the stigma, the pollen tube growing through the intervening tissues into the ovary, which contains the ovule. The seed (i.e., an ovule containing an embryo) is thus enclosed within the ovary, and the latter structure develops into the **fruit** of the flowering plant. The pistil is composed of one or more units called **carpels.** The latter structures are considered to be reduced branching systems that have fused, or modified leaves (i.e., megasporophylls). If this structure bore ovules at the margins, an inrolling and a fusion of the margins could produce a structure like a pistil with the ovules enclosed. Figure 27-2 represents the manner in which this may have occurred over long periods of evolution. Certain primitive plants have sporangia borne on the upper surfaces of modified structures. Although an intermediate partially folded (or rolled) condition has not been found in primitive forms or in fossils, anatomical investigations have produced evidence that pistils may have evolved in this fashion. Also, the whole series exists during the development of pistils in many living species. A simple pistil is composed of one carpel, as in the garden pea (*Pisum*), whereas a **compound pistil** consists of two or more fused carpels, as in the lily (*Lilium*); both conditions are represented in Figure 27-3. As with perianths and stamens, the size, shape, number, and kind of pistils vary, depending upon the flower under discussion. The pistils collectively are termed the **gynoecium.**

Figure 27-1 has been referred to as representing a complete flower, which certainly implies the existence of incomplete flowers. The sepals and petals are not directly concerned with reproduction and are **accessory** parts of the flower, while the **essential** parts are the stamens and pistils. An **incomplete** flower lacks one or more of the floral organs, as in the grasses (Family Gramineae), which contain no petals. An **imperfect** flower lacks one of the essential organs. Flowers may be **staminate,** as the tassel flowers of corn (*Zea mays*), or **pistillate,** as the flowers of corn that form the ears lower down on the stem. *Zea mays* produces both kinds of flowers on the same plant, a **monoecious** condition. A plant such as willow (*Salix*) is **dioecious,** because staminate and pistillate flowers are produced on separate plants of that species.

Figure 27-2. Suggested manner in which the pistil may have evolved. A simple pistil is shown in **(c),** and a compound pistil in **(d)**; the latter could readily be formed by the fusion of carpels.

(a)   (b)   (c)   (d)

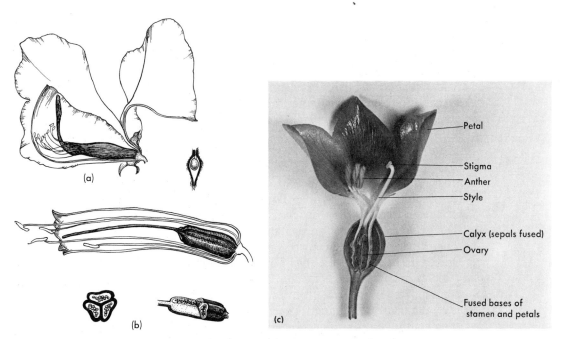

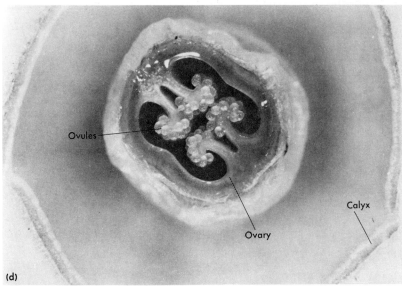

Figure 27-3. (a) Garden pea flower with a simple pistil. The pistil is also shown in section to expose one of the seeds. (b) Lily (*Lilium*) flower with a compound pistil. Lower sectional views of the ovary with seeds. (c) *Orpheum fruticosum* flower with fused floral parts (×4). (d) *Orpheum fruticosum*, cross section of ovary of compound pistil (×100). (Courtesy of Peter Jankay.)

27.3 Diversity in Flowering Plants

Anyone who has taken even a cursory glance around a flower garden or at flowering plants in the fields and woods is immediately made aware of the tremendous diversity of such organs. Certain structures, which have been discussed as parts of a generalized and complete flower, may be missing or greatly reduced.

477

But in addition to this, parts may be fused, shapes may be irregular, or the insertion (attachment) of structures may be various. The manner in which flowers are arranged on an axis, such clusters being termed **inflorescences,** also varies considerably in the number of flowers, the number of branches, and the way in which the individual flower stalks are attached. A few examples may help to emphasize the great differences that obtain from flower to flower.

Position of the Ovary

Most flowers have ovaries that are termed **superior;** i.e., they are above the place of attachment of the other floral parts. Figure 27-4(a) represents a **hypogynous** flower with such an ovary as could be found in the buttercups (*Ranunculus*), grasses (*Poa*), and morning glories (*Ipomoea*). In **perigynous** flowers the basal portions of the perianth and stamens are united and form a cuplike structure or floral tube around, but distinct from the superior ovary; the stamens, petals, and sepals appear to be borne on the rim of the cup within which the superior ovary is located (Figure 27-4[b]); such flowers are found in the apple (*Malus*), rose (*Rosa*), and cherry (*Prunus*).[1] In **epigynous** flowers there is also a floral tube, but it is united with the wall of the ovary, so that the other floral parts appear to be attached at the top of the ovary (Figure 27-4[c]). The ovary is said to be **inferior;** such flowers are found in the *Iris,* honeysuckle (*Lonicera*), and composites (*Aster, Zinnia, Helianthus,* or sunflower, *Taraxacum,* or dandelion).

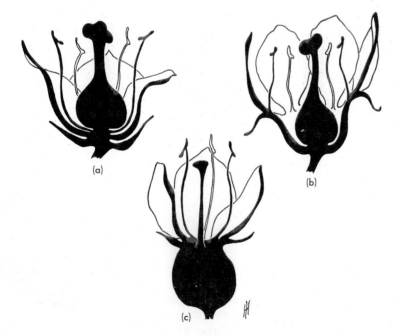

Figure 27-4. Position of the ovary. **(a)** Hypogynous flower; the ovary is superior. **(b)** Perigynous flower, superior ovary. **(c)** Epigynous flower with the ovary inferior; the basal portions of the floral parts fuse around the ovary.

[1] The genus *Prunus* includes plums, peaches, and apricots in addition to cherries; all are perigynous.

Parts Fused or Irregular

In many kinds of flowers some of the floral parts may be fused to form tubular or bell-shaped structures, or the members of one whorl of parts may be more or less united with members of another whorl. The bluebell (*Campanula rotundifolia*) has fused petals as well as fused sepals; in *Hibiscus* the filaments of the stamens also are fused, forming a tube around the pistil; in the Compositae (sunflower family) the anthers are fused as a cylinder around the pistil, but the filaments are free. The most common examples of irregular flowers, those that are not radially symmetrical, can be found in the snapdragon family (Scrophulariaceae), the bean family (Leguminosae[2]), and the orchid family (Orchidaceae). Many other examples could be given, but those listed should be sufficient to indicate the great diversity within flowers.

Wind Pollination vs. Insect Pollination

As comparisons are made with regard to flower structures, basic similarities become evident among wind-pollinated flowers and among insect-pollinated flowers. Also evident are certain basic differences between these two groups of plants. Such similarities and differences particularly concern the flower, its structures and numbers.

In general, wind-pollinated species tend to bear many, rather small, inconspicuous flowers. The petals are frequently much reduced and not colorful, whereas the stamens are quite numerous and conspicuous; both the stamens and pistils are well exposed. By the very nature of the pollinating agent, those plants that happened to evolve the type of flowers mentioned would tend to reproduce and survive more readily. Relatively large amounts of pollen are advantageous because of the rather inefficient mechanism involved. Large, showy structures would be of no advantage and could be of considerable disadvantage by forming wind barriers, which would interfere with pollen distribution. The oak (Fagaceae), willow (Salicaceae), and grass (Graminae) families are all wind-pollinated. In general, wind-pollinated flowers are considered to be more primitive and the insect-pollinated ones more advanced.

Those plants that are insect-pollinated tend to have flowers in which the petals are quite colorful and conspicuous and in which nectaries are present. Both these structures attract a variety of insects. Frequently the presence of fused petals forces the insect to crawl toward the nectary, which ensures its brushing along stamens and pistils. Examples of insect-pollinated flowers are found in the bean, snapdragon, orchid, and mint (Labiatae) families.

The sunflower family (Figure 27-5) is rather interesting because it is insect-pollinated and yet produces large numbers of small flowers. However, the flowers are arranged in clusters; each "sunflower" or "daisy" is not a single flower but one such compact cluster. The outer flowers, or ray flowers, each have one conspicu-

[2] Included in the Leguminosae are peas (*Pisum sativum*), beans (*Phaseolus vulgaris*), peanuts (*Arachis hypogaea*, alfalfa (*Medicago sativa*), clover (*Trifolium*), sweet peas (*Lathyrus odoratus*), and so forth.

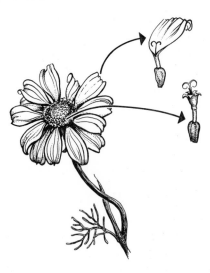

Figure 27-5. Composite (sunflower family) inflorescence. The upper right figure shows a ray flower enlarged, and the lower right figure shows a disc flower enlarged.

ous strap-shaped petal, which attracts insects, since the group together is conspicuous. The inner flowers, or disc flowers, are small and tubular. Thus, no one flower is really very attractive, but the cluster of flowers, termed a **head** in this case, creates a very pleasant appearance. Because of the compactness of arrangement, one visiting insect pollinates many flowers.

There are many fascinating adaptations of flowers to their pollinators. Those plants that are pollinated by moths, for example, tend to have white flowers with a heavy fragrance; such flowers tend to open at dusk, the time when moths are flitting about. Some of the odors emanating from flowers are not very pleasant to humans, but flies are attracted readily to such aromas as decaying matter, dung, carrion, putrefaction—odors produced by certain flowers. One such plant has the appropriate common name of ''skunk cabbage'' (*Symplocarpus*). There is even an orchid flower (e.g., *Ophrys*), which resembles the female species of wasp; considerable pollination occurs as the male wasp moves in hopeful activity from flower to flower. In the tropics especially there are flowers that are pollinated by animals other than insects; bats and hummingbirds are frequently important pollinators in these regions, the former being attracted mainly by odor and the latter by color.

Self-pollination and Cross-pollination

Flowers may be variously modified in ways that ensure either self-pollination or cross-pollination. The latter type is much more advantageous,[3] and the greater number of plants have flowers that are usually cross-pollinated. However, self-pollination does occur in oats (*Avena*), wheat (*Triticum*), and violets (*Viola*). In some plants, such as the willows (*Salix*) and poplars (*Populus*), cross-pollination must occur because they are dioecious, with the stamens and pistils occurring on

---

[3] Review Chapter 18 on genetics, if necessary.

different plants. In others the same flower contains both male and female parts, but these mature at different times, thus preventing self-pollination, or the stamens and stigmas occur at different levels.

Dicotyledoneae and Monocotyledoneae

The Anthophyta are divided into two classes, the Dicotyledoneae and the Monocotyledoneae, on the basis of differences within the flower and seed structures as well as within the leaf and stem structure. As indicated by the name, the dicotyledonous plants are those in which two cotyledons (see Section 27.8) are present in the embryo. In addition to this, the flower parts are in whorls of four or five or their multiples, the leaves are net-veined (reticulate venation), the vascular tissue of the stem is in the form of concentric cylinders or in bundles that form a ring, and a cambium is present. The monocotyledonous plants have but one cotyledon in the embryo, the flower parts are in threes, leaf venation is usually parallel, vascular bundles are scattered in the stem, and a cambium is almost always lacking.

Obviously, then, from the previous paragraph, the clearest distinctions between the two classes are in embryo and flower structures. The other characteristic differences will apply to almost all plants, but exceptions to these are somewhat more frequent. Some palms, for example, have leaves in which the venation is certainly not parallel, even though these plants are monocots. Although the distinction between a monocot and dicot is not as sharp and clear as had been thought earlier, they are very useful taxonomic groupings.

27.4 Primitive and Advanced Flowers

The great differences that occur among the Anthophyta are a result of evolutionary processes. Although this topic will be discussed in considerable detail in the next chapter, the evolution of the flower will be mentioned at this time while the involved structures are fresh in mind. This section presents such a well-illustrated example of evolution within a group that it should be reviewed after the chapter on evolution has been completed.

The general trend of evolution of the flower has been toward a gradual shortening and compaction of the axis (receptacle). Primitive flowers (e.g., buttercup, *Ranunculus; Magnolia*) have an elongated receptacle, whereas advanced flowers (e.g., sunflower, *Helianthus;* dandelion, *Taraxacum;* orchid, *Cattleya*) have a very short, blunt receptacle. This shortening has resulted in a number of changes in the structure and arrangement of the flower parts, changes that can be observed quite readily when examining a variety of flowers.

Change from Helical to Cyclic Arrangement of the Flower Parts

In the primitive *Magnolia* flower, the parts are attached in the shape of a helix (i.e., like the thread a screw); the stamens, for example, are of an indefinite number attached in a helical fashion to the axis. In the advanced *Iris* flower, the parts are arranged in a whorl or cycle; the stamens are definite in number (three), and all are attached at the same level.

Reduction in the Number of Cycles

The most advanced flowers always have only four cycles of parts, that is, one each of sepals, petals, stamens, and carpels (pistils). *Trillium* or *Lilium* (lily) have flowers of a moderate degree of advancement; all the parts are arranged in cycles, but the stamens are in two cycles.

Union of Petals

In primitive flowers the petals are separate, whereas they are fused or joined in most advanced flowers. The bluebell (*Campanula rotundifolia*), for example, has flowers in which the fused petals (a **sympetalous corolla**) form the distinctive feature from which the name is derived. The fusion of members of one whorl with those of another is also considered to be an advanced character; in snapdragon (*Antirrhinum*) and *Phlox* the stamens are fused to the petals.

Union of Carpels

In primitive flowers, such as the buttercup, each carpel forms a pistil. In advanced flowers, such as the lily, several carpels fuse to form a single pistil. In many of the most advanced types, the outer sets of parts seem to be inserted at the top of the ovary, the epigynous condition mentioned in the previous section.

Irregularity of Parts

In primitive flowers all parts of a set, such as the petals, are alike. In some of the advanced flowers, as in the bean family (Leguminosae), the various petals are quite dissimilar. The general arrangement of floral parts may be that of **radial symmetry,** a circular plan with parts of equal size equally spaced in position; or **bilateral symmetry,** a right-and-left plan with parts that vary in size and are unequally spaced. The latter type, as in sweetpeas (*Lathyrus*), is considered a more advanced condition than the former, as in roses and petunias.

27.5 Development of Megaspores and Megagametophytes

One to many **ovules** are produced in an ovary, and each is attached to the tissues of the ovary by a small stalk, the **funiculus.** The integuments (**inner** and **outer**) develop along with the megasporangium and eventually surround it completely except for the tubular micropyle (Figure 27-6). The single, large, diploid megasporo-

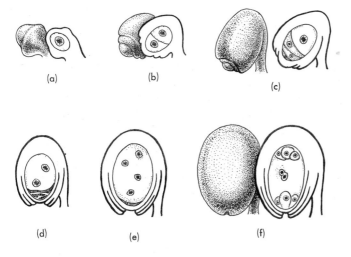

Figure 27-6. Development of the mega-gametophyte of flowering plants. **(a), (b), (c)** The megasporocyte is undergoing meiosis as the integuments develop around the megasporangium; only one megaspore remains functional. **(d), (e)** The nucleus of the functional megaspore is undergoing mitotic divisions. **(f)** The mature megagametophyte has developed within the ovule. (From *The Plant Kingdom,* by W. H. Brown, copyright 1935 by Ginn and Co., Boston.) **(g)** Photomicrograph of lily ovule with megasporocyte (×300). (Courtesy Carolina Biological Supply Company.)

(a)

(b)

(c)

(d)

(e)

(f)

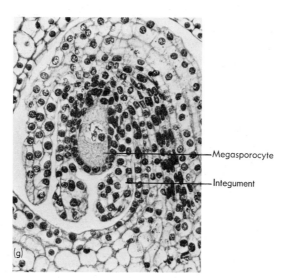

—Megasporocyte

—Integument

(g)

cyte (megaspore mother cell) undergoes meiosis, which results in the formation of four haploid megaspores. Three of these megaspores degenerate, while the nucleus of the remaining one undergoes mitotic divisions that result in eight haploid nuclei being located within the enlarged megaspore. (Some variations occur in different flowers, such as *Lilium,* but the structures subsequently discussed in this section are basically alike.) Three of these nuclei migrate toward the micropylar region, three migrate to the opposite end, and two remain near the central region of the megaspore. Membranes and thin delicate walls eventually form around each of the nuclei at the extremes but not around the central nuclei. As a result, the mature megagametophyte (female gametophyte) consists of several cells.

Figure 27-7 represents an enlarged ovule sectioned to show the internal struc-

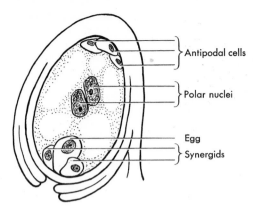

Antipodal cells

Polar nuclei

Egg

Synergids

Figure 27-7. Mature ovule (longitudinal section) showing the structures of the megagametophyte.

tures. One of the cells at the micropylar region is the egg, or female gamete, and the other two are **synergid cells.** Because it is frequently difficult to differentiate the egg cell from the other two, the three cells collectively are sometimes referred to as the **egg apparatus.** The two nuclei in the center are **polar nuclei,** and the three remaining cells are **antipodal cells.** It is quite obvious that the megagametophyte of the flowering plants is reduced much beyond that of any other group of plants. The antipodal and synergid cells merely degenerate and disappear after fertilization occurs.

In many texts the terms *nucellus* and *embryo sac* are used to refer to the megasporangium and megagametophyte, respectively. These are terms that were utilized for such structures before the realization that they were homologous with structures in lower (simpler) plants. In this text the more descriptive terms are used to emphasize homologies and comparable structures.

27.6 Development of Microspores and Microgametophytes

As the stamen matures, the cells of the central portion of anther become quite distinct from the peripheral cells; the former are the microsporocytes (microspore mother cells) (see Figure 27-8). As a result of the disorganization and absorption of surrounding cells, eventually many such microsporocytes are located within a chamber, the microsporangium, and each anther is composed of four microsporangia. Each diploid microsporocyte undergoes meiosis, producing four functional haploid microspores that lie freely in the microsporangium as a result of further gradual extension of the surrounding sterile tissue. The microspores undergo a mitotic division, but no walls separate the two resultant cells, which are both retained within the wall of the microspore. The cell that is usually smaller is the **generative cell,** the larger one is the **tube cell;** the entire structure is a **pollen grain,** or young microgametophyte.

At this stage of development, the anther usually splits open (dehisces) along a vertical line between the two pollen chambers on each side of the filament, thus releasing pollen grains. The generative cell eventually divides to form two sperm.

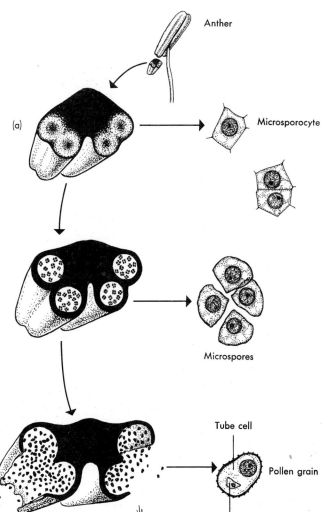

Anther

Microsporocyte

Microspores

Tube cell

Pollen grain

Generative cell

Figure 27-8. **(a)** Development of microspores and pollen grains in an anther. The generative cell will eventually divide to form two sperms (male gametes). **(b)** Immature anther of lily (*Lilium*), transverse section (×30). (Courtesy of Robert Gill.) **(c)** Lily anther with pollen grains (×25). (Courtesy Carolina Biological Supply Company.)

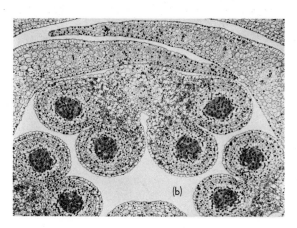

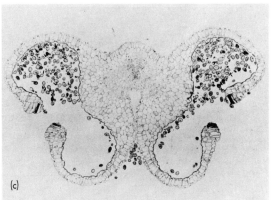

27.7 Pollination and Fertilization

Pollination, or the transfer of pollen from the anther to the stigma, is accomplished in a variety of ways but most usually by wind currents or by insects. The surface of the stigma is frequently covered with a rather viscous secretion to which the pollen grains adhere. The tube cell produces a long extension, the **pollen tube,** which projects through one of the thin areas of the pollen grain wall and grows down through the tissues of the stigma, style, and ovary toward the ovule. Although the pollen grain contains some food material, most of the food utilized in the growth of the pollen tube undoubtedly is obtained from the tissues of the pistil. The generative cell divides mitotically, producing two male gametes, or sperm; at this stage of development, the structure may be considered to be a mature male gametophyte (Figure 27-9). The sperm are carried within the pollen tube to the ovule, where entrance is usually through the micropyle.

The pollen tube penetrates the megasporangium to the egg and ruptures, thus liberating the sperm within the female gametophyte. One of the sperm fuses with the egg; fertilization (syngamy) thus occurs after pollination. The two terms are not synonymous although they are frequently used as though they were. The resultant diploid zygote undergoes many mitotic divisions as it develops into a multicellular embryo. The second sperm fuses with *both* polar nuclei. This *triploid* (3N) nucleus undergoes many mitotic divisions: eventually the resultant triploid nuclei are delimited by cell walls and form a multicellular food storage tissue, the **endosperm.** The antipodal cells and synergid cells of the female gametophyte and the tube cell of the male gametophyte disintegrate (see Figure 27-10). The term *double fertilization* is sometimes used to emphasize the fact that two sperm are involved (one fusing with the egg cell and the other fusing with both polar nuclei), a condition restricted to the Anthophyta.

27.8 The Seed

As in the Coniferophyta, the ripened ovule containing an embryo is a **seed.** However, an important distinction exists with regard to the food storage tissue. In the conifers the enlarged female gametophytic tissue functions in this manner, while

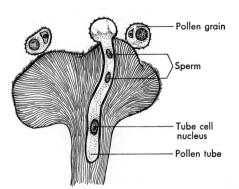

Pollen grain

Sperm

Tube cell nucleus

Pollen tube

Figure 27-9. Pollen grains on a stigma. One pollen tube is growing through the tissues of the stigma and style; the sperm are carried within this tube.

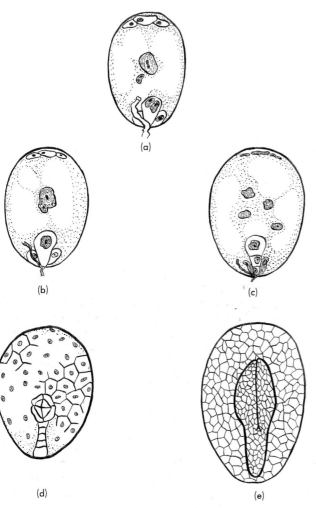

(a)

(b)   (c)

(d)   (e)

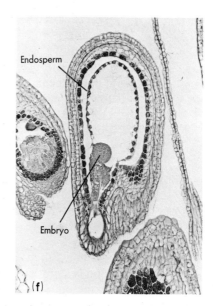

Endosperm

Embryo

(f)

Figure 27-10. Fertilization and the development of the embryo and endosperm. **(a)** One sperm is fusing with the egg in the lower portion of the figure as the other sperm is approaching the fused polar nuclei; the pollen tube is degenerating. **(b)** Fusion of one sperm with polar nuclei forming the triploid endosperm nucleus. **(c)** Endosperm nucleus has divided twice mitotically; zygote developing into the embryo. **(d)** Formation of cell walls around the nuclei of the endosperm. **(e)** Embryo (with two cotyledons) surrounded by the multicellular, triploid endosperm tissue. **(f)** Photomicrograph of developing *Capsella* embryo (×100). (Courtesy of Robert Gill.)

in the flowering plants an entirely new type of tissue develops. Not only is this a new type of tissue with regard to its method of development, but it is also composed of cells that have three sets of chromosomes (triploid, 3N). The chromosome number of the endosperm cells actually varies depending upon the number and type of polar nuclei that are produced and that fuse with the sperm. It is beyond the scope of this text to discuss the ten basic types of endosperm. It should be noted, however, that the chromosome number may be 2N, 3N, 5N, or higher. We will limit our discussion to the triploid condition, which is the most common. As the embryo develops, it receives nourishment from the endosperm. In many seeds most, if not all, the endosperm has been utilized by the developing embryo, so that the mature seed contains nothing but remnants of this triploid tissue. This is true of most dicotyledons. However, the endosperm persists as a food storage tissue in many monocots. It is highly developed in the grasses (see, for example, Figure 27-12), and is of economic importance as a food source.

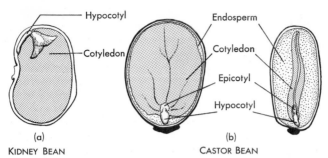

Figure 27-11. Longitudinal sections through seeds. **(a)** Kidney bean. The endosperm has been utilized and food is stored in the cotyledons; one cotyledon has been removed so that the hypocotyl and epicotyl are visible. **(b)** Castor bean. The endosperm is the bulky storage tissue; two views to show the sheetlike cotyledons.

The integuments grow and harden, forming the **seed coat.** Although the micropyle is closed as a result of integumentary growth, in most seeds the location of this area is visible as a slight depression. The **hilum** is a scar on the surface of the seed left where the seed broke from the stalk. Figure 27-11 represents longitudinal sections through a kidney bean seed in which the endosperm has disappeared and a castor bean seed in which the endosperm forms the bulk of the seed. In both types the embryo, or immature plant, is readily distinguishable. The seed actually consists of tissues of the original generation as well as those of the new generation, although some of the tissues are usually digested away by this time.

The embryo consists of a short axis with one or two attached cotyledons. The portion above the point of attachment of the cotyledons is the **epicotyl,** which becomes the shoot system of the new plant as the seed germinates. The lowest portion of this axis is the **radicle,** which becomes the root system. The area between the epicotyl and the radicle is the **hypocotyl,** a transitional zone of stemlike structure, which elongates in some seeds during germination. The **cotyledons** are usually food storage organs, which digest and absorb food from the endosperm and may function as leaves after germination. In seeds such as the kidney bean, food is stored in the cotyledons; in those similar to the castor bean, food is stored in the endosperm. Both the kidney bean and castor bean are dicotyledonous plants; in Figure 27-11(b) the two cotyledons are clearly shown.

In monocotyledonous plants only one cotyledon is present. The embryo of some of these, as in members of the grass family (e.g., *Zea, Avena,* and *Triticum*), has an additional structure, called the **coleorhiza,** covering the root tip and a tubular leaf sheath, the **coleoptile,** enclosing the young primary shoot. These structures are shown in Figure 27-12. In the corn grain the seed coat is fused with the ovary wall, and so the grain is really a fruit. The single massive, shield-shaped cotyledon remains permanently in the grain in direct contact with the large endosperm tissue. Upon germination, the radicle and epicotyl rupture through their enclosing tissues.

Seed Germination

The other parts of the seed may enlarge greatly as a result of absorbing water, but the embryo is really what is growing when a seed germinates. As is true of

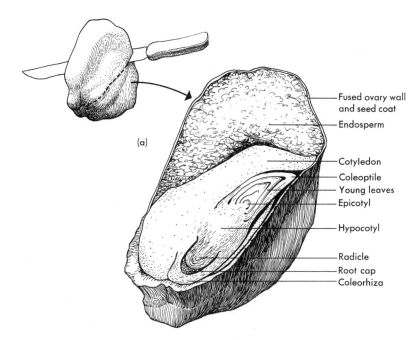

(a)

- Fused ovary wall and seed coat
- Endosperm
- Cotyledon
- Coleoptile
- Young leaves
- Epicotyl
- Hypocotyl
- Radicle
- Root cap
- Coleorhiza

Figure 27-12. (a) through (d) Corn (*Zea mays*) grain dissected to show structures. The single cotyledon almost encloses the embryo axis (b). (e) Photomicrograph of corn grain, longitudinal section (×10). (Courtesy Carolina Biological Supply Company.)

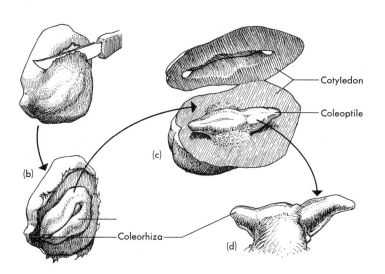

(b)

(c)

(d)

- Cotyledon
- Coleoptile
- Coleorhiza

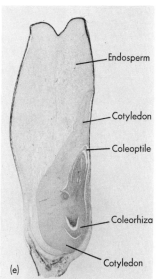

(e)

- Endosperm
- Cotyledon
- Coleoptile
- Coleorhiza
- Cotyledon

most living organisms, growth will not occur unless moisture, oxygen, and a suitable temperature are available. Sufficient food and minerals are stored in almost all seeds, so that these factors do not limit germination. They are, however, additional essential factors that must be present for continued growth of the seedling once it has become established. A source of light for photosynthesis enables the young plant to manufacture its own food as soon as leaves have developed.

As water is absorbed by a seed, the inner tissues swell more rapidly than the

489

seed coats, resulting in a rupture of the latter. The penetration of water also results in tissues becoming hydrated, and enzyme activity is tremendously enhanced. Foods stored in the cotyledons or the endosperm are digested and utilized as the embryo grows. The radicle emerges first, grows rapidly downward through the soil, and develops into the root system. In some seeds the hypocotyl elongates, carrying the cotyledons above the surface of the soil, where they may function as green leaves. This is especially true of plants, such as the castor bean and squash, where the cotyledons are very leaflike; the cotyledons of the garden bean function much more as storage organs than as leaves, even though they are exposed upon germination (Figure 27-13). In other seeds, such as the pea, the hypocotyl does not elongate, and the cotyledons remain in the soil (Figure 27-14). The

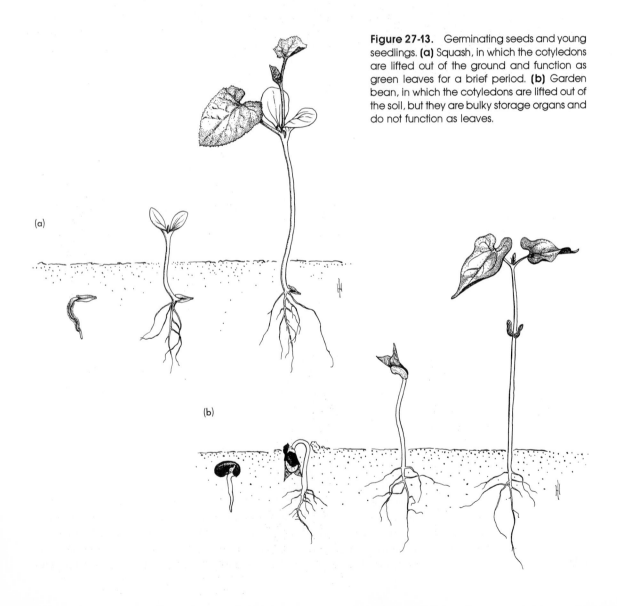

(a)

(b)

Figure 27-13. Germinating seeds and young seedlings. **(a)** Squash, in which the cotyledons are lifted out of the ground and function as green leaves for a brief period. **(b)** Garden bean, in which the cotyledons are lifted out of the soil, but they are bulky storage organs and do not function as leaves.

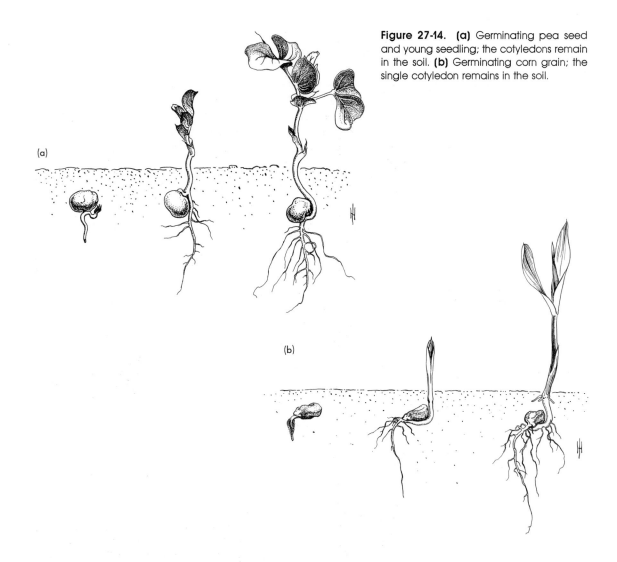

Figure 27-14. (a) Germinating pea seed and young seedling; the cotyledons remain in the soil. (b) Germinating corn grain; the single cotyledon remains in the soil.

(a)

(b)

epicotyl rapidly develops into the aboveground shoot system of the new plant. The plant is now established and able to absorb water and minerals as well as to manufacture its own food. The cotyledons eventually wither and die.

27.9 Seed Dormancy

The seeds of many kinds of plants, but not all, will not germinate immediately even when placed in an environment conducive to growth, a condition known as **dormancy.** The seeds are viable, however, and after a period of a few weeks to a few months, germination will occur. Factors that bring about dormancy are related to the seed itself, and the dormant period is one in which gradual changes take place that eventually allow for growth of the embryo. This condition is particularly advantageous in enabling progeny to withstand adverse environmental conditions. For example, if seeds germinated as soon as they were mature, many young seedlings in northern climates would be exposed to injuriously low winter temperatures. A similar situation would obtain in areas where rainfall is seasonal; seedlings might be exposed to conditions of dangerously low rainfall if they germinated immediately. Seeds are much more resistant than are seedlings and would withstand adverse conditions during dormancy, germinating later under favorable environmental conditions.

In some seeds (e.g., alfalfa, vetch, clover) dormancy results because the seed coat is impermeable to water; in others (e.g., many composites and grasses), the seed coat is impermeable to oxygen; in still a third group (e.g., many common weeds, such as mustard), the seed coat is mechanically resistant and prevents expansion of the enclosed tissues. In all such cases where the seed coat is involved, gradual changes result in the seed coat eventually becoming permeable or less resistant. Damage to the seed coat, as by abrasion, will have the same effect and may occur when the seed is blown about or washed about among rough soil particles. Once such changes have occurred in the integuments, the seed will germinate quite readily if conditions are favorable (as indicated in the previous section).

In some types of seeds, dormancy is not a result of the permeability or resistance of the seed coat, and the embryo will not grow even if the seed coat is removed. Such seeds are considered to have dormant embryos. In some cases (e.g., iris, red oak, ash), inhibiting substances have been shown to be present, and these materials gradually disappear during the period of dormancy. A series of rains, or possibly one heavy rainfall, may leach inhibitors from the seeds and thus break dormancy. In others the specific cause of dormancy is not known, although the environmental conditions that break dormancy have been ascertained in many instances. The seeds of apple, peach, dogwood, and pine require a period of low temperature of from two weeks to three months before the seeds will germinate; a number of grass seeds require alternating temperatures to break dormancy. One of the most interesting treatments to increase the percentage of germination has been discovered in the case of some varieties of tobacco and lettuce seeds. Exposure to light greatly increases the number of seeds that eventually germinate (see also Section 29.14). In all the treatments indicated, the seeds must be moist for the treatment to be effective. In some cases plant growth regulators may break dormancy of seeds (see Section 29.11). It is not yet possible to explain all the factors that are involved in dormancy or the breaking of dormancy. Much additional investigation is essential in this relatively unknown phase of botany.

27.10 The Fruit

As the ovule develops into a seed, the ovary portion of the pistil sometimes increases enormously in size as it develops into the **fruit,** a ripened ovary bearing one or more seeds. The seeds are enclosed and not naked as was true of the seeds in the Coniferophyta. The stimuli that bring about such growth of the ovary are secretions from the pollen tube and from the developing seeds. (The possible nature of these secretions will be discussed in Chapter 29.) The tremendous growth entailed is made more meaningful when one realizes that a peach, a string bean, a pea pod, and a watermelon all develop from ovaries no more than 2 to 10 mm in diameter, hundreds of times smaller than the mature fruit. This is a considerable drain on the food supply of the plant, and vegetative growth frequently ceases as the fruits mature.

Pinching off, or pruning, flower buds will result in larger flowers and fruits developing from the remaining buds as a result of the extra food supply available to the latter. Such thinning out is a common practice among amateur gardeners as well as among professionals. In fact, if perennial plants are not thinned, flower and fruit production may be quite irregular from year to year. A year in which a very large yield is obtained is quite likely to be followed by a year with a relatively poor yield. The amount of food utilized in fruit production during the first year results in such little vegetative growth or in the storage of such low food reserves that the development in the subsequent year is greatly curtailed. Besides pruning, various chemical sprays can be used to bring about flower and fruit drop with its concurrent thinning. Commercial fruit tree growers, especially, utilize a spray program for this purpose (see Section 29.11).

Fruit Types

Many kinds of fruits that develop in various ways are to be found. They may be composed of one or more pistils, with or without including adjacent parts of the flower, and dry or fleshy; they may or may not be dehiscent. The great diversity in fruits makes it difficult to devise a classification that will include them all. However, a few of the more common fruits are described below, with examples of each. Remember that, botanically speaking, the term *fruit* is very specifically defined, while *vegetable* is not; the latter may or may not also be a fruit.

Fruits are generally divided into four large groups, which may each then be further subdivided: simple, aggregate, multiple, and accessory fruits. A **simple fruit** is one that develops from a single ovary. Most Anthophyta fruits are of this type. An **aggregate fruit** is formed from a group of ovaries produced in a single flower, as in raspberries and blackberries. A **multiple fruit** is composed of the ovaries of a cluster of flowers that are borne on the same stalk, as in the mulberry, fig, and pineapple. An **accessory fruit** consists of one or more ovaries together with additional tissues, usually the calyx or receptacle or both, as in the strawberry, apple, and pear.

The pistil of the flower shows great diversity, and fruits also could be expected to be of various kinds. Simple fruits are frequently subgrouped on the basis of the characteristic appearance of the mature **pericarp,** or ovary wall, which sometimes becomes differentiated into three distinct layers: the **exocarp** (outermost), **mesocarp,** and **endocarp** (innermost). A **berry** (Figure 27-15) is a fruit in which the pericarp becomes soft and fleshy at maturity (although the outer layers are often skinlike), as in the grape, tomato, citrus fruits (lemon, orange, and so forth), and date. In a **drupe** (Figure 27-16) the exocarp is a thin skin, the mesocarp is the fleshy

Figure 27-15. Grape (*Vitis*), a typical berry; the pericarp is soft and fleshy.

pulp, and the endocarp is extremely hard (the pit), as in the cherry, plum, peach, and olive. A **legume** (Figure 27-17) consists of a single dry (or leathery) carpel, which splits along two sutures, as in the bean, pea, clover, vetch, and acacia. A **capsule** (Figure 27-18) consists of dry fused carpels, which open in various ways, as in the lily, poppy, and tulip. An **achene** is a dry, indehiscent, one-seeded fruit formed from a single carpel, as in the sunflower and dandelion. A **nut** is similar to an achene but is formed from fused carpels and has a very hard pericarp, as in the oak, chestnut, and beech. The **grain,** or **caryopsis,** is a dry one-seeded fruit in which the seed coat is fused with the ovary wall, as in corn, wheat, oats, and other grasses.

Fruit classification is in just as much a state of flux as is general plant classification. No one system is completely adequate in all respects. For example, although a blackberry is usually considered to be an aggregate fruit, a small portion of that fruit consists of the pulpy receptacle or stem tissue. Some botanists would prefer the term *aggregate-accessory* in this instance. For most purposes, however, the classification utilized in this text is quite suitable. Remember that not all fruit types, but merely the most common ones, have been included. Remember also that common names are not necessarily botanically correct. The student might consider, in the light of this statement, which of the following is actually a berry: blackberry, strawberry, mulberry, or grape. The term *nut* is used just as indiscriminately: peanuts are legumes; pine nuts are seeds; almonds, coconuts, pecans, and walnuts are all drupes.

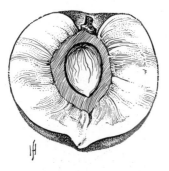

Figure 27-16. Peach (*Prunus*), a typical drupe. The exocarp is a thin skin, the mesocarp is fleshy, and the endocarp is extremely hard.

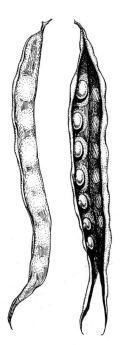

Figure 27-17. Bean (*Phaseolus*), a typical legume. It is a single dry carpel which splits along two sutures.

27.11 Seed Dispersal

The tissues of the seed and the fruit increase the survival possibilities of the progeny by the protection they afford to the embryo. In addition to this, various modifications of the seed, and frequently the fruit, facilitate the dissemination of the embryo. Such seed dispersal may result in the seed germinating in a more suitable environment than that in which the parent plant is located. The new environment might be less suitable, of course, but in either case the progeny are distributed at some distance from their point of origin. This in itself is advantageous, helping to

Figure 27-18. *Yucca*, a typical capsule. It consists of dry fused carpels.

prevent overcrowding and the attendant deleterious effects of shading and competition for water and minerals.

Although the production of large numbers of seeds aids in the propagation of a species, this factor should not be assumed to be the most important criterion for the successful establishment of a plant. The really important factors are the ability of the plant to grow in diverse environments and the facilities of the plant for dispersal to these various environments. Bermuda grass (*Cynodon dactylon*), for example, will spread with distressing rapidity by means of runners. Even if the flowers are trimmed by mowing a lawn before seeds are mature, this will not retard the progress of this grass to any great extent. In fact, although many flowers are produced, very few seeds actually mature; vegetative propagation is the rule in this plant.

Wind Dispersal

One of the most important agents in the dissemination of plants is wind. In some instances the seeds are so light that they can be carried great distances. This is true of the orchids, whose seeds are so tiny as to appear like specks of dust; several hundred thousand weigh approximately 1 gm. In other plants the seeds or

Figure 27-19. Examples of plants that are dispersed by the wind. *Upper left:* Hairy seed of milkweed (*Asclepias syriaca*). *Lower right:* Hairy fruit of dandelion (*Taraxacum officinale*); individual fruits are shown being blown away from the dense cluster which is typical of this plant.

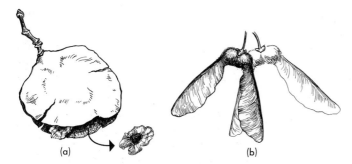

Figure 27-20. Examples of plants that are dispersed by the wind. **(a)** Winged seed of *Jacaranda ovalifolia*. **(b)** Winged fruit of maple (*Acer*).

fruits have hairlike appendages or wings that tend to keep them suspended in the air, often for considerable periods of time. The seeds of the common milkweed (*Asclepias syriaca*) have hairy appendages arising from the seed coat, and the fruit of the dandelion (*Taraxacum officinale*) bears a hairy parachute formed by the modified calyx (Figure 27-19); the hairs have a similar function although a dissimilar origin. One also discovers winged seeds and winged fruits. In the former the wings are outgrowths of the seed coat, as in members of the Bignonia Family (Bignoniaceae; for example, *Jacaranda ovalifolia*) (Figure 27-20[a]). The fruit of the maples (*Acer*) has two wings, each of which grows from the back of a carpel (Figure 27-20[b]); the seeds of pine (see Section 26-2) are also winged.

Animal Dispersal

The other agent of most importance in the spread of seed plants through the world is the animal population, including man. The fruits of some plants have hooks or spines which become entangled in the hair of various animals (or in the clothing of humans) and then are carried about, eventually dropping off as the animals brush against objects (Figure 27-21). The seed may then germinate if the environment is suitable. Some seeds or fruits may be carried considerable distances. even though they do not possess such spiny protuberances, by being located in mud or earth which adheres to the feet of animals, especially wading birds. The mud may be shaken loose or drop off as it dries.

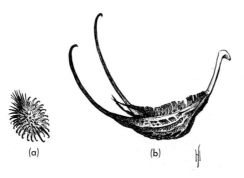

Figure 27-21. Examples of plants that are dispersed by animals. **(a)** Cocklebur (*Xanthium*). **(b)** Fruit of unicorn plant (*Martynia*).

Many fruits are edible, such as grape (*Vitis*), cherry (*Prunus serotina*), and *Pyracantha* berries. These attract birds and other animals, which may spit out the pits or swallow them. The seeds are usually resistant to digestive enzymes and pass through the animal virtually undamaged. In fact the action of digestive acids and enzymes on the seed coat may even enhance germination (see Section 27.9 regarding dormancy). The fecal material excreted with the seeds may actually be beneficial to the growth of the resultant seedling. Remember that bird manure, usually from chicken farms, is an excellent source of nitrogen when used as a fertilizer. Birds quite likely are responsible for carrying seeds over natural barriers, such as water. Nuts and nutlike fruits are carried about by rodents and are frequently abandoned without being eaten. Oak (*Quercus*), hickory (*Hicoria*), and beech (*Fagus*) are usually scattered in this fashion.

Man in his travels throughout the world has played an exceedingly important role in the dissemination of plants. He has purposely carried with him the crop plants, forage plants, and ornamentals he prefers. However, he has also, usually unknowingly, carried seeds and fruits of weed plants along with him. Such weeds, in competition with crop plants, result in tremendous economic losses. Millions of dollars must be expended each year in an attempt to control weed growth, and additional millions of dollars are lost as a result of decreased crop yields caused by weed competition. In addition man has often brought in diseased plants resulting in such disasters as the almost complete loss of the American chestnuts (*Castanea dentata*) by blight disease and the decimation of the elm (*Ulmus americana*) population by the Dutch elm disease. The fungus causing chestnut blight was inadvertently introduced into the United States about 1904 on planting stock brought in from the Orient, where only mild infections occur on the native species. Neither the American nor the European chestnut were resistant, and the disease spread rapidly; within 35–40 years the chestnut had pretty well disappeared. In a similar situation, the Dutch elm disease fungus was apparently brought to the United States on elm logs from Europe in 1933. By 1970 the disease had spread through most of the Northeast and Midwest and destroyed most of the trees in these areas. Disastrous introductions of this sort brought about the establishment of quarantine laws. However, unwanted pests still occasionally get through.

Additional Dispersal Mechanisms

Water acts as an agent of dispersal for some kinds of plants. The seeds and fruits of various aquatic plants float and may be carried about by water currents. Rainfall may wash seeds about, and receding floods may strand seeds in likely spots, where germination then occurs. If these are seeds of a nonaquatic plant, they will not remain viable if submerged in water too long. However, most seeds are capable of germination after submersion of one to several days. Probably the most outstanding example of a water-dispersed plant is the mangrove (*Rhizophora mangle*). The fruits of this tree are carried great distances by ocean currents, and the plant is now well distributed along tropical shores. The fruit of the coconut palm (*Cocos nucifera*) is also exceedingly well adapted for distribution by sea currents. Because of the air trapped between the fibers in the mesocarp layer of the fruit wall, coconuts float easily and sea water almost never comes in contact with the seed. Germination occurs quite readily if the fruit is washed up on shore.

A few types of plants have modifications that result in a mechanical seed dispersal. The drying-out of various layers of the ovary wall at different rates may result in such strains that a very rapid rupture of some of these tissues results in the forcible ejection of seeds. This frequently occurs in the fruits of the violet (*Viola*), sorrel (*Oxalis*), and witch hazel (*Hamamelis*). In the squirting cucumber (*Ecballium*) osmotic pressures build up as the inner tissues of the fruit ripen until the tensions are such as to forcibly extrude the mucilaginous contents and the seeds through a small rupture in the leathery outer layer of the fruit. These forces, which develop osmotically or upon drying, are frequently sufficient to cast seeds several feet away from the parent plant.

Tumbleweeds, such as *Amaranthus albus* and *Salsola kali* var. *tenuifolia,* effect seed dispersal in an interesting manner. The bushy plant breaks off near the surface of the soil and is rolled along open ground by the wind. Seeds and fruits are dislodged during this tumbling process, which may carry the plant for considerable distances.

27.12 Importance of the Division Anthophyta

The flowering plants are undoubtedly the dominant members of our present-day flora and are found in almost all land and shallow water environments. Man's most important crop and forage plants—the plants of greatest economic value—are Anthophyta. The major uses of wood have been noted in Section 8.6, and the reader should review these.

Man's basic food supply depends upon the cereals (rice, *Oryza sativa;* wheat, *Triticum vulgare* and *T. durum;* corn, *Zea mays;* oats, *Avena sativa;* rye, *Secale cereale;* and barley, *Hordeum vulgare*) and grasses. Most of the cereals are eaten directly, but they may also be fed to livestock, which are then eaten. The forage grasses, of course, are consumed indirectly by man. Of those plants just mentioned, rice is used by more people than any other food; probably half the world's population, mainly in the Asiatic countries, consumes this plant. Cereals, especially barley and corn, are also frequently used in the production of alcohol and the distillation of alcoholic beverages.

The reader might also review Section 10.6 for mention of various fibers and chemicals obtained from flowering plants. Of the former, the most important are undoubtedly cotton, linen (flax), and hemp. Many synthetic fabrics are now produced, but most of these utilize cellulose (a plant product) as a base. The following are but a few of the materials obtained on a commercial basis from flowering plants: alcohol, acetone, acetic acid, tars and resins, tannins, oils, gums, and dyes. Rubber and cork are obtained from certain trees: *Hevea brasiliensis* and *Quercus suber,* respectively. Various drugs and beverages are also obtained from flowering plants: quinine, morphine, strychnine, digitalis, and ephedrine; tea, cocoa, and coffee.

Many other plant parts than those already mentioned are useful to man. Entire volumes have been written concerning edible seeds, fruits, rhizomes, tubers, and so forth; this brief discussion is intended merely to emphasize the economic importance of the Anthophyta and not to list all the uses to which these plants have been put. Chapter 2 presents a brief discussion of different ways in which humans make use of various plants and plant parts, but the reader should refer to a text on economic botany for a more extensive coverage and greater details.

27.13 Evolutionary Advantages Featured in the Anthophyta

As in the Coniferophyta, the gametophytes develop within protective tissues and are not directly exposed to the environment. However, the Anthophyta are in a more favorable condition with respect to the female gametophyte and the embryo than are the conifers. The ovules of the former plants are enclosed within an additional mass of tissue, the ovary, which affords protection to the developing gametophytes and eventually the embryo, as well as frequently aiding in the dissemination of the mature seed. Just as the seed habit is considered to be a major evolutionary advance, so is the evolution of the fruit considered to be exceedingly significant. The production of a fruit, the various modifications that aid in seed dispersal, and the presence of vessels in the conducting tissue are factors that were important in establishing the Anthophyta as the dominant plants of today. Some of the modifications leading to vegetative reproduction (e.g., runners, rhizomes, tubers, bulbs) are somewhat important, but sexual reproduction and its resultant variations are much more significant when examined from an evolutionary viewpoint.

Only one egg is present in an ovule of the flowering plant. This ensures a greater supply of food for the developing embryo, since the single embryo thus has no competition, as was the case in the conifers. In the latter several archegonia and eggs may be produced by the female gametophyte.

The evidence is not clear as to the ancestors of the flowering plants. They probably evolved from a primitive group of conifers, but additional evidence must be obtained before any definite statement may be made.

27.14 Comparison of Life Cycles

In Figure 27-22 is represented the generalized life cycle of a flowering plant. This should now be compared with the life cycles of representatives from the various groups of the plant kingdom that have been studied.

Certain general observations should be clear when examining these life cycles. An ever-increasing dominance of the sporophyte phase is evident in the more complex plants, correlated with a general reduction in size of the gametophyte and its eventual dependence on the diploid sporophyte. As a result of these changes, the delicate gametes and gametophytes are more and more protected in the advanced plant groups. The dissemination of a species depends more and more upon

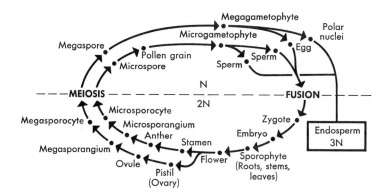

Figure 27-22. Diagram of life cycle of a flowering plant.

the distribution of fruits and seeds and the movements of spores, especially microspores or pollen grains. These meiospores are much hardier than gametes. The large sporophyte development is thus a great advantage—many meiospores are produced.

Summary

1. Anthophyta produce specialized structures, the flowers, in which enclosed seeds develop. They are the dominant members of the world's flora.

2. A complete flower has the following parts: sepals, petals, stamens, and at least one pistil. Microspores are produced in the anther portion of the stamen; the stalk is termed the filament. The pistil consists of the stigma at the tip, and elongated style, and the enlarged ovary at the base; one or more ovules are contained within the ovary. In some plants one or more of the floral parts may be absent.

3. A single megasporocyte in the ovule develops into four megaspores as a result of meiosis; three of the megaspores degenerate. Mitotic divisions of the functional megaspore result in the production of the megagametophyte: an egg cell, two polar nuclei, synergid cells, and antipodal cells.

4. Many microsporocytes in the anther produce microspores by meiotic divisions. The microgametophytes that develop from these spores are transported to the pistil in a variety of ways (pollination). Two sperm are produced as the pollen tube grows down the tissues of the pistil.

5. One sperm fuses with the egg, forming a zygote; the second sperm fuses with both polar nuclei, eventually forming the endosperm (a triploid storage tissue). The zygote develops into a multicellular embryo.

6. The ovule enlarges, the integuments harden, and the resultant structure is the seed. The seed may be dormant for a time. Upon germination, the cotyledons may be carried up and out of the soil. The cotyledons may also function as leaves for a short time.

7. As the ovule develops into a seed, the ovary increases enormously in size and forms the fruit. Flowers and fruits are of many different kinds.

8. Seeds and fruits may be variously modified, a factor that frequently assists in seed dispersal. Wind, animals (including humans), and water are the most important agents of dispersal.

9. The Anthophyta are divided into two classes, the Dicotyledoneae and Monocotyledoneae. The former have two cotyledons present in the embryo, the flower parts are in multiples of four or five, the leaves are net-veined, and a cambium is present. The monocotyledonous plants have one cotyledon in the embryo, the flower parts are in threes, leaf venation is usually parallel, and a cambium is usually lacking.

10. Primitive flowers have the following general characteristics: helical arrangement of floral parts, separate petals and carpels, and regularity of structures. Advanced flowers usually have a cyclic arrangement of parts, fused petals and carpels, and irregular or dissimilar parts.

11. The flowering plants, in addition to their aesthetic values, are our most important food and forage plants. Many fibers are obtained from the Anthophyta; so also are cork and rubber. Drugs, beverages, and other commercial products of various sorts (alcohol, acetone, tars, resins, dyes, and so forth) are also obtained.

12. The enclosing of the seed within the fruit increases the possibility of survival in diverse environments.

Review Topics and Questions

1. Discuss, with the aid of a word diagram, the general life cycle of a flowering plant with regard to alternation of phases, general chromosome number, fertilization, and meiosis.
2. Indicate the similarities between the life cycle of a flowering plant and that of a moss, a fern, and a pine.
3. Indicate the differences between the life cycle of a flowering plant and that of a moss, a fern, and a pine.
4. Describe the development in a flowering plant of a sperm, starting with a microsporocyte. Indicate where each structure develops and whether it is haploid or diploid.
5. Describe the development of the megagametophyte, starting with the megasporocyte. Indicate where each structure develops and whether it is haploid or diploid.
6. Diagram a complete flower and label all parts.
7. Define the following: carpel, incomplete flower, imperfect flower, monoecious, endosperm, epicotyl, hypocotyl, and fruit.
8. Describe in detail what happens when a dry (viable) seed is placed underground and watered. Assume that you have a garden bean seed, and conclude your discussion with the fully established seedling.
9. Discuss at least four mechanisms for seed dispersal, indicating the advantages that accrue to plants having such mechanisms.
10. Define the following and give an example of each: simple fruit, accessory

fruit, berry, nut, hypogenous flower, perigynous flower, and epigynous flower.

11. Compare the general structure of a wind-pollinated flower with that of an insect-pollinated one.

12. Discuss the functioning of those structures that have made it possible for the Anthophyta to obtain their prominent position in the flora of the world.

13. List at least ten ways in which the flowering plants are important economically.

Suggested Readings

BIERHORST, D. W., *Morphology of Vascular Plants* (New York, Macmillan, Inc., 1971).

BURNS, G. W., *The Plant Kingdom* (New York, Macmillan, Inc., 1974).

FOSTER, A. S., and E. M. GIFFORD, JR., *Comparative Morphology of Vascular Plants,* 2nd ed. (San Francisco, Freeman, 1974).

SCHERY, R. W., *Plants for Man,* 2nd ed. (Englewood Cliffs, New Jersey, Prentice-Hall, 1972).

Evolution

28.1 Evolution

The idea that all organisms are descended from past organisms from which they differ more or less is one of the broadest concepts in biology and has influenced many fields outside of biology. **Evolution** most simply means descent with change and implies that all organisms are more or less related one to another; the "change" really refers to changes in the genetic composition of a population (non-inheritable changes are not involved). Presenting this idea in a logical way, we shall first examine the general evidence that leads to the acceptance of an evolutionary concept. Then we can consider several theories that eventually led to the modern formulations of this process. Finally, then, we shall be able to examine more specifically evolution within the plant kingdom in the light of the discussions of various plant groups in the preceding chapters.

The term "evolution" embodies the idea of an orderly series of changing events related in a succession so that preceding events influence succeeding ones. This is not really a strange concept, and we speak of the "evolution of an idea" or the "evolution of the train" (as a means of transport). The basic idea is one of change in an orderly fashion; one suggestion or advance paving the way for the next. The concept of *organic evolution* or *biological evolution* reflects a similar situation in that the first living organisms were very simple in structure, and that various evolutionary changes brought about increasing structural complexity and diversity that culminated in the eventual appearance of the more complex or advanced forms. It is accepted that organisms are still evolving although human life spans are so short as compared to evolutionary events that one usually is not aware of such changes. However, humans frequently speed the process by selective breeding. Common examples are the many breeds of dogs and the various breeds of pigeons that are now available as compared to those present early in recorded history. In both cases there has been an extreme radiation of different types with humans selecting which shall survive.

28.2 Geological Evidence

The earth has undergone a tremendous physical evolution. In some areas land masses have risen from the sea and in others have sunk beneath the oceans. Iceland, for example, is a volcanic plateau of rather recent origin, probably 7 to 10 million years old. On the other hand, the southern tip of the Scandinavian peninsula has depressed about 2 m during the past 300 years, while the northern portion has raised somewhat. The 1960 calamitous earthquakes in Chile are examples of earth movements; so also is the sudden appearance of the massive volcano Paricutin in Mexico during 1943. In the Chilean earthquake some layers of rock were moved upward almost 10 m (about 33 ft.) Similar movements occurred in the more recent Alaskan earthquakes of 1964. The processes that caused changes in the past are causing them today, but human time scales are too short for recording the many changes that have occurred since the earth's beginning. Estimates of the age of the earth vary from two-and-one-half to at least five billion years, depending upon the method of estimation, but five billion is the consensus at the present time.

The most accurate determinations of geological ages are based upon known rates of decomposition of radioactive materials. Such breakdown occurs at a uniform rate and is independent of temperature, pressure, light, or other environmental factors. By determining the ratio of radioactive materials to their breakdown products, we can estimate the length of time during which the original material underwent disintegration. One method is to measure the accumulation of lead, the stable product derived from the disintegration of radioactive uranium, in igneous rocks. Such rocks are formed from molten materials, and it is presumed that they are the oldest rocks of the earth's crust, having been derived from molten materials formed fairly early in the history of the earth. The half-life of radioactive uranium is four-and-one-half billion years; that is, one half of the remaining uranium atoms will be converted to lead every four-and-one-half billion years. Assuming that only uranium atoms were present when the rock was formed, the ratio of lead to uranium atoms gives an approximation of the age of the rock. Some difficulties with this technique have yet to be worked out; so our estimates tend to be a little conservative. However, the earth very likely is older than five billion years. In fact, each new estimate pushes the age farther and farther backward. Fossils are not found in igneous rocks, but they are often associated with dated igneous rocks in such a way that the age of the layer in which they are both found can be estimated.

A current theory held by many geologists proposes that the surface of the earth is composed of gigantic, closely fitting, plates of rock floating on an inner molten core. These plates are moving from place to place as a result of internal pressures. In an area between two contiguous plates the plastic mantle rises from below and pushes the plates apart; with this movement the edge of one plate is moved against another and it is raised or lowered as one slides over or under the other. The continents move about because they are on such plates. The idea of **continental drift** was proposed over 50 years ago, and aroused much controversy but little acceptance. However, significant new data and the theory of *plate tectonics* during the past few years, have now resulted in rather widespread acceptance of this idea. It is suggested that the continents have moved apart and together over long periods of time. The matching configurations of South America and Africa on their Atlantic Ocean sides had been noted by geologists for ages although there was little evidence, until recently, that the continents had drifted apart.

These changes of land movement and drift emphasize the tremendous environmental variations to which living organisms must have been subjected. Land masses were connected at one time, separated at others; sea levels rose and fell, as did the land in various areas; and if continental masses drifted from the south polar regions to their present locations, as has been suggested, considerable climatic changes must have occurred. Some of the different kinds of organisms could well have been better adapted to a new environment than were the parental types. This does not imply that the environmental variations *caused* the adaptations in the organisms. It is considered that environmental factors are *selective* or *directive* forces rather than causative forces of evolutionary change. If variations in organisms occurred, the environment could well determine which would survive.

28.3 Paleontological[1] Evidence

The concept of evolution is probably best substantiated by the existence of fossils. Basically, a **fossil** is any trace of a former living thing and may be of several types:

1. Intact specimens or parts of them. These are sometimes found embedded in amber, in tar pits, or in localities where decay has been prevented (e.g., in bogs that are relatively toxic). Plant remains have been found in the frozen muck of boreal regions after thousands of years without decay because of the low temperatures. The wood of some trees contains substances, such as tannin, which protect it from bacterial action. The dried and hardened resin of gum-bearing trees may be preserved as amber. Some of the specimens, which became entrapped when the gum was soft, are preserved in fossil ambers from the Tertiary (about 50 million years ago) and are perfect in every detail. In these rather rare instances plant parts are preserved in essentially their original state.

2. Molds, casts, or impressions. The shape of the organism is retained, but none of the original plant (or animal) material remains. Rapid burial in fine-grained sediments or volcanic ash greatly facilitated the formation of imprints of various plant parts, and these impressions remained after the sediments had been compacted into solid rock. Even though the original plant tissues have disappeared, much of the basic structure is often clearly evident.

3. Compressions. These consist of carbonized plant material more or less in the original shape but greatly compressed.

---

[1] Paleontology is the study of ancient life.

4. Petrifications. The plant parts are infiltrated or replaced to a great extent by mineral substances; the structure is preserved. Frequently mineralization results in the total replacement of the original material by minerals (commonly silica and calcium carbonate) carried by water, and the details of both external and internal structure are reproduced with amazing fidelity.

5. Accumulated products. These are structures, such as reefs, composed of the secretions of living organisms, or incomplete decay products, such as oil and coal. On the basis of present knowledge, such structures are accepted as indications of the activity of living organisms, or at least that living organisms were present in those areas at some time.

The finding of such remains or indications of the one-time presence of living organisms enables biologists to deduce much concerning the changes that have occurred during past eras. The age of the fossil is determined by the location of the rock stratum in which it is found and/or by radioactive measurements. In general, the deeper the stratum, the older the fossil; earth upheavals or lowerings must be interpreted with care, of course. The radioactive measurements are made using rock samples or portions of the fossil itself (carbon-14 dating). Whereas the rock strata indicate more relative ages, radioactive dating presents us with more nearly an exact time scale.

Carbon-14 dating is applicable to carbon-containing material and covers the time range of about 100 to 50,000 years; it is an exceedingly useful method for determining the age of biological material. This technique depends upon the fact that the ratio of carbon-14 (the radioactive isotope) to carbon-12 (the normal carbon atom) is constant on earth and in living organisms. The former atoms disintegrate at a constant rate with a half-life of 5570 ± 30 years, but new carbon-14 atoms are continually produced in the atmosphere when high-energy cosmic rays collide with nitrogen atoms. When an organism dies, it is no longer obtaining new supplies of carbon-14 (by way of photosynthesis or by utilizing the products of photosynthetic organisms). The carbon-14 present in the dead organism breaks down at the known rate. When the ratio of radioactive to nonradioactive carbon is determined, the time since the death of that organism can be determined fairly accurately, at least for materials less than 50,000 years old. Because the uranium-lead method is useful only for very ancient materials, millions of years old, other methods of dating utilize isotopes of fluorine and argon to fill the gap of time between these two techniques.

In some cases the fossil records are quite complete, while in others they are not. This is not too strange when one considers what happens to organisms when they die; decay occurs, or the dead organism is even more rapidly consumed by various animals (e.g., jackals, vultures, maggots, rodents, and various grazing ones). An organism (or portion of it) must be protected quickly from such a fate in order for it to be preserved. Such protection may occur in swamps and bogs because of a low oxygen content or toxic materials that decrease or prevent bacterial action, in rapidly silting streams and lakes, in tar pits, and in glaciers; in these areas animals cannot reach the dead organisms.

One finds on examining fossils that the flora and fauna of the earth are constantly changing; new types of plants and animals arise, flourish, and sometimes become extinct. The general sequence of evolutionary change from organisms of very simple structure to more complex organisms is well substantiated; the older remains are the more primitive. The evidence for the presence of bacteria and some of the more primitive algae is mostly indirect because of their extremely small size and simplicity of structure. Table 28-1 is a comparative chart, based upon available knowledge, of geological and paleontological history and indicates the organisms that were present during the various periods.

Table 28-1 Geological time scale

Eras	Periods	Millions of years from present (approximate)	Major events in the evolution of plant life as shown by fossil record
Cenozoic	Quaternary	Present to 2	Extinction of many trees through climatic changes. Increase in herbaceous flora.
	angiosperms late Tertiary	25	Dwindling of forests; climatic segregation of floras. Rise of herbaceous plants.
	early	65	Development of many modern flowering plant families. Rise and worldwide extension of modern forests.
Mesozoic	Cretaceous	136	Flowering plants gradually become dominant; some modern types represented. Cone-bearing plants decline.
	Gymnosperms Jurassic	190	Earliest known flowering plants. Cycads and conifers dominant; primitive cone-bearing plants disappear.
	Triassic	225	Increase of cycads and conifers.
Paleozoic	Permian	280	Waning of arborescent club mosses and horsetails. Early cycads and conifers.
	Carboniferous	345	Extensive coal-forming forests of giant club mosses, horsetails, and primitive cone-bearing plants. Bryophyta.
	Devonian	395	Early vascular plants: psilophytes, primitive club mosses, horsetails, ferns, and first seed plants. Early forests of arborescent club mosses. Possibly Bryophyta.
	Silurian	430	Algae dominant. First direct evidence of land plants.
	Ordovician	500	Marine algae dominant. Possibly first land plants.
	Cambrian	570	Some modern algal groups established.
Proterozoic		570 to 1500	Bacteria and simple algae.
Archeozoic		1500 to 5000	No fossils known; possibly unicellular algae.

Source: Victor A. Greulach and J. Edison Adams, *Plants, An Introduction to Modern Botany* (New York, Wiley 1962), as adapted from Eames and with the permission of the publisher; time scale modified in accord with "Geological Society Phanerozoic time scale 1964" (*Quart. J. Geol. Soc.,* London).

28.4 Biogeographical Evidence

Plant geographers frequently point out that many types of plants are widely distributed and that others are quite restricted in occurrence. Also, different kinds of plants are found growing in areas that are widely separated but have similar environments. Such observations indicate that the environment alone cannot explain distributions of organisms. If areas containing similar species are isolated from each other by various barriers, such as mountain ranges or bodies of water, these organisms are quite likely to evolve along divergent lines and become very different from one another. The occurrence of variants is a haphazard phenomenon, after which the environment will act as a selective agent in perpetuating those individuals that are best able to survive under the existing conditions. The evidence suggests that dispersal and colonization are followed by isolation and evolution in the new environments.

Charles Darwin was greatly impressed by the flora and fauna of the Galápagos Islands, a group of small islands about 600 mi west of Ecuador, South America. Here he found certain organisms (e.g., the giant tortoise and marine iguana) not found elsewhere; he also found that tortoises from different islands in the group were distinguishable. This was an example of areas only 50 to 60 mi apart that had differing organisms, a clear case of divergent evolution. However, Darwin's investigation of the finches was what provided him with his first insight as to the mechanisms involved in the origin of species. Although they were basically alike, he found 13 distinct species of these birds varying in both beak structure and habit —some fed on fruits, some ate seeds, and others hunted for insects. He noted that they all showed a marked relationship to those of South America. They had to be descendants of a common ancestor, which had migrated from the mainland. The absence of other land birds as competitors enabled the finches to evolve in many directions—directions that would not be available to them in a populous area, where the finches are restricted to very narrow niches. For example, if by an accident of heredity a finch became able to crack open and live on seeds, it would have the advantage of a supply of food in addition to insects. In time, and especially if finches with slight differences were isolated on different islands so that there was little interbreeding between populations, various species with different habits and different appearances would emerge. As a result of his observations, Darwin became aware of the importance of both environment and competition as selective factors in evolution.

28.5 Taxonomic Evidence

Plants and animals are classified mainly with regard to their similarities and dissimilarities of structure. The kinds of organisms thus become arranged in various groupings that indicate relationships. The greater the structural resemblances, the more closely related are the organisms. A dog and a wolf are more closely related than a dog and an antelope; peas and beans are more closely related than peas and roses. If organisms are closely related, they must have a common recent ancestor.

Those organisms that are quite dissimilar can be related only very distantly to a common ancestor ages ago. Relationship (common ancestry) is the simplest valid general explanation of similarity of organisms. Occasionally, when parallel evolution is involved, some difficulty may be encountered in the interpretation of similarities and relationships. This will be discussed later (Section 28.12).

Related organisms have similar basic structural patterns. One example of this is the flower, which is characteristic of one large group of plants (Division Anthophyta). Subdivisions of this group exist with regard to the particular kind of flower, but the members of this group are certainly more closely related to each other than they are to plants that bear cones rather than flowers. The homology of vertebrate appendages (e.g., wing bones of birds and finger bones of man) is another example of structural relationships. Even more basic than these examples is the cellular structure of living organisms.

28.6 Embryological Evidence

In many cases the early development of organisms may be very similar, even though the mature individuals are quite different. The early stages of a moss gametophyte (the protonema) and a fern gametophyte (prothallus) are algalike in appearance and tend to suggest that algae were ancestors to these plants. More vivid evidence of embryological similarities can be found among animals. In the very early stages of development, when only a roundish mass of cells is present, one animal embryo is almost impossible to distinguish from another. Even somewhat later, when the head and other regions are clearly visible, many animal embryos look very much alike. At this stage, the human, pig, chick, and shark have many points of similarity. The human embryo even has gill lobes and a tail, both of which disappear before birth. These various similarities during the development of organisms certainly are indicative of relationships between them.

28.7 Evidence from Comparative Physiology

Just as structural patterns exist, so do basic chemical patterns in organisms. Photosynthesis, chlorophyll, respiration, cytochromes, adenosine triphosphate (ATP), deoxyribonucleic acid (DNA)—the presence of these and similar materials and processes attests to general relationships among organisms. In a more restrictive sense, we may consider the resins, such as turpentine, of the pines (Pinaceae), the essential oils of the mints (Labiatae), and the latex found in the spurges (Euphorbiaceae); these materials are less widespread than, for example, chlorophyll and indicate closer relationships.

As is the case with animals, certain proteins are characteristic of plant groups. The more nearly alike the proteins, the closer is the relationship. Conversely, plant groups that are not closely related tend to have quite different protein characteristics. Analyses of various pigments, such as the flavonoids, have also been useful in indicating relationships. In fact comparative phytochemistry is a recent field of study that has revealed some interesting correlations between series of chemical compounds found in groups of plants. Such investigations will probably play an ever-increasing role in the understanding of evolution and classification.

28.8 Evidence from Genetic Studies and Domestication

Genetic studies have provided a source for evolutionary change. The hybrids between species are occasionally sufficiently self-fertile to become new species. Evidence for this can be found in the California pines. A cross between *Pinus jeffreyi* and *P. coulteri* produces mostly sterile hybrids, which have no future. The cross between *P. jeffreyi* and *P. ponderosa* produces some fertile hybrids, but they are poorly adapted. The cross between *P. jeffreyi* and *P. ponderosa* var. *scopulorum* probably resulted in the production of a successful (fertile) hybrid now known as *Pinus washoensis,* which grows near Lake Tahoe. This is apparently a new species that arose as a hybrid.

New species have also been shown to arise from a multiplication of the original basic chromosome number, usually as a result of the failure of one cell division during meiosis. The diploid gametes then produce a tetraploid zygote. Such multiplication of basic chromosome sets occurs naturally or may be induced by colchicine, and the individuals are polyploid.[2] Polyploids usually look different from the original plant and are usually not capable of interbreeding with it; thus, a new species may arise. Polyploids are oftener larger and more desirable than diploids. Many cultivated lilies, petunias, snapdragons, and blackberries are polyploid.

In the genus *Clarkia* are several well-substantiated cases of fertile hybrid species that developed when polyploid individuals arose. For example, *C. similis* combines gene sets from *C. modesta* (8) and *C. epilobioides* (9); the numbers in parentheses refer to the normal gametic chromosome number. The hybrid, having a diploid number of 17, would be sterile; with an uneven and nonhomologous number of chromosomes, pairing (synapsis) could not occur. A chance doubling of the chromosome number resulted in the fertile hybrid with the even number of 34; the gametes, of course, contain 17 chromosomes.[3] Similar results possibly may be obtained with other sterile hybrids by using colchicine to double the chromosome number.

Domestication is really evolution under more or less controlled conditions, with man, instead of the natural environment, determining which individuals will survive and reproduce. This was mainly a hit-or-miss proposition until the 1900s, when the field of genetics opened up tremendous possibilities in plant and animal breeding. Without an understanding of genetics, the procedure was merely to keep the seeds of the "best" plants, hoping that such seeds would develop into equally good offspring. In many instances this was quite successful, especially

---

[2] There are various kinds of polyploids: triploids, tetraploids, pentaploids, and so on.

[3] Additional information concerning *Clarkia* may be found in "The Genus Clarkia" by Harlan Lewis and Margaret Ensign Lewis, *University of California Publ. in Botany,* **20** (No. 4), 241–392 (1955).

over rather long periods of time. However, if several pairs of genes or multiple alleles or dominance is included, the progeny will be quite diverse. The concept of inheritance being determined by particles (genes) enabled one to predict results, to estimate possibilities, and to plan an accurate series of breeding programs. The results since 1900 have been astounding.

Plant-breeding programs, with their genetical basis, are generally concerned with developing plants that have one or more of the following characteristics: (1) increased yields (quantity), (2) better quality, (3)shorter growing seasons, (4) disease resistance, and (5) the ability to grow in diverse environments. During the past 50 years, for example, the yield of wheat (*Triticum aestivum*) has increased from 20 bushels per acre to 100 (maximum about 200), and that of corn (*Zea mays*) from 30 bushels per acre to 150 (maximum about 250). Some of these increases are caused by modern agricultural practices, but most of the increases are a result of new varieties. The many varieties of apples are good examples of man's efforts to produce quality products. At first the improvement of the apple (*Malus sylvestris*) consisted largely of the selection of chance seedlings, and this is even true today, the Golden Delicious being a prime example. The Cortland variety, however, is an example of the results of controlled breeding, which is being carried on with significant results. European grape varieties based on *Vitis vinifera* were unsuccessful in North America because of a lack of resistance to low temperatures in the north and various diseases in the south. Hybrids between native American species and the European grape are of better quality than the native grapes and retain the resistance of the native types.

Possibly one of the best examples of the use of a breeding program is the work that has been done with the resistance of wheat to a fungus disease. The wheat plant is one of the most important food plants in the world, especially in the Western hemisphere. Unfortunately, the yield is greatly reduced by various diseases, with wheat rust (organism: *Puccinia graminis* var. *tritici*) the major culprit. The only economically feasible method of controlling this disease is by using resistant wheat varieties. In many agricultural experiment stations, wheat-breeding programs are a continuing project, because the rust organism produces new varieties as a result of sexual reproduction or as a result of gene mutation. Varieties of wheat that are resistant one season may be attacked by new rust varieties within a few years. Here, then, is an example of manmade evolution attempting to maintain an advantage over natural evolution.

Domestication has resulted in the production of various species from a common ancestor. The wild cabbage (*Brassica oleracea*) is the plant from which cauliflower, kale, kohlrabi, and brussels sprouts have been produced, in addition to the cabbage itself. One might also list the many kinds of dogs, horses, and chickens that have resulted from man's efforts. In all these, conscious selection by man has preserved those types in which he was interested. Darwin himself was impressed by this great variety of domestic plants and animals. He studied, among other things, variations in the breeds of domestic pigeons. By careful selection over many years, pigeon fanciers had developed numerous strange and exotic types, all of which originated from the common pigeon, like the kind still living in many cities.

28.9 Lamarck's Theory

Jean B. Lamarck was the first (1802) to suggest that all species had descended from other fewer species by gradual changes. He realized the importance of close and distant relationships as a result of his studies of comparative anatomy and the difficulties he encountered in clearly defining a species. He understood evolution as a dynamic process, continually occurring and leading to increasing complexity and specialization; that there was a progression extending from the simplest to the most complex. These were accurate and excellent ideas, but his theories as to the method of evolution have not been accepted.

Lamarckian evolution is frequently termed evolution through the transmission, or lack of it, of characteristics gained or lost by use or disuse—the inheritance of acquired characteristics. As new needs or wants arise, these tend to cause new parts or organs to develop. The frequent use of a structure tends to improve its functioning or bring about its further development and so produces a variation that is transmitted to subsequent generations. Similarly, disuse tends to cause structures to cease functioning and eventually disappear. The environment changes the needs and requirements of an organism, and this stimulates the development of new parts, the greater development of existing parts, or the loss of parts no longer used. In this way variations great enough to be considered a new species are built up. Lamarck emphasized the development of structures *in order to* accomplish some end. The long neck of the giraffe must have been acquired by generations of giraffes stretching for foliage in trees; long legs must have been developed in certain animals in order for them to escape enemies; keen sight must have developed so that the animal could find its prey.

Experimental evidence was readily accumulated that refuted Lamarck's ideas. Acquired characteristics are not inherited as long as they have no effect on the genes. No one denies that organs or structures are affected by use or disuse. The environment does, of course, greatly influence the development of an individual, but such development is not inherited. The children of weight lifters are not born with bulging muscles; cutting off the tails of mice for many generations does not produce progeny without tails.

One of the simplest experiments to demonstrate environmental influences on the individual and to refute Lamarckianism is to grow plants of the same genotype under different environmental conditions: sufficient versus insufficient water, light versus shading, sufficient nitrogen versus deficiency of nitrogen, and so forth. Many differences in size, shape, and color will result. If these plants are capable of producing seed (self-pollination), seeds from the different-appearing plants, grown under uniform conditions, will develop into plants that are all the same. The genetic makeup alone determines the *capabilities* of the organism, while the environment may influence the actual growth of an individual. A similar situation obtains in humans existing on a diet insufficient in certain vitamins, for example. The individual is capable of normal growth if he receives the required nutrients; without the essentials his growth and development are abnormal (e.g., a lack of vitamin D results in rickets; a lack of vitamin C results in scurvy).

In recent years T. Lysenko, a Russian plant breeder, claimed to have produced

513

heritable acquired characteristics. We cannot in this book detail the mass of evidence in opposition to this claim.[4] Suffice it to say that independent workers in other countries have never been able to duplicate Lysenko's experiments. In fact, all the data that such investigators have obtained are opposite to those of Lysenko. Needless to say, his claims have not been accepted.

28.10 Darwin's Theory

Charles Darwin was a naturalist who correlated his immense powers of observation with an idea gained from Thomas Malthus' ''Essay on Population'' (1798) in devising a concept of evolution that is accepted in modified form by biologists today. Darwin observed the many variations of plants and animals and collected many of them, particularly during his voyage around the world aboard H.M.S. *Beagle* from 1831 to 1836. (This was partially discussed in Section 28.4.) His ideas are frequently incorporated in the phrase, *the theory of natural selection, or the survival of the fittest.*

Actually Darwin's initial interest in life of the earth's past was stimulated by the work of Charles Lyell. In the 1830s Lyell published one of the classical works of geology. He pointed out that variations in the topography of the earth could have come about by the action of existing forces; valleys could be carved by flowing rivers, mountains could be worn down, and coast lines could be altered by wave action. He described evidence for concluding that these and other changes had occurred throughout the ages. In general, then, the earth has had a long history of change. In reading this work Darwin, as a biologist, wondered about life that may have existed thousands of years ago and whether it was similar to the kinds that we now have.

Darwin' observations led him to the idea that inheritable variations of all kinds and of various degrees, but mostly small, occur in all organisms. He acknowledged Malthus' statements that the reproductive potential of any organism is almost limitless, leading to overproduction or an overabundance, and that the actual number of organisms was not increasing. A pine tree or a dandelion, for example, may produce hundreds of seeds; a million or so may be produced by a single orchid; even humans have a tremendous reproductive capacity, women having been known to bear 12 to 15 children or more during a lifetime. Since it is apparent that no one group of organisms swarms uncontrollably over the earth, there must be keen competition among individuals and a high death rate. Because organisms are exposed to various adverse forces under natural conditions, those individuals best fitted (adapted) to the particular environment are most likely to survive. This survival is based upon the better-adapted individuals leaving proportionately more offspring, thus increasing the prevalence of the favorable variations. As generation follows generation, the better-adapted individuals will form an increasingly greater proportion of the population. The continued operation of this natural se-

---

[4] Conway Zirkle, *The Death of Science in Russia* (Philadelphia, Univ. of Pennsylvania Press, 1949). This is an excellent discussion of the effect of political control on the progress of science.

lection over long periods of time produces forms sufficiently different from the original to be called new species.

Darwin's theory of evolution consists of several important points. (1) *Variation*: individual organisms of the same species are not exactly alike but differ among themselves. (2) *Overproduction*: more offspring are produced than can possibly survive. (3) *Competition*: these individuals must compete with each other in a struggle for existence. (4) *Survival of the fittest:* the best-adapted individuals survive. (5) *New species originate*: this natural selection of the best-fitted individuals over many generations produces new forms of life.

In Darwinian terms those giraffes that happened to have longer necks could obtain more food than their short-necked brethren and thus would have a better chance of surviving in that environment. The difference in approach between this statement and that by Lamarck is so important it should be stressed. Lamarckian thought is that the long neck evolved *in order to* guarantee survival, while the Darwin concept implies that survival was possible *because* the long neck *happened* to evolve. In other words, many variations occur at random in any population of organisms, and the environment determines which variants are suitable and thus likely to be successful in the production of offspring. In this instance, ancestral giraffes probably had necks that varied in length and these variations were hereditary. Competition for food led to the survival of the longer-necked offspring at the expense of the shorter-necked ones; eventually only long-necked giraffes survived. The environment is the selective agent for changes that have occurred in the genetic composition of a population. Of course, many gradations of adaptability exist.

Darwin gave full credit to Lamarck and others for initiating the theory of evolution, but his own careful observations added such a tremendous amount of detail and such thorough documentation that he is really considered to be the person who developed most thoroughly the basis of our modern concept of evolution. His concept of the method of selection was quite different from Lamarck's, of course. However, Alfred R. Wallace, entirely independently of Darwin, reached the same conclusion concerning natural selection. In fact, Darwin's *The Origin of Species by Means of Natural Selection* (1859) was published as a result of the article Wallace submitted. Since his voyage Darwin had been compiling his data and writing —he had shown a rough draft of this material to various friends as early as 1844— but had spent about 20 years in the effort. With the arrival of Wallace's paper, friends persuaded Darwin to submit a summary of his evidence along with the article by Wallace to a meeting of the Linnaean Society (one of the foremost scientific societies of the day) for publication in its *Journal*. The more massive amount of data accumulated by Darwin and the careful and thorough exposition of his ideas, as well as the fact that his project had been well underway since about 1838, earned for him fame throughout the world, whereas Wallace is known mainly to biologists (and not to all of them, either). Independent discovery of basic ideas is nothing new. If the data are present, the theory will eventually be suggested by someone. Newton did not manufacture gravity, and Einstein did not produce relativity. These were basic ideas that would have had to be developed eventually. Fortunately, the world produces intelligent men who perceive concepts more readily than do most of us.

Darwin's ideas concerning evolution, one must realize, do not explain the *source* of variations. He assumed that variations occurred and that natural selection took place among these. We are indebted mainly to de Vries for an understanding of how variations develop.

Darwin's theory of evolution aroused rather violent controversy, especially on religious grounds, but it was gradually accepted. Some biologists were almost immediate converts while others were eventually persuaded as more and more data accumulated. This theory provided a unifying concept which explained otherwise puzzling facts. For example, the fossil record with its appearance of progressively more complex types in later rocks, and the extinctions that took place frequently, now fell into place in this theory.

28.11 De Vries' Theory

In 1901, Hugo de Vries, a Dutch botanist, proposed his mutation theory of evolution as a result of his studies of the inheritance of a species of evening primrose (*Oenothera lamarckiana*). Although the patterns of heredity were generally orderly and predictable, he observed that individuals with new types of flowers occasionally and suddenly developed, that these changes were inherited, and that some of the new individuals were sufficiently different from the parental type to be considered new species. The sudden changes that resulted in new kinds of plants were called mutations. Later evidence supported the idea that mutations are chromosomal changes, thus influencing hereditary characteristics.

One of the important aspects of sexual reproduction is the production of variability in the offspring as a result of recombinations of different kinds of genes. The origin of a different gene, or a new allele, as a sudden unpredictable change of an existing gene is termed **mutation.** Neither Darwin nor Mendel had any idea of such changes; they just assumed that variations existed. The term *mutation* is sometimes used to refer to any chromosomal change, while the term *gene mutation* is used in referring to changes in genes. The changes observed by de Vries were actually cases of polyploidy and chromosome aberrations, but many cases of gene changes were subsequently investigated. The more common practice is to use the term *mutation,* as in this text.

Usually mutations are of minor effect, recessive, and harmful, although occasionally dominant and major changes may be brought about. That mutations tend to be harmful and recessive is not really strange: the living organism is a highly complex system, and even small changes may disrupt a whole sequence of events —one nonstandard bolt may prevent an entire machine from functioning properly. Also, life has existed for a considerable length of time. Mutations have occurred throughout this period, and *natural selection* by environmental conditions would have resulted in perpetuating the more favorable of these changes. Mutations occurring at the present time have probably occurred in the past and have been selected *against.*

The gene may be pictured as a specific and definite portion (i.e., length) of a DNA molecule, one of a chain of such molecules that comprise the essential part of a chromosome (see Sections 4.5 and 18.20). Mutations could then consist of rearrangements of the atoms of the molecule as a result of internal or external forces. X-rays, infrared rays, cosmic rays, ultraviolet light, heat, and various chemicals serve to increase the rate of natural mutations. In the absence of such external forces, mutations probably occur at the rate of one gene per million genes each generation, although this will vary considerably with regard to different organisms as well as with different genes. A particular gene may vary in several ways, resulting in multiallelic forms, or the mutant may revert back to the original. The changes are at random, as are the genes that may mutate. Artificial methods (e.g., X-rays, mustard gas) are utilized to increase the rate at which mutations occur, so that they may be studied more readily, but such treatments do not change the kind of mutations that occur, nor do they enable one to predict the mutant that arises; the occurrence is still at random. (Radiations and mutations have a direct correlation: a little radiation produces few mutations, while much radiation produces many mutations.)

Some mutations are not inherited. Some are lethal and cause the death of the cell that has mutated. As another example, the navel (seedless) orange is a result of a mutation in a bud. Obviously, since seeds are not produced, such a mutation cannot be transmitted to progeny. However, vegetative progagation by means of cuttings or grafts serves to perpetuate such desirable plants. Somatic mutations, those occurring in vegetative cells, are not inherited but may be visible as variations in flower color or shape, leaf size, and so forth. These may not be advantageous to the plant (the seedless orange, for example), but they may be desirable to man, who aids in their survival. This is also true, of course, with inherited mutations that would die out except for man's interference. For his own benefit man has frequently perpetuated variations that could not compete under natural conditions and would have been eliminated. Many of our important crop plants are probably in this category. Seed dispersal in *Zea mays* (maize, or corn) would be very restricted, except for man, because of the firm attachment of the grain to the "corncob" and the presence of the "husk" leaves covering the fruiting structures (or "ears"). If the cob should fall to the ground, its grains could all germinate together, producing a competing mass of seedlings. If man did not cultivate and tend a wheat field, how many of his plants would survive? How many such wheat plants are seen in an untended field next to a wheat field? Man's desires do not necessarily involve selecting plants for uncontrolled competition; he protects his plants from competition.

The idea suggested by de Vries was actually a modification of Darwin's theory of natural selection. The latter emphasized an unbroken, gradually merging series of variations. De Vries believed that new forms, often sufficiently distinct to be called new species, appear suddenly, repeatedly, and unpredictably in the ordinary course of reproduction. These new forms (mutants) breed true. That is, in reproduction they produce organisms like themselves. Mutations are of all kinds and of various degrees and frequently result in quite large and distinct changes. The competition resulting from overproduction produces a natural selection (as suggested by Darwin), which results in the survival of those mutants best fitted to the environment. Actually, de Vries' theory merely emphasizes the importance of mutations as a fundamental process in evolutionary change; it offers no explanation of the cause(s) of these mutations.

28.12 The Modern Concept of Evolution

The considered judgment of biologists today is that from a single original form of life, or from a relatively few original forms, of very simple organization, other forms of greater complexity and diversity have in some manner evolved; that from these forms still others, mostly of greater complexity but sometimes of simpler organization, have evolved in the same manner; that this process has continued at varying rates but without interruption from the time of the beginning of life and is still going on; and that all organisms, both extinct and living, both plant and animal, have arisen in this way.

Studies of plant fossils have emphasized this evolutionary change from simple organisms to complex ones. The oldest layers of rocks, at least 3 to 5 billion years old, have no evidence of plant remains (see Table 28-1), but in rock strata of the Proterozoic Era, which was about 1 billion years ago, evidence of bacteria and algae has been found. During the Paleozoic Era, besides the algae, the first true land plants and eventually seed plants evolved. Great beds of coal were formed during the Carboniferous period of this era, probably 300 million years ago, from the bodies of plants (primarily ferns, Lycophyta, Sphenophyta, and primitive cone-bearing plants) that had existed in enormous numbers at that time. Various geological changes converted masses of plant material into the compressed form we know as coal. Later, during the Mesozoic Era, the conifers flourished, the flowering plants appeared, and the conifers began a marked decline. Although the first flowering plants appeared about 150 million years ago, their rapid and diverse evolution is shown especially in the fossil specimens found in the Cenozoic layers; during the era of 60 million years or so, the flowering plants emerged to attain their present dominant place among the flora. All this information has been obtained by a careful study of rock layers and the fossils they contain. Those fossils located in rock levels of more recent origin are found to be more complex; the older the fossil, the simpler the organism.

Although progressive evolution is the most conspicuous trend, examples exist of **retrogressive evolution** in both the plant and animal kingdoms. The parasitic mode of life, wherein structurally simple organisms have evolved from more complex ancestors, is considered to be of the latter type; so also is the possible derivation of fungi from algal ancestors as a result of loss of chlorophyll. Such changes are side branches along the main trunk of evolution and have not appreciably altered the direction of evolutionary advances in the plant kingdom as a whole. **Parallel evolution** refers to the common evolutionary pattern of different plant groups and may result in plants that are morphologically similar. Parallel evolution of this type is sometimes called **convergent evolution** and is shown clearly in the xeromorphic characteristics of the cactus and some members of the spurge (euphorbia) family; *Agave* (amaryllis family) and *Aloe* (lily family) have evolved similarly (Figure 28-1). The morphological characteristics of the vegetative structures are so similar that it is almost impossible to distinguish the members of these several families without a careful examination of flowers. This common pattern of xeromorphism among the members of widely separated families growing in similar, or nearly identical, environments is one of the most striking examples of convergent evolution.

Figure 28-1. Two plants of similar appearance, but members of different families; parallel evolution. *Left:* Cactaceae. *Right:* Euphorbiaceae. (Courtesy of Peter Jankay.)

Sequence of Events in Evolution

If all members of any one group were born alike, survival would be a matter of chance. But they are not all alike, as Darwin emphasized. Variations are produced by chance (unpredictable) mutations, and these variations are recombined at random by sexual reproduction. Many of the variants will be less fitted to the existing environment and will die, but others may arise that are better adapted and will survive and may completely replace the parental type by leaving more offspring. In the latter case, the parental type becomes an extinct group. If mutations are not lethal, the changes are handed down to the progeny. Thus, if the environment gradually changes, a few of the new genetic combinations may be better adapted to the different environment than was the original organism, and more of the new types will survive. If environmental changes are sufficiently prolonged or severe, the original plant will become extinct, and the new type will become normal. This is environmental selection of the mutant type over the original parental type, and the latter becomes a smaller and smaller part of the population and may even become extinct. At the margins of its geographic range and ecologic tolerance, a species is subjected to the most evolutionary pressure. If the proper mutations occur, the species may be able to occupy new territory. By the continued preservation over long periods of time of those individuals with the most favorable variations, living organisms become amazingly well adapted to their environment.

It is important to remember that evolution involves *selection,* with environmental factors, reproductive tendencies, and competition furnishing the pattern of selectivity. Evolution is an ongoing process that can take many directions, but these various selective factors funnel the genetic changes in certain directions. Many variations are simply lost because they are not suitable for one reason or another; such organisms are poor competitors, leave fewer offspring, and gradually disappear from a population. Other variations are perpetuated because they have been successful; they leave more offspring.

One might consider the evolution of a species like producing a book from scattered type picked up at random. At first, this latter process appears to be impossible. However, if we select only words as they form and discard groups of type that do not form words, the construction of a book is only a matter of time. We have selected groups of type (words) that are suitable. In a similar way mutations scattered at random throughout populations are selected by environmental and other forces. The evolution of *some* species is no miracle. It is only when an attempt is made to specify a particular direction of evolution that difficulties arise. After all, somebody wins the daily double at horse races. The miracle occurs if a previously specified individual wins.

An example of selection by environmental conditions within the last 100 years can be seen in the change of the peppered moth (*Biston*) from its normal light color to a black coloration. Occasional dark mutants were first collected about 100 years ago in Great Britain. Since the moths rest on tree trunks during the daytime, the dark ones are much more easily seen by birds and eaten. Thus, they become eliminated, even though they are hardier than the normal. In the smoke-grimed industrial areas, however, the "normals" no longer match their surroundings as well as do the "blacks," which thus have a selective advantage. As a result the dark forms have now almost entirely replaced the light forms in England, whereas the light forms are still the predominant members of the population in Scotland and other rural areas. This situation is completely understandable in terms of our evolutionary concepts.

Isolating two or more segments of an interbreeding population encourages their evolution into distinct species. When variations occur anywhere in the original population, they can spread entirely through that population by means of sexual reproduction. When such variations occur in isolated segments, they cannot spread to other segments. Similar mutants are not likely to occur at the same time in the various segments; also possible is that the environments of the isolated segments will be somewhat different, thus favoring different variants. If the period of isolation is sufficiently prolonged, the original species may evolve into two or more species. So many hereditary differences will have accumulated during this period that the genes of the various forms will not combine readily. Natural selection will tend to intensify the differences, distinct species will evolve, and interbreeding may eventually become impossible. Isolating mechanisms may be geographical (oceans, mountain ranges, and so forth), temporal (differences in flowering time or reproductive cycles), or genetic (incompatibility of gametes, sterility).

Optimum conditions for evolution appear to be small, semi-isolated populations. Such populations are isolated enough to evolve independently and differently and yet close enough to receive occasional "transfusions" of new variability from adjacent populations.

28.13 The Pattern of Evolution

A schematic diagram (Figure 28-2) of evolution emphasizes the fact that this is not a straight-line process but resembles more the branches of a tree—thus the term *family tree*. Variations may persist for a short or long time, depending upon possible environmental shifts and other factors. In the diagram, the original organism **A** of an ancient age gives rise to **B, C,** and **D. B** becomes extinct and gives rise to nothing, whereas **D** persists unchanged to the present time. **C,** on the other hand, eventually evolves into two different groups, **E** and **F.** Group **E** gives rise to **G, H, I,** and **J,** while group **F** produces **K** and **L. G** persists unchanged. **H** gives rise to **M** and **N. I, J,** and **K** become extinct, while **L** evolves into **O.** The groups living at the present time are **G, M, N, O,** and **D;** the other groups are found only as fossils. The extinct groups arose as mutants or variants that were not well adapted to the environment, or a change in the environment may have resulted in their becoming unsuited, or they may have died out for other reasons. In any event, they left no living descendants. In this representation **M** and **N** are closely related, with a recent common ancestor (**H**), and are rather similar morphologically, though different enough to be distinct species. **M** and **D** are very dissimilar, although related distantly through an ancient common ancestor, **A. D** and **O** might look very similar because of similar adaptations, but they are only distantly related (**A**). This could be an example of convergent evolution (e.g., *Agave* and *Aloe*). Only a small portion of the total number of individuals evolved exist at any one time, the remainder having become extinct. Without extinctions, classification would be extremely difficult because of the lack of gaps in the range of variations.

When a new basic series of adaptations, such as the flower, arises and becomes established, the group remains small during establishment and early development. Many variations on the original structure are evolved as the group becomes dominant and exploits many environments. A group may decline as a new group when a better basic adaptation arises. The group may also decline if it becomes overspecialized and cannot adapt to changes in the environment.

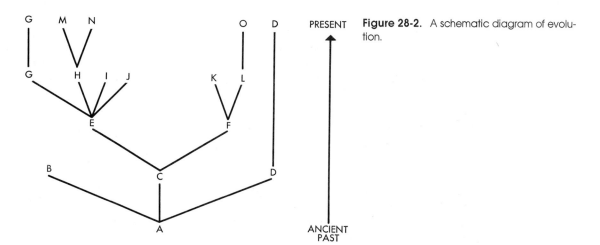

Figure 28-2. A schematic diagram of evolution.

The *Yucca* is potentially overspecialized. The styles of *Yucca* (Spanish bayonet) flowers are hollow, with the stigmatic surface at the lower end of this tube. Pollination is accomplished when the female Yucca moth (*Pronuba*) collects a ball of pollen from the stamens and stuffs this pollen down the tubular style. The moth then drills a hole through the ovary with her ovipositor (egg-laying device) and lays eggs among the ovules. The moth larvae develop within the ovary and feed upon the developing ovules. Enough ovules are present so that some are almost always left as seeds. The larvae bore out of the ovary and form cocoons in the soil; the adult emerges at about the time that the *Yucca* flowers. Note that the *Yucca* can be pollinated only in this fashion and the moth larvae can use only *Yucca* ovules for food; thus, the plant and the animal share a complete interdependence. Such overspecialization is quite likely to result in extinction of both organisms. That is, the demise of one species is sure to result in the extinction of the other.

A few of the older species usually persist after a new group becomes dominant. If an organism finds itself in a suitable environment, it may remain unchanged to the present. The blue-green algae are probably such primitive organisms that arose ages ago.

28.14 Evolution in the Plant Kingdom

Evolution can be discussed from two points of view. First, we can indicate that there is a gradual change from the simple to the complex by the accumulation of additional traits as we examine representatives from the primitive bacteria to the modern flowering plants. A second approach is to discuss each general advantageous characteristic separately, indicating the manner in which the simple and then the complex plants solved the problem of survival.

In the first approach we may list important descriptive features of an organism that differentiate it from simpler types but are also partially descriptive of all the more complex types. In other words, the traits listed for one type of organism would be found in all the more complex types, but not in any of the simpler forms. If the great groups of plants are arranged in order of complexity starting with the simplest, we find that, in general, any group will display the principal characteristics of all the simpler groups above it, together with certain new characteristics peculiar to itself. Hence in the Anthophyta we found all the fundamental characteristics of all the lower groups. Such a listing of characteristics affords an excellent overall view of the increasing complexity of organisms when comparing types that were present early in the period of time encompassing life on earth with those that were present more recently. Now that various groups within the plant kingdom have been covered, we can discuss more readily some of the specific modifications and adaptions that form the basis for our concept of evolution.

The second approach is more general, in that we select an advantageous feature, such as the protection of reproductive parts, and examine the various types of organisms with an attempt to ascertain with what degree of success they have accomplished the task.

Increasing Complexity in the Plant Kingdom

Remember that the characteristics that are listed occur in most organisms within the group but not necessarily in all. Also, some groups have a range of organization from simple to complex within the group itself.

I. Schizophyta (bacteria). These are considered to be the most primitive of living plants, and they have certain traits that are characteristic of plants in general:

1. Cytoplasmic membranes and cell walls.
2. Growth and cell division.
3. Metabolism.

II. Chlorophyta (green algae). Actually, most of the algal groups have the new characteristics enumerated below in addition to the three traits above. The green algae are utilized as the type organism because they are more closely related to the flowering plants.

4. Chlorophyll and associated pigments in chloroplasts, photosynthesis.
5. Multicellular organization, a nucleus in each cell.
6. Sexual reproduction, meiosis, fusion of gametes (syngamy).
7. Alternation of phases, diploid and haploid.

III. Bryophyta (moss and allies). Even though these plants are considered to be a side branch of the plant kingdom, they do indicate a position more advanced than the algae and less complex than the ferns. These are the simplest of the land plants and have the following features:

8. Epidermis with a cuticle and stomata in the sporophyte.
9. Multicellular sex organs, antheridia and archegonia.
10. An embryonic stage that is protected by tissues of the preceding generation.

IV. Pterophyta (ferns). The additional features found in the ferns are significant for a terrestrial existence:

11. Independent sporophyte.
12. Leaves, stems, and roots as part of the sporophyte.
13. Vascular tissue.

V. Coniferophyta (pines and associated forms). These plants add to the features that are advantageous to existence on land:

14. Heterospory and the associated two types of gametophytes.
15. Very large sporophyte and dependent gametophytes.
16. Pollination.
17. Seeds.

VI. Anthophyta (flowering plants). These plants possess all the traits mentioned above, plus others that indicate that the flowering plants are truly the most complex plants:

18. Vessels.
19. Ovary, which matures as the fruit.
20. Flower.

The list of fundamental features could be expanded, depending upon the wishes of the individual making a similar list. All the specific characteristics of the various plant groups were not utilized, nor was every plant group afforded a position in the listing. For example, we could quite logically list "heterospory" under the Lycophyta and then follow with the conifers. However, the principle is the same, and a simplified scheme was utilized.

General Evolutionary Trends

After the discussion of plant groups in preceding chapters and the listing of specific features of plant groups to indicate evolutionary accumulation or complexity in the foregoing pages, we can now indicate how certain problems of existance were solved (or partially solved) by different plant groups.

Protection of reproductive parts. Sexual reproduction is exceedingly important to an evolutionary process because of the variations that are produced in the offspring. In considering the life cycle of any plant, one discovers that the gametes are the most sensitive or the least resistant structures with regard to environmental factors. These sex cells never develop thick walls. To do so would be very disadvantageous (why?), and such plants would soon die out; they would not be able to compete successfully with other plants. The gametes, whether isogametes or egg or sperm, are bounded only by a thin membrane but by no other protective structure. As a result, their survival would be enhanced if they were retained within a gametangium (or similar structure) as long as possible before fertilization, or if fertilization took place within such a structure. If necessary, the student should review the various plant groups discussed in the rest of this section.

I. CHLOROPHYTA

The green algae are utilized because they probably gave rise to the complex land plants.

1. *Ulothrix.* This plant is selected as one of the simplest multicellular organisms exhibiting sexual reproduction. Motile isogametes are liberated into the surrounding water. No protection exists prior to and at fertilization (syngamy), except for the brief period during which the gametes are developing in the gametangium. The isogametes and the resulting zygote are at the mercy of the environment.

2. *Oedogonium.* In this plant one discovers oögamy. Only the sperm are really exposed to the environment as they swim to the egg, which is retained within the wall of the oögonium. Not only is the egg somewhat protected by a cell wall, but it

also contains a considerable amount of stored food. As a result, the zygote is adequately nourished as well as protected by the oögonium wall; fertilization occurs within the oögonium.

II. BRYOPHYTA

3. The moss plant is one of the simplest in which multicellular sex organs, the antheridia and archegonia, are found. The importance of such sex organs is that the gametes are protected by a layer of cells rather than a single cell wall, as in *Oedogonium*. The sperm are, of course, still liberated and swim to the egg, but the latter remains within its protective jacket of cells. The zygote and resulting embryo are also sheltered for a time.

III. PTEROPHYTA

4. The ferns are quite similar to the mosses, except that the base of the archegonium is embedded within the vegetative portion of the gametophyte in the former. The egg is contained within this basal portion, and the additional tissues are undoubtedly beneficial to its chances of survival.

IV. CONIFEROPHYTA

5. In pines, the development of ovules and the seed habit resulted in tremendous advantages to the gametes and to the progeny. The gametes are protected by the gametophyte, and the reduced gametophytes are in turn protected by the tissues of the sporophyte. The female gametophyte and its eggs are never really exposed. The male gametophyte, but not the male gamete, is exposed for a short time but is protected by a rather thick wall. The sperm is actually brought to the egg; the male gamete no longer swims to the egg. In addition to this protection of the gametes, the young plant (embryo) is sheltered within the seed. Dormancy also becomes possible with seed production, and this is again an advantageous condition for the new generation.

V. ANTHOPHYTA

In the flowering plants one finds a situation similar to that in pines, plus the added protective feature of ovules enclosed within ovaries. The gametophytes are further reduced and protected. Many fruits not only afford additional protective layers but also aid in dispersing the seed.

Increased diversity. The simplest plants, the types that were first present on earth, are unicellular and exhibit little diversity.[5] The plants that evolved later show increasing diversity, a multicellularity resulting in specialization and a division of labor. Without such specialization large land plants would not evolve. These plants require supporting tissues and conducting tissues, as well as others. One type of cell or one type of tissue alone could not possibly be sufficient for

[5] This is always a debatable point. *Euglena* (of the Euglenophyta) has considerable diversity as compared with *Staphylococcus* (of the Schizophyta), even though both are unicellular. However, our simplified discussion is still valid. We must merely decide where to draw the line in the comparisons. We could just as easily discuss evolution within a group, such as a single division or class, as in the broader view of a kingdom.

these diverse functions. Compare, for example, the conducting and supporting cells of the xylem with the food manufacturing cells of the leaf mesophyll. Elongated cells with thick walls are essential in a supporting function, whereas such structures would be detrimental to cells carrying on photosynthesis. Gaseous exchange in the latter, for one thing, makes it imperative that the cells be thin-walled and that there be much intercellular space, a condition efficiently handled by the leaf mesophyll. The most complex plants, then, are multicellular and furthermore consist of specialized types of cells that perform certain functions and not others; the summation of the interrelated functions of the various tissues is the resultant whole plant. This specialization results in the efficient performance of the various functions of the plant, including reproduction.

I. Chlorophyta

1. *Chlamydomonas.* This plant is a single-cell type in which no great specialization is possible. Relatively few gametes and spores are produced.

2. *Ulothrix* and *Oedogonium.* These are multicellular plants having both vegetative and reproductive cells. The plants are larger than *Chlamydomonas;* many more gametes and offspring are produced, and this provides more material for natural selection and evolution. No longer does one type of cell carry on all functions of the organism. The sporophyte is unicellular, however.

II. Bryophyta

3. The mosses produce a fairly large multicellular gametophyte and also a more specialized sporophyte. Both these structures have localized areas devoted to the production of gametes or spores, providing large numbers of offspring and the possibility of variations. Also, the central cells of the gametophyte are slightly modified as conducting cells. This increases the efficiency of food and water transport, without which a small size or a water habitat is mandatory.

III. Pterophyta

4. In ferns the large sporophyte with its vascular tissue and many meiospores results in a dispersal of the species to various environments. Spores are more resistant than gametes and are better able to withstand the difficulties inherent to dissemination. Even if large gametophytes with vascular tissue had evolved in some groups, they probably would not have been nearly as successful as those plants in which diversification was emphasized in the sporophyte. Natural selection would undoubtedly favor the sporophyte development because adverse environmental conditions would be more effective against the rather delicate and susceptible gametes than against the rather resistant meiospores.

IV. Coniferophyta

5. In pines the exceedingly large sporophyte results in the production of enormous numbers of meiospores, and only the microspore (pollen) is exposed. The differentiation of separate male and female gametophytes provides a greater efficiency of gamete formation. The emphasis upon sexual reproduction provides additional material for variations in offspring. A well-developed root system and

vascular tissue are also advantageous to the survival of the sporophyte, which in turn protects the gametophytes.

V. ANTHOPHYTA

6. The flowering plants provide highly efficient mechanisms of pollination, and sexual reproduction is more certain. The leaf structure is very effective in exposing photosynthetic cells to those factors that are essential to food manufacture. Vascular tissue in these plants is more efficient than in the pines as a result of the development of vessels in the xylem. The flowering plants are thus the most successful of plants, and one finds that they inhabit almost every environment of the earth.

Retention of water. This is a problem restricted to land plants. Without some means of water retention, land plants could not exist; water would evaporate, and the individual cells would rapidly die.

I. BRYOPHYTA

1. Moss plants can exist on land because of the cuticle and cuticularized meiospores that have evolved. The cuticle greatly retards water loss from exposed surfaces. The lack of organized vascular tissue and the necessity of a film of water for fertilization (why?) prevented the mosses from developing to any large size.

II. PTEROPHYTA

2. The development of vascular tissue, well demonstrated in the ferns, was the big evolutionary step that made possible the existence of large land plants. Although the cuticle retards water loss, it does not stop transpiration completely, especially with the presence of stomata in the epidermis. Diffusion of water (osmosis) through parenchyma-type cells is far too slow to supply aboveground portions of a plant with water, except in the case of small plants such as the mosses.

III. CONIFEROPHYTA AND ANTHOPHYTA

3. These land plants have various modifications and adaptive structures that make existence possible in rather dry areas: thick cuticles, sunken stomata, water storage tissues, reduced leaf surfaces and so forth.

Summary

1. Evolution means descent with change and implies that all organisms are more or less related to each other. All organisms are descended from past organisms. This concept is based upon much evidence of various sorts.

2. The vast physical changes occurring during the earth's history have resulted in so many and such radical environmental changes that organic evolution must have occurred for living organisms to have survived.

3. The discovery of fossils, traces of former living organisms, emphasizes the

general sequence of evolutionary change. The older fossils are basically of much more simple structure than are the more recent fossils or the dominant organisms of the present.

4. The wide distribution of some plants, the restricted distribution of others, and the different kinds of plants growing in similar environments that are widely separated all suggest that variations are haphazard phenomena that are acted upon selectively by the environment.

5. Basic similarities and differences between plant groups are found by a study of embryological development and chemical patterns in living organisms. Recently, genetical studies have shown that new species may arise as hybrids between existing species or as a result of the multiplication of the original basic chromosome number.

6. Lamarck's ideas concerning evolution were based upon the inheritance of characteristics that were gained or lost by use or disuse. Darwin's observations led him to suggest that of the variants that arose, those that survived were the ones best fitted to the environment. This did not explain the source of variations. De Vries suggested that variations arose as sudden unpredictable changes in existing genes. Biologists basically agree that mutations bring about changes and that these changes, and variations in general, are perpetuated and expanded by sexual reproduction. The complex plants and animals of the present are descended from much simpler organisms that existed ages ago, some of which may still exist.

7. Optimum conditions for evolution appear to be small, semi-isolated populations. Such populations are isolated enough to evolve independently and differently and yet close enough for occasional interbreeding to occur.

8. In examining the various groups of the plant kingdom, one is impressed with the increasing complexity of those plants in the higher groups, which are of relatively recent origin, as compared to those plants of ancient origin. Three evolutionary trends of great importance that have reached their highest evolution in the flowering plants are protection of reproductive parts, diversity, and retention of water.

Review Topics and Questions

1. Discuss what is meant by the term *evolution*. Indicate at least five kinds of evidence supporting the concept of evolution.
2. Compare Lamarckian and Darwinian evolution. Discuss the kind of evidence that tends to refute some of the ideas of Lamarck. What suggestions by Lamarck are acceptable to our modern concept of evolution?
3. What was the contribution of de Vries to the concept of evolution? In what way did his ideas differ from those of Darwin? In what ways did they agree?
4. What is meant by "retrogressive evolution"? Give an example.
5. Discuss why semi-isolated, but not completely isolated, populations afford optimum conditions for evolution. What is the significance of "isolation"?
6. Why is "plant breeding" sometimes called "manmade evolution"?
7. Compile a descriptive list showing how the haploid meiospores, produced as a

result of meiosis, are afforded greater and greater protection with the increasing complexity of plant groups. Starting with the simplest, list each plant group and describe the structures that afford protection to the spores.

8. Discuss the causes of evolution.

9. What are fossils? Describe the various kinds of fossils. Why is the fossil record incomplete? (In other words, why are some groups not found as fossils? Why isn't the record of ancient relationships more exact? Why aren't we certain, from examing fossils, which organisms are descended from which primitive ones?)

10. What is the significance of sexual reproduction as far as evolution is concerned?

11. Since the bacteria are considered to be extremely primitive organisms, why have they not been replaced by more complex groups?

12. Bearing in mind the general similarities among the life cycles of mosses, ferns, and flowering plants, describe (with the aid of word diagrams, labeled diagrams, or general descriptions) the evolutionary trends shown in the following: (a) relative sizes of the gametophyte and sporophyte phases, (b) dependence and independence of the two phases, and (c) dependence of the sexual process on water.

Suggested Readings

DE CAMP, L. S., "The End of the Monkey War," *Scientific American* **220,** 15–21 (February 1969).

DOBZHANSKI, T., "The Genetic Basis of Evolution," *Scientific American* **182,** 32–41 (January 1950).

EISLEY, L. C., "Charles Darwin," *Scientific American* **194,** 62–72 (February 1956).

KETTLEWELL, H. B. D., "Darwin's Missing Evidence," *Scientific American* **200,** 48–53 (March 1959).

LACK, D., "Darwin's Finches," *Scientific American* **188,** 66–72 (April 1953).

RYAN, F. J., "Evolution Observed," *Scientific American* **189,** 78–83 (October 1953).

STEBBINS, G. L., *Processes of Organic Evolution,* 2nd ed. (Englewood Cliffs, New Jersey, Prentice-Hall, 1971).

29
Growth, Development, Flowering, and Plant Movements

29.1 Growth

The term *growth* refers to two basic factors of a plant's development: (1) an increase in the number of cells as a result of cell division, and (2) an irreversible enlargement and differentiation of existing cells accompanied by an increase of cellular components. No sharp distinction exists between these phases of growth, but to consider them separately is useful. Remember that when a cell divides, even if the volume of the two resultant cells does not total more than that of the original, some increase in cellular components always occurs. The new cell wall that forms during cell division and the duplicated chromosomes that are produced are examples of additional components that were not present in the original cell. However, the greatest increase of cell material occurs during the enlargement and maturation of the cell.

Meristematic tissues, cell division, and general areas of cell enlargement have already been discussed in previous chapters. In those discussions growth was considered on the basis of an increased length and, sometimes, increased diameter of the plant as a result of tissue development. It was also stated that for growth to occur, various factors are required: food as a source of energy and a source of basic materials for synthesizing cellular components, minerals, water, a suitable temperature, and usually oxygen. In addition to these, investigations over the last 40 to 60 years have shown that growth regulators are also necessary.

29.2 Growth Regulators

The term **growth regulators** is used in referring to those organic compounds, other than nutrients, present in minute quantities, that promote, inhibit, or otherwise modify growth, development, or differentiation of plants. A **hormone** is a growth regulator that is produced in one part of the organism and elicits a response at some other location in that organism. In other words, the hormone that is essential for growth is produced in one area, while the actual growth response occurs elsewhere in that plant. In animals hormones are produced in and secreted from discrete glandular[1] structures, but such glands are not present in plants. The term growth regulator can include both the natural (endogenous) and the synthetic compounds that have been found to modify plant growth.

The **auxins** constitute an important class of plant growth hormones, and the principal naturally occurring one that has been extracted from plant material is **indole-3-acetic acid,** or **IAA.** All auxins are basically similar, and they are all organic acids. Although not all have been examined, a sufficiently great variety of plants have been found to contain auxins so that botanists agree that auxins are probably universally present, with the principal centers of auxin synthesis being apical meristems, buds, young leaves, and young flowers.

One of the characteristics of growth regulators is that they exert regulatory effects upon metabolism when present in only minute quantities. This makes very difficult the extraction of sufficient material for chemical analyses, and the existence of various hormones is frequently inferred from the occurrence of various physiological reactions rather than from the isolation of a specific compound. More information has been gathered with regard to auxins than with regard to other hormones, and the former will be discussed in some detail in the next section.

If plant tissues are cut, various parenchyma cells at the site of injury usually become meristematic, forming a **callus** (a mass of undifferentiated parenchyma cells) from which a cork cambium develops in the outer layers and eventually cork is produced. If the freshly cut tissue is immediately rinsed with water, very few cell divisions occur. If some tissue is removed, ground into a paste, and smeared onto the rinsed (cut) surface, cell divisions do occur. Some substance diffuses into the intact cells and stimulates relatively differentiated cells to divide again. In at least one case, a compound called **traumatic acid,** which acts like a **wound hormone,** has been extracted. Several such wound hormones may exist, but additional investigation is necessary to clarify the situation.

In recent years **cytokinins** (or **phytokinins**) and **gibberellins** have been undergoing considerable investigations as additional plant growth regulators. The former, of which **kinetin** and **zeatin** have been identified, appear to be general growth regulators. Cytokinins have been obtained from yeast, coconut milk, apple fruit extracts, and many other plant materials. They may be important in maintaining protein contents in the plant although their primary influence is the stimulation of cell division; the name is derived from *cytokinesis.* The gibberellins were first observed in connection with the bakanae disease of rice in Japan. *Gibberella fujikuroi,* the causal fungus, produces several gibberellins, of which the most familiar is **gibberellic acid.** The best-known effect of gibberellin is increasing the elongation of stems. In fact, the Japanese investigators noted that diseased seedlings grew unusually tall before they died and referred to such as "foolish seedlings" (bakanae). Before any specific substances were isolated, it was known that the fungus could be grown in a liquid culture medium, and that such liquid contained fungal secretions that induced symptoms of the disease when sprayed on healthy plants. Eventually, at least six gibberellins were isolated from such cultures, and

---

[1] A gland is an organ of secretion, such as the adrenal, thyroid, and pituitary glands in humans.

532

Ch. 29 / Growth,
Development,
Flowering,
and Plant
Movements

they are usually referred to as GA_1, GA_2, GA_3, etc. Procedures were quickly developed for producing large quantities of gibberellins (**GA**), which could then be used to investigate various physiological effects of this class of growth regulators. It soon became apparent that gibberellins are found in various plant parts of many kinds of plants, and that they are, indeed, naturally occurring plant growth regulators. Subsequent investigations indicated that the gibberellins have a multiplicity of effects on flowering, germination, and dormancy. These various facets will be discussed in subsequent sections.

The growth and differentiation of a plant appear to be under the influence of delicate chemical balances maintained by the interaction between growth regulators. The situation is not at all clear, possibly because there *is* such an interaction. It is not certain whether an applied growth regulator brings about an observed response as a result of its own activity or because the basic balance is shifted. Many investigators are convinced that there are a variety of growth regulators, and that more and more of them will be identified as techniques become more sophisticated. At present, our information is pretty well limited to the three types just mentioned: auxins, gibberellins, and cytokinins (Figure 29-1). However, although much less is known about them, two other growth regulators (ethylene and abscisic acid) will also be discussed.

29.3 Auxins

As is true so frequently, the presence of a growth factor or a hormone was postulated before auxin was isolated, just as the vitamins in the human diet were known to be essential before any of them had been isolated. The presence of such ma-

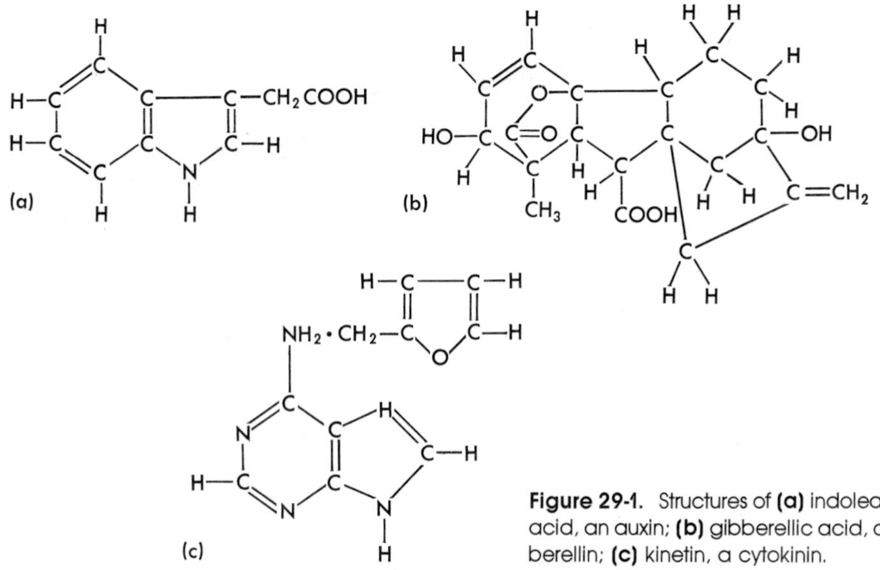

Figure 29-1. Structures of **(a)** indoleacetic acid, an auxin; **(b)** gibberellic acid, a gibberellin; **(c)** kinetin, a cytokinin.

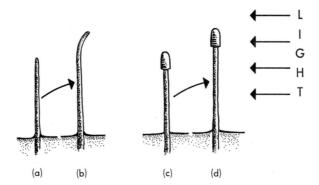

Figure 29-2. Darwin's experiment. **(a), (b)** Control seedlings bend toward a light source. **(c), (d)** Seedlings with lightproof caps over the tips do not bend.

terials is inferred from the physiological reactions that occur in the presence, or absence, of certain tissues or extracts. Darwin, as early as 1880, had shown that seedlings bent toward the light only if the tips were exposed and that such bending resulted from greater growth on the nonilluminated side of the stalk. He exposed grass seedlings to a unilateral source of light and noted that they curved toward the light source. When he covered the tip of the seedlings with lightproof caps, no bending took place. The control seedlings, those without coverings, responded as before (Figure 29-2). This indicated that some influence must be transmitted from the tip downward. Although no further explanations were forthcoming until 1910 and later, in the light of our *present* understanding of hormones the work of Darwin would imply the presence of hormone—a material produced in the tip of the seedling that elicits a response (curvature, or differential elongation) in the stem below. Such a material would fit well into our present definition of a hormone.

Boysen-Jensen's investigations (1910–1913) stimulated other botanists to explore further this field of plant curvature toward a source of light.[2] He decapitated the seedling, placed a bit of gelatin on the stump, replaced the tip on the gelatin, and found that curvature toward light ensued just as with intact seedlings (Figure 29-3). The substance diffused from the tip through the gelatin to the stump. If the tip is not replaced, curvature does not result.

Most of the information about plant hormones has been obtained from experiments using the coleoptile of grasses, chiefly oats (*Avena*). The **coleoptile** consists of a leaflike, cylindrical structure with a conical top, which encloses the developing leaves of the germinating seed. Early in the development of the seedling, the parenchyma cells of the coleoptile stop dividing, and all subsequent growth is by cell elongation. The coleoptile serves as an excellent organ for the study of the cell-enlargement phase of growth.

Paal (1918) demonstrated that if a freshly severed coleoptile tip is replaced on the stump eccentrically, more growth results on that side, even though this is done without illumination (Figure 29-4). All of the experiments just cited were in agreement with the idea that a growth-promoting substance was produced in the co-

---

[2] The layman "explains" the bending of a plant toward a source of light by stating that the plant is seeking the light it requires for food manufacture (photosynthesis). This is, of course, not an explanation. No plant consciously seeks anything.

534

Ch. 29 / Growth,
Development,
Flowering,
and Plant
Movements

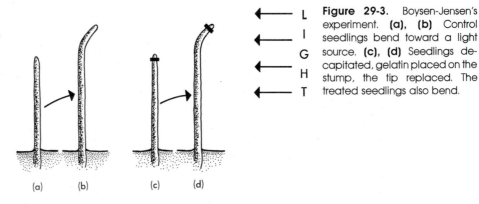

Figure 29-3. Boysen-Jensen's experiment. **(a), (b)** Control seedlings bend toward a light source. **(c), (d)** Seedlings decapitated, gelatin placed on the stump, the tip replaced. The treated seedlings also bend.

leoptile tip and diffused to the cells below, where it stimulated normal growth even in the dark. Light was not necessary for its production, although unilateral light might cause the substance to move toward the dark side. This would then give a response similar to that found by Paal, where cells elongated only on the side receiving a supply of the growth-promoting substance from the eccentrically placed coleoptile tip.

Went was especially impressed by the work of Boysen-Jensen and Paal and was convinced that a growth hormone must be involved. He devised experiments that combined aspects of these other workers as well as additions of his own. A number of technical difficulties had to be solved, but eventually Went showed in 1928 that a growth substance diffused into agar[3] if excised tips are placed thereon for a short time. He then cut the agar into small blocks, which, when placed eccentrically on decapitated coleoptiles in the dark, caused curvatures of the coleoptiles; they could be used in place of the severed tips that Paal had used. Controls utiliz-

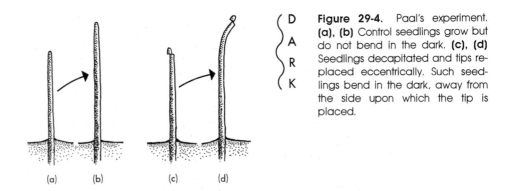

Figure 29-4. Paal's experiment. **(a), (b)** Control seedlings grow but do not bend in the dark. **(c), (d)** Seedlings decapitated and tips replaced eccentrically. Such seedlings bend in the dark, away from the side upon which the tip is placed.

[3] Agar is a polysaccharide that attains a gelatinlike consistency when dissolved in hot water and allowed to cool.

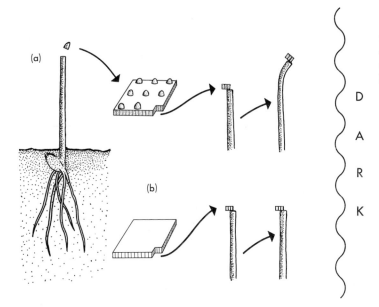

(a)

(b)

D
A
R
K

Figure 29-5. Went's experiment. **(a)** Coleoptile tips are placed on agar, and later small blocks of this agar are placed on one side of cut coleoptiles; bending of the coleoptiles results. **(b)** Plain agar blocks do not elicit a bending response.

ing plain agar, elicited no response (Figure 29-5). Also, during his investigations, Went discovered that this growth regulator moved only from the tip (or apex) toward the base of coleoptile tissues even if the segments were inverted. The mechanism responsible for such *polar transport* is not known, but it is an attribute of living cells. If the tissues are under anaerobic conditions or are treated with respiratory inhibitors, polar transport is negated.

One of the most important aspects of this work by Went was his demonstration that the degree of curvature was proportional, within limits, to the concentration of the hormone in the agar blocks. For example, if two coleoptile tips were placed upon a block of agar, the curvature was twice that caused by a similar block of agar upon which only one tip had been placed. This presented a *quantitative* test for auxin, and it is still the basis for determining the concentration of these growth substances in tissues. No chemical test is sensitive enough to determine the presence of plant growth hormones in the minute amounts that elicit responses. Thus, the biological techniques must be utilized. The term **biological assay** (or **bioassay**) refers to this type of analysis where the response of living material is used to test the effect of biologically active substances. Actually, not until 1934 was this plant growth hormone isolated and found to be indole-3-acetic acid (IAA). Since then, it has been possible to use the bioassay to determine the amount of IAA in a tissue by comparing the curvature brought about by an extract of that tissue with the degree of curvature brought about by a known quantity of indoleacetic acid under standard (controlled) conditions of temperature and humidity. It has been calculated that 1 mg of IAA is capable of bringing about a 10-degree curvature in 50 million decapitated coleoptiles. This number would make a row of coleoptiles, standing closely side by side, nearly 50 mi in length. A similar 10-degree curvature in one coleoptile would be brought about by approximately 2×10^{-11} gm of IAA.

535

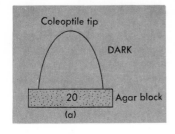

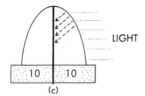

(c)

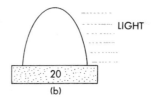

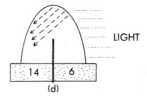

Figure 29-6. The lateral displacement of IAA in unilaterally illuminated coleoptile tips. Numbers on the agar blocks represent the degree of curvature when those blocks are used in the standard *Avena* coleoptile curvature test.

29.4 Phototropism

The investigations mentioned in the previous section really resulted because botanists were curious as to why a plant bent toward a source of light. The cells on the nonilluminated side were soon shown to elongate more than those on the illuminated side. Auxins stimulate such elongations. Went was able to show that unilateral illumination of coleoptile tips leads to a transport of auxin from the light side to the darker side, and that light did not cause the destruction of auxin. As shown in Figure 29-6, the same amount of auxin is collected in agar blocks beneath excised coleoptile tips regardless of whether they are kept in the dark or illuminated (**a** and **b**). If the coleoptile tips are split and the two halves separated by a thin piece of glass (or similar material to prevent diffusion between the halves), about the same amount of auxin is collected from the illuminated and darkened halves (**c**). However, when the two halves are only partially separated at the base (**d**), more auxin diffuses from the darkened half than from the illuminated half. Such results indicate that unilateral light induces a lateral movement of IAA toward the darkened side. Note that the total amount of auxin collected from the coleoptile tips was about the same in spite of different treatments. Thus, light did not cause destruction of auxin.

The curvature of a stem toward a source of light results from the differential distribution of auxin; the higher concentration of auxin on the darker side stimulates greater elongation, and the stem bends toward the other side. Growth responses resulting from external stimuli are termed **tropisms.** Since the stimulus in this case is light, the term used is **phototropism** (photo = light). Considerable evidence supports the suggestion that light stimulates a transverse, or horizontal, diffusion of auxin toward the darker side. The mechanisms responsible for this movement of IAA are not known.

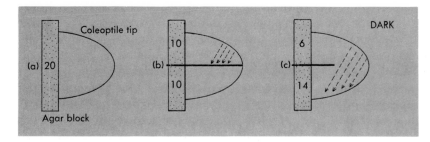

Figure 29-7. The lateral displacement of IAA in horizontally oriented coleoptile tips. Numbers on the agar blocks represent the degree of curvature when those blocks are used in the standard *Avena* coleoptile curvature test.

29.5 Geotropism

For a great many years, biologists had observed that when a plant is placed in a horizontal position, the stem curves upward, the shoot eventually resuming its normal upright position. Once phototropism was explained on the basis of auxin concentrations, **geotropism,** the growth response caused by gravitational stimulus, was soon shown to result from a similar asymmetrical distribution of auxins in the plant. The total amount of auxin diffusing into an agar block from a horizontally oriented coleoptile tip is the same as that from a vertically placed tip (see Figures 29-6[a] and 29-7[a]). If the coleoptile is divided into completely separated halves, similar quantities of IAA move into the two agar blocks (Figure 29-7[b]). However, if the coleoptile is partially split into an upper and a lower half, more IAA diffuses out of the lower half (Figure 29-7[c]). Thus, considerably more auxin accumulates in the lower side as compared with the upper side of a coleoptile or stem that has been placed in a horizontal position. This stimulates a greater elongation of those cells on the lower side, and the coleoptile or stem tends to grow upward to an erect position. The lateral movement of radioactive auxin in such horizontal stems has been detected.

The foregoing explanation is satisfactory in explaining the negatively geotropic response of shoots; it is not sufficient for an understanding of the positively geotropic response of roots in bending toward the source of stimulus (down, in this case). Fortunately, the work of Thimann (Figure 29-8) demonstrated that shoots

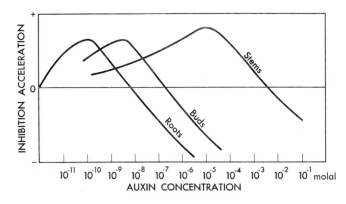

Figure 29-8. The inhibition and growth promotion of different organs as a function of auxin concentration. (From *The Action of Hormones in Plants and Invertebrates*, by K. V. Thimann, New York, Academic Press, Inc., 1952.)

538

Ch. 29 / Growth,
Development,
Flowering,
and Plant
Movements

are accelerated in their growth by concentrations of auxin that inhibit root growth; the zone of elongation of the root is more sensitive to auxin than the similar region of the stem. Auxins accumulate in the underside of both roots and shoots placed in a horizontal position. While this induces a more rapid growth of the underside of stems, the concentration of auxin is such as to inhibit growth of the underside of roots. In roots the cells on the upper side are not inhibited and elongate more than the lower ones; curvature is downward. As a result, stems tend to grow up and roots tend to grow down, both as a result of asymmetrical distribution of auxin. Again, the mechanism responsible for such movements of IAA are not known.

29.6 Mechanism of Auxin Action

The manner in which auxin brings about cell elongation has not yet been clarified. However, investigations have suggested various possibilities. IAA has been shown to increase the plasticity of the cell walls of treated coleoptiles, and this is quantitatively similar to the effect of this auxin upon the growth of the coleoptile. With this increase in plasticity of the walls, cells stretch more easily in response to turgor pressure that develops as water is absorbed. Changes in plasticity of cell walls probably involve reactions catalyzed by enzymes, and IAA does stimulate RNA synthesis. The suggestion is that IAA acts upon DNA to influence the production of the mRNA, which codes for the synthesis of enzymes that make the cell walls expand easier. This possible effect of auxin on enzyme synthesis is undergoing considerable investigation. One major difficulty with this hypothesis of auxin action is that the elongation of a cell as a response to IAA treatment occurs so rapidly that there does not appear to be sufficient time for enzyme synthesis and so forth to take place. Localized membrane effects are also being studied because such changes may bring about very rapid responses. It is quite possible, of course, that auxins may bring about their effects by more than one mechanism.

Growth stimulation by auxins requires metabolic energy, as can be shown by the inhibition due to respiratory poisons. Recent evidence indicates that IAA greatly increases oxygen uptake due to oxidative phosphorylation in isolated mitochondria of corn (maize). The formation of citric acid from oxaloacetic acid and acetyl coenzyme A in the mitochondria is more effective in the presence of IAA. If this proves to be true of other auxins and with other plants, growth stimulation could result from an increased energy supply (ATP synthesis).

It has been shown that auxin applications stimulate ethylene production from plants, and that some of the responses obtained with auxins can be duplicated by exposing the plants to ethylene. Therefore, it has been suggested that responses attributed to auxin may actually be due to ethylene. This also is still under investigation. (See also Section 29.9.)

Auxins have many effects upon plant growth and development. The manner in which they bring about such effects is only now being elucidated, and it is not yet clear what the specific mechanism is or whether several mechanisms are involved. In any event, the usual sites of auxin formation in vascular plants are meristems and enlarging tissues (e.g., the apical meristem, and young expanding leaves). Active polar transport is then involved in moving the auxin from these areas to sites where growth responses are elucidated.

29.7 Gibberellins

These compounds are more complex than the auxins (Figure 29-1). Various of the gibberellins have been found in diverse plant groups, including flowering plants, conifers, ferns, and *Fucus* (one of the brown algae, Phaeophyta), and they are probably of universal occurrence.

Gibberellin will not cause curvature in the *Avena* coleoptile as auxin will. Lateral transport of the former occurs very rapidly, and the compound is rather uniformly distributed in the coleoptile; elongation, but not curvature, will result. This is one way of differentiating auxins from gibberellins. The basic test for the latter utilizes dwarf maize (corn). These plants are very stunted in growth as compared with normal maize plants. Increasing concentrations of gibberellin applied to the dwarf maize result in increasing growth responses as a result of cell enlargements and division; $\frac{1}{1000}$ of a microgram (1 microgram = one millionth of a gram) of gibberellin can be detected in this manner. These maize plants do not respond to auxin treatments, and normal maize plants do not respond to gibberellin treatment. Similar responses have been found in other dwarf mutant plants (e.g., peas and beans). It is likely that such plants lack gibberellin, caused possibly by the lack of an enzyme involved in gibberellin synthesis, and treatment with this growth regulator produces a normal plant by supplying a material that was lacking.

Gibberellins have specific effects different from those of auxins, but these effects also may result from influences upon the formation of specific mRNA and enzyme molecules. In germinating barley grain, as well as other grain, gibberellin released from the embryo induces an increased production of α-amylase (an enzyme that hydrolyzes starch) and proteases (protein-digesting enzymes) in the outermost cells of the endosperm (the aleurone layer). Using various inhibitors, it was indicated that the gibberellin acts by causing formerly repressed genes to become active, resulting in new mRNA and then new enzymes. These enzymes are responsible for digesting and releasing storage products of the endosperm, which are then available to the developing embryo. This action of gibberellin (influencing mRNA synthesis) is similar to the way in which auxin is thought to act. Once again the suggestions are stimulating, but more research must be done to clarify the way in which growth regulators might act.

Gibberellins and auxins frequently bring about similar responses in plants. These and other growth regulator responses will be discussed in subsequent sections. The main sources of GA are the leaf primordia in the apex, rather than the meristem itself, and the root system. GA biosynthesis also occurs in the embryo, fruits, and seeds. Translocation is without polarity and takes place quite readily throughout the plant. In fact, the ease with which movement occurs is probably the reason that GA does not cause *Avena* coleoptile curvature.

29.8 Cytokinins

The best-known cytokinin, **kinetin,** is more complex than the auxins but not as complex as the gibberellins (see Figure 29-1). It is primarily involved in cytokinesis although it has been shown to influence a variety of growth processes. Cell division takes place in an orderly sequence of DNA synthesis, mitosis, and then cytokinesis. There is evidence that IAA is mainly involved in the first two, and that the last step is controlled by kinetin or other cytokinins. These materials have been found in coconut milk, yeast extracts, various fruits, and actively dividing tissues; they are considered to be widely distributed in higher plants. Nine naturally occurring cytokinins have been reported in the literature, as well as at least 100 synthetic cytokinins, and most of these compounds are derivatives of adenine. **Zeatin,** which has a molecular structure similar to kinetin, has been isolated from corn (**Zea mays;** maize), and most of the cytokinin activity in coconut milk is attributed to zeatin or its derivatives.

Cytokinins promote cell expansion in many tissues, but inhibit such growth in others. They have no effect on dwarf maize (or other dwarf plants) or on coleoptile curvature. The diagnostic test for cytokinins consists of culturing tobacco pith (parenchyma) cells in a nutrient solution. If kinetin, or similar compounds, and auxin are added, an acceleration of cell division is evidenced; no response occurs with kinetin, auxin, or gibberellin alone. The initiation of leafy shoots from a callus derived from tobacco pith can be obtained by increasing the ratio of kinetin to auxin. If the kinetin-to-auxin ratio is lowered, roots are formed. The callus can be retained in an undifferentiated state provided the proper balance of auxin and kinetin is maintained, along with various nutrients. This is a beautiful example of the control of differentiation by interactions of growth regulators.

If an excised leaf is floated on water plus kinetin, it stays green, succulent, and healthy for a relatively long time. Similarly treated leaves, but without kinetin, shrivel and die rather shortly. We can even treat one portion of a leaf with kinetin and cause it to remain green and healthy, while the other portion withers and dies. This peculiar response to kinetin is called the Richmond-Lang effect. If amino acids are added to the water in which treated leaves are floating, these amino acids tend to accumulate only in the treated portion of the leaves. Protein synthesis appears to be increased by kinetin treatment, and at the same time protein degeneration is decreased. Recent work indicates that kinetin may delay protein degeneration, and thus delay senescence, by its ability to suppress formation of the enzyme protease.

Evidence obtained during the past few years has implicated cytokinins in the control of almost all phases of plant growth. In some instances at least, they interact with auxins. Cytokinins appear capable of determining the orientation of cell enlargement induced by auxins, and interact with auxins to promote either cell division or cell enlargement depending upon other physiological factors. They influence tissue differentiation and embryo development in a number of plants. Some of the growth effects caused by cytokinins are possibly a consequence of the ability of these compounds to promote the transport, accumulation, and retention of metabolites in tissues and organs; they could be important regulators of nutrient flow between plant parts, or even influence movement of other regulators (such as auxin). It has been shown that applied cytokinins are incorporated into transfer RNA. This may indicate that they influence the synthesis of certain specific tRNAs. The suggestion has been made that cytokinins, and other growth regulators, may act as gene derepressors (see Section 18.20).

29.9 Ethylene

Ethylene ($H_2C = CH_2$) is a volatile gas formed by the incomplete combustion of carbon compounds, such as coal and petroleum, and its damaging effect on plants has been observed for over 60 years. Only recently, however, has it been realized that ethylene is a natural plant growth regulator. Although the biosynthesis is not clear, various plant tissues and organs produce ethylene. The amount present is usually very small, less than 0.1 ppm, but ripening fruits synthesize rather high concentrations. In fact ethylene speeds ripening in most fruits; a ripe apple put in with nonripe ones will trigger ripening of the latter. (The "rotten" apple in the barrel?)

Ethylene seems to modify growth by inhibiting stem elongation and stimulating transverse expansion so that the stem appears swollen. It also accelerates the abscission of leaves, stems, flowers, and fruits. In general ethylene appears to increase senescence of plant organs, but much more investigation will be required to clarify the situation.

Auxin applications can stimulate ethylene production by factors of from 2 to 10, and investigations are now proceeding to ascertain whether some of the various growth responses usually attributed to auxins are, in effect, really responses to increased ethylene. Irrespective of whether this turns out to be the case or not, the mechanism of ethylene action remains of prime consideration. There is considerable evidence that ethylene brings about alterations in the amounts or effectiveness of various enzymes, and this may be accomplished through RNA-directed protein synthesis. Some investigators have suggested that ethylene acts by bringing about the synthesis of a new mRNA, which then directs the synthesis of a specific enzyme. However, some of the responses to ethylene occur too rapidly to be accounted for by the synthesis of new protein (enzyme). An ethylene-stimulated release of enzymes from a bound form has been reported. Thus, the enzymatic changes induced by ethylene may not involve enzyme biosynthesis; the effect may be to convert enzymes from an inactive (i.e., bound) to an active form. Obviously, the mechanism of action of this growth regulator is no more clear than that of the others already discussed. It is anticipated that future investigations will clarify the situation.

29.10 Abscisic Acid

The discovery of and then the extensive investigations regarding growth stimulating substances soon brought about speculation as to the possibility of growth inhibitors. It was clear that plants did not grow uniformly during the year; that

542

Ch. 29 / Growth,
Development,
Flowering,
and Plant
Movements

growth, in fact, frequently ceased during cold or dry seasons. Were growth inhibitors, as well as growth promoters, involved? Ethylene (see Section 29.9) was not really considered at first, even though many of its effects were being investigated, because it was known as an externally supplied material and not recognized as a natural growth regulator for many years. Investigations dealing with growth cessation actually included a variety of possible aspects of the problem as pursued by different groups of plant physiologists. One approach sought answers to the question of dormancy. For example, why were buds dormant during the cold of winter? The advantages of such a situation are obvious; dormant (nongrowing) buds are relatively resistant as compared with buds that are growing and producing leaves and/or flowers. What is the mechanism or effective agent, however, that causes such buds to cease growth after primordia (see Section 7.1) have developed? Other investigators sought answers to the question of senescence and leaf fall (or abscission).

By 1950, it had been demonstrated that dormant buds contained large amounts of unidentified growth inhibitors that declined as dormancy was broken. A dormancy-inducing substance, termed *dormin,* was isolated and found to induce bud dormancy when applied to woody plants growing under vegetative conditions. Its chemical structure was not identified until the mid-1960s, when it was shown to be identical to that of another substance capable of causing senescence and leaf abscission. This latter substance was originally termed *abscisin II,* but it was later agreed that abscisin II-dormin should be named **abscisic acid (ABA);** the structure is shown in Figure 29-9.

Abscisic acid is of widespread occurrence in plant tissues and usually interacts with other growth regulators in an inhibitory manner. The auxin-induced growth of *Avena* coleoptiles is negated by ABA, as is the GA-induced synthesis of hydrolytic enzymes in barley aleurone cells. These inhibitions are reversed by increasing the concentrations of IAA and GA respectively. If ABA is applied to a leaf, the treated areas yellow rapidly (i.e., senescence, as indicated by chlorophyll breakdown), an effect which is opposite to that of the cytokinins. The inhibitory aspects of ABA appear to be general and include inhibition of seed germination in addition to the other effects just mentioned.

As is the case with other growth regulators, the mechanism of ABA action is not really known. There is considerable evidence that indicates that ABA influences the nucleic acid (DNA-RNA) and protein-synthesis systems. One suggestion is that RNA synthesis is depressed because of inhibition of translation (see Section 18.20). It has also been shown that ABA brings about a decrease in both DNA and RNA levels, and it has been suggested that this may result from increased nuclease activity (i.e., increased activity of the enzymes involved in the breakdown of nucleic acids). As with the other growth regulators, the rapidity of some plant

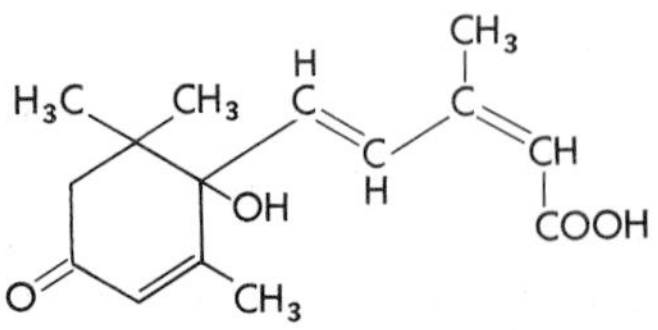

Figure 29-9. Abscisic acid.

responses to ABA raises questions about whether alterations in the nucleic-acid-directed protein synthesis system are involved. It is hoped that future research will clarify the situation.

29.11 Effects Attributed to Growth Regulators

Most of the discussion thus far has concerned the influence of various growth regulators individually. It has become quite evident that great ramifications and interactions are involved as regards processes influenced by IAA, GA, CK, ethylene, and ABA. In addition to this, many chemicals that do not occur in plants have been found to act as growth regulators. As a result, commercial preparations of plant growth regulators frequently are not naturally occurring products but compounds that are more active or that may be obtained more readily and with less expense than the naturally occurring materials. The following discussion will indicate some of the interrelationships, overall effects, and the usefulness of manipulating growth regulator concentrations by spray programs.

Growth Inhibition and Dormancy

Almost everyone knows that removing the terminal bud of a shoot results in a bushier plant. This is commonly done with such garden flowers as chrysanthemums. The bushier appearance results from the growth of lateral buds, which are normally inhibited by the presence of a terminal bud. This influence of a terminal bud in suppressing the growth of lateral buds is termed **apical dominance.** If the excised terminal bud is replaced by agar blocks containing auxin, the lateral buds will not develop. If the auxin-containing blocks are removed, the dormant axillary buds start to grow; controls of plain agar do not inhibit such growth. Although the mechanism of lateral bud inhibition is not clear, auxin produced in the terminal bud is certainly involved. However, cytokinin applications can remove the auxin inhibition; and gibberellin stimulates lateral bud growth in some plants and enhances apical dominance in others. Ethylene has been shown to suppress lateral bud growth (an effect relieved by CK), and it is possible that IAA may function in part by influencing ethylene biosynthesis. The most likely interpretation is that bud growth or inhibition depends upon a balance between the various growth regulators.

In the previous discussion of seed dormancy (see Section 27.9) it was mentioned that one of the factors frequently involved in this phenomenon is the presence of an inhibitor. The abscisic acid contents of dormant seeds have been found to be higher than nondormant seeds, and applications of ABA have been shown to prevent seed germination. Although other inhibitors may be present, ABA is the only one for which data have been obtained. Auxin treatments have no effect upon seeds but gibberellins, cytokinins, and ethylene have all been shown to break dormancy and stimulate seed germination. Thus, one again discovers an in-

544

Ch. 29 / Growth,
Development,
Flowering,
and Plant
Movements

teraction among growth regulators. The various concentrations and ratio of amounts appear to be the important factors in determining plant response.

Natural periods of dormancy are associated with high contents of inhibitors in buds and ABA again is the principal growth regulator responsible. The environmental signal that induces dormancy is the short photoperiods that occur in late summer, and many trees can be brought out of dormancy by long photoperiods. It is most common, however, for buds to be brought out of dormancy by exposure to low temperatures; ordinarily this would occur during the winter, and the tree then "leafs out" in the spring. Dormant buds are much more resistant structures than are young leaves and stems; the latter would be damaged quite readily by low winter temperatures. The period of dormancy results in such young leaves and stems not being produced until after the winter has passed. Those plants, where such a system happened to evolve, would have an advantage and tend to survive more readily in cold seasonal climates.

Application of ABA induces bud dormancy and this can be reversed by the addition of GA or CK. Ethylene has also been reported to break dormancy. Phloem sap, obtained by allowing aphids to feed on the plant (see Section 8.1), contains significantly more ABA under short photoperiods than under long ones. Also, ABA gradually decreases during chilling treatments and GA content increases; in some woody plants CK increases during emergence from dormancy. Whether a bud becomes dormant or continues growth is regulated by the interactions of inhibitory influences by ABA and promotive influences of GA and other growth regulators.

Abscission

Plants and their parts develop pretty much continuously from germination until death, and the deteriorative processes that naturally terminate the functional life of a structure are collectively called **senescence.** Aging in perennial plants results in the abscission of leaves as senescence progresses; leaves may fall more or less simultaneously at the end of a growing season (the deciduous condition), or there may be a more spasmodic leaf fall (the evergreen condition; leaves last one to seven years but eventually abscise). Injury or the removal of the leaf blade accelerates petiole abscission. A similar senescence and abscission of flowers occurs, and fruits also drop as they ripen.

Auxin has been shown to inhibit the separation of abscission layers (see Section 10.2 and Figure 10-9) in the petioles of leaves, flowers, and fruits. The natural abscission that occurs appears to be a result of decreased auxin production as senescence progresses. For example, if the leaf blade is removed, petiole abscission soon follows as a result of diminished auxin supply (usually available from the blade). However, if auxin is applied to the cut petiole, abscission is considerably delayed. It is apparently not the actual auxin concentration that is important, but the concentration gradient across the abscission zone (i.e., the relative concentrations of auxin on the stem side and leaf side of the zone). If the auxin concentration is high on the leaf side and low on the stem side, no abscission occurs. When

545

Sec. 29.11 /
Effects
Attributed to
Growth
Regulators

the gradient disappears or is slight (as would occur when the leaf ages and the auxin content of the blade decreases), there is abscission. If the gradient is reversed (high on the stem side and low on the leaf side), abscission is accelerated.

Investigations during recent years have made it clear that the situation is more complex than originally considered. A frequent consequence of progressive senescence is ethylene production, and this greatly accelerates abscission. Ethylene stimulates the production of cellulase and probably pectinase, hydrolytic enzymes that bring about the separation and degradation of cells in the abscission zone. Although abscisic acid, gibberellin, and cytokinin can each promote abscission, the major growth regulators of dehiscence seem to be IAA as an inhibitor and ethylene as a promoter. In general, IAA production decreases and that of ethylene increases during senescence; the result is abscission.

Synthetic auxins, such as naphthaleneacetic acid, at about 5 to 10 parts per million (ppm), are used in dusts or sprays at harvest time to delay the premature drop of certain fruits, especially apples. Preharvest drop of apples often causes large losses to growers, because the crew of pickers cannot work around the orchard rapidly enough or because the apples fall before they are sufficiently ripe. The McIntosh variety of apple is especially susceptible to premature fruit drop and is one in which auxin spray programs are very important.

It is also possible to bring about thinning of fruit crops by using auxin sprays at relatively high concentrations. For example, naphthaleneacetic acid applied at full bloom in a concentration of 50 ppm causes flower drop and will reduce fruit set in apples by about 75 per cent; at 100 ppm, the entire harvest will be eliminated. Thinning is advantageous; the fewer remaining fruit will be larger (more food is available to each fruit) and yield will be more uniform from year to year. Care must be utilized, of course, that the concentration of spray material is not so high as to damage the harvest. The opposite effects of low and high concentrations of auxins on abscission are reminiscent of such effects on cell elongation, as pointed out in our discussion of geotropism (see Figure 29.8).

Parthenocarpy

The development of fruits from unpollinated flowers occasionally occurs, as in banana, navel orange, seedless grape, and seedless grapefruits. Such **parthenocarpic** fruit can be produced in some plants by treating the pistils with auxins, usually indolebutyric acid, gibberellins, or cytokinins. Although edible seedless fruit have been produced in this manner in strawberry (*Fragaria*), eggplant (*Solanum melongena* var. *esculentum*), cucumber (*Cucumis sativus*), and squash (*Cucurbita maxima*), the most satisfactory results have been obtained with tomato (*Lycopersicon esculentum*). Under natural conditions most fruits are produced by the development of the ovary brought about by auxin introduced from the pollen or by auxin synthesized in the ovary or ovules as a result of a stimulus introduced from the pollen. Treatment of flowers with growth regulators merely provides a supply of these materials without the presence of pollen.

546

Ch. 29 / Growth,
Development,
Flowering,
and Plant
Movements

Root Initiation

In many kinds of plants, stems, leaves, and even roots may be severed from the parent plant and used for propagation. This vegetative propogation by the use of cuttings is frequently very useful in perpetuating desirable characteristics, such as disease resistance, quality, yield, and vigor. The offspring are identical with the parent plant; no possibility exists of genetic change, because sexual reproduction (gamete fusion) is not involved. Also, with cuttings much more rapid growth is obtained than with seed; a saving of time from two to four years in obtaining a mature plant may be realized. This is quite significant to the nurseryman and gardener.

Treatment of stem, leaf, and root cuttings with any one of several plant growth substances frequently results in a stimulation of root formation. Auxin-treated cuttings root more rapidly and have a greater profusion of roots than the untreated; ethylene and abscisic acid have similar effects. Dipping cuttings into dusts or solutions of such substances as indolebutyric and naphthaleneacetic acids, however, does not necessarily ensure the production of roots. Auxin is only one of several factors that may be limiting the process, and careful attention must be given to water, light, temperature, humidity, and the general condition of the cuttings. Also, some kinds of plants are difficult to root and do not respond to hormone treatments. Gibberellin and cytokinin generally inhibit the formation of adventitious roots, although the latter does stimulate rooting in some plants.

Cell Division

Growth regulators are involved in cell division also. Auxin, cytokinin, and gibberellin have been shown to stimulate cell division whereas abscisic acid causes an inhibition. The influx of cambial activity during the spring appears to be caused by the increasing production of auxin by terminal meristems, buds, and developing leaves, all of which show increased activity at this time of year. Gibberellins may also influence cambial activity.

Weed Control

As work with growth regulators progressed, some of the naturally occurring substances were synthesized in the laboratory, and eventually new compounds with similar properties were synthesized. The latter are not found in plants, but may be cheaper, more effective, or easier to manufacture or use. By this time, scientists had found that whereas low concentrations of some of these compounds would stimulate growth, larger amounts were toxic and killed the plant. Beside the effects already noted, these materials had great influences on cellular metabolism, high concentrations appearing to be disruptive of general metabolism.

Some of the synthetic growth regulators affected certain types of plants more than others, thus suggesting the possibility that weeds might be eradicated by *selective* sprays. Chemical weed killers had been in use for many years, but these

. were compounds that generally killed all vegetation; they were directly toxic to the cells with which they came in contact. Some of these all purpose vegetation-destroying chemicals are still used in special cases: ferous sulfate, sulfuric acid, arsenical preparations, aromatic fractions of fuel oils, and so forth. Differential killing with these materials is basically a result of differences in the permeability of the cuticle; if the material penetrates to the cell, the cell is destroyed. Herbicidal plant regulators, on the other hand, are highly effective at concentrations that do not directly destroy cells with which they come in contact. The killing appears to be caused by greatly increased respiration, an acceleration of cell enlargement, abnormal cellular proliferation, and a disruption of phloem tissue as a result of excessive development of parenchyma in the phloem region. These hormone herbicides are selective: many broad-leaved plants (dicotyledons) are extremely susceptible, while narrow-leaved monocotyledonous plants are highly resistant. This selectiveness has not yet been explained.

Fortunately, most of the common weeds are broad-leaved. This makes possible the use of something like 2,4-D (2,4-dichlorophenoxyacetic acid) to destroy weeds without injuring grasses. Such regulators are transported throughout the plant and are very effective in eradicating plants like dandelion (*Taraxacum officinale*) and morning glory (*Ipomoea purpurea*), which have extensive root-storage systems. The chemical weed killers mentioned in a previous paragraph would destroy the aboveground portions, but sprouts would subsequently emerge from the undestroyed roots. Materials like 2,4-D are transported within the plant as are the natural growth regulators. Thus, their effect is produced throughout the plant, including the below-ground portions. Not all auxins can be used as weed killers; indoleacetic acid and indolebutyric acid, for example, are not effective.

Some of the synthetic herbicides are effective defoliants. By application of such compounds, for example, it is possible to defoliate cotton plants before harvesting to avoid plugging the mechanical cotton harvester with cotton leaves.

An interesting note is that this practical application of one phase of botanical research was totally unexpected. No one could have guessed that an investigation concerned originally with phototropism would eventually lead into the extremely important field of weed control. Because certain investigators were interested in a fundamental plant phenomenon, the practical problem of killing weeds is much nearer a solution. This is really the way in which many practical problems become solved; fundamental research with the general aim of increasing knowledge is always the foundation for the solution of practical problems.

29.12 Flowering

In previous chapters the general structure, growth, and development of the flowering plant were discussed. All these plants undergo a period of vegetative development before flowers are produced, and the length of this period varies greatly. Some trees may grow for years before any flowers develop, whereas flowers appear on some annual herbs in a few weeks from the time of seed germination. Since about 1920, the stimulus that initiates flowering has been the subject

548

*Ch. 29 / Growth,
Development,
Flowering,
and Plant
Movements*

of considerable investigation and speculation. About this time Garner and Allard began publishing the results of their investigations, which demonstrated that flower production in many plants was influenced by day length. Subsequently, other investigators added temperature as another environmental factor that affected flowering.

Photoperiodism

Flowering plants can conveniently be arranged in three general groups (Table 29-1) with regard to the duration of the light period (**photoperiod**) required for flowering: (1) "long-day" plants, which will flower only if exposed to day lengths longer than a critical amount, (2) "short-day" plants, which will flower only if the duration of the light period is shorter than a critical length, and (3) "day-neutral" plants, which appear to be unaffected by day length and flower readily over a wide range of photoperiods. Varietal differences exist with regard to the critical day length, and a "short-day" species may flower if the light exposure is less than 13

Table 29-1 Examples of well-known plant species that are short-day, long-day, or day-neutral with regard to their flowering behavior

Species	Common name	Critical day length (hours)
Short-day plants[a]		
Xanthium pennsylvanicum	Cocklebur	15 to 15.5
Cosmos sulphureus var. *Klondike*	Cosmos	12 to 13
Nicotiana tabacum var. *Maryland Mammoth*	Tobacco	13 to 14
Chrysanthemum indicum var. *Queen Mary*	Chrysanthemum	14 to 14.5
Euphorbia pulcherrima	Poinsettia	12 to 12.5
Long-day plants[b]		
Anethum graveolens	Dill	11 to 14
Hyoscyamus niger (annual variety)	Henbane	10 to 11 (22°C)
Spinacea oleracea	Spinach	13 to 14
Hibiscus syriacus	Rose mallow	12 to 13
Day-neutral plants[c]		
Lycopersicon esculentum	Tomato	
Zea mays	Corn	
Fagopyrum esculentum	Buckwheat	
Antirrhinum majus	Snapdragon	
Nicotiana tabacum var. *Java*, etc.	Tobacco (most varieties)	
Capsicum annuum	Chili pepper	
Cucumis sativus	Cucumber	

[a] Short-day plants flower only when the illumination is shorter than the critical day length.
[b] Long-day plants flower only when the illumination exceeds the critical day length.
[c] Day-neutral plants flower under a wide range of day lengths.
Source: James Bonner and Arthur W. Galston, *Principles of Plant Physiology* (San Francisco, Freeman, 1952), by permission of the publisher.

hours or 15 hours or 17 hours, depending upon the variety used. This is true, for example, in soybean (*Glycine soja*): the Biloxi variety flowers if the day length is less than 13 hours and remains vegetative if the day length is prolonged beyond this amount; for the Peking variety the critical period is 15 hours, and it is 17 hours for the Mandarin variety. All these soybean varieties will flower at about the same time if the day length is 12 hours or less. Under these conditions, the duration of light is less than the critical period for all the varieties. Differences in flowering would be noticeable only when the day lengths are prolonged.

Another complicating factor with regard to flowering is that both "short-day" and "long-day" plants may flower when exposed to a 12-hour photoperiod. This appears to be rather strange but is readily explained. The critical day length of many "short-day" plants is about 14 hours (for cocklebur it is actually about 15 hr; see Table 29.1), while it is approximately ten hours for many "long-day" plants (note henbane in Table 29.1). Thus, a 12-hour day length is less than the critical for the one group and more than the critical for the other group, and both flower. Obviously, then, "short-day" and "long-day," when used in reference to flowering, do not imply a particular length of time. The terms refer to the fact that flowering occurs if the photoperiod is shorter or longer than a critical amount. The number of photoperiodic cycles required to induce flowering differs from one species to another and frequently from one variety to another within a species. In a few types, one or two exposures to the correct photoperiod is sufficient to induce some flowering, although a greater number of flower primordia will be initiated with a greater number of photoperiodic induction cycles. With some plants flowering will occur at any day length, but more rapidly at long days; with others flowering occurs more rapidly at short days.

Some plants like carrot, cabbage, and henbane (*Hyoscyamus*) produce a whorl of leaves (a rosette, the leaves develop but the internodes do not elongate) and then the flower stalk emerges from the center after an appropriate stimulus. Flowering is usually induced by exposure to long days (i.e., long photoperiod), to cold (as in biennial plants), or to both. Gibberellin treatment can substitute for such stimuli in many plants which produce rosettes; as a result, the stem elongates (frequently termed *bolting*) and the plant flowers. Recently it has been reported that cytokinin application can substitute for long-day photoperiods in at least some plants.

Although photosynthesis and the resultant accumulation of food reserves are essential to flowering, and a certain degree of maturity is usually required before plants will flower, the photoperiodic stimulus that brings about the initiation of flower primordia does not function through the photosynthetic process. The light intensity that is sufficient to produce flowering may be so low as to be unimportant photosynthetically. Also, while red wavelengths are very effective, blue wavelengths have the least influence on flowering. Besides, "short-day" plants can be prevented from flowering by prolonging the day length. The same treatment would increase the amount of photosynthetic material produced by these plants.

As Hamner first pointed out, the terms *short-day* and *long-day* are somewhat misleading. For this reason the two terms have been used within quotation marks. He found that if *Xanthium* (cocklebur, a "short-day" plant) is exposed to short photoperiods, flowering is inhibited when the *dark* period is interrupted by a brief

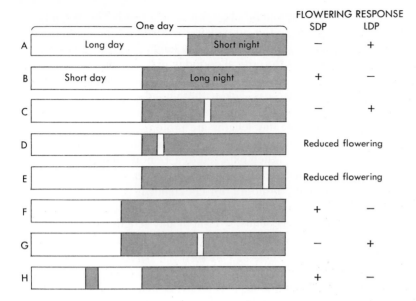

Figure 29-10. Diagram to show the effect on "short-day" and "long-day" plants of various photoperiods, and interruptions of either the dark or light periods. See text for discussion. SDP = short-day plants. LDP = long-day plants.

flash of light. Later investigations showed that "long-day" plants flowered when exposed to a similar regime (i.e., short days followed by an interrupted long night). Plants have been exposed to a variety of photoperiods and manipulations of the light and/or dark periods. Figure 29-10 represents the situation as it occurs generally in most "short-day" (SDP) and "long-day" (LDP) plants. Plants in **Group C** receive the same total amount of light and dark exposure as the plants in **Group B,** but the interruption of the dark period results in the SDP not flowering and stimulates LDP to flower even though **Group C** is on a short-day photoperiod. If the flash of light (actually a five-minute flash of low intensity is sufficient for many plants) occurs near the beginning or near the end of the dark period, as in **Groups D** and **E,** flowering is reduced. Even when the night is lengthened and the day shortened, a brief flash of light inhibits flowering of SDP and stimulates flowering in LDP (see **Group G** and compare with **F**). Note that a period of darkness during the daytime, **Group H,** has no effect on flowering. Clearly it is the length of uninterrupted dark, not light, that is important. Although the terms *short-day* and *long-day* are so well established in the literature that they will undoubtedly be retained, we should reexamine their meanings and redefine them. For a "short-day" plant to flower, the *continuous* dark period must exceed a certain length; in the case of a "long-day" plant the longest dark period must be shorter than a critical length. Because our 24-hour days include periods of light and dark as the sun rises and sets, a "short-day" plant flowers when undergoing short days and long nights.[4] We must merely remember that both the light and dark periods are important.

[4] Some botanists have suggested using "long-night" rather than "short-day" and "short-night" instead of "long-day." This is probably more logical. In either case, the understanding of photoperiodism, rather than the peculiar terminology, is important.

The geographical distribution of plants is governed in part by their photoperiodic responses. Since the length of the daylight period increasingly varies from season to season in latitudes increasingly north or south of the equator, "short-day" plants cannot compete too well beyond 50-degrees latitude, because the growing season in such regions is restricted to the time of year during which long days prevail. Also, sexual reproduction of plants requiring longer than a 13-hour light exposure cannot be accomplished in the tropics. The "day-neutral" group of plants can grow in either locality. In the temperate zones all three groups of plants flourish because of the varying day lengths, but they flower during different seasons. In general, the "long-day" plants bloom in late spring or early summer, while the "short-day" species bloom in early spring, late summer, or early autumn. Of the latter group, those that grow rather rapidly and can withstand chilly weather will bloom early; the majority, however, tend to bloom during the latter portion of the growing season. Once they are mature, the "day-neutral" plants may flower throughout the year if the weather is favorable for their growth.

Photoperiodic responses are not limited to plants; the reproductive activity of a number of invertebrate and vertebrate animals has been shown to be controlled by day length. The basic mechanisms responsible for such responses are not known and probably are different in plants and animals, but the time-measuring system in both instances involves a recognition of changing day length and night length with the season. It is interesting to note the diversity of organisms that are capable or responding to changes in the photoperiod: algae, ferns, flowering plants, mites, snails, fish, birds, and sheep. It is very likely that this list can be lengthened almost yearly as results of investigations are published. In any event the synchronous timing of sexual reproduction provides a means of securing cross-fertilization and a great variety of genetic recombinations by which new adaptations are made possible.

The Temperature Factor

Investigations have demonstrated that temperature influences flowering in some kinds of plants. Most biennials will flower only after exposure to relatively low temperatures, the condition that exists between their two seasons of growth. During the first season, growth is normally vegetative. After overwintering, these plants produce flower stalks and eventually seeds. For example, celery (*Apium graveoleus*) remains vegetative at 16° to 21°C. If the plants are exposed to temperatures of 5° to 16°C for 10 to 15 days, flowers and seeds will be produced. Thus, the biennial can be converted to an annual by manipulating the environment. Such induction of flowering by low-temperature treatment of seeds or plants is called **vernalization.** Occasionally, weather conditions are such that late cold spells in the spring may cause celery to go to seed during its first season of growth. This is a loss as far as the farmer is concerned, since he is interested in the stalks (petioles) and not in the seed. Gibberellin treatment can substitute for the cold-temperature stimulus in rosetted plants. Cytokinins also can replace the requirement for low temperatures in some plants.

The low temperature stimulus is perceived by the apical meristem in most

552

*Ch. 29 / Growth,
Development,
Flowering,
and Plant
Movements*

plants, and this alters the morphological expression of growth. In those plants where a low-temperature treatment of the seed is effective in bringing about flowering of the plant after germination and growth, experiments in which the embryo was excised have shown that the apex of the embryo is the site of perception. High temperature, following the low-temperature treatment, nullifies the flowering effect. This suggests that a product formed during the cold experience may be metabolized away if high temperatures follow too soon for the product to establish its effect. In a few plants, such as *Hyoscyamus niger,* the transmission of a vernalization stimulus across a graft union has been demonstrated. This has led some investigators to postulate a hypothetical substance, termed *vernalin,* as responsible for the transmission. No such substance has been isolated, however.

Some species, such as lettuce, rice, and China aster, are caused to flower by high temperatures. The White Boston variety of lettuce (*Lactuca sativa*), for example, is stimulated to flower by relatively high temperatures. If the temperature remains below 15°C, heads, but no flowers, will be produced; at 16° to 27°C, no heads will be produced, and the plants will flower rapidly.

Night temperature, especially, appears to have an influence in reinforcing the effect of the photoperiod. In general, the number of flower primordia increases if the night temperature during photoperiodic induction is high rather than low.

29.13 Flowering Hormone

Most botanists are convinced that flower initiation is a result of a hormonal influence, and the tentative name **florigen** has been assigned to this unknown material. Even though the hormone has not been isolated, a vast amount of evidence that it exists has been gathered. Although the evidence is somewhat indirect, this situation was also true of the vitamins before they were isolated.

The young, fully expanded leaves are the structures that perceive the photoperiodic stimulus, which is then transmitted to the growing point. This was demonstrated by exposing the buds to one photoperiod and the leaves to another. Flowering resulted only when the leaves were exposed to the correct light duration. In some plants, when one branch is exposed to an induction period, flowering occurs throughout the plant, indicating that some material was transmitted from the one branch to the rest of the plant. The stimulus is also transmitted through a graft union. These data all imply the presence of a hormone—a material produced in one part of the organism (i.e., the leaves) and which elicits a response in other parts (i.e., the meristems). Several investigators have shown that extracts from leaves of flowering *Xanthium* (SDP) were capable of inducing flowering when applied to the leaves of *Xanthium* plants maintained under vegetative (long-day) conditions. The material has not been purified or identified and the results are still preliminary, especially because there is considerable variability in the success of these experiments. This extract is without effect on long-day plants, and similar extracts have not been obtained from long-day plants. The difficulty probably arises from the fact that the material is used up during the flowering process and from the probability that extremely minute amounts are effective. It is also possible that the active substance(s) may be labile.

In spite of the fact that "short-day" and "long-day" plants flower under differing photoperiods, the hormone responsible for flower initiation is in all likelihood very similar in these two groups of plants. If a "short-day" variety is grafted to a "long-day" variety, flowering can be initiated in both plants under either type of photoperiod. The hormone must be quite similar although the internal mechanism that utilizes this material probably differs substantially.

Further investigations with regard to florigen promise to be of tremendous value. A future where man can govern whether or not a given plant will flower is not too inconceivable. In fact, this can be done with many plants at the present time. For example, chrysanthemums normally bloom in the fall, as do many "short-day" plants. They can be made to bloom in midsummer by covering the plants during the afternoon or morning, thus shortening the days and lengthening the nights. With florigen and "antiflorigen" compounds of the future, the same effect may be obtained by the use of economical sprays.

29.14 Phytochrome

Recent investigations have shown that inhibition of flowering, by interrupting the dark period in "short-day" plants, can be caused by red light of a wavelength of 660 nm (actually orange-red in color). This effect of red light can be counteracted by exposure to far-red light of a wavelength of 730 nm. This is a reversible photoreaction. The fact was obvious to investigators that some kind of a pigment must be involved, since light energy cannot be effective unless it is first absorbed. (A substance that absorbs visible light is a pigment.) In 1959 Borthwick and Hendricks finally demonstrated the presence of phytochrome.

Phytochrome appears to be a protein with a chromophore, or colored group, and the latter is closely related to phycocyanin (a pigment found in certain algae). It occurs in two possible forms: P_r is the form of phytochrome that absorbs red light (660 nm), and P_{fr} is the form that absorbs far red light (730 nm). These forms are readily interconverted as follows:

$$P_r \underset{730 \text{ nm}}{\overset{660 \text{ nm}}{\rightleftarrows}} P_{fr}$$

When P_r is exposed to red light, it is converted to P_{fr}; when P_{fr} is exposed to far-red light, it is converted to P_r. In the dark a slow change from P_{fr} to P_r takes place. Complete interconversion does not occur because there is considerable overlap in the absorption spectra, especially in the 600–690 nm region; so both forms are usually present. It is the ratio (or the relative amounts) of the two forms that varies, and this depends upon the wavelength of light. The P_{fr} form is considered to be the active form. In "short-day" plants P_{fr} inhibits flowering, and this is the form that gradually decreases in amount during the dark period. In "long-day" plants P_{fr} promotes flowering.

We now can suggest a possible explanation for the results obtained with *Xanth-*

554

*Ch. 29 / Growth,
Development,
Flowering,
and Plant
Movements*

ium (SDP) when the dark period is interrupted with a flash of light (Figure 29.10). At the end of the light period, the phytochrome is mainly in the far-red-absorbing form. P_r is converted to P_{fr} more rapidly than the reverse conversion of P_{fr} to P_r. Thus, even in sunlight or incandescent light (which contain both red and far-red wavelengths), there is more P_{fr} than P_r. Then follows a slow conversion of P_{fr} to the red-absorbing form during the subsequent dark period. The removal of P_{fr} releases the inhibitory effect and allows the formation of a flowering substance (FS, probably florigen). When enough FS has been produced, flowering is initiated. If the dark period is interrupted with red light, P_r is forced back to the P_{fr} form and flowering is inhibited. If the exposure to red light occurs early in the dark period, sufficient dark period still remains to convert some P_{fr} to P_r and build up some FS; flowering will be reduced but not inhibited. If the exposure is near the end of the dark period, enough of the flowering substance has already been formed to provide for some flowering; FS is not affected by light. Figure 29-11 represents the interconversions of phytochromes and the resultant flowering response. The influence of phytochrome in the flowering of "long-day" plants has not been worked out as well as with "short-day" plants. Very likely the far-red-absorbing form of phytochrome promotes the formation of FS in "long-day" plants.

The situation is still somewhat confusing because of the conflicting and contradictory results obtained by different investigators with various plants. It is possible that there are several forms of phytochrome, and possibly intermediates in the conversion of P_{fr} to P_r. The available evidence seems to indicate that flowering may be controlled by an array of stimuli and inhibitors acting together in the plant. Phytochrome may be involved in setting the stage for the synthesis of some regulating principle. Further data are necessary; our ideas may have to be modified considerably in the future as investigations proceed.

Phytochrome is also involved in seed germination in some plants. For example, moist lettuce seeds require an exposure to light for germination (this exposure usually need be for only a few seconds); in the dark they will not germinate. If exposed to red light, they germinate; if exposed to far-red light, they will not germinate. Actually, a range of red and far-red light is involved, with the maximum responses at 660 and 730 nm, respectively. The same is true for the influence of light on flowering. Plants may be exposed to red and then far-red illumination. The wavelength of the last exposure determines the response of the plant.

Phytochrome apparently is involved in a variety of processes in plants, including stem elongation and leaf expansion. It has been suggested that phytochrome effects are brought about by repressing or activating genes that code for specific mRNA molecules, and thus specific enzymes necessary to cause the biochemical

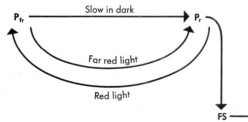

Figure 29-11. The possible participation of phytochrome in the flowering response of "short-day" plants. P_{fr} = far-red-absorbing form of phytochrome. P_r = red-absorbing form of phytochrome. **FS** = flowering substance (florigen?).

changes leading to the observed responses. However, some of the responses mediated by phytochrome appear to be too rapid for such a mechanism. It is likely that changes in membrane properties are responsible for the rapid effects, and that the former mechanism may determine long-term morphological events. Future investigations will undoubtedly expand our knowledge of this interesting pigment.

29.15 Photomorphogenesis

By **photomorphogenesis** is meant the control exerted by light over growth, development, and differentiation of a plant, independent of photosynthesis. Other environmental factors (e.g., water and mineral availability) also influence growth and development, but this is in a quantitative way by determining the maximum attainable size of the plant. The light effect with which we are concerned at this time is in addition to its influence on food production (photosynthesis). Phototropism is one example already discussed (Section 29.4), and the influence of light on seed germination and leaf expansion has been mentioned briefly. These latter ef-

Figure 29-12. Pea seedlings, eight days old. *Left:* Germinated and grown in light. *Right:* Germinated and grown in dark; etiolated. (Courtesy of Peter Jankay.)

556

Ch. 29 / Growth,
Development,
Flowering,
and Plant
Movements

fects, and a few others, deserve a bit more extensive treatment now that the phytochrome system has been presented.

Many dicotyledonous plants, if germinated and grown in the dark, have an appearance significantly different from those grown in the light (Figure 29-12). Such plants are said to be **etiolated.** They have poorly developed leaves (scalelike leaves), elongated stems with an apical hook, and are yellowish rather than green. Exposure of these plants to various wavelengths of light has shown that far-red light duplicates the effects found with plants grown in the dark, while red light results in responses similar to those with light-grown plants. Remember the P_r-P_{fr} interconversions? The phytochrome system is implicated in these photomorphogenic effects.

Stem Elongation

If seedlings are grown in far-red light, the stem is greatly elongated as would be the case if they were grown in the dark. Red light has the opposite effect; elongation is not as rapid (similar to plants grown in the light). It is considered that the presence of P_{fr} inhibits elongation and that its conversion to P_r allows elongation to occur. This situation may be beneficial to the seedling. Seeds in general germinate underground (dark), and the rapid elongation speedily results in the seedling emerging where light is available for photosynthesis. Until this occurs, the seedling is dependent upon food stored in the seed for its growth. The faster the plant reaches a source of light, the better its chances of producing food (photosynthesis) before the supply in the storage tissues is exhausted. Remember that food materials provide substrates used in synthesizing complex cellular components and also provide the source of energy that makes such syntheses possible. Continued rapid elongation would result in a weak, spindly plant easily damaged by wind action or collapsing as its weight increases.

Leaf Expansion

Dicotyledonous seedlings tend to have tiny scalelike leaves if grown in far-red light, and larger expanded leaves if grown in red light. Leaf expansion, thus, occurs when phytochrome shifts to the P_{fr} form. This may be beneficial to the plant. Most of the substances required for growth are used for stem elongation and are not utilized in leaf expansion until the plants are tall enough to intercept greater amounts of sunlight. This enables considerable stem growth from the available stored foods without some of that food being siphoned off to the production of leaves that would not really function for a time. Also, if leaf expansion commenced with germination, penetration of the seedling through the soil might be hindered and the leaves might be damaged by abrasive soil particles. Expansion above ground (light) would be advantageous in providing a greater photosynthetic surface.

Apical Hook

Etiolated dicotyledonous seedlings, or those grown in far-red light, have a stem that is recurved in the subapical region to form a hook. This hook readily penetrates the soil, actually leading the apical region where the scalelike leaves are produced. In the presence of light, or red light, the hook unfolds and the stem straightens. The advantage to the plant might be in preparing a pathway through the soil. The tender, delicate apical meristem and leaves are not *pushed* through the soil by the elongating stem; they really are *pulled* through with the hook region leading the way. It is less likely for the leaves and apical regions to be injured by abrasive action in this fashion.

Chlorophyll Formation

Light is essential for chlorophyll formation in most plants. As was mentioned previously, etiolated plants are yellowish rather than the green of plants grown in light because chlorophyll is not formed in the former. Plants exposed to far-red light behave in a similar manner. Exposure to light, or red light, results in the rapid production of chlorophyll. This response of the plant to light is probably advantageous in that food reserves and other materials are used to support stem elongation until the time when the plants are tall enough to be exposed to light. At this stage in growth, photosynthesis could occur if chlorophyll were present and thus supply a source of food materials. Plants in which such a mechanism happened to develop would tend to have a better chance of surviving. They would reach a source of light rapidly, owing to the rapid stem elongation in the dark, and chlorophyll would be produced primarily at the time when it could be functional. If a portion of the stored material was utilized in chlorophyll formation, elongation might be less and the plant might not reach a height sufficiently tall to expose its leaves to light.

Seed Germination

In some plants, after seeds have imbibed water, their germination is stimulated by red light, an effect that can be reversed by far-red light. This red/far-red relationship again implicates the phytochrome system. In some species soaking the seeds in gibberellin overcomes the light requirement. Red light increases the gibberellin content of soaked pea seeds, while far-red light reduces the level of this growth regulator. One effect of gibberellin is to increase the production of certain digestive enzymes in seeds, which would enhance germination by providing a usable food supply. It is possible that the phytochrome system may be involved in gibberellin production, but more investigation is needed to demonstrate whether this is the general situation with regard to seed germination as stimulated by light.

It is possible that the failure of seeds to germinate unless they are exposed to

558

*Ch. 29 / Growth,
Development,
Flowering,
and Plant
Movements*

light may be an ecological advantage. Only a fraction of the seeds present in the soil may be disturbed and exposed to light in any given season. If an unfavorable growing season resulted in death of those plants that developed, other ungerminated seeds would still be available for germination in subsequent years. Germination after light exposure would also tend to assure the availability of light to those seedlings that do grow.

29.16 Résumé

No plant growth regulator functions individually. The interactions of many regulators undoubtedly govern the growth, development, and differentiation of a plant. These interrelationships, although essential to the normal functioning of complex organisms, greatly increase the difficulties encountered in attempting to study individual growth regulators.

Table 29-2 presents some of the known characteristic effects of auxins, gibberellins, cytokinins, abscisic acid, and ethylene. In some aspects these growth regulators are similar; in others they are quite different. Data are conflicting in some instances in that one group of investigators finds a stimulation while another group discovers an inhibition. This may result because experimental procedures may not be exactly similar, different plant species are used, the amounts of growth regulator may differ, and so on. At times, data may be suggestive but not conclusive and some possible effects may simply not have been examined yet. Investigations are continuing and expanding; new information is being reported almost daily; and it is quite likely that other plant growth regulators will be discovered in the future.

Table 29-2 Some of the characteristic effects of growth regulators

	Auxins	GA	CK	ABA	Eth
Cell elongation	+[a]	+	+	−	−
Cell division	+	+	+	−	−
Parthenocarpy	+	+	+		
Lateral bud growth	−	+, −	+	−	−
Root initiation	+	−	+, −	+	+
Abscission	−	+	+	+	+
Seed germination	0	+	+	−	+
Flowering of rosetted nonvernalized biennials and rosetted "long-day" plants	0	+	+	−	
Richmond-Lang effect	0	0	+		
Polar transport	Yes	No	No	No	
Avena curvature test	+	0	0		
Dwarf maize test	0	+	0		

[a] + refers to stimulation, − refers to inhibition, 0 refers to no effect. GA = gibberellins. CK = cytokinins. ABA = abscisic acid. Eth = ethylene.

Plant movements are usually too slow for direct observation but the results are easily noticed. The opening and closing of flowers, the unfolding of buds, the bending toward light, and the twining of tendrils are examples of movements that have all been described many times in terms of the beginning and final conditions with little, if any, comment about the intermediate positions of the plant structures or the mechanisms involved. The use of time-lapse photography has made it possible to observe such movements in considerable detail. As a result of these and other investigations, plant movements have been classified in various ways, but all such classifications are simply an aid to understanding. There are overlaps and not all movements fit neatly into the given categories.

In general, movements of plant parts can be classified, according to the mechanism involved, as turgor movements or growth movements. **Turgor movements** are caused by differential changes in turgor and size of the cells as a result of the gain or loss of water, and are easily reversible. **Growth movements** are irreversible and due to differential growth rates of cells in different parts of the affected organ.

Turgor Movements

The effective cells are often different from ordinary cells and may be concentrated in certain areas. The rolling of leaves of many grasses in dry weather is caused by loss of water of the bulliform cells, which form longitudinal rows in the epidermis. This movement and the opening and closing of stomata have already been mentioned (Sections 15.4 and 15.5). The drooping and folding up at night (sleep movements) in some plants are caused by turgor changes in the cells of the pulvinus at the base of the leaf or leaflet. The **pulvinus** is a thickened portion of the petiole, composed primarily of parenchyma cells and relatively large intercellular spaces and a central strand of vascular tissue. Water passes into or out of the cells more freely on one side of the pulvinus than on the other. This unequal movement of water causes unequal enlargement or shrinkage (turgor responses), and a consequent movement of the petiole and blade. The drooping of leaves or leaflets of the sensitive plant (*Mimosa pudica*) when touched results from differential changes in turgor of the cells in the pulvini (Figure 29-13). The mechanisms bringing about such turgor changes have not yet been clarified.

Growth Movements

Growth movements may be spontaneous and self-controlled, **autonomic,** or they may be induced by external stimuli, **paratonic.** The latter may be further classified as tropisms or nasties (or nastic movements). **Tropisms** are responses to stimuli that come chiefly or wholly from one direction. **Nasties** are responses that are independent of the direction from which the stimulus is received; the stimulus may be uniform or nondirectional.

Figure 29-13. Sensitive plant, *Mimosa pudica*, showing the normal condition of the leaves and the closed condition after being touched. (Courtesy of Peter Jankay.)

Autonomic movements. The growing tip of a young stem moves slowly back and forth in a more or less helical fashion owing to alternately changing growth rates on opposite sides of the organ. This is usually termed **nutation** (Figure 29-14). The nutation of twining vines is exaggerated and is strongly accentuated when a solid object is touched. Thigmotropism (see later) is probably involved at this time.

Paratonic movements. (1) T R O P I S M S. Geotropism and phototropism have already been discussed (Sections 29.4 and 29.5). However, leaves as well as stems are often phototropic. In many instances the petioles of leaves bend in such a way that the blades are oriented at right angles to the source of illumination with a minimum of overlapping, fitting into patterns termed **leaf mosaics** (Figure 29-15). Such an arrangement results in little shading and a maximum use of available light.

Thigmotropism is a growth movement in response to contact with a solid object and is most readily apparent in tendrils of climbing vines (see Figure 8-6). The immediate reaction is probably due to turgor changes, but this is followed by increased growth on the side opposite to the point of contact so that the tendril coils

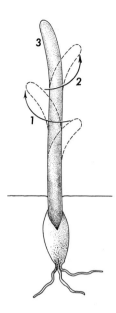

Figure 29-14. Nutation. The broken lines indicate the back-and-forth motion of the stem tip during growth. (From *Plant Form and Function*, by G. J. Tortora, et al., New York, Macmillan, Inc., 1970, reprinted by permission.)

around the supporting object. Tendrils do not respond to liquids, not even to heavy liquids such as mercury; nor do they respond to perfectly smooth solids. A rough or uneven surface is required for reasons that are not clear.

Examples of **chemotropism,** the response to chemicals, are the growth of the pollen tube through the style to the ovule and the bending of *Drosera* "tentacles" around an insect (Figure 14-7). The latter movement is partly thigmotropic, but the continuous chemical stimulus of a trapped insect causes the tentacles to remain curved; recovery within a day occurs if the object is not digestible.

(2) NASTIES. The growth movements of bud scales, young leaves, and petals are examples of nasties. These organs grow faster on the lower side when young, resulting in an upward and inward bending, which encloses the tip of the axis. The

Figure 29-15. Leaf mosaic formed in response to light by some plants.

562

Ch. 29 / Growth,
Development,
Flowering,
and Plant
Movements

unfolding of the bud occurs later when growth is more rapid on the upper surface. The principal stimuli appear to be the photoperiod or temperature to which the bud is exposed. Some investigations indicate that these growth changes may be caused by some balance or ratio between growth inhibitors (abscisic acid?) and growth stimulators (gibberellins?). Sleep movements of some plants are growth responses rather than turgor movements. Fluctuations of light and temperature produce nastic movements in many flowers. Tulip and dandelion (*Taraxacum*) flowers tend to be open during the day but close at night, while some species of *Oenothera* (evening primrose) open in the evening.

The closure of leaves of *Dionaea muscipula* (Figure 14-8) is a rapid turgor response followed by a differential growth response, which may be considered to be a combination of thigmonasty and chemonasty. Closure may be brought about by touch, but continued closure depends upon further chemical stimulation with nitrogenous substances from the insect body.

In spite of much investigation, very little is known regarding the mechanisms that bring about plant movements. Even the differential distribution of auxin, which results in phototropic or geotropic responses, is only a partial explanation. What *causes* such distribution? Why do certain cells lose water, and thus turgidity, when stimulated by touch or light? Why does contact result in differential growth in some plants? These are all unanswered questions.

Summary

1. Growth consists of an increase in the number of cells and an increase of cellular components. Growth regulators are organic compounds that promote, inhibit, or otherwise modify growth, development, and differentiation of plants. A hormone is a regulator that elicits a response at some distance from where it is formed.

2. Known growth regulators are auxins, gibberellins, cytokinins, ethylene, and abscisic acid.

3. Auxins are identified by their effectiveness in causing curvature of oat coleoptiles. They are important in phototropism, geotropism, inhibition of lateral buds, inhibition of abscission, fruit development, stimulation of root initiation, and stimulation of cell division and growth.

4. Gibberellins are identified by their effectiveness in causing elongation of dwarf maize plants. They are influential in stimulation of lateral buds, seed germination, fruit development, inhibition of root initiation, and stimulation of cell division and elongation.

5. Cytokinins are identified by the tobacco pith test. They bring about stimulation of lateral buds, seed germination, cell division, and cell elongation; and they delay senescence.

6. Ethylene and abscisic acid are general growth inhibitors; that is, inhibition of cell elongation, cell division, and lateral bud growth. They accelerate the abscission of various plant parts and increase senescence.

7. Some of the synthetic plant growth regulators are useful as selective herbicides.

8. In many plants, flowering is induced as a result of a photoperiodic stimulus; in others the stimulus is high or low temperature. For convenience, we divide plants into three groups on the basis of their response to photoperiodic stimuli: (1) "short-day," (2) "long-day," and (3) day-neutral.

9. "Short-day" plants flower only if the day length is shorter than a critical length and if the dark period is continuous. "Long-day" plants flower only if the light period is longer than a critical length. Day-neutral plants flower over a wide range of photoperiods.

10. Certain wavelengths of light influence the conversion of phytochrome from one form to another. These conversions appear to be influential in flowering and in other aspects of photomorphogenesis.

11. Plant movements result from differential changes in turgor or growth. Growth movements may be spontaneous or induced by external stimuli. The latter type includes tropisms and nasties.

Review Topics and Questions

1. Distinguish between growth regulators and hormones.
2. Describe an experiment that could be used to determine whether or not auxin is produced by the ovary of an apple flower. Explain how the condition of the ovary may influence your results.
3. Explain the mechanism that brings about the geotropic response of stems and roots.
4. Describe in detail how you could prove that a differential concentration of auxin is responsible for phototropism.
5. An investigator isolates a growth factor from cucumber seedlings. Describe in detail what he should do in order to determine whether this cucumber factor is an auxin, gibberellin, or cytokinin.
6. Discuss at least five ways in which auxins influence plant growth and development. Do the same for gibberellins, cytokinins, ethylene, and abscisic acid.
7. Differentiate between "short-day" and "long-day" plants. What are day-neutral plants? During what time of the year do these three types of plants usually flower? How does day length (photoperiodism) influence plant distribution?
8. Describe an experiment that demonstrates the importance of the dark period in flower induction.
9. Discuss the evidence that indicates the existence of a flowering hormone.
10. Discuss the manner in which phytochrome may participate in flower induction of "short-day" plants.
11. Why does a brief flash of red light during the dark period prevent flowering in "short-day" plants? How does this response vary if the red flash is given at different times during the dark period?

564

Ch. 29 / Growth,
Development,
Flowering,
and Plant
Movements

12. Compare the growth and development of plants germinated in the dark with plants exposed to light as they germinate. Discuss possible advantages to plants of such differences.
13. Indicate the various kinds of movements that plants may undergo and discuss whether or not such movements may be advantageous to the plant.

Suggested Readings

GREULACH, V. A., "Plant Movements," *Scientific American* **192,** 101–106 (February 1955).

LEOPOLD, A. C., AND P. E. KRIEDEMANN, *Plant Growth and Development,* 2nd ed. (New York, McGraw-Hill, 1975).

SALISBURY, F. B., "Plant Growth Substances," *Scientific American* **197,** 125–132 (April 1957).

SALISBURY, F. B., "The Flowering Process," *Scientific American* **198,** 108–114 (April 1958).

VAN OVERBEEK, J., "The Control of Plant Growth," *Scientific American* **219,** 75–81 (July 1968).

WILKINS, M. B. (ed.), *The Physiology of Plant Growth and Development* (New York, McGraw-Hill, 1969).

Our Environment

30.1 Environment

The term *environment* is commonly used simply to indicate the surroundings in which we find ourselves situated. However, this really means that we should consider all the living and nonliving things that are about us, but such tremendous complexities and details are beyond the intent of this book. It entails discussions of the effects of environmental factors upon individuals and upon groups of individuals as well as the interactions of these various organisms. **Ecology** is such a study of the interactions of organisms with one another and with their environment. Throughout the text, this facet of botanical studies has been mentioned and emphasized. How are plants distributed and why? What are the factors that determine which kinds of plants survive, and what sort of groupings there are among organisms? Such discussions may proceed from individuals to populations to communities to ecosystems or be restricted to any of these groupings. A **population** is a group of organisms of the same species occupying a particular area, and a **community** consists of *all* the plants and animals in that area (i.e., the latter is made up of populations). Any interacting system of living organisms and their environment comprises an **ecosystem.**

When associations of plants were first studied, it seemed pretty clear that environmental factors influenced plant distributions and that plant growth in turn brought about changes in the environment. As a result, over many years, various plants would invade an area, flourish, die, and be replaced by others, until finally the vegetation would develop a discrete association best able to exist in that environment. It was recognized that various associations overlapped and blended into each other at their boundaries, of course. However, further and detailed investigations and measurements of all the species and their importance in the community made it apparent that the idea of discrete associations was an oversimplification. Associations were shown to be complex and largely continuous patterns of populations. As one moves along the earth's surface, the environment changes gradually in a continuum and so do the species; instead of discrete associations, there is a vegetational continuum wherein the many kinds of communities integrate continuously as combinations of species correspond to the pattern of environments. And yet, botanists survey the general patterns of plant growth and describe principal terrestrial **biomes,** those large complex groupings of organisms that have a recognizable vegetational structure; for example, the *grasslands* of temperate and somewhat dry areas are a biome.

Within each biome there is a continuum of communities and, thus, gradual changes between the biomes. Then how is it possible to discuss vegetational types, or biomes, within an area? These are really not discrete entities, but the *dominant* vegetation occupies large areas and is distinctive enough to enable us to distinguish between one biome and another. Although these life-zones intergrade and are continuous with one another, and distributions of the major plant species overlap broadly, the basic idea of biomes is a useful concept. It is similar to the acceptance of colors as a useful concept even though the spectrum of wavelengths of light is known to be continuous. Later we will discuss some of the more important plant associations as recognized by many botanists.

The discussions in this chapter will be concerned with general biomes or vegetation types of North America since they are indicative of the various environments available to humans. Although animals will be mentioned only occasionally, it should be borne in mind that it is the vegetation that basically determines animal distributions and numbers by providing food and shelter. Plant succession will also be considered in order to incorporate a general understanding of the development of vegetation and the series of changes that occur when vegetation is disturbed. The chapter will conclude with a brief consideration of how humans use and misuse the environment, especially the land and water.

The environment consists of physical and biological factors that interact to produce habitats for organisms. Physical factors may be subdivided as light, water, soil, atmosphere, temperature, topography, and fire. The biological factors include green plants, nongreen plants, and animals. We have already seen how some of these factors are important to plants and animals.

Photosynthesis provides food and oxygen and is influenced by light, temperature, water, and soil (minerals). Light also frequently influences flowering and thus seed production (reproduction). Plant growth in general occurs within rather restricted ranges of temperature and water supply, the ranges varying somewhat with the plant involved. For example, as has already been mentioned, grasses tend to have lower water requirements than most broad-leaved plants, and some plants can grow at higher temperatures than others. Topography, the physical characteristics of an area, may influence such factors as temperature and rainfall. In general in the northern hemisphere, south-facing slopes tend to be warmer and drier than north-facing slopes because the former are exposed to more direct rays of the sun. Mountain ranges intercept rain clouds, and thus their windward side tends to be moist and the leeward side is usually drier. Warm air has a greater water-holding capacity than cold air. The moisture-laden air rises when it reaches the intercepting mountain ranges, expands, becomes cooler, and loses its capacity for holding water. As the air passes over the mountains, it falls and becomes warmer. It is able to hold more water, and rainfall tends to be low or scarce. All these factors, and others, interact to produce the vegetation types that comprise the most noticeable aspect of our environment.

Throughout the text, an attempt has been made to indicate the influence of various environmental factors upon plant processes in order to present reasons for plant distributions and the concomitant animal distribution. It is expected that the reader is acquainted in a general way with this material.

30.2 General Vegetation Types

The environment determines the vegetation that will develop in an area. Although our discussion is restricted to North America, similar vegetation develops in other areas of the world under similar environments. Temperate grasslands occur in similar climates in Europe, Asia, South America, and Africa as well as in North America. The specific plant species may vary but the general growth forms are comparable, and this is really what is meant by vegetation types. Those patterns found from Panama in the south to the Arctic in the north can also be found in other regions provided the environmental conditions are similar. As another example, one might mention the southern coastal part of California, which has a Mediterranean climate. That is, the climate is very similar to that found in the Mediterranean region of Europe and North Africa. The vegetation *types* or *biomes* in the two regions are quite comparable although the species comprising the plant communities differ considerably. In many areas, especially with intensive agriculture, the natural vegetation has been removed. The following discussions, however, refer to the original and natural types.

The categories used in classifying vegetation types vary considerably depending upon the amount of detailed subdividing that one wishes to include. This section will consider general main categories only (Figure 30-1). No matter what groupings are used, there is no sharp line of demarcation; one vegetation type merges with its neighbor in the region between them. For example, where one region may consist of forest and the next region a grassland, the boundary between them consists of decreasing trees and increasing grass as one moves in the direction of the latter. In fact, a **savanna** is really an area where scattered trees are found along with grasses. It is usually considered a transition zone. However, it should be noted that sharp boundaries are sometimes found where fire and other harsh environmental factors are involved.

Tundra

The **tundra** (Figure 30-2), extending through the northernmost region of North America, is an area characterized by the absence of trees and the presence of relatively few species of plants. The growing season is very short (two to three months), frost may occur at any time, and **permafrost** occurs at a depth of a few inches to a few feet (a meter or so). This term refers to a soil in which the water is continually frozen. Precipitation is low, but so are evaporation and runoff; the region is relatively flat and poorly drained. The upper layers of the soil thaw during

Figure 30-1. Distribution of major vegetational types in North America.

Figure 30-2. Alpine tundra. (United States Forest Service photograph.)

the summer and are rather spongy and saturated with water. The long days (or continuous days) and relatively high temperatures permit rapid growth during the brief growing season of 5–10 weeks.

The plant cover consists of grasses, sedges, small herbs, low shrubs, mosses, and lichens; annual plants are rare. "Reindeer moss" is actually a fruticose lichen several cm high. Toward the north vegetation becomes more sparse and is predominantly mosses and lichens. To the south the tundra gives way to the boreal forest.

In mountainous areas a similar vegetation is found at elevations above the **timberline** (i.e., the highest limit of tree growth). This is frequently referred to as **alpine tundra** in contrast to the **arctic tundra** of the preceding paragraphs; low temperatures and short growing seasons are common to both. The former has better drainage, gusty winds, and shorter days but with relatively high light intensities. Mosses and lichens are found to a lesser extent in the alpine tundra. Farther south the alpine tundra is found at progressively higher elevations, but it may even be found in modified form in the tropics, usually above about 12,000 feet (4 kilometers).

Boreal Forest

The **boreal forest** (Figure 30-3) is dominated by conifers and is frequently called the northern coniferous forest. It forms a wide belt of small trees (usually less than

569

20 m tall) across Canada and Alaska. It is an area that is cool and moist, and has a growing season of three to four months and long, cold winters. Rainfall averages 50 cm (about 20 in.) per year, evaporation is low, and most of the land is not very hilly. Thus, there are many lakes, slow streams, and extensive bogs. Decomposition is slow in the cold climate and lakes frequently become filled in with peat, the organic matter produced primarily from sphagnum mosses, although sedges may also be involved. In some northern countries peat from such bogs is used for fuel after first being dried. Many gardeners also take advantage of the water-storing properties of peat and use it as a valuable soil additive. White spruce (*Picea glauca*) and balsam fir (*Abies balsamea*) are the most important trees, but jack pine (*Pinus banksiana*) also occurs. In the more moist areas tamarack (*Larix laricina*), black spruce (*Picea mariana*), willows (*Salix* spp.), and poplar (*Populus* spp.) are found. Relatively few species occur in the rather cold areas of the boreal forest and tundra as compared with warmer regions. However, the northern limit of tree growth is probably influenced more by the lack of high enough temperatures during the growing season than by the extreme cold. Permafrost is also a factor by limiting root growth and penetration.

Eastern Forests

The eastern forests extend from the Atlantic coast halfway across the United States, beyond the Mississippi River, to the eastern boundary of the grasslands. They are limited in the north (where they give way to the boreal forest) by progressively lower mean temperatures and shorter frost-free periods, in the west by decreasing rainfall, and in the south by increasing temperatures and low soil fertility. These forests are classified as three general types: hemlock-hardwood forest, eastern deciduous forest, and southeastern coniferous forest.

Hemlock-hardwood forest. This forest (Figure 30-4) extends from New England west, across southern Canada and northern United States, to the Great Lakes with a southerly projection in the Appalachians reaching to Georgia. The dominant trees are hemlock (*Tsuga canadensis*), beech (*Fagus grandifolia*), and maple (*Acer saccharum*). Various other coniferous trees (such as pine, spruce, and fir) and deciduous trees (such as birch and basswood) are also present but in fewer numbers. The shrub and herb understory usually blooms in early spring before the trees leaf out. Once leaves are produced there is a fairly complete crown cover and considerable shading at the ground level. Winters are cold and summers are warm and humid; precipitation ranges from 75 to 100 cm (30 to 40 in.) per year, decreasing toward the west. This forest can be considered as transitional between the boreal coniferous forest and the temperate broad-leaved forest.

Figure 30-4. Hemlock-hardwood forest. (United States Forest Service photograph.)

Figure 30-5. Eastern deciduous forest. **(a)** Summer condition. **(b)** Winter condition. (United States Forest Service photograph.)

Eastern deciduous forest. This forest (Figure 30-5) covers a rather broad area (south of the hemlock-hardwood forest) from the East Coast to the grasslands west of the Mississippi River, but not extending into the coastal plains of the Atlantic states (south of New Jersey) or the Gulf states. The climate is strongly seasonal: summer is hot and humid; winter is cool or cold; rainfall ranges from 75 to 125 cm (30 to 50 in.) per year. This forest is variable with beech, maple, and basswood (*Tilia*) of the northern portion giving way to the dominant oaks (*Quercus*) and hickory (*Carya*) in the southern part. Numerous other tree species form a less prominent portion of the forest.

Chestnut trees were once quite common in the eastern portion, but they have been destroyed by a severe fungus disease. Chestnut blight was first noticed in 1904 and has just about eliminated the chestnut tree from our flora. The pathogen (*Endothia parasitica,* an ascomycete) is native to China where it is not a serious parasite. When it reached the United States on imported nursery stock, it found a very susceptible host in the native chestnut (*Castanea dentata*) and spread with tremendous rapidity. No control measures have been successful, and the completeness of destruction wrought by the disease is without parallel in the annals of plant pathology.

Southeastern coniferous forest. The coastal plain of the Atlantic and Gulf states, south of New Jersey and west to the border of Texas, is dominated by pine forests (Figure 30-6). The sandy soil is low in minerals and poor in water-holding capacity, which is not as disadvantageous to pines as to broad-leaved trees, which prefer a soil with more humus. Repeated fires are also an important factor in the maintenance of the pine forest. Some of the species sprout again from just below the surface, which rapidly replaces the trees lost by burning, and fire hastens the

Figure 30-6. Southeastern coniferous forest. (United States Forest Service photograph.)

opening of some cones. The seedlings of some species of pine are much more fire-resistant than seedlings of broad-leaved species. Broad-leaved species also tend to have small branches and leaves lower down on the trunk than is typical of most pine trees. Periodic low fires frequently damage the former but do little damage to the latter. If protected from fire, the pines usually give way to an oak-hickory climax. The most abundant pines are pitch pine (*Pinus rigida*), loblolly pine (*P. taeda*), slash pine (*P. elliotii*), and longleaf pine (*P. palustris*). Some of these yield the turpentine, rosin, and pulpwood of commerce. Beech and oak are found in more moist areas; bald cypress (*Taxodium distichum*), gum (*Nyssa*), and ash (*Fraxinus*) are found where the water table is high.

Western Coniferous Forests

These forests reach from western Canada down into the United States, where the Pacific Coast forests stretch along the coast to the center of California and the Rocky Mountain forest extends along the Rocky Mountains to Central America. The two prongs of the coniferous forests are separated by the Great Basin, a cool desert area in the United States.

Pacific Coast forests. These forests include the conifer associations along the coast and into the Cascade and Sierra Nevada mountain ranges. It is an area that is more moist than that occupied by the conifers of the Rocky Mountain region but with summer drought. However, because these two forests merge in the north, they are usually treated as one grouping with two southerly diversions. The

573

Figure 30-7. Douglas fir (*Pseudotsuga menziesii*). (United States Forest Service photograph.)

coastal region has 125 to 375 cm (50 to 150 in.) of precipitation, cool summers, mild winters, and a long growing season; the Sierra-Cascade region has 75 to 250 cm (30 to 100 in.) of precipitation, warm summers, cold winters, and a short growing season. The characteristic trees in the former region are Douglas fir (*Pseudotsuga menziesii,* Figure 30-7), western hemlock (*Tsuga heterophylla*), Sitka spruce (*Picea sitchensis*), western red cedar (*Thuja plicata*), and the redwoods (*Sequoia sempervirens*) in California. Western white pine (*Pinus monticola*), ponderosa pine, Jeffrey pine, incense cedar (*Calocedrus decurrens*), white fir (*Abies concolor*), sugar pine (*P. lambertiana*), and the big tree (*Sequoiadendron giganteum*) occur at higher elevations in the Sierra-Cascade Mountains. At still higher elevations, lodgepole pine (*P. murrayana*), whitebark pine (*P. albicaulis*), mountain hemlock (*Tsuga mertensiana*), and alpine fir (*Abies lasiocarpa*) are the dominant trees. An elevational zonation is quite pronounced throughout the region. The short summers at higher elevations and the summer drought at lower elevations tend to be disadvantageous to broad-leaved trees in most of this region.

574

Rocky Mountain forest. This coniferous forest extends along the Rocky Mountains from British Columbia to Central America and produces much of the commercial timber used in the United States. Prominent elevational zonation is apparent, with ponderosa pine (Figure 30-8) giving way to Douglas fir and then to spruce (*Picea engelmanii*) and alpine fir (*Abies lasiocarpa*) at higher elevations.

Grassland

Forests give way to grasslands as rainfall decreases, and this is the situation as one travels west from the Atlantic Coast of the United States. The entire central portion of the United States, and extending into part of southern Canada, consists of an immense grassland (Figure 30-9). There is a zonation within this area with tall grasses (about 2 m high) in the eastern region giving way to short grasses in the more arid western region to the Rocky Mountains. This area is now heavily used

575

for agriculture and grazing, and most of the original vegetation has been almost entirely replaced as a result of man's use.

The original tall grassland is now the intensively cultivated Corn Belt with wheat farther west reaching into the short-grass zone. Most of the latter area, where bison once roamed in such enormous numbers, furnishes natural grazing for cattle and sheep. The soil is dark and fertile and exceedingly productive, limited only by rainfall, where it has not been damaged by erosion resulting from overgrazing and overfarming (see Section 15.8 for a discussion of the Dust Bowl). Rainfall varies from a yearly average of 60 cm (about 25 in.) in the tall-grass region to only 25 cm (about 10 in.) in the western short-grass area. The grassland region is wider (east to west) in the north than in the south, reflecting the effect of temperature on evaporation and transpiration (and thus on water requirement).

There are other smaller grasslands in North America, one of which is the Central Valley of California. Here overgrazing and farming (under irrigation) have pretty well destroyed the native vegetation, which has been replaced by weedy annual grasses accidentally introduced from Europe.

Broad Sclerophyll Vegetation

Most of California from the Pacific Ocean to the western foothills, and extending into northern Baja California (Mexico), has a Mediterranean type of climate with water being the critical factor during most of the growing season. Summers are long, hot, and dry while winters are short, mild, and moist (25 to 90 cm of rainfall per season; about 10 to 35 in.). The dominant species are woody with small, firm, hard, evergreen leaves. Two general associations are present, broad sclerophyll (sclero = hard; phyll = leaf) woodland and chaparral.

Broad sclerophyll woodland. In the more moist and cooler areas scattered trees are found (Figure 30-10), resembling a savanna rather than a forest, although dense thickets of oaks may also occur. The characteristic trees are evergreen oaks (*Quercus wislizenii, Q. agrifolia*), California laurel (*Umbellularia californica*), and madroño (*Arbutus menziesii*).

Chaparral. The chaparral community (Figure 30-11) is dominated by evergreen shrubs (1 to 3 m tall with firm, small leaves) and occurs in the drier habitats. It is subject to frequent fires, and many species sprout freely from the base; within a few years after a fire the vegetation is back to normal. The dominant species are chamise (*Adenostoma fasciculatum*), manzanita (*Arctostaphylos* spp.), *Ceanothus* spp., scrub oak (*Q. dumosa*), and toyon (*Heteromeles arbutifolia*).

Figure 30-10. Broad sclerophyll woodland. (United States Forest Service photograph.)

Figure 30-11. Chaparral in southern California. (United States Forest Service photograph.)

Deserts

Desert regions have low rainfall and a high rate of evaporation with high day temperatures usually alternating with low night temperatures. This creates a difficult situation for plant growth, and various adaptations that enable certain types of plants to survive in such habitats have already been discussed (see Section 15.5). Although some of the soils are fertile, irrigation is required for successful farming. Even this is not without peril because of the accumulation of salts as a result of the high evaporation rates.

Cold desert. In the United States between the Cascade-Sierra ranges on the west and the Rocky Mountain range to the east lies the Great Basin. Rainfall is 25 cm (about 10 in.) per year because the prevailing westerly winds drop most of their water when rising over the more western mountains; the area is in a gigantic rain shadow. The winters are usually long and cold with frequent light snow, unlike the deserts farther south. The cold desert covers Nevada, western Utah, and parts of Idaho, Wyoming, Oregon, and California. It is an area of sagebrush (*Artemisia tridentata*) and scattered grasses (Figure 30-12), although overgrazing has caused a decrease in the latter. Various weedy annual grasses have been introduced. Greasewood (*Sarcobatus vermiculatus*) and shad scale (*Atriplex*) are common in saline soils. An open woodland of piñon pine (*P. monophylla*) and juniper (*Juniperus* spp.) occurs at higher elevations where moisture is more abundant.

Warm desert. Moisture is even more of a problem in the warm desert than in the cold desert with the former receiving about 12 cm (about 5 in.) of rainfall per year. The summers are long and hot, and the winters are mild with occasional frost in some areas but little if any snow. This area covers southern Nevada,

Figure 30-12. Cold-desert vegetation; dominated by sagebrush (*Artemisia tridentata* and related species). (United States Forest Service photograph.)

southeastern California, southern Arizona, and extends into Mexico. The most characteristic plant is the creosote bush (*Larrea tridentata*), which occurs widely spaced (5 to 10 m) with much bare ground in between. Other shrubs, such as bur sage (*Franseria dumosa*) and ocotillo (*Fouquieria splendens*), are also present. Scattered in various locations are the century plant (*Agave americana*), saguaro cactus (*Carnegiea gigantea*), and many other cacti (Figure 30-13). Small trees such as paloverde (*Cercidium*), mesquite (*Prosopis*), ironwood (*Olneya*), and smoke tree (*Daleaspinosa*) are found in watercourses and other areas that are not quite so dry as the others. Scattered oases with abundant year-round water support groves of palms. Ephemeral annual herbs, which have a very short life cycle, may form a colorful display in the spring if the winter rains have been sufficient.

Tropical Forests

Warm temperatures and high rainfall (125 to 875 cm; 50 to 350 in.) evenly distributed throughout the year provide a hospitable climate in which the diversity of species greatly exceeds that of any other area. The dominant trees are very tall flowering evergreens with many herbaceous plants growing as epiphytes on the trunks and branches (Figure 30-14). This frost-free forest barely reaches the tip of Florida but covers much of Mexico and Central America. Where the rainfall is more seasonal, the trees are deciduous during the dry season, and *tropical savanna* or *thorn forest* is produced. In the former, the vegetation is dominated by tall grasses, which often burn during the dry season, and scattered large trees. In the latter, the vegetation consists mainly of small, often thorny trees. In southern Mexico and Central America a tropical forest is present on the Atlantic Coast and tropical savanna or thorn forest on the Pacific Coast.

579

Farming in the tropical forests has been by the traditional cut-and-burn (or

Figure 30-13. Warm-desert vegetation; dominated by creosote bush (*Larrea divaricata*). (United States Forest Service photograph.)

milpa) method. A small area is cleared by cutting and burning, cultivation follows for a few years, and then the area is abandoned as yield declines. The heavy rainfalls leach minerals out of the soil, and they are not replaced by litter as they are beneath the normal forest canopy. Because of the high moisture and high temperatures, much leaching and decomposition occur and these soils are not very fertile. Forests are supported because minerals are replenished by the decomposition of litter.

30.3 Plant Succession

Not only are plants affected by environmental factors, but their own growth and development involve interactions between them that serve to bring about changes in the habitat. Plants shade each other, frequently release toxic products, and

compete for minerals and water. As liter decomposes and humus content of the soil increases, the water-holding and mineral-holding capacities of the soil increase and the structure of the soil changes (see Chapter 17). As the habitat is modified, the types of plants that comprise the community are also subject to change. New species invade the area, which had been inhospitable to them previously, and supplant the original plants. Several successive replacements may occur in this fashion. Such a replacement of one plant community with another in a given site over a period of time is known as **succession.**

As plants compete in an area, some species gradually become dominant and form the greatest portion of the community. Eventually a stage is reached wherein a recognizable community is established that succeeds itself; succession tends toward stability. This is known as a **climax.** The vegetation types discussed in the previous section (e.g., forest, grassland, tundra) are the climaxes in their respective areas. They represent the final stage of succession and are the plants best adapted to the environmental conditions of the area; no other combination of species is successful in replacing the climax community under the present environmental conditions.

Just as there are various environments, there also are various kinds of succession. **Primary successions** are those that begin in areas that have not previously been inhabited by plants, such as bare rock or open water. Once plants become established, succession may be interrupted or a climax destroyed by various forces, such as fire, overgrazing, or abandoning agricultural lands. **Secondary successions** are those that begin on soil after the natural vegetation has been destroyed. A few examples of succession will be discussed.

Bare Rock Succession

Plant succession on a bare rock surface (Figure 30-15) ordinarily begins with crustose and foliose lichens, which grow when water is available and remain dormant during dry periods. There is no soil at this time and thus no reservoir of water that could be utilized between rainfalls. Only the lichens, which require little water to begin with and can withstand considerable desiccation, are capable of surviving in this situation. The mechanical action of the growing lichens (and possibly organic acid production) breaks up rock particles, and their rough surface configuration serves to catch blowing dust. Accumulation of such debris and the

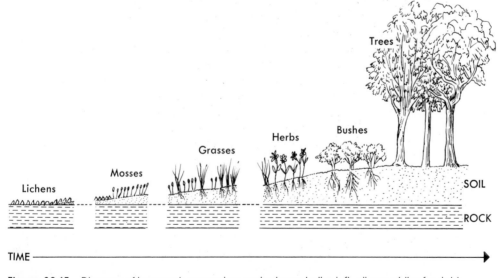

Figure 30-15. Diagram of bare rock succession; early stages to the left, climax at the far right.

decay of lichens initiate the formation of thin soil layers. This small amount of soil and humus increases the moisture content of the area slightly (water-holding capacity is increased) and fruticose lichens and mosses may gain a foothold. These stages may last a very long time; growth is very slow because of the general water deficiency. As the moss mat develops, shading and competition for water and minerals usually eliminate the lichens. As more windblown debris and decaying plant parts accumulate, the soil layer increases.

Rock crevices tend to fill in rather rapidly with wind-borne materials, so that soil conditions are distinctly better from the start and development is not so slow as compared with the bare surface. These crevices commonly begin with a higher type of plant, such as mosses or even herbs. The shoots of these pioneers project from the crevices and intercept material that tends to build up the soil above the level of the crevice. The basic result is similar to that just discussed in that the important factor is the development of soil, which then supports the growth of larger plants.

As the water-holding capacity of the area increases, small annual and perennial herbs, including grasses, become established. More soil develops by the disintegration of the rock and the decay of roots, stems, and leaves. The water-holding capacity is increased still further, and the soil is enriched with nutrients from the decay processes. This enables shrubs to develop, which tend to shade out most of the herbaceous plants but themselves give way to trees. The type of forest that develops depends upon the general environmental conditions (e.g., rainfall, temperature). In fact, the final stage may be grasses or shrubs in drier regions.

The critical factor in this succession is the formation of soil, which provides a reservoir of water that plants can absorb throughout the growing season. Some of the rainfall is not used immediately, or lost as runoff, but is stored in the soil from which it may be absorbed later. The soil also provides anchorage for plants and a supply of minerals. As more and more soil accumulates, larger and larger plants can be supported. These gradually crowd out the earlier stages, mainly by shading although toxic products may also be involved. Each stage in the succession prepares the way for the next. The final stage usually consists of the most mesophytic types that can be supported in that environment; the prevailing climax is adapted to the climate of the area.

Open-Water Succession

The first plants to invade a new body of open water are algae and these are soon followed by submerged aquatic vascular plants, such as *Elodea* and *Myriophyllum*. Inorganic sediment (eroded soil) carried in by drainage water and organic remains gradually cause the pond to become shallower. Nearer the shore, aquatics with floating leaves, such as water lilies, root in the sediment and tend to shade out the submerged plants. The invasion of emergent aquatic vascular plants (e.g., cattail, *Typha;* bulrush, *Scirpus;* sedge, *Carex*) near the shoreline speeds up the filling process. These overtop and crowd out the floating-leaved aquatics. The continued accumulation of debris and sediment results in the pond becoming ever more shallow and the gradual encroachment of the shoreline toward the center of the pond (Figure 30-16). Moisture-loving shrubs and trees invade along the shore (e.g., willow, *Salix;* alder, *Alnus*).

As the pond decreases in size, each vegetation zone moves in toward the center. The amount of dead material settling from above exceeds the rate at which it can be decomposed by decay bacteria, and the pond fills in at an increasing rate. The taller plants shade out the others, and terrestrial vegetation gradually occupies the area as the pond fills. Eventually the climax forest (e.g., beech-maple) develops, with a small central marsh possibly remaining as the only remnant of the pond. Depending upon environmental conditions, the climax may be grassland rather than forest or a conifer association.

All lakes and ponds gradually fill with mineral sediments and organic remains. This may take place in a few decades or thousands of years depending upon the extent of the lake, its depth, the rate of inflow, the rate of plant growth, and other factors. Each stage of succession prepares the way for the next stage.

Secondary Succession

The orderly progress of plant succession may be interrupted by natural forces or by man's intervention. Once a soil has developed, the natural vegetation may be destroyed by fire, grazing, lumbering, or cultivation. If the soil is not destroyed by erosion as a result of removal of the original vegetation, it is suitable for rapid restoration. This revegetation tends to occur more rapidly than the original succession, which depends upon the slow processes of soil formation. Such revegetation is termed **secondary succession.**

Fire. Nearly all terrestrial vegetation is susceptible to burning if the plants are spaced closely enough to carry fire. Fertility may be improved, at least temporarily, by the nutrients released in the ash and by the decaying roots of those plants that have been killed. The available soil moisture status may improve as a result of reduced transpiration. However, there is always some loss of fertility due to volatilization of nitrogen, leaching losses, erosion, and runoff.

After a fire, there is usually an influx of weedy species, which produce copious quantities of well-disseminated seed. These share the habitat with survivors of the fire and with sucker shoots that come up from surviving subterranean organs of woody and herbaceous plants. If the area has not been damaged too severely by erosion, succession is rather rapid and original climax species eventually return to their dominant position.

In some regions fire-resistant species remain the dominant portions of the vegetation as a result of periodic burning although the climatic regime would support other species that are fire-sensitive. The latter are kept in subordinate roles or excluded altogether by the frequent burning. This is apparently the situation in the southeastern portion of the United States where the coniferous forest is dominant. It is also likely that the eastern deciduous forest would extend farther west, into the present grassland, except for periodic fires. Many grasses can mature and produce seed within two years or even the first year in the case of annuals, whereas

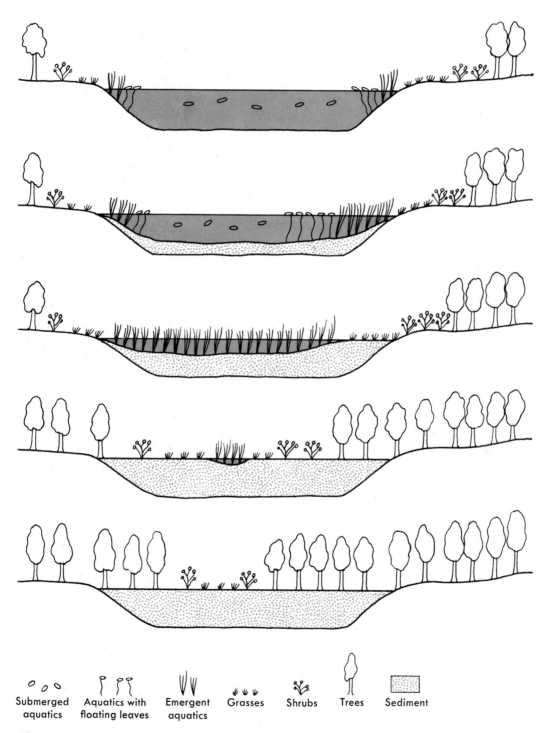

Submerged aquatics	Aquatics with floating leaves	Emergent aquatics	Grasses	Shrubs	Trees	Sediment

Figure 30-16. Diagram of open water succession; early stages at the top, climax at the very bottom.

trees require considerably longer. Recurrent fires would place trees at a reproductive disadvantage.

Abandoned fields. In regions where the natural fertility of the soil is low or where rainfall is low, fields are often cultivated for a time and then abadoned when yield decreases. Secondary succession begins as soon as tillage stops. The first plants to invade are usually annual weeds, which produce large amounts of seed and have previously been kept at a reduced level by cultivation. In a few years the annual weeds give way to perennial herbs. In the more moist areas, shrubs and trees eventually replace the low-growing species. In the Piedmont of North Carolina successional studies have shown vegetational changes of this sort. The original vegetation was a forest consisting mostly of deciduous trees, such as oaks and hickories, which was cleared by the European settlers to make room for their crops. As a result of extensive cultivation and harvesting, nitrogen supplies in the soil became depleted and farms frequently were abandoned. The first plants to invade such fields are mainly herbaceous weeds and annual grasses, followed by perennial grasses. At about this stage, pine seedlings begin to appear, grow slowly at first, but overtop the herbaceous plants between five and ten years after abandonment. Shading makes the area unsuitable for further herb growth, and eventually prevents reproduction of the pine whose seedlings require considerable light for successful growth. (Review *compensation points,* Section 14.1, if necessary.) Seedlings of oaks and hickories are more tolerant of shade than are pine seedlings, and the former also have deeper root systems. These deciduous trees grow upward and eventually overtop and shade the pines. Within 80 to 140 years as the pines mature and die off, they are replaced and the old farm field becomes an oak–hickory forest much as it was originally. This forest is in equilibrium with the climate, and is the climax vegetation for the region. In the drier grassland areas, low fertility of old cropland, resulting from man's removal of parts of crop plants, may delay natural revegetation. Fifty to one hundred years may be required for natural redevelopment of the climax grasses.

Grazing. Moderate grazing is not detrimental to range vegetation and may actually increase productivity. Grazing and browsing tend to remove senescent shoot tissues, which are a drain on the production of the young and vigorous tissues, vegetative reproduction is stimulated, and the amount of litter is decreased. The accumulation of too much litter is detrimental to germinating grasses and small grasses. Heavy grazing, or overgrazing, reduces the photosynthetic capacity and food reserves, dwarfs the root system, and results in weakened plants. There is a decrease in abundance of the more palatable plants and an increase in the less palatable ones until the livestock is forced to eat the latter. Weedy annuals then become common. If the plant cover is diminished too much, rainwater tends to run off quickly rather than sinking in. Flooding is exaggerated and streams may dry up in summer because of a lack of water reserves in the soil. Erosion is of course increased with such runoff.

Moderate grazing, alternating with a "rest" period of no grazing, will retain a parcel as productive grazing land. If an area has been overgrazed to the point where most forage grasses have been depleted and annual weedy species have invaded, it may take many years for restoration of the original vegetation type.

30.4 Land Utilization

Man has utilized land continuously throughout his existence. Early man simply collected plant materials and caught animals as he needed them, and there was no concerted effort toward agriculture or animal husbandry. These latter were activities that gradually appeared as man learned by trial and error. As societies and cultures developed, land began to be used in a more formal way. At the present time, we can consider this usage in terms of farming, grazing, lumbering, and recreation.

Humans have preferences for certain plant materials just as do most animals. Humans differ from most animals, however, in being capable of planting and caring for those plants they deem most suitable to their needs. As a result, much land is utilized in cultivating plants that are not necessarily good competitors—in fact they usually are poor competitors. They are successful because of man's intervention in controlling unwanted types (or weeds). Breeding programs develop varieties suitable for man's use but which usually would have a difficult time surviving without man.

Farming has frequently resulted in exceedingly detrimental effects upon soil productivity because harvesting removes those minerals that the plant has absorbed. Recycling, an important facet of nature, does not occur. This is, of course, the basic reason for fertilizer applications. As less new land becomes available for cultivation, owing to increased population pressures, it is not possible to abandon farms and move to new areas. In many areas of the world fertilization is not practiced (frequently for economic reasons). Crop yields gradually decrease as soils are depleted.

Rotation of crops is a useful tool in alleviating some of the effects of mineral removal. Plants vary in requirements and in the absorption of minerals, so that varying the types of crops that are grown in an area may decrease the rate at which a particular mineral becomes deficient. Rotation including a legume, such as clover or alfalfa, is especially beneficial in maintaining nitrogen contents (see Section 20.6). Plowing under those plant parts that are not harvested will return minerals to the soil and improve soil structure by the addition of humus.

It should also be borne in mind that crop plants grown in rows expose considerable amounts of soil to the erosion effects of water and wind—much more so than a natural mass of vegetation covering that soil. Plow furrows, especially, provide pathways for flowing water. Contour plowing is utilized for this reason; furrows at right angles to drainage flow greatly retard runoff and decrease erosion. Windbreaks of trees or shrubs can decrease the power of the wind and are useful in erosion control. If care is used in applying these methods of cultivating, productivity can be maintained.

Much land is used for grazing purposes. In this case also tremendous amounts of damage have been done throughout the world. Recycling is interfered with by the removal of livestock, and overgrazing destroys the vegetative cover. The resultant erosion removes the upper, most productive layers of the soil. As in any land use, the factor to bear in mind is that plants are essential for soil development and to anchor soil in place. Grazing, as well as farming, can be continued as long as the methods used allow for regeneration of the plant cover.

Forest lands supply us with lumber for building and with wood that is used in the production of paper, turpentine, and many other materials. Unfortunately, man usually has considered the supply of trees to be unlimited. Only in recent years have any real attempts been made to manage forests on a sustained-yield basis. In some areas at least the more sensible methods of block cutting and selective cutting are used, depending upon the trees involved. Douglas fir (*Pseudotsuga menziesii*) is not shade tolerant as a seedling, and its place in the forest would rapidly be appropriated by shade-tolerant species if selective cutting were practiced. Harvesting is done by cutting all marketable trees in a block of 100 acres or less with ample standing timber left between the blocks. Seeds blow in from the surrounding forest. The new stand is thinned periodically to provide growing space and to remove undesirable trees. In about 40 years the trees in the first cutover block are mature enough to provide a seed source for neighboring blocks. In a few more decades this first block can be harvested again. The block-cutting technique has the inherent difficulties of increased water runoff and decreased support of the animal population in that area. Thus, if it is possible, selective cutting is preferable.

Young red spruce will grow in the shade of older trees, and harvesting is done by selecting marketable trees for cutting. The remaining trees provide seed, and additional trees are cut as they mature. With care the forest can provide a continuous supply of trees. Natural reseeding is not completely adequate and it is usually combined with aerial, hand, or machine seeding. In many forests that are being managed to provide a sustained yield, young trees are planted from plantation stock.

Many land areas are used for a variety of recreational purposes. The increasing interest of people in hiking and camping has emphasized the importance of establishing national, state, and local parks to retain land in an undisturbed condition. Many such areas are available although most people who are interested in "nature" realize that additional areas should be preserved before it is too late. Once the natural vegetation has been changed by man's intervention, it is almost impossible to recover the original condition.

Unfortunately, a great deal of land has been removed from recreational or productive use or from aesthetic appreciation by being covered by suburbs, roads, shopping centers, parking lots, and so forth. Concrete and asphalt have covered millions of acres of potentially productive land at about two million acres per year after World War II. The characteristics that make land desirable to a farmer (level area, accessibility, lack of rocks) also make it attractive to the developer. It is estimated that at least 40 million acres of land in the United States has been covered in this fashion; probably more than 15 per cent of the cropland. In some areas the situation is much worse than in others. The eastern seaboard of the United States is almost a continuous urbanized mass from Washington, D.C. to Boston, Massachusetts (a distance of about 805 km or 500 miles); and 20 per cent of California's best agricultural land has been lost. The best farmland is often associated with cities, which expand into this prime land. On a national scale about 70 million acres is considered to be such prime agriculture land, and about one half of this has been urbanized. Much of the thousands of acres of land that remains is not usable by farmer or urbanite. It is not merely the amount of land that is being removed from agricultural use that is a matter for concern but the fact that much of this is the best (most productive) farmland. It is pretty well impossible to reclaim such land. One can only hope that greater ecological planning will invade our system of development.

30.5 Water Use

The largest water consumer is industry, using about 225 billion gal per day. A large paper mill, for example, uses more water each day than does a city of 50,000 inhabitants. The next greatest single use of fresh water in the United States is for the irrigation of approximately 44 million acres of land, about 7 per cent of the total cropland. This is about 180 billion gal per day, or almost one half of the freshwater we use annually. Use of water in and around the home varies from about 150 gal per person per day in homes with running water to about 15 gal per person per day for homes without that convenience. In the United States the average is about 40 gal, although only a small portion of this is actually consumed. In ancient villages the total water used for all purposes averaged about 3 gal per person. To compare this latter figure with our technological society, we should really calculate the per capita consumption of all water in the United States and not just home use. On this basis, the average daily use of water for all purposes in the United States is approximately 1500 to 2000 gal per person. The per capita average daily use of water was only 600 gal in 1900 and is expected to be 2300 gal by 1980. It is this tremendous consumption that is cause for alarm.

The Geological Survey estimated that an average of approximately 4300 billion gal of water are released daily over the United States through precipitation (about 30 in. annually for the entire country), and that about 70 per cent returns to the atmosphere through evaporation and transpiration. The rest contributes to the formation of ponds and streams or percolates as groundwater into the earth. It is this component that may be used for domestic, industrial, or agricultural purposes, with surface waters (e.g., ponds, streams) satisfying about 80 per cent of man's needs. It is estimated that lakes and rivers in the United States contain about 650 billion gal of water.

It should be borne in mind that water *use* and water *consumption* usually are not the same. For example, of the 225 billion gal per day of water that industry uses, less than 20 billion gal per day are actually consumed. Most of the water is used for cooling or to remove wastes. In fact, except for irrigation, no major water use consumes more than a small fraction of the total water used. The yearly demand for water in the United States is about 115 billion gal. If all the water now in reservoirs, water supply systems, and waste systems were efficiently recycled, the annual demand and losses to the atmosphere would easily be replaced by annual precipitation. In terms of total amounts, there is plenty of water. However, it is not always sufficient in a particular area nor is it always suitable for use when it is available. Industrial discharges, sewage, and salt accumulation may make a water supply unfit for use by humans, other animals, and plants. The greater the consumption of water, the greater the necessity for purifying, reclaiming, and recycling the water supply.

Various lakes and streams are used for recreational and commercial fishing. There are over 35 million licensed fishermen in the United States. With the increasing population and the increase in leisure hours and mobility, sport fishing is growing enormously in popularity. The pollution effects mentioned in the previous paragraph would seriously curtail this use of water, again emphasizing the need for conservation and recycling.

Water can also be used to produce electric power by channeling water flow to drive a turbine that generates electrical energy. Water impounded behind a dam may be used in this fashion, as may water flowing over a falls. However, only a small portion of the power needs of most countries is now met from hydroelectric sources.

Summary

1. The environment consists of physical and biological factors that interact to produce habitats for organisms. The vegetation that develops is determined by such interactions.

2. In North America the vegetation may be classified as several natural, and convenient, types. The tundra, in the northernmost regions, is characterized by the absence of trees and the presence of low-growing plants. The boreal forest, primarily in Canada, is dominated by conifers. The hemlock–hardwood forest extends across southern Canada and northern United States from the East Coast to the Great Lakes. South of this forest is the eastern deciduous forest, which extends westward to the grasslands. The coastal plain of the Atlantic and Gulf states, to the border of Texas, is dominated by pines that comprise the southeastern coniferous forest. Grasslands are found in the entire central portion of the United States and extending into part of southern Canada. The western coniferous forests reach from western Canada down into the United States, where the Pacific Coast forests stretch along the coast to the center of California and the Rocky Mountain forest extends along the Rocky Mountains to Central America. Most of California from the Pacific Ocean to the western foothills, and extending into northern Baja California (Mexico), has a broad sclerophyll vegetation: broad sclerophyll woodland in the more moist and cooler areas, and chaparral in the drier habitats. The cold desert of the United States, with its sagebrush and scattered grasses, lies between the Cascade-Sierra ranges on the west and the Rocky Mountains to the east. The warm desert, with its creosote bushes and scattered cacti, covers parts of southwestern United States and extends into Mexico. The tropical forest vegetation reaches the southern tip of Florida and covers much of Mexico and Central America.

3. The replacement of one plant community with another in a given site over a period of time is termed plant succession. Primary successions are those that begin in areas that have not previously been inhabited by plants, such as bare rock

succession and open water succession. Secondary successions are those that begin on soil after the natural vegetation has been destroyed, as by fire or grazing.

 4. Man utilizes land for farming, grazing, lumbering, and recreation.

 5. Water is used for irrigation, industrial processes, domestic consumption, fishing, and recreation.

Review Topics and Questions

1. It has been stated that the term *environment* refers to a very complex situation. Briefly indicate the complexities that are involved.
2. What is meant by "the timberline"? What is responsible for its presence?
3. Discuss possible reasons for the northern limitation of the boreal forest.
4. Discuss possible reasons for the establishment of grasslands, rather than some other vegetation type, in the central part of the United States.
5. Why is there no sharp line of demarcation between the eastern forests and the grasslands in the United States?
6. Discuss the basic modifications of the environment that result in successional stages of plant growth, starting with (a) bare rock and (b) open water.
7. What is meant by "secondary succession"?
8. Discuss the effect of farming on soil productivity.
9. The per capita average daily use of water in the United States is about 1500 gal. Obviously only a small amount is actually consumed by humans. Indicate the types of uses that are responsible for this high utilization.

Suggested Readings

BILLINGS, W. D., *Plant, Man, and the Ecosystem,* 2nd ed. (Belmont, Calif., Wadsworth, 1970).

ETHERINGTON, J. R., *Environment and Plant Ecology* (New York, Wiley, 1975).

GLEASON, H. A., and A. CRONQUIST, *The Natural Geography of Plants* (New York, Columbia University Press, 1964).

KORMONDY, E. J., *Concepts of Ecology* (Englewood Cliffs, N.J., Prentice-Hall, 1969).

OWEN, O. S., *Natural Resource Conservation: An Ecological Approach* (New York, Macmillan, Inc., 1971).

WAGNER, R. H., *Environment and Man* (New York, Norton, 1971).

WHITTAKER, R. H., *Communities and Ecosystems,* 2nd ed. (New York, Macmillan, Inc., 1975).

Populations and Their Problems

At various times throughout this text, the dependence of all organisms upon green plants for a food supply has been emphasized. The few chemosynthetic autotrophs that can manufacture their own food are insignificant as far as a supply of food for other organisms is concerned. In order to understand more fully the basic relationships involved, we will undertake a discussion of factors that influence the numbers of individuals within a given population.

31.1 Population Increase

To most biologists, and to many other individuals as well, the most serious problem facing us is the tremendous increase of the human population: not merely the large population that now exists on the earth, but the rate at which this population is increasing each year. Malthus was one of the first to point out that all populations have a tremendous capacity for increase because of the reproductive potential of organisms. Each plant or animal is capable during its lifetime of leaving large numbers of offspring upon the earth. Consider, for example, the enormous numbers of spores produced by puffballs, the number of seeds (or fruits) released from a single small dandelion plant, or the large quantity of sperms and eggs secreted by spawning salmon. If all the spores or seeds survived or if all the eggs were fertilized and survived, each of these organisms (in pairs, possibly) is capable of producing progeny in such quantity that the earth would soon be cluttered with their descendants. A similar situation would result in the case of organisms that reproduce more slowly, except that there would be a longer interval before these progeny would overrun the earth. Elephants reproduce rather slowly, starting at about 30 years of age and sometimes continuing until the age of 90. But even at this slow rate, if all progeny reached their full life span, the descendents of a single pair would number over 18 million in 750 years. At the other end of the scale, one could examine certain bacteria that divide about every 30 minutes under ideal conditions. In 24 hours, a single bacterium could give rise to 2^{48} descendants if they all survived. This would be more than 281 trillion, weighing ap-

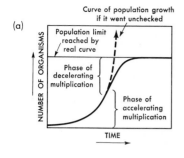

(a)

Curve of population growth if it went unchecked

NUMBER OF ORGANISMS

Population limit reached by real curve

Phase of decelerating multiplication

Phase of accelerating multiplication

TIME

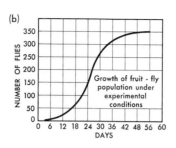

(b)

NUMBER OF FLIES

350
300
250
200
150
100
50
0

Growth of fruit - fly population under experimental conditions

0 6 12 18 24 30 36 42 48 56 60
DAYS

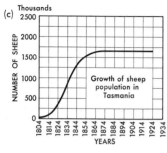

(c)

Thousands
2500

NUMBER OF SHEEP

2000
1500
1000
500
0

Growth of sheep population in Tasmania

1804 1814 1824 1834 1844 1854 1864 1874 1884 1894 1904 1914 1924 1934
YEARS

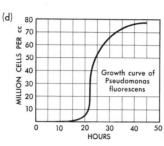

(d)

MILLION CELLS PER cc

80
70
60
50
40
30
20
10

Growth curve of Pseudomonas fluorescens

0 10 20 30 40 50
HOURS

Figure 31-1. Curves of population growth. **(a)** The broken line represents the potential growth of a population; the solid line represents the actual growth. **(b), (c), (d)** Growth curves of actual populations. ([a] and [c] from *Life, An Introduction to Biology*, by G. G. Simpson, C. S. Pittendrigh, and L. H. Tiffany, New York, Harcourt, Brace & World, 1957. [b] from *The Biology of Population Growth*, by R. Pearl, New York, Alfred A. Knopf, 1930. [d] from *Physiology of Bacteria*, by Otto Rahn, copyright 1932, Blakiston Div., McGraw-Hill, used by permission.)

proximately 245 gm or 9 oz; an average-size bacterium, such as *Escherichia coli*, weighs about 8×10^{-10} mg. Although the number of individuals is represented by an impressive figure, the weight of such a large number of bacteria may not appear to be very significant. However, if this rate of reproduction and survival continued for a mere 30 days, the weight of this mass of bacteria would be more than 118 million kg (118,000 metric tons or 130,000 U.S. tons). Possibly this is more impressive.

Populations obviously do not increase continually at the rate of which they are capable. Eventually, the rate decreases until the population stabilizes, because the number of individuals that can be supported by an environment is limited. Figure 31-1(a) represents the potential growth of a population as a broken line and the actual growth as a solid line. Figures 31-1(b), 31-1(c), and 31-1(d) are growth curves for actual populations; similar curves could be obtained from data of other populations. The limiting factors that bring about the stabilization of a population are no food supply and its resultant starvation, predators, parasites, accidents, and climatic conditions. Organisms retain the ability to reproduce rapidly, but only a small fraction of the offspring actually survive because of various limiting factors.

31.2 Human Population

The problem of most vital concern to us is the future of human populations. Demographers are those individuals concerned with the vital statistics (births, deaths, and so forth) of populations although the term is usually used in reference to those concerned with human population. Many demographers have pointed out that the number of humans has been increasing with leaps and bounds, especially during recent generations. Since records either were not kept or were not very accurate before the 1800s, the estimates of early populations are not as valid as one would wish. Various ways of making such estimates have been suggested. For example, man requires about 2 sq mi of good hunting territory to support himself. Therefore, before the shift from food collecting to food producing (agriculture), man required much land for his survival. The entire earth, on the basis of available hunting territory, could have supported probably no more than 8 to 10 million people. Archeological and geological investigations place other limitations on population estimates—in what areas is there evidence of man's existence, what areas probably had a suitable climate, and so forth. Realizing the variabilities and inadequacies of such data, demographers are rather careful to be conservative in their estimates. By using many different kinds of evidence from various sources, however, they can make estimates that are probably fairly accurate.

On the basis of various kinds of information, Warren Thompson suggests the following data as an indication of population growth. Estimating that 1 million people lived on earth 10,000 years ago and 600 million in 1700, the average yearly increase was only 0.64 of a person per 1000 of the population. From 1700 to 1950, a period for which the figures are understandably more accurate, the population increased to 2.5 billion; this is an average annual increase of 5.5 persons per 1000. The rate of increase is accelerating and it takes fewer years for the population to double (Table 31-1). If the 1976 rate continues, doubling will occur in 35 years (i.e., eight billion by 2011 A.D.). The factor primarily responsible for this sharp increase is the relatively rapid drop in the death rate, as a result of modern medical and sanitary facilities (Table 31-2). Also, the average life expectancy has increased (Table 31-3). Even now, however, the life expectancy at birth of at least one half of the present population is about 50 years, compared to 70 to 73 years in countries where modern preventive medicine has been applied (Table 31-4); consider what the population increase would be if such preventives were in universal use.

The rate of growth of the human population is more than 2 per cent per year (as of 1975). Although this figure appears to be very small, no sustained population growth, even at a very low rate, could be maintained for very long. For example, a rate of one half of 1 per cent maintained for another 2000 years would leave less

Table 31-1 Increase in human population

Date (A.D.)	Estimated population (millions)	Doubling time (years)
1650	500	1500
1850	1000	200
1930	2000	80
1950	2500	—
1970	3500	—
1976	4000	46

Table 31-2 Decrease in death rates

| From 1920–1924 to 1957 | | United States | |
		Year	Deaths per 1000
British Honduras	56%	1900	17.0
Chile	57%	1950	9.6
Costa Rica	54%	1960	9.4
Japan	64%	1975	8.9
Taiwan	67%		

than 4 sq yd of the earth's land surface for each person. At the 1975 rate of increase, it would take less than 500 years for this situation to obtain. The question is not merely one of food supply, but also of living space.

The immediacy of the problem becomes impressive if one examines the percentage increase per year and the doubling time in years for that increase. Table 31-5 gives the annual rate for selected countries, while Figure 31-2 shows the approximate doubling time in years for these rates. Although the annual rate of increase varies somewhat, the 2 per cent increase annually for the world as a whole means a doubling of the world population in about 35 years. The resulting competition for available resources and space becomes more striking with the passing years.

Demographers indicate birth and death rates in terms of the number of births or deaths for every 1000 people. An annual birth rate of 40 per 1000 means that each year the population is increased by 40 for every 1000 people who were present the year before. If 250,000 people live in an area with such a birth rate, the population will be 260,000 the following year if all survived (250,000 × 40/1000 = 10,000 births). Costa Rica recently announced an annual growth rate in excess of 5 per cent; this would result in a doubling of population in about 14 years. A number of countries were experiencing a population increase of a bit over 3.5 per cent (1975), which would bring about a doubling in about 20 years (e.g., Algeria, Pakistan, Syria, Venezuela). The increase of world population between 1950–1970 was

Table 31-3 Length of life for Americans

| Year | Age | |
	Male	Female
1900	46.3	48.3
1930	58.1	61.6
1960	68.4	72.1
1970	67.1	74.8
1975	68.2	75.9

Table 31-4   Life expectancy (1975)

Country	Age
Netherlands, Sweden	75
Australia, U.S.A.	72
Japan, U.S.S.R.	70
Argentina	69
Mexico	55
India, Pakistan, Cambodia	46
Guatemala	40
Guinea, Ivory Coast	35
World	55

about two times the size of the entire world population in 1650; and there are as many people in China today as in the whole world in 1800.

In Ceylon, that large island off the southeast coast of India, the population has doubled since World War II as a result of malarial control. Birth rates have remained fairly steady, whereas death rates were greatly decreased by spraying malarial areas with DDT (a potent, residual insecticide). Similar actions aimed toward disease prevention and further efforts to establish greater sanitary facilities and cleanliness will undoubtedly bring about great decreases in death rates in many areas. The world death rate of 25 per 1000 in 1935 dropped to 14.5 by 1975 and is expected to fall to 12.7 by 1980. The resultant population increases will be truly astounding, even more so than they have been already.

The present world population is over 4 billion, and it is increasing at a rate that will add in the neighborhood of 1 billion people in the next 10 years. This increase is more than the total present population of the Western Hemisphere. In 1975 over

Table 31-5   Annual rate of increases in population for selected countries, 1975

Country	Percentage increase
Venezuela	3.5
Ecuador	3.4
Brazil	2.8
India	2.1
China	1.7
Japan	1.3
U.S.A.	1.1
U.S.S.R.	0.9
Norway	0.7
Sweden	0.4
World	2.0

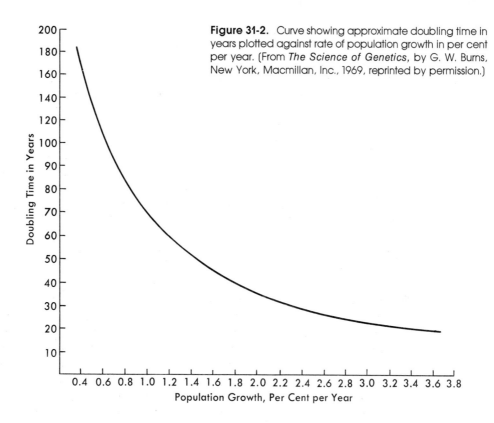

Figure 31-2. Curve showing approximate doubling time in years plotted against rate of population growth in per cent per year. (From *The Science of Genetics*, by G. W. Burns, New York, Macmillan, Inc., 1969, reprinted by permission.)

88 million people were added to the world, at a rate of almost three per second (which amounts to more than a whole new United States every three years), and most of these drew their first breath in Asia. The question that surely must arise in everyone's mind is "How will they obtain food?" For example, total land under cultivation in India is about 80 per cent that of the United States, but total crop yield is only 40 per cent that of the United States, while the population of India is almost three times greater.

Besides the 170 new mouths to feed each minute, there are 170 new bodies to clothe and shelter; 170 new brains to educate; 170 new inhabitants needing their share of land, water, jobs, medical care, and so forth. The world population may reach almost eight billion by 2000 A.D. if the present rate continues—a doubling in 35 years from the four billion in 1976 at a 2 per cent increase. Most demographers are anticipating a dropoff in the rate, and this is already occurring in North America and western Europe where the change has been from about 1.6 to 1 per cent over the past ten years. Some demographers even estimate that the world growth rate has actually fallen a bit below 2 per cent and that it may drop to 1 per cent by 1985, which would lower the estimate of total population to 5.4 billion by 2000 A.D. Unfortunately, however, declining death rates in some developing countries have not yet been matched by decreases in birth rates, and the annual rate of increase for the world may be as high as 2.2 per cent (as estimated by some demographers). Usually population estimates tend to be conservative. For example, in

1955 the U.N. population office was estimating that the annual increase in 1965 would be 35 million and in 1965 it was actually about 55 million. At present (1977), the world's population increases every two weeks by an amount greater than the number of people living in the city of Los Angeles, and every month over seven million people must be accommodated. (See Figure 31-3.)

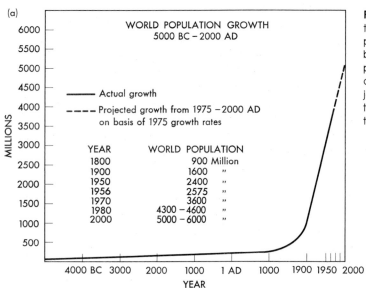

Figure 31-3. Curves of human population growth in various areas. **(a)** World population growth. The size of the interval between 1900 and 2000 has been disproportionately lengthened to give a clearer presentation of past and projected population growth during the twentieth century. **(b)** Growth of population in continental United States.

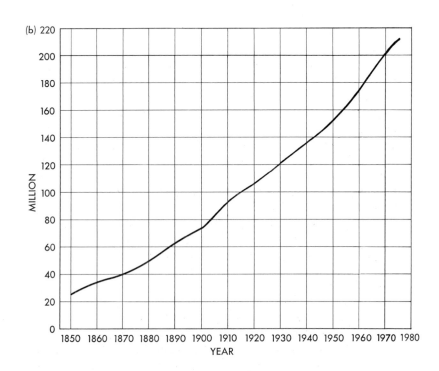

(c)

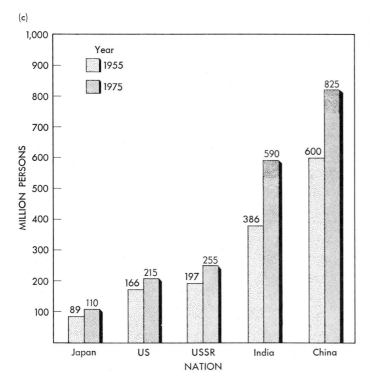

Although the population problem is most serious in the Asiatic countries, this should not make us lose sight of the situation in the United States. In April 1960 the population of the United States reached 180 million; it was 185 million by December 1961, 200 million in 1968, 213 million in 1975, it may pass the quarter-billion mark by 1980, and probably pass 350 million by the year 2000. These figures may seem somewhat insignificant when compared with the 1980 estimates for China at 1100 million and India at 660 million. However, one must also bear in mind what these population increases mean in terms of land in the United States.

In 1950 there were 12.5 acres of land per individual in the United States, including that in cities, highways, mountains, deserts, and marshland. That figure shrank to 8.8 by 1975 and will probably shrink to 5.7 acres per individual 25 years later. The United States Department of Agriculture estimates that there were 3.1 acres of cropland per individual in 1950, 2.2 in 1975, and that this will fall to 1.6 acres by 2000. These are figures that concern biologists and others. We can continue to produce more food per acre for quite some time, but there is a limit. Instead of farm surpluses, we will have to strain our mighty farm machine to keep our standard of living as high as it is today. And what of the rest of the world? Rapid improvements in agricultural techniques have increased food supplies faster than the arithmetical increase suggested by Malthus, but he was right in principle. There are limits to the number of persons that the Earth can support.

31.3 The Food Problem

According to nutritional experts of the World Health Organization, each individual requires approximately 2500 calories per day to be adequately fed. The term *calorie* refers to energy requirements and is used in a general way to indicate dietary needs. However, calories alone do not really measure nutritional needs. If the calories are predominantly provided by cereals and potatoes, a deficiency of essential vitamins, proteins, and minerals is very likely. Fewer than 2500 calories per day is certainly insufficient, even though a greater caloric intake may not be qualitatively sufficient; therefore, the caloric content of foods can be utilized as an indication of an absolute minimum. At the present time, by this standard, more than one half of the world's population is undernourished. According to the 1955 report of Political and Economic Planning (PEP), *World Population and Services,* the number of underfed approximates 98 per cent in Asia, 93 per cent in Africa, 80 per cent in South America, and 44 per cent in Europe. Other estimates indicate that probably 20 per cent of the people in the United States are also undernourished. The situation has improved somewhat because food production has increased by an average rate of 2.4 per cent per year, or slightly more than the population increase over the last two decades; but the increase has been exceedingly slight in the developing countries where the problem is greatest. There has been little real improvement in the diet of those who are in chronic nutritional need. And this is without taking into consideration *malnourished* people, those that do not eat the right kinds of food (especially with regard to lack of protein in the diet).

Food production has increased in all countries of the world, but in many of them the increase has been less than the percentage increase of the population. In Pakistan, for example, a 20 per cent boost in food production since World War II has been canceled out by the great population increase there. Today that country has 10 per cent less food per person than 15 years ago. Egypt's Aswan High Dam is expected to increase arable land by 30 per cent (about 1.2 million acres). But by the time the dam was finished in 1970, Egypt's population had soared by slightly more than 30 per cent. Even more modern agricultural techniques cannot keep pace with such a situation. In fact, A. E. Staley of the Stanford Research Institute stated, "Despite all the vaunted technological and economic progress of modern times, there are probably more poverty-stricken people in the world today than there were fifty years ago."

It is not merely the increase in numbers of people that makes the food problem a difficult one. It must be borne in mind that the amount of food required increases steadily from the time of birth to about 19 years of age. Ivan L. Bennett, for example, estimated that it would require 20 per cent *more* food in 1975 than in 1965 to *maintain* the low level of nutrition of the Indian population if *no new* children were added during this ten-year period. To elevate the diet to the minimal standard of 2500 calories would require a 30 per cent increase. In terms of wheat, these caloric increases represent 30 to 40 million tons, more than twice the amount of our largest annual shipment to India and about the increase obtained from favorable trends in Indian agriculture. However, the population of that country increased at more than 2 per cent each year. Thus, no improvement in the standard of living could occur even with a 30 per cent increase in agricultural production. In fact if the world population keeps expanding at the present rate, within 100 years it would be necessary to increase food production about eight times if present standards are to be maintained—and yet this is inadequate for over one half of the present number of people. In other words, we need almost an immediate doubling of food production to reach an adequate diet for the present population. Such rapid and large increases have not been achieved in the past.

In many areas lack of conservation measures and overcultivation without regard for a sustained-yield basis have resulted in the gradual loss of many formerly productive lands. Members of the Soil Conservation Service have estimated that over 280 million acres (113 million hectares) of farm and grazing land have been ruined or seriously damaged in the United States and that erosion is active on an additional 775 million acres (314 million ha). The development of good topsoil is a process that takes thousands of years, while its destruction may take place in a very few seasons as a result of poor land management. C. E. Millar[1] of Michigan State College points out (in 1948) that erosion has seriously curtailed average crop yields in his state, and that 25 per cent of Michigan land has lost 5 to 7.5 cm (2 to 3 in.) of topsoil. He indicates that a few crops, such as wheat and potatoes, have had an increase in yield but that the average yields of crops most generally grown are about the same as they were 60 years ago. These yields should be much greater as a result of the advances in the agricultural sciences, such as the greater use of commercial fertilizers, new varieties of plants, more irrigation, and more effective control of diseases and pests.

Similar erosion problems and the resultant impoverishment of the land have occurred in every country of the world. Large areas of forest, farm, and grazing lands are no longer useful. About 17 per cent of the land now cultivated on Earth is subject to such rapid erosion that it may well be lost for food production; about 6 million hectares (14.8 million acres) of land are being lost each year. The saddest feature of such destruction is that man has caused the entire situation by a lack of an appreciation of natural laws. This is one of the reasons that an understanding of the basic principles of plant distribution is so important.

31.4 Food Sources

Estimates vary, of course, as to the amount of land now under cultivation, the amount necessary to maintain a population at an adequate level of living, and the amount of land that could yet be utilized. Assuming rather conservative estimates, probably $2\frac{1}{2}$ billion acres (1 billion ha) are now under cultivation and another 1 billion acres could be brought into cultivation. With the present agricultural tech-

[1] Mentioned in Stanley A. Cain, "Natural Resources and Population Pressures," *Bios,* XXI, 4 (December 1950).

niques, approximately 2 acres (0.8 ha) per person are required to sustain an adequate level of living. This means that the present world population probably could be supported if we make certain assumptions, which may or may not be valid.

First, we must assume that some mechanism could be devised to distribute food in such a manner that areas with a surplus would supply areas with a deficiency. This is similar to farmers supplying city dwellers. However, in the latter case the urbanites supply a variety of useful articles (e.g., cars, tractors, furniture, electrical appliances, and so forth) to the farmer in return for this food. Would similar possibilities of exchange exist between nations? At the present time, very few nations are able to fill their own needs with regard to food supplies; most nations import foods. England, a great industrial nation, imports approximately 40 per cent of its food. This indicates the illusion of those who believe that industrialization of nations will solve food problems. If all nations increased their industrial potential to the height enjoyed by England, the United States, and a few others, to whom would such nations turn for their food supplies? At the present time, only the United States, Canada, Argentina, New Zealand, and Australia do not need to import food, and the last named is not in too good a position. One possibility is the export of minerals, such as iron ore, or oil in return for food. Such an exchange cannot continue indefinitely of course.

A second assumption in our attempt to support the world's population is that everyone will be willing to accept an "adequate level of living." Assuredly, this means raising the standards of the majority of people. Just as assuredly, it would mean the lowering of standards for most of the people in at least North America and Western Europe. Human nature is such that we will heartily support the raising of standards in general—our own as well as our neighbors'—but we will usually not support the raising of our neighbors' standards at the expense of our own. Probably very few individuals will be willing to accept any substantial lowering of their own level of living.

The third basic assumption is probably the most serious. Our *present* world population could be supported. Therefore, we must assume that the population of the world will not increase substantially or that food production will increase at the same rate that the population does. But populations are increasing tremendously, even though the rate is only approximately 2 per cent for the world as a whole. If the death rates in most of the world are reduced to those of North America or Western Europe, the rate of increase would be close to 3 per cent. Within a relatively short period of time, certainly during this century, there would be 6 billion people and only 3.5 billion cultivatable acres—an almost impossible situation with our present techniques. Also, as we have already noted, food production in general is barely keeping pace with population increases, and not even that in most developing nations.

It should be emphasized that *protein deficiency* is the most important aspect of the food problem. Milk, meat, and eggs are pretty well restricted to the diets of North Americans and West Europeans. Hundreds of millions on earth never see such delicacies and receive what proteins they do get from plant foods. Plant proteins, however, are usually low in certain amino acids that are essential to humans. A diet almost exclusively of rice, or other grains, would tend to be deficient in lysine, and probably methionine and tryptophan also. These are three of the eight amino acids that humans must obtain from food.

Available Land

Much of the additional land that could be brought under cultivation consists of rather dry areas, including deserts, and tropic areas with their excessive rainfall. In the former, the expense of irrigation would vary considerably, depending upon the distances from a source of water and upon feasibility of converting seawater to freshwater. However, as food supplies become increasingly difficult to obtain, the question of expense will become less important, and undoubtedly more and more of these relatively dry areas will be utilized. In many areas of the southwestern United States, irrigation has made possible the production of tremendous crop yields. A difficulty, which should not be overlooked, is that irrigation of areas where evaporation is high usually results in an increase of salt content of the soil. Water contains dissolved salts, and as the water evaporates, the salts are left behind. This increase in salt content eventually results in poor plant growth and reduced yield.

Cultivation of the tropics poses an entirely different problem. Here the difficulty is not moisture but the absence of minerals. The average individual finds the fact almost impossible to believe that most of the soils in tropical rain forests are impoverished—sadly depleted or lacking in minerals. Hundreds of years of rains and warmth have weathered and leached such soils to the point where mineral reserves are almost exhausted. Then what supports the luxuriant forest vegetation? The answer lies in the accumulation and decay of humus. This humus is continually replaced by the shedding over the ground of tons of leaves. As these leaves decay and participate in the formation of humus, mineral elements are released to the soil. This continues as the humus is decomposed. The forests thus are supported because they return, or *recycle,* considerable quantities of minerals to the soil. Once the trees are removed, the rapid decomposition of humus (which always occurs in such moist, warm areas) results in the production of minerals, which are washed or leached out of the soil by the following rains. No replenishing of the humus takes place because crop plants are harvested and at least a considerable portion removed; soils rapidly deteriorate to the point where they will no longer support much growth. A former North American agricultural attaché in Brazil remarked, "Put an Iowa farm in the Amazon valley and the soluble salts would be gone from its soil in two years, its humus in three." In order to maintain productivity in this area a tremendous program of fertilizer application would have to be undertaken. This would be exceedingly expensive because of leaching, a result of the heavy rainfall, which would remove much of the applied minerals. Humus, with its enormous water-holding and mineral-holding capacity, greatly retards such leaching, but the humus is no longer there.

In many areas of the moist tropics, land has been cleared for the production of corn, beans, sweet potatoes, yams, and rice. For a few years yields were substantial, until the humus gave out. The fields were then abandoned, and gradually the forest reclaimed the clearings. At the present time, the natives practice a system of shifting cultivation; as one field gives out, they move to a new area. This cannot be continued indefinitely. However, tropical forest areas can be used for the production of tree crops. Oil, rubber, palm, and cacao could be cultivated successfully by making use of an understanding of the basic reasons why trees are successful and removable crops are not.

Many areas that are not now cultivated present a tremendous problem as far as transportation is concerned. For such an area to be useful, it must be possible to transport food from the region in which it is produced to the region where it will be consumed. In order to provide such useful transportation facilities, enormous amounts of money must be expended. The African and South American continents, immense lands that could be utilized more fully, present unique problems. In Africa, the geographical characteristics are such that most of the rivers do not afford the ease of transportation found on the Mississippi or the Amazon; waterfalls and rapids are constant barriers. On the other hand, the cataracts with their cascading waters can be used to produce hydroelectric energy. Just recently has advantage been taken of this situation. Also, more and more roads and railroad lines are being constructed. If all these factors could be correlated with correct land usage, productivity within the African continent could be increased tremendously. But one must bear in mind that much of the energy that is obtained would be most useful if used in the production of nitrogen fertilizers. No amount of hydroelectric energy or transportation facilities can replace the minerals that are so sadly lacking in most of the moist areas of this great land.

In the Amazon basin of South America, that other great land mass to which people turn hopefully as a possible food-producing area, the flatness of the terrain produces problems quite distinct from those of Africa. The Amazon River and its great tributaries are used extensively for transportation; in fact, almost the entire population and its agriculture are concentrated adjacent to these rivers. During seasons of high water, motor launches can travel at least 3500 miles up the Amazon from the coast without encountering any natural obstacles. During seasons of low water, travel is greatly curtailed; sand bars and tortuous avenues prevent any large vessels from navigating the river in many areas. The extreme flatness of this basin also brings about severe flooding when the torrential equatorial rains pour enormous volumes of water into the various tributaries from the Andes mountains to the sea. Farmers must then await the receding of flood waters before the next crop can be grown. No simple solution to this flooding problem seems to exist. Iquitos, Peru, is 2300 miles from the Atlantic Ocean but only 350 ft above sea level; 1000 miles from the Atlantic, the land is only 125 feet above sea level. Dams would not suffice; tremendous dikes and levees are needed, and it is doubtful whether they could be built high enough in that relatively flat area to contain the tremendous amounts of water that flow to the ocean (estimated as 64 billion gal, or 250 billion liters, pouring into the sea every second). Roads and railroads are needed to service those areas not immediately adjacent to rivers. This would be an immensely difficult undertaking. It could be done, but will it be done in time?

31.5 The Standard of Living

The grade or level of subsistence and comforts in everyday life enjoyed by a community will be influenced by a number of factors, all of which may conveniently be considered under one of three categories: (1) land area, (2) cultural management, and (3) population intensity.

The term **land area** is used to indicate that land which is used, or can be used, productively. Industrial and residential land is usually included because the products of such land may be exchanged for food. However, as the demand for food increases, agriculturally productive land clearly becomes increasingly important. In the long run, then, we are really speaking of land that is suitable for cultivation or the raising of livestock.

The term **cultural management** is a more inclusive one. Here we must refer to various methods of agriculture, science, industry, medicine, and social organization that enable humans to derive greater benefits from nature than would otherwise be possible. This is really what places man ahead of other animals, and modern man ahead of ancient man. This is a facet of man's intelligence. Because of such management, the city dwellers can be supported by the relatively few people who remain on farms and ranches in an industrialized country like the United States. Transportation and storage facilities make possible the use of advanced agricultural techniques in producing much more from land than could be utilized by the people working that land. The building of farm machinery is essential to such advanced techniques and is part of the exchange between city and farm.

Population intensity refers to the actual number of individuals in a particular area. The rapidity with which such numbers are increasing has already been discussed. This factor has the greatest influence on the standard of living, and the effect is inversely proportional to the number of people.

The standard or level of living of any people can be raised by increasing the land area or improving the cultural management of the community in question. Unfortunately, the earth is finite, limiting the possible expansion of land area. As we have seen, not much more land is available for cultivation. Also questionable is whether the additional land placed in cultivation during any one year will counterbalance the productivity loss resulting from erosion during the same period. In the future, man will need to examine more thoroughly the possibility of obtaining additional food supplies from the oceans of the world (see the next section). Also unfortunate is that the techniques and facilities with which man obtains his wants and necessities are not advancing rapidly enough. In comparison with man's achievements previous to this century, amazingly rapid advances, especially within the last few decades, have been made; however, even such relatively rapid advances have not kept pace with our problems of erosion. In spite of tremendous strides in the agricultural sciences, yields have been increased in only some crops, whereas overall yield has not been substantially increased. If our present rate of soil destruction continues, the situation will become worse, because our best lands are already in use. Much less erosion is needed to ruin a poor soil than to ruin a soil that was originally quite productive. Also, much of our best agricultural land is being used as sites for cities, buildings of various sorts, and roads. The expansion of these facilities removes millions of acres of farmland out of production each year.

If the land area available to a community does not increase and if the cultural management increases but slowly, then the increase in population will result in a lowering of the standard of living. Except for possibly Ireland and France, the population of every country is expanding at a rate of 1 to 3 per cent annually. At the present time, there are many countries in which the population intensity is such that any possible improvement in the level of living is immediately canceled by the increase in numbers even at the low rate of 1 per cent. After all, 1 per cent of 1 million is 10,000; a country of 200 million people with a 1 per cent annual increase (such as the United States) will have 202 million the next year. An enormous advance in cultural management is necessary to raise the standards of the existing population, as well as to provide for an additional 2 million people. Even the rapid advances of the last 20 years have not managed to do this, and whether advances in the future can be rapid enough is questionable.

The problem really revolves about finding food and facilities for the peoples of the world. As the Roman philosopher Seneca wrote in 49 A.D., ''A hungry people listens not to reason, nor cares for justice, nor is bent by any prayers.''

31.6 Optimistic Possible Solutions to the Population Problem

The previous portions of this chapter have been couched in a rather pessimistic vein, and justifiably so, in the author's opinion. This opinion is shared by many individuals, scientists and nonscientists alike, but certainly not by everyone. Opinions vary from the extreme pessimists who believe the matter is hopeless to those wonderful optimists who are certain that everything will work itself out advantageously. I hope that I am somewhere between these extremes and will discuss what I am convinced must be done to alleviate the difficult situation now existing in the world.[2] The heading ''optimistic possible solutions'' has been used simply because of the colossal difficulties that must be faced in implementing these solutions even under the best of conditions, and because of the grave possibility that we have started too late in trying to solve the population—food supply problem. Of course, natural laws will eventually solve the problem for us, but I am sure that no one favors starvation, even for others.

The goal for any procedures to be discussed as possible solutions to the population problem is to raise the level of living. This is much more difficult than lowering the death rate. Wiping out disease by medical means and by improved sanitation entails expenditure of time and money but is certainly within reach. Raising living standards is highly complex and involves education, science (including agriculture, industrial techniques, and so forth), sociology, religion, and politics. These are interrelated problems, with difficulties on every side. This is why immediate steps are so essential.

Education

If one feature is common to just about all problems, it is education. No procedure could be functional, even without considering its possible success, unless

---

[2] These discussions do not take account of the difficult political situations existing in the world at the present time. Such situations are entirely out of my sphere of capabilities, and I do not wish to become embroiled in political arguments. I will present my carefully considered opinions as to methods for solving the ''population–food supply'' question. The implementation of some of these solutions may be very difficult, if not impossible, at the present time, because of political factors. I am well aware of this.

607

Sec. 31.6 /
Optimistic
Possible
Solutions to the
Population
Problem

educational facilities are vastly increased. This is really the key to any solution. Such education must be worldwide in scope. The activities of any one community could be negated if a neighboring community were not similarly engaged. An excellent example is derived from the Dust Bowl situation of the Midwestern United States. Farmers utilizing good land management were often handicapped by the smothering soil blowing toward them from adjacent lands. If all of the individuals in this area had been aware of the basic reasons that resulted in the development of the Dust Bowl, all more likely would have adopted more sensible practices.

One of the basic difficulties that must be faced is that probably a majority of the people in the world are not even aware that a problem of food supply exists. Too frequently people assume that a mere redistribution of the world's supply is all that is required to alleviate the suffering of untold millions. With education, the problem can be recognized and attacked on many fronts. But how does one approach an illiterate population? Without a tremendous increase in literacy, failure is certain. A majority of the people cannot be reached unless use can be made of the written word. The estimated illiteracy varies from less than 2 per cent of the population to at least 80 per cent in some countries. The first step, then, must be to increase literacy. After this, explanations and reasons and solutions must be discussed thoroughly. All this must be done immediately.

The following discussions of procedures that might help to solve or at best alleviate much of our difficulties will assume that education of the people is being emphasized.

Increased Cultural Management

This complex category includes many technical advances that could be utilized now and suggestions that lend promise for the future. As was mentioned previously, any increase in the facilities included under cultural management would assist in improving a standard of living. Again, it is mandatory that achievements not be unilaterally applied.

Agricultural production. Many different techniques are used to produce food from the land, either as crops or as livestock. A great increase in production should be possible by the diligent application of modern technology and practices to all land areas. Fertilizers, machinery, and hybrid seed are only a few examples of factors that would influence production favorably. Modern agricultural practices also include crop rotation, fallowing, and the use of legume plants where possible. These are items that would bring about immediate results.

We must also think in terms of long-range plans. Plant- and animal-breeding programs must be vastly increased to provide organisms that can be utilized more efficiently in the diverse environments of the world. Basic research in genetics and physiology must be emphasized. Such investigations are the foundation for an understanding of plant and animal growth, distribution, and interactions. We cannot emphasize too strongly that any basic research in the biological sciences (which necessarily overlaps other fields, such as chemistry, physics, psychology, and sociology) must be pursued with vigor. Although the data and concepts that result do not always appear to have practical applications, this should not alter our efforts. Any knowledge is important. Also, much basic research frequently has future applications of considerable significance.

There are two excellent examples where basic research resulted in important applications that could not have been predicted: weed killers and nylon. The former resulted from investigations that started with studies as to the curious curvature of plants toward light and led to the discovery of plant growth hormones, to chemically synthesized growth regulators, and finally to weed killers, after studies concerning respiratory effects of such materials were undertaken. Nylon resulted from investigations concerning polymerization, the chemical formation of large molecules from smaller ones; some of the polymers had threadlike characteristics. The discovery of chemical weed killers (such as 2,4-D) and nylon could not have been anticipated. They did not even exist when the basic research was started.

An increase in agricultural production refers to the increase per land area and not to an increased production per person. For example, the use of machinery could enable one man to do the work of five; if the yield remains the same, however, the only advantage obtained is that four men can relax while one works. The real need is to produce a greater sustained yield per acre. Most of the agricultural techniques have been developed as a result of research in temperate zones, and similar sorts of investigations will probably be necessary for the tropics and semitropics in order to utilize these areas most efficiently. Methods that are suitable in one climatic regime are not always suitable in others. Breeding programs, especially, would be exceedingly important in developing more productive organisms.

Conservation. The term *conservation* is most frequently used synonymously with the slogan, "prevent erosion and save wildlife." Much more is involved. Erosion control is certainly essential, but the building of dams to achieve this is only a partial measure. The most important feature in the struggle to retain our topsoil is to keep a vegetative cover on the ground, especially at the headwaters of our various streams and rivers. Grasses, shrubs, trees, and even weeds deflect the pounding force of rain, and their roots hold soil particles against the forces of wind and flowing water. Unfortunately, all the dams we may build not only hold water but also collect silt. As more and more silt collects, the effectiveness of the dam declines. Figure 31-4 is a photograph of a silt dam built upstream from a reservoir dam (in the mountains just north of Santa Barbara, California). The purpose of the smaller dam was to decrease the amount of silting in the reservoir, and there are several of these dams. As can be seen, the silt dam *was* rather efficient; the accumulated silt has reached the top of the dam. Of course, this dam no longer serves its function. Gibralter Reservoir, downstream from the area in the photograph, is approximately 30 per cent filled in with waterborne soil after about 50 years. The task of removing such debris and silt from dams is prohibitively expensive, even if possible. It would also be necessary to find some place to put the silt that is removed. It cannot merely be dumped in the vicinity where future rains can wash it back behind the dam or downstream to the next dam. Figures 31-5, 31-6, and 31-7 indicate the damage that results as soil is washed away by water, and many areas are much more severely damaged than those shown in these photographs.

Conservation also entails sewage disposal and the reuse of water, with the

Figure 31-4. Mono Dam in Los Padres National Forest, California, completely filled with silt. *Top:* Upstream view; *bottom:* downstream view. (United States Forest Service photographs.)

609

Figure 31-5. Gullies cut into rolling land by rainfall. (Courtesy of United States Department of Agriculture, Soil Conservation Service.)

sludge being used as fertilizer. Complete sewage disposal plants will be commonplace in the near future, as the demand for water keeps expanding and the supply continues to diminish. This is not to imply that the amount of rainfall is any less than it was. The greater the built-up areas, the less soil is available to act as a sponge in retaining water. The soil that remains after the topsoil is removed (by erosion) is coarser and has low water-holding capacity. Water tables are lowering in many areas as a result of excessive use, and many wells are going dry. All these factors result in less and less water being available.

The amount of water utilized in an industrial nation such as the United States is considerable. Although an individual *consumes* probably no more than 2 liters or qt per day, it has been estimated that the average household uses about 40 gal per person per day for *all* purposes. (One gal is equal to about four liters.) Industrial and irrigation requirements are much greater, of course. Table 31-6 summarizes the total amount of water used in 1950 and in 1975. The demand for water increases markedly with the progress of industrialization. For example, in terms of the daily use per person of all water in the United States, the figures are 600 gal in 1900, 1230 gal in 1950, 1500 gal in 1960, and 1900 gal in 1975. A large steel mill uses about 250 million gal of water per day; a large paper mill uses more water than a city of 50,000 people does. In addition to finding such a great supply of water, we are also faced with the problem of pollution from industrial wastes. If we are to solve the problem of water supply, it is essential that we eliminate wasteful use, purify and reuse water, and convert saltwater to freshwater.

Conservation should also be thought of in terms of studying the possible uses of

Figure 31-6. Severe gully erosion. (Courtesy of United States Department of Agriculture, Soil Conservation Service.)

industrial by-products. In certain areas of Florida, orange rind, a waste product of the orange juice industry, is used as a dietary supplement for dairy cows. The enormous variety of products now obtained from the peanut plant (*Arachis hypogaea*) exemplifies the possibilities involved: (1) oil for cooking and for salads, margarine, soap, and lubricants, (2) the pressed cake for stock feed (exceedingly high protein content), (3) peanuts themselves or peanut butter as food, and (4) foliage as fodder. Much more research should investigate possible uses of by-products and waste materials.

Nothing has been said about conservation in terms of aesthetic appeals and the preserving of the wonders of nature. I do not mean to slight the importance of such conservation goals, but this phase of conservation involves other considerations than the basic problem of food supply and will not be covered here although I consider it a valid and important goal.

Food chains. Under this category are included two general items: (1) the shortening of existing food chains, and (2) the utilization of foods not now consumed or consumed only in relatively small amounts. Many of our feeding habits will probably have to be reorganized as food becomes less available as a result of increased

Figure 31-7. A passing rainstorm produces severe erosion. (Courtesy of United States Department of Agriculture, Soil Conservation Service.)

people to feed. More and more plant food will be consumed directly and less animal food will be utilized; this, of course, refers to plants that are edible to humans. Any such elimination of the "middleman" in a food chain greatly increases the amount of original food value that is available to humans. (See Section 14.3.)

Traditional agriculture has biological and environmental limits and, as the human population continues to increase, there is more and more of a stimulus to the search for new food supplies. Two excellent possibilities have come to light: yeasts and algae. In some localities yeast plants are already being utilized as a dietary supplement because of their relatively high protein content and their supply of B vitamins. The sugar and nitrogen necessary for yeast growth can be obtained fairly inexpensively from acid digestions of cellulose or starch and from ammonium salts. With these materials and an initial inoculum of yeast cells in large tanks, tons of yeast plants per day can be obtained. Yeast cells will double their weight within a period of a few hours, and the final amount of yeast will depend upon the initial inoculum, the amount of food and minerals, the temperature,

613

Sec. 31.6 /
Optimistic
Possible
Solutions to the
Population
Problem

Table 31-6 Water requirements of the United States

	Billions of gallons per day	
	1950	1975
Municipal and rural	17	20
Industrial	80	225
Irrigation	88	180
Total	185	425

and the concentration of oxygen. One conservative estimate is that ten fermenting tanks could produce at least 10 tons of yeast per day, starting with about 500 lb of yeast per tank. This would be equivalent to approximately 150 head of cattle or 80 pigs per day. Taste or flavor is not being considered. Yeasts do not have an unpleasant taste, and they can be mixed with other foods. Also, there are varieties of yeast that produce the natural flavor of meat (*Torula utilis*) and the flavor of bacon. Additional flavors and condiments can certainly be utilized.

Proteins are important to humans because of the amino acids that are obtained when such compounds are digested. Of the 20 amino acids required by the human body, man must obtain 8 in his food because his body does not synthesize them. The proteins of most common plant foods, however, generally lack lysine and frequently are poor in one or two other amino acids (i.e., tryptophan and methionine); this is true of the cereal grains. Yeasts have a well-balanced variety of amino acids and a high content of lysine, which makes them a useful complement to the cereals. With this in mind, Alfred Champagnat has been growing yeast on low-grade petroleum plus some mineral salts to supply nitrogen, phosphorus, and potassium; it utilizes the least valuable constituents (the waxes) and leaves a better-grade oil suitable for diesel engines and domestic heating. The yeast is separated from all traces of petroleum and dried; the purified concentrate is a white powder with little odor or taste. Laboratory animals have thrived on this protein source, and it is now being fed to livestock. In 24 hours 1000 lb (454 kg) of yeast cells will gain 5000 lb, 2500 lb of which are edible protein; a 1000-lb steer can synthesize at most 1 lb of protein in the same period. A number of sophisticated versions of the food have been prepared with various flavors. There seems to be no reason why it could not be used by humans.

At Purdue University, a team of investigators has discovered a way in which to overcome the lysine inadequacy of corn (maize) grain. These geneticists found that in one kind of corn the mutant gene Opaque-2 shuts down production of zein, a non-nutritious protein, leaving material free for the production of proteins containing lysine and tryptophan. This has inspired a worldwide search for similar mutant genes in other major grain crops (wheat, rice, millet, sorghum) that would fill out their protein complements. Such grain would be of great importance in those areas where protein in the form of meat is seldom obtainable—Asia, Africa, Central and South America.

The International Rice Institute has developed new rice strains that increase production by 65 to 85 per cent, require a shorter growing period, which allows the production of three crops per year, and are dwarf plants. Because these short plants do not become top-heavy and collapse, there is less preharvest loss of grain. Rice production in Pakistan can be used as an example to indicate the success of these new strains. In 1967 Pakistan was receiving rice shipments from the United States. In 1968, after growing the "miracle rice," Pakistan had eliminated its rice deficit and began exporting rice the next year. Unfortunately, the population increase caught up and there no longer is any surplus.

Various algae, especially marine forms, are now utilized for food or as cattle fodder. However, the most promising alga for future use is *Chlorella,* a unicellular member of the Chlorophyta. The environmental conditions under which *Chlorella* is grown can be varied to alter the composition of each cell. The protein content can be increased from 50 per cent to 88 per cent of the total dry weight, and fat content from the normal 7 per cent to 75 per cent. Each hectare devoted to algal culture can produce up to 90 metric tons (dry weight) or 40 times the protein yield of a hectare of soybeans and 160 times the hectare yield of beef protein. The taste of fresh *Chlorella* is similar to that of somewhat oily broccoli, while the bright green powder (dried algae) resembles lima beans. However, artifical flavorings could be added, or other algae might be found that will have a more suitable taste. Investigators have already prepared alga bread, rolls, noodles, soup, ice cream, and a sauce that tastes like soy sauce. The addition of *Chlorella* to rolls causes an increase of 20 per cent in protein, 75 per cent in fat, and a considerable enrichment of the vitamin A and C content. A tablespoonful of *Chlorella* powder is equivalent in food value to a 1-oz piece of steak, a comparison that emphasizes the possibilities of utilizing this plant as a dietary supplement. For example, mixing algae powder with tea produces a brew as nourishing as beef broth without materially affecting the taste.

At the present time, preliminary designs indicate that algae could be produced at a cost of about 35 cents a pound. This is low compared to the price of meat, but high when compared to that of vegetable proteins like soybean meal or yeast at about 10 cents a pound. However, the cost might be lowered with changes in technology; estimates actually range down to 6 cents a pound. Besides, expense may become meaningless as food becomes ever more scarce. The suggestion has also been made, and tried with some success, that algae be grown in sewage-disposal ponds. The sewage requires large supplies of oxygen for the bacterial action that brings about rapid decomposition, and the algae supply this oxygen (photosynthesis). Considerable amounts of carbon dioxide, nitrogen, and other minerals on which algae thrive are made available as the sewage decomposes. When heat-dried or cooked, the sewage-pond algae are edible. If humans are aesthetically insulted by such food, the algae could be used to feed cattle; and this has been done with excellent results. Many other uses can be found for this algal material. The oils may be extracted and used for fuel. Algae contain vitamins, oils for paints, sterols to make cortisone and other hormones, pigments for dyes, and probably hundreds of other items that might be useful.

Fish protein concentrate (FPC, or fish flour) is another potentially powerful weapon against protein deficiency; the protein value is similar to that of meat.

615

Sec. 31.6 /
Optimistic
Possible
Solutions to the
Population
Problem

FPC is made by shoveling whole trash fish (skates, sea robins, dogfish, hake, and so forth) into a hopper, grinding them up, washing with solvents to deodorize, and finally extracting and dehydrating the material to a tasteless and odorless powder containing at least 70 per cent protein. The United States Food and Drug Administration has approved fish flour as being safe and sanitary, fit for human consumption. This and the "petroleum protein" mentioned previously, because of their neutral flavors, can fit into local diets as additives in foods that people already accept and eat (e.g., soups, tortillas, and bread). They would not have to overcome the traditional habits and prejudices against new foods.

Some foods can be disguised to make them acceptable. A high grade of protein can be produced from soybean meal, but it is bland and unappetizing; and so it is processed to make the material more palatable. The protein material is made into a batter about the consistency of honey and then forced through tiny holes, becoming extremely fine, colorless, odorless, and tasteless fibers when coagulated in an acid bath. The fibers are flavored, colored, and packed with binder into any form desired; vitamins and minerals or other supplements may be added. The strength and diameter of the fiber can be regulated so that the "chewing texture" can be similar to a fish filet, hamburger, sliced ham, steak, chicken, or any other texture desired.

Humans have also never really utilized food from the oceans as fully as might be done. Almost all the annual fish haul is from the Northern Hemisphere, and yet conditions in parts of the Southern Hemisphere are equally suitable for fish. Circulation patterns of ocean currents, upwelling of waters from deep layers, and proximity of land are factors that result in the availability of mineral nutrients upon which phytoplankton[3] thrive. These plants are the "grasses of the sea" and provide food for zooplankton; both are food sources for various fishes, and the food chain in all its ramifications can eventually support humans to a much greater extent than it is now doing. Food supplies from the sea can probably be tripled, especially if humans are willing to consume a greater variety of fish than is now the case. Also, aquaculture has been used successfully for years in Asia to grow oysters, certain fish, and some seaweeds (algae). This could easily be expanded and improved; one estimate is that such cultivation of marine organisms could result in at least a fivefold increase in the present (1977) annual production of 1.8×10^6 metric tons of fish and shellfish.

Industrial potential and trade. Future industrial expansion of most nations should emphasize power supplies and raw materials. The exchange of factory products for agricultural products is a thing of the past. So many industrialized nations (e.g., United States, Great Britain, Germany, Japan, Canada) now exist that additional nations similarly oriented will result in surpluses of industrial products, with little hope of exchanging such materials for essential food supplies. As nations with food surpluses gradually build up their industries, with whom will other nations trade for food? Energy and power supplies, however, will be required in ever-increasing amounts as populations and industries increase, espe-

---

[3] The drifting microscopic algae and the microscopic animals that feed upon them are called *plankton;* the plants are the phytoplankton, the animals are zooplankton.

cially if living standards can be raised. Fossil fuels, such as coal and oil, are being depleted at a fantastic rate; expanding economies may very well result in the complete utilization of all known coal and oil supplies within 200 to 300 years. Within this span of time the need is incumbent upon us to find new energy sources. At the present time, atomic energy appears to be the only possible solution. However, so many technical difficulties are involved that industrial uses of such energy may not be available soon enough. The simple matter of shielding or protecting humans from atomic radiations may preclude utilizing atomic engines in many instances. For example, automobiles and airplanes would require such heavy shielding that they cannot be built at present. Disposal of radioactive waste is another difficult problem. In order to allow additional time to search for solutions to such problems, atomic energy likely will be used at first exclusively to create electrical energy. Similarly, water power in many nations could be harnessed to provide electrical energy. With these new sources of electrical energy available, coal and oil could be transported to those areas where electricity is not suitable. This would conserve fossil fuels and might extend the supply for an additional 200 to 300 years.

Industry also requires a tremendous amount of raw materials, such as iron ore, tungsten, magnesium, boron, uranium, and many others. The spread of industrialization and modernization will increase the rate at which raw materials are consumed, and many of these are nonrenewable. Per capita consumption of raw materials, other than food, is ten times as high in the United States as in the rest of the world. With increasing technological progress throughout the world, there will be increasing shortages of copper, iron, lead, tin, and other metals so essential to industry. Vast supplies of these materials exist in many underdeveloped countries of the world. Such nations should increase their industrial potential, with mining as the primary goal. This would provide a basis for trade.

Industrialization as such does not solve the basic problem of providing sufficient food for a population. However, industrialization is essential for improving standards of living. This results not only from the material things that industry provides but also from the increased education that must necessarily accompany such expansion. A well-educated society typically tends to grow slowly for people are both aware of population control and motivated to practice it. The importance of education has already been discussed.

Population Stability

Undoubtedly the only real solution to the population problem is the maintenance of a stable population. Even if our cultural management increase could keep pace with our population expansion, which has not been possible thus far, populations will have to be stabilized eventually or there will be "standing room only" on the earth. At the present rate of increase, people standing shoulder to shoulder would cover the earth's land surface in another 700 years. Obviously this will not happen. Unfortunately, however, if populations are not stabilized in some humane way, starvation, war, and disease will do the job in an inhumane way.

Figure 31-8 represents a situation in nature that might be considered to exem-

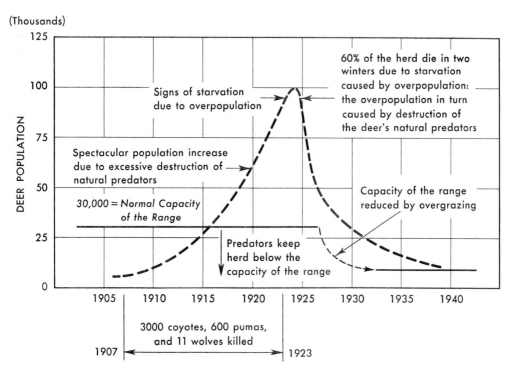

(Thousands)

125

100

75

DEER POPULATION

50

25

0

Signs of starvation
due to overpopulation →

60% of the herd die in two
winters due to starvation
caused by overpopulation:
the overpopulation in turn
caused by destruction of
the deer's natural predators

Spectacular population increase
due to excessive destruction of →
natural predators

30,000 = Normal Capacity
of the Range

Capacity of the range
reduced by overgrazing

↓ Predators keep
herd below the
capacity of the range

1905   1910   1915   1920   1925   1930   1935   1940

3000 coyotes, 600 pumas,
and 11 wolves killed

1907 |←――――――――――――――→| 1923

Figure 31-8. The history of the Kaibab deer. The dashed line is the graph of population numbers of the deer. (From *Life, An Introduction to Biology*, by Simpson, Pittendrigh, and Tiffany, New York, Harcourt, Brace & World, 1957.)

plify this situation with which humans are faced. The original deer population on the Kaibab Plateau in Arizona was fairly stable as a result of predation and was well below the capacity that the vegetation could support. A campaign of extermination against the predators resulted in a greatly decreased death rate for the deer. (A similar result has been obtained in human affairs by advances in medicine and sanitation, which have brought about rapidly declining death rates over recent decades.) Naturally, the deer population increased tremendously, and far beyond the number that could be supported by the available vegetation. Starvation rapidly decimated the deer population until it stabilized at a relatively low figure. The range had been seriously damaged by overgrazing and could now support fewer animals than was originally possible. Unfortunately, the rapidly increasing numbers of humans have also been destroying soils for many generations, resulting in fewer really productive acres than was once the case. The future is not pleasant to contemplate in the light of examples from nature.

Methods of stabilizing populations are highly complex and involve all phases of human existence. Sociological, religious, psychological, and political aspects of the problem must be studied. In general, the birth rate of a population tends to decrease as economic development, industrialization, and education increase. Differences in birth rates between countries are consequences of social rather than biological factors; reproductive potential does not vary, but social customs do.

Late marriages, contraceptive practices, less dependence upon hand labor and thus less need for large families are among those factors that influence birth rates. The Scandinavian countries have a fairly stable population. Ireland and France have stable populations. Even Italy now has birth rates little above replacement levels in spite of relatively high birth rates in the south. Although the problem of population stability is a difficult one, it can be solved. One can hope that the solution is reached in time. Only with a stable population can we maintain a standard of living that we might consider adequate. With an increasing population, no matter how slow the rate, the future will bring malnutrition, starvation, disease, and poverty.

Increased Land Area

Enough has been said with regard to the availability of land to indicate that some increase is possible. Every effort should be made to utilize all available land. This utilization must be based upon sound land-use practices. Our knowledge of plant distributions, agricultural technology, environmental influences on soil fertility, genetics, pathology, storage, and transportation must be correlated for the most efficient methods of production for the area in question. The practices useful under some environmental conditions are not necessarily suitable in other climatic regimes. Surveys should be undertaken in an effort to ascertain whether production could be increased above present yields and how best to achieve a *sustained-yield* situation. Most of our efforts will be time consuming and expensive but necessary.

31.7 In Retrospect

If we could utilize all known conventional agricultural techniques, including supplemental irrigation of land now under cultivation, we might be able to double food production. If new lands were brought into cultivation in a similar fashion and if the food sources of the seas were utilized to a greater extent, food production might be increased fourfold or fivefold. Improved plant- and animal-breeding programs would further increase this gain. Optimistically, it might be possible to increase total food production, including new foods, by a factor of eight or ten. With such an increase, a world population of probably three or four times the present number could be supported at an adequate dietary level. We assume, of course, that transportation problems and political difficulties could be overcome, and frequently traditions also. Harrison Brown is willing to estimate the much higher figure of 50 billion. However, his estimates are based upon 1 billion acres of algae farms, for one thing. If this food source is not utilized, his estimate of the population would be closer to 10 billion. This figure is approximately that which I would estimate as the maximum that could be supported.

However, one important factor remains—the space that is available for the population. Urban development continually removes land from cultivation, and this reduces food production. More important is the actual living space for an individual. No population can increase indefinitely, no matter how slow the rate. As Karl Sax so aptly put it, there would soon be "standing room only." Even if it became possible to migrate to new planets, the solution would not be obtained. Even a low 1 per cent increase of an estimated 6 billion people in the year 2000 would mean 60 million added per year. It would take a million spaceships carrying 60 passengers each just to stay even, one ship leaving every 31 seconds. Even the most optimistic estimates of food production cannot solve the problem of available space. This problem can only be solved when the population stabilizes.

Summary

1. All populations have a tremendous capacity for increase because of the reproductive potential of organisms. Such a rapid increase does not continue indefinitely, and eventually each population stabilizes. The factors limiting the size of a population are food supply, predators, parasites, and climatic conditions.

2. The human population is expanding very rapidly and has not reached stability except, possibly, in a few areas.

3. The large numbers of humans in the world and the continuing increase in numbers have created a serious food shortage. Expansion of modern technology to all countries and to all usable land areas would probably make possible the feeding of the present population of the world at an adequate level of nutrition.

4. The standard of living depends upon the land available, upon modern technology (especially agricultural technology) and knowledge, and inversely upon the number of individuals who must be supported. The amount of land cannot be increased indefinitely, and technological advances have not been rapid enough in the past to keep pace with expanding populations.

5. The basic problem revolves around finding food, clothing, shelter, and other facilities for the peoples of the world. Many steps could, and should, be taken to alleviate this situation. Universal education, especially with regard to literacy, is the most important step and the one upon which all others depend. The initiation of modern agricultural methods in all areas, including conservation methods, should be undertaken immediately. New food sources must be utilized, and the length of food chains must be reduced. The industrial potential of various areas must be exploited more fully. Of utmost importance is the study of methods to stabilize populations. Even if food could be made available to an infinitely large population, the problem of space for the individuals of that population would remain.

Review Topics and Questions

1. Explain why no population continues to increase in numbers at the rate at which it is capable.

2. Discuss why the human population has increased much more rapidly during the last 200 years than it has in all of the preceding years.
3. Discuss the factors that govern the standard of living of a population. What is meant by "standard of living"?
4. Explain why the soils of the moist tropics are usually not very productive after the first few years. Why can crops be grown successfully for the first few years?
5. Discuss the factors that you consider to be the most important to solving the population problem.
6. Why will a shortening of food chains enable us to feed more people with the available food?
7. Why are yeasts and algae considered to be excellent possibilities as new food sources?
8. Since industrialization will not solve the problem of food shortages, what is the importance of a country becoming industrialized? What form of industry would you suggest for those countries that are now primarily agricultural? If your answer varies with the location of the country, explain why.

Suggested Readings

BORGSTROM, G., *The Hungry Planet* (New York, Macmillan, Inc., 1965).

BRITTAIN, R., *Let There Be Bread* (New York, Simon & Schuster, 1952).

BROWN, H. S., *The Challenge of Man's Future* (New York, Viking, 1954).

BROWN, L., *In the Human Interest* (New York, Norton, 1974).

CHAMPAGNAT, A., "Protein from Petroleum," *Scientific American* **213,** 13–17 (October 1965).

CHRISPEELS, M. J., and D. SADAVA, *Plants, Food, and People* (San Francisco, Freeman, 1977).

EHRLICH, P., A. H. EHRLICH, and J. HOLDREN, *Ecoscience: Population, Resources, Environment* (San Francisco, Freeman, 1977).

OSBORN, F., *Our Plundered Planet* (Boston, Little, Brown, 1948).

OSBORN, F., *The Limits of the Earth* (Boston, Little, Brown, 1953).

OSER, J., *Must Men Starve?* (New York, Abelard-Schuman, 1957).

SAX, K., *Standing Room Only* (Boston, Beacon, 1955).

TURK, A., J. TURK, and J. T. WITTES, *Ecology, Pollution, Environment* (Philadelphia, Saunders, 1972).

VOGT, W., *People* (New York, Bartholomew House, 1961).

"The Human Population" *Scientific American* **231,** 30–182 (September 1974). The entire issue.

Pollution

32.1 Pollution, Population, and Technology

The term **pollution** refers to the contamination of soil, water, or the atmosphere that lessens their value to man or other living organisms. Of all organisms man is the greatest polluter. Only man does not recycle materials but removes them, destroys without need, and tends to ignore the basic relationships between organisms and their environment. All organisms bring about changes in the environment, but without man's intervention an equilibrium stage eventually is reached, which we have previously referred to as a climax. Man, however, brings about such disruptive changes that the environment usually deteriorates and frequently quite drastically. Man is not usually in equilibrium with his environment because of his disregard for recycling, which is so essential to an environmental equilibrium.

An agricultural society is generally acquainted with the interactions of the environment and the organisms it supports. But increasing numbers of people exert pressure for increased production of food. The resultant overgrazing and overcultivation, especially of watersheds, and the general abuse of the soil create problems of erosion and destruction of productivity. This has occurred many times during recorded history. For example, Mesopotamia, the semiarid land between the Tigris and Euphrates rivers of the Near East, had a thriving agricultural economy dependent upon irrigation waters before the birth of Christ. Attempts to feed more and more people resulted in heavy grazing pressure, deforestation, and cropping of slopes, which brought about erosion of uplands where the rivers had their sources. Accumulated sediment clogged irrigation canals and required constant labor to keep them open. Because of periodic revolutions or invasions, irrigation canals were left unattended at times and gradually deteriorated beyond repair. Today this area produces probably no more than one sixth of the agricultural yield that once had been obtained.

An increase in people means an increase in all types of environmental pollution and an accelerated depletion of natural resources. This is accentuated in a technological society. For example, although the United States has less than 8 per cent of the population of the world, it is responsible for over 50 per cent of the total consumption of raw materials. Most of our demands upon natural resources have no correlation with need. Our consumption of cars, television and radio sets, air conditioners, dishwashers, containers of various sorts, and so forth does not stem from need. Such items use enormous amounts of nonrenewable mineral resources as well as renewable resources that could be put to better use. Even renewable resources may become scarce if used at a rate greater than that of renewal. For example, many packaging materials are produced from wood pulp obtained from trees that are renewable. However, our obsession with packaging utilizes wood at an enormous rate with very little being recycled. The various uses of lumber provide an additional drain upon wood supply. The United States uses about 450 lb of pulp wood and paper per person per year, which is four times the lifetime consumption of a person in an underdeveloped country. Renewal is simply not keeping pace.

Our discussion of various forms of pollution will encompass the most conspicuous and most deleterious aspects and will concern primarily the United States. It should be borne in mind, however, that similar situations exist in other countries. The degree of pollution depends upon the number of people, the area involved, and the technological aspects of the society. As one reads the following sections concerning the pollution that we have been spreading so thoroughly throughout our environment, it might be of interest to note that during the past decade we have spent about 30 billion dollars for agricultural subsidies, about 45 billion dollars for space explorations, and only about 5 billion dollars on water pollution abatement.

32.2 Litter

Litter is a disorderly accumulation of carelessly discarded objects, and people living in the United States deposit an incredible amount of litter in their environment. We annually discard about 48 billion cans, 26 billion bottles, 30 million tons of paper, 4 million tons of plastics, 100 million tires, 7 million cars, and smaller amounts of various other items. Degradable items eventually decompose and present less of a problem than the nondegradable ones, of course. A tin can, which is really steel covered with a very thin layer of tin, will rust away in a few years. An aluminum can, on the other hand, is practically indestructible. It can be reused as scrap, but it is so readily discarded that few are actually collected for recycling. The replacement of the degradable plain brown bag by nondegradable plastic bags in many stores also has accentuated the problem of littering.

Frequently it is less expensive to buy a new item than to repair an old one; so the latter is thrown out. Much of our clothing, toys, and dinnerware is now made of microbial-resistant plastics and cannot be disposed of easily. This all contributes to our mass (mess?) of litter.

Garbage is classified as waste resulting from growing, preparing, cooking, and serving food. Over 5 lb of garbage per person per day is discarded in the United States. In New York City alone approximately 5 million tons of garbage are collected yearly. The primary methods of disposal in most localities are the open dump and landfill. The former are infested with rats and flies, are potential sources of diseases, often catch fire, and trash frequently blows about. In an efficiently run landfill garbage is compacted and covered with earth. This decreases chances of fires, decreases possibility of blowing trash, and attracts fewer vermin. Obnoxious seepage from the buried trash may pollute water supplies, however. Incineration is frequently used for garbage disposal, but this usually pours large quantities of soot, fly ash, and smoke into the atmosphere. When properly designed and operated, though, incinerators can cope with large amounts of combustible trash: the health problem is eliminated; the volume of solid waste is reduced about 80 per cent, and the residue is inert and readily handled. Government studies indicate, unfortunately, that 94 per cent of the nation's dumps and 75 per cent of all incinerators are inadequate to handle the solid wastes that are accumulated at the rate of about 3.6 billion tons per year.

32.3 Water Supplies

The amount of water used and its sources have already been discussed (Section 30.5). We are now concerned with the quality of that water and the ways in which various water supplies are being polluted and frequently made unfit for consumption or other uses. Water can be purified before being consumed, although this may be quite expensive, but this does not alleviate the situation with regard to plants and animals that are dependent upon such water for their existence, especially aquatic organisms.

Sewage

A survey in 1966 revealed that raw sewage from 11 million people living in 1342 communities is being discharged into streams and lakes with no treatment whatsoever, and that 17 million people in 1337 communities require new or enlarged sewage systems. This nauseous situation is dangerous because pathogenic organisms are capable of transmission by sewage discharge and stream flow for many miles from their point of origin. The causative organisms of typhoid fever, cholera, polio, infectious hepatitis, and other diseases have been found in various streams and rivers. An outbreak of waterborne infectious hepatitis in 1961 resulted from the victims eating clams harvested from Raritan Bay off the New Jersey coast. The clams had taken the virus into their bodies during the process of filtering food from sewage-contaminated waters. However, infectious agents do not survive for very long outside the host organisms, and retention of sewage for treatment will usually eliminate them. This is assuming that the treatment is efficient and thorough, including chlorination of the final effluent, because some infectious agents have been shown to survive various treatment processes. Their numbers are much reduced, as compared to untreated sewage, but safety resides only with complete elimination.

Even if pathogenic organisms do not create a problem, sewage has an extremely deleterious effect upon the water ecosystem into which it is discharged. The organic matter is immediately acted upon by bacteria in the streams and lakes. Such bacterial action is important in recycling essential elements, such as nitrogen and phosphorus, which would otherwise be locked into organic compounds and not available for further use. However, this bacterial activity consumes oxygen dissolved in the water. If the sewage load is too great, the level of oxygen may be decreased to the point where game fish and other organisms cannot survive. Undesirable forms, such as *Tubifex* worms, insect larvae, and trash fish, become dominant. With decreasing oxygen content, decomposition becomes more and more anaerobic, and bacterial activity produces increasing amounts of foul-smelling hydrogen sulfide gas.

The net effect of sewage pollution is to impair recreational and aesthetic values severely, and the possibility of exposing a population to disease organisms. The only solution is a greatly expanded system of effectively processing sewage so that the effluent is reusable. It is possible with present methods to treat sewage so that the effluent water is usually clearer and more pure than the original tap water in the home. This reclaimed water could be returned to the reservoirs and then to distribution lines, or at least used for irrigation as it now is in some cities. This would entail a considerable expense, approaching 30 billion dollars by some estimates, but is an absolute necessity if we are to prevent further deterioration of our environment and repair some of the damage already done. One suggestion that would be likely to speed the development of sewage treatment plants is to insist that water supplies be obtained downstream and discharges be made upstream.

Industrial Discharges

Various industries use tremendous quantities of water as a solvent, a cleaning agent, a coolant, a waste-removal agent, or a mineral extractant. The industrial wastes are frequently discharged without treatment, and industrial pollution of water rapidly occurs. Dyes and stains, foul odors, oily scum, and solid residues are readily in evidence around most industrial areas.

Organic wastes, which bring about a lowering of the oxygen concentration in waters as mentioned before, include blood and discarded tissues from slaughterhouses, hair from tanneries, sulfite liquors from paper pulp mills, and peelings and other materials from fruit and vegetable-processing plants. If the wastes themselves are not toxic, fish and other organisms may die from lack of oxygen. Heavy-metal salts, such as zinc and copper, may be directly toxic to aquatic vertebrates at a few parts per million and cause killing of fish.

The effective treatment of industrial wastes is complex and expensive. However, some industries have reduced pollution by converting their wastes into commercially valuable by-products. In any event, treatment must be implemented if pollution is to be stopped. The cost would undoubtedly be passed on to the consumer by way of price increases, but this would simply mean that those using the product must pay the expense of maintaining a clean water supply. If this is not done, the expense of treating the water supply will have to be borne by all individuals in the area and not just those consuming that particular product. This is a decision that should be made as soon as possible. That is, should each industry discharge only treated water or should the water be treated after discharge and before consumption? In the author's opinion, the former is the only logical decision. Treatment after discharge will not help fish and other organisms living in those waters. It also seems rather logical that those responsible for polluting should be responsible for removing the pollutants. The environment belongs to all of us, and no one should have the right to damage what is not his alone.

Most of the water used in industry is for cooling purposes. The water is warmed, of course, as it circulates through heated equipment and is returned to the stream, which warms up. Heat pollution is usually undesirable because less oxygen dissolves in warm water than in cold water, and less desirable fish may replace game fish. Also, with less oxygen there would be less activity of aerobic bacteria. The result could be an unsightly, foul-smelling stream. Several industries have installed cooling towers where the heated effluent water can cool. In Oregon the warm water discharged from a paper mill was sprayed on fruit trees and prevented frost damage. Other beneficial uses of such waters may be devised after additional study.

Fertilizers and Pesticides

Plant growth is frequently limited by deficiencies of various mineral elements, especially nitrogen, phosphorus, and potassium. Other things being equal, the more such nutrients are available, the faster and more luxuriantly will plants grow. These three elements are prevalent in sewage, industrial wastes, and agricultural fertilizers. As more and more fertilizers are used to increase crop yield, greater quantities wash into streams by runoff water and leaching by way of underground sources. Probably 10 to 25 per cent of fertilizers are lost in this way (i.e., are not absorbed by the crop plants that were being fertilized) and burden aquatic ecosystems. With this increase of minerals, aquatic plants frequently produce explosive growths. Such aquatic vegetation has become an increasing nuisance in streams, ponds, lakes, and irrigation canals. Dense blooms of algae may cause death of submerged plants by decreasing sunlight penetration. They may interfere with swimming and boating and frequently impart a foul taste to the water. Death of this lush growth creates anaerobic conditions as it decomposes, resulting in the release of odorous hydrogen sulfide gas. This has been discussed previously.

Lakes whose waters are relatively rich in plant nutrients, such as phosphates and nitrates, are termed **eutrophic.** They have much decaying organic matter and usually show reduced oxygen concentrations. **Oligotrophic** lakes are those whose waters are low in plant nutrients, usually highly oxygenated, and have relatively small amounts of decaying organic material. The process of eutrophication is enormously accelerated by man, especially in a sophisticated technological society. For example, one of the most overfertilized, or overpolluted, lakes in the world is Lake Erie, one of the Great Lakes. It is shallow and has a large number of population centers along its shore. Much sewage and effluents from various industries are discharged into the lake. The agricultural land between the cities adds insecticides, herbicides, and fertilizer. Because of the small volume of Lake Erie, this amount of discharged material cannot be diluted or decomposed rapidly enough to prevent pollution. The more desirable fish, such as pike, whitefish, and sturgeon, have practically disappeared and have been replaced by catfish, carp, and smelt, which are better able to withstand such polluted waters. Whole square miles of lake surface are frequently covered by dense mats of algae. Swimming is impossible in most areas because of decaying vegetation and quantities of sewage in the water. This misuse of Lake Erie emphasizes the far-reaching effect of pollution. If water resources are to be available in the future, we must realize that water should not be viewed as a free waste disposal system. While the supplies of water are pretty well unlimited, the ability of water to dispose of wastes is not. Whether our polluted waters, such as Lake Erie, can ever be restored to full usefulness is uncertain. What is certain is that they cannot be restored unless discharges are treated to prevent, or at least greatly decrease, the addition of organic matter, phosphates, and nitrates to these waters.

Many pesticides used since World War II, such as DDT, are not readily biodegradable and may remain intact for many years in either water or soil. Terrestrial and freshwater systems have become polluted, and such pesticides are being cycled into the marine environment. DDT, for example, is soluble in fats and accumulates in fatty tissues. It has been found in marine fish oils and even in fatty tissues of seals and penguins from the Antarctic. This material is not readily metabolized and, therefore, tends to build up higher and higher concentrations in a food chain. Remember that each link in a food chain consumes considerable quantities of the next lower organism: the pyramid of protoplasm or numbers (see Section 14.3). The organism at the top of a food chain accumulates a much higher concentration of the pesticide than was originally applied.

It should be emphasized that pesticides are extremely beneficial in many ways. Man basically establishes a **monotypic agriculture,** that is, fields composed of a single species where originally there existed a natural ecosystem including a variety of species. The removal of grassland in order to cultivate wheat or corn, the removal of forests to provide space for farms, and even the development of a front lawn are all examples of monotypic agriculture. In such a biologically simplified ecosystem, pests spread with devastating rapidity. Their natural controls, such as predators and parasites, are usually not present, whereas their host plants are present in dense concentrations and are rather uniform in their susceptibility, which provides an easy pathway for the dissemination of the parasite. The southern corn blight, which reduced overall crop yield about 15 per cent within one year (1970), is an excellent example of such a situation; as is the late blight disease of potato (see Section 21.5).

Insects, bacteria, fungi, and weeds reduce crop yields up to 50 per cent or more at times and cause a loss of billions of dollars. In order to reduce such losses, enormous amounts of pesticides are used in the form of insecticides, fungicides, and herbicides. If care is used in their application, the concentrations of such pesticides are not injurious to crop plants or animals (other than insects of course). The difficulty arises if materials are used that are not readily biodegradable. These may accumulate, through food chains, to a dangerously toxic level. This is the real reason why the use of DDT, and similar chlorinated hydrocarbons, is being discouraged.

An investigation by E. G. Hunt and A. I. Bischoff of the California Department of Fish and Game illustrates the disruption of an ecosystem that may result from using some pesticides. Clear Lake, about 90 miles from San Francisco, has been a favored waterfowl nesting area for many years. DDD, a chlorinated hydrocarbon, was applied to the lake in 1949 to eradicate a tiny gnat that formed dense cloudlike swarms that were irritating to fishermen, hunters, and other vacationists. The concentration of the insecticide was only 0.02 part per million (ppm), and two additional applications were made in 1957. Within a few years the bodies of dead fish, gulls, ducks, geese, and grebes began to litter the beaches. The investigators found that the DDD was originally absorbed by algae and then transferred with its molecular structure intact to successive links in the food chain. Grebes, a diving bird, became poisoned by way of the algae-crustacean-fish links in the food chain of which they were the terminal link. Five parts per million of DDD were found in crustaceans, several hundred ppm in fish that eat crustaceans, and 1600 ppm in the gonads of the fish-eating grebes, an eighty-thousand-fold increase over the original concentration or 0.02 ppm. Apparently the insecticide severely impaired reproduction of the grebes.

Whether persistent pesticides, such as DDT, are a threat to human health is still a matter of controversy. Many biologists, ecologists, and medical men are exceedingly concerned about the situation, even though no one has died directly because of DDT properly used. DDT has been found in milk and in body fat of humans, and it is quite likely that DDT is present to some extent in all inhabitants of the United States; probably about 12 ppm in the fat tissue. This level appears to have no effect, but many people are not convinced that long-range exposure to this amount of DDT will not ultimately result in serious damage. Certainly birds and fish have eventually been killed by DDT sprayed to control insects. As long as there is any doubt about the role of DDT in human metabolism, there is a risk. We are not obliged to use DDT, and this doubt has led to the banning of DDT in many areas.

Irrigation and Salt Accumulation

Many arid areas are productive only as a result of irrigation. The soils are fertile, and only a lack of rainfall limits plant growth. Such areas, however, are in regions of low humidity and usually rather high temperatures. As a result, much evaporation and transpiration occur, which effectively remove much water and concentrate the minerals in the remaining water. Salt gradually accumulates and presents a problem because of its deleterious effect upon plant growth.

Seed germination is particularly sensitive to high salt concentrations, and seedlings are more susceptible to injury than well-established plants. Most crop plants, however, are rather susceptible to salt injury. The reduction in growth that results from increasing salt concentrations in the soil is partly the result of reduced water absorption caused by osmotic effects in the root environment. The primary effect, however, results from the uptake of abnormally high concentrations of salts. Although the effects of high salt content on metabolic processes are not well understood, various investigators have reported decreased hydration of proteins including enzymes, suppression of RNA and DNA accumulation, decreased photosynthesis, and changes in respiratory rates. An excess of salt appears to bring about decreased growth because of protoplasmic injury rather than from desiccation.

32.4 Air

With a few exceptions, such as the anaerobic bacteria, all organisms require oxygen. Although green plants produce this oxygen, it is the oxygen supply in the air that is the immediate source for most terrestrial organisms. Aquatic plants and animals obtain oxygen that is first dissolved in water of course. Irrespective of the structures and mechanisms involved in obtaining oxygen, the organism is also exposed to any other gases that are present in the atmosphere. The problem of air pollution arises when gases, or particulate matter, build up to concentrations that are deleterious to plants and/or animals. There are differences in susceptibility to various pollutants, and we will discuss only those materials that appear to be most troublesome.

We have now reached the point where our factories, smelters, refineries, automobiles, home fires, and municipal dumps are spewing out more than 150 million tons of pollutants annually into the air we breathe—almost one-half million tons a day. Factories and electric-power generating plants fire some 50 million tons of fly ash (particulate matter) and 26 million tons of sulfur oxides into the skies. Fluorides, nitrogen dioxide, and ozone are additional pollutants that cause difficulties. In California crop damage from air pollution is greater than 100 million dollars annually, and the total cost of air contamination is considered to be in the billions of dollars for the United States as a whole. This includes crop damage, deterioration of various fabrics and materials, and health costs. But how does one put a price on aesthetic and recreational values? What of the beauty that disappears with the death of a forest from sulfur dioxide fumes, or the ugliness that appears in the form of discolorations to buildings and damage to ancient art objects of marble and sandstone?

Air pollution is mainly the result of things burning, such as gasoline, coal, oil, trash. The curve of dirty air follows fairly closely the curve of national wealth—more cars, more power plants, more factories, more trash. The biggest single source of air contamination, gasoline consumption in motor vehicles, has more than quadrupled in the last 20 years. Antipollution devices are now installed on all new cars, but the air is getting worse as a result of population increases, more vehicles, more production, and more trash. The problem is aggravated by the increasing complexity of the materials combusted and the increasing numbers of uses to which fires are put.

Sulfur Dioxide

Varying amounts of sulfur compounds are present in coal and oil, and sulfur dioxide is released in the smoke as these fuels are burned. Additional sulfur dioxide is released during the refining of sulfide-containing ores, such as iron, copper, and zinc. Sulfur dioxide enters plants principally through the stomata and causes disintegration of chloroplasts, plasmolysis, and eventual death of mesophyll cells. Some plants, such as barley, are damaged by concentrations of 0.1 to 0.3 ppm sulfur dioxide, which cause decreased photosynthesis and general growth suppression. There are instances where entire plant communities have been destroyed by prolonged exposure to sulfur dioxide.

The classic illustration of sulfur dioxide destruction is the Copper Basin area, a southern Appalachian valley, of Tennessee. Before the fumes from a copper smelter were eliminated, about 7000 acres of dense deciduous forest were completely denuded and another 17,000 acres reduced to sparse grass. Most of the topsoil, exposed to rain and water flow without protection from vegetation, washed away leaving a desertlike area in the midst of lush forest.

Sulfur dioxide in a humid atmosphere greatly accelerates rusting and corroding of metals. It also brings about rotting of wool, cotton, and leather and the erosion of marble, sandstone, and limestone. In many cities of Europe, ancient statuary has been damaged more in the past 50 years than in all the years since they were produced. Air pollution in London results from burning huge quantities of coal and oil, and sulfur dioxide is the main contaminant.

Fluoride

Flouride toxicity has greatly intensified with the vast expansion of the aluminum and phosphate reduction industries, which use minerals high in fluorides. Exposure of various woody plants, such as aspen, grape, citrus, and pine, to as little as 5 parts per billion of fluoride reduced leaf size 25 to 35 per cent; tree height, crown volume, and total green weight were also reduced. Fluoride caused a significant reduction in the number of fruits and the average weight per fruit in tomato. Typical symptoms in most plants consist of chlorophyll destruction and reduction in leaf size, although death of cells occurs if exposure is prolonged or if fluoride concentration is too high. Fluorides appear to act as cumulative poisons; even low concentrations eventually kill.

Nitrogen Oxides

Any combustion process that produces high temperatures in the presence of nitrogen and oxygen will yield nitrogen oxides, especially nitrogen dioxide. Fuels used in heat and power generating contribute significant amounts, but about 70 per cent of the nitrogen dioxide originates from automobile exhausts. The atmosphere usually contains about 0.1 ppm nitrogen dioxide, but concentrations in the Los Angeles area of southern California have exceeded the "alert" level of 3.0 ppm. Depending upon the plant, 10 to 200 ppm for one to eight hours causes tissue collapse, bleaching, death of cells, and up to 100 per cent defoliation. There may also be a cumulative effect: concentrations that failed to cause injury during three successive days of fumigation brought about death and collapse of cells on the fourth day.

PAN

Industrial expansion and population increases in the unpolluted Los Angeles basin of southern California during the 1940s resulted in the clear air being replaced by a dirty brown haze, which became known as "smog." This was really an incorrect use of the term, but common usage over a relatively long period will undoubtedly establish a new definition. **Smog** was originally used in London to denote fog that had become mixed and polluted with smoke. The phenomenon in Los Angeles involves neither fog nor smoke.

Hydrocarbons and oxides of nitrogen emitted from automobile exhausts react in the presence of sunlight to yield the toxic gases that cause plant damage, eye irritation, and reduced visibility. Nitrogen dioxide is split by ultraviolet radiation of sunlight into nitric oxide and atomic oxygen. The latter can combine with molecular oxygen to form ozone or react with some of the unburned hydrocarbons to form various oxidation products of which **peroxyacetyl nitrate (PAN)** is the principal constituent causing damage to plants.

The rather inefficient internal combustion engines of automobiles emit tons of nitrogen dioxide and unburned hydrocarbons. About 20 per cent of the hydrocarbons emitted are a result of evaporation from gas tanks and carburetors, about 30 per cent from the crankcase vent, but most are vented to the atmosphere through the tailpipe. Combustion adds carbon monoxide and carbon dioxide as well as the nitrogen dioxide. Every 1500 cars daily discharge into the air an estimated 3.2 tons of carbon monoxide, 600 lb of hydrocarbon vapors, and 200 lb of nitrogen oxides, plus smaller amounts of other chemicals that depend upon the various additives in the gas (e.g., tetraethyl lead, detergents, lubricants, antirust compounds, antiicing compounds). All these pollutants, and the additional reaction products formed in the atmosphere, comprise the mixture that should be identified as *photochemical smog.*

PAN at sublethal concentrations of about 0.01 ppm brings about reduced yield of sensitive plants, such as tobacco and petunia; lesions develop with exposure to 0.1 ppm. Chloroplasts are readily destroyed. Enzyme activity, respiration, photosynthesis, ion absorption, and carbohydrate and protein syntheses all may be impaired by PAN concentrations far lower than necessary to produce death. Trees exposed to naturally polluted air in the Los Angeles area have up to 30 per cent more leaf drop, and yields are often only half of those in clean air. Chemical analyses show that PAN concentrations in southern California regularly range from 0.01 ppm on average days to 0.5 ppm on "smoggy" days. Highly sensitive crops can no longer be grown in the Los Angeles area, where a noticeable haze now develops about 200 days out of every year.

Ozone

Although the natural concentration of ozone is about 0.02 ppm, it may range as high as 0.5 ppm in some urban atmospheres. In such areas ozone concentrations are increasing at an alarming rate as a result of the action of ultraviolet light on nitrogen dioxide, as mentioned above. Concentrations in and around cities almost daily exceed the 0.05 ppm at which plant damage is expected; in Los Angeles 0.15 to 0.35 ppm is common and 1.0 ppm has been reached.

After ozone enters the leaf through the stomata, it disrupts membrane structure, increases permeability, destroys enzymes and organelles, increases respiration, and reduces photosynthesis. Even sublethal ozone concentrations may decrease photosynthesis by 40 to 70 per cent. In some leaves not visibly injured, respiration increases greatly if fumigation is continued over an extended period of time and depletes carbohydrate reserves. In addition to its effects on plants, ozone is very reactive and attacks rubber causing it to crack and decompose.

Human Health

It is believed that air pollutants contribute to or aggravate such acute and chronic respiratory diseases as bronchial asthma, sinusitis, emphysema, and bronchitis. They are also strongly suspected of aggravating cardiac and circulatory ailments, and some medical scientists believe that long-term, low-level exposure reduces resistance to other diseases. Although a direct relationship between air pollution and human disease has not been proved, there is much circumstantial evidence to indicate this. Even if unclear air does not hurt human health, it often makes life miserable—eye irritation, offensive smells, decreased visibility, and damage to property.

When pollution increases substantially, there is an accompanying increase in the death rate. The most devastating example of this occurred in London in 1952. A sooty fog settled over the city, reducing visibility to near zero. This smog lasted four days and hospitals became crowded with hundreds of patients suffering from pneumonia, bronchitis, and heart disease. During the week of the smog and the week that followed, London's normal death rates had been exceeded by 4000. Similar episodes have occurred in other urban areas with sulfur dioxide implicated as the irritant mainly responsible. There are also some studies that indicate that air pollution accounts for a doubling of the bronchitis mortality rate for urban, as compared to rural, areas. Evidence from European studies shows an association between lung cancer and air pollution. While individual studies may be criticized on the grounds that none managed to provide controls for all causes of ill health, the number of studies and the variety of approaches are persuasive. Basically, one concludes that there is an important association between air pollution and various morbidity and mortality rates.

Estimates of the dollar benefit of pollution abatement are difficult to make. In a recent article in *Science,* L. B. Lave and E. P. Seskin suggest that the total annual health cost that would be saved by a 50 per cent reduction in air-pollution levels in major urban areas, in terms of decreased morbidity and mortality, would be over two billion dollars. They also state that this is surely an underestimation of total costs. Psychological and aesthetic effects were not considered nor were additional costs associated with the effects of air pollution on vegetation, cleanliness, and the deterioration of materials. It is obvious that costs of air pollution total billions of dollars.

Summary

1. Pollution is the contamination of soil, water, or the atmosphere that lessens their value to living organisms. Pollution increases drastically as the number of humans increases and is accentuated by technological developments.

2. Litter is a disorderly accumulation of discarded objects. Humans discard vast quantities of material, including garbage.

3. Humans contaminate water supplies by discharging into them untreated, or partially treated, sewage and various industrial wastes and coolants. Considerable amounts of agricultural fertilizers and pesticides eventually drain into our water and frequently are accumulated and concentrated in various food chains. Irrigation of arid areas results in increased salt concentrations of soil and water to the detriment of plants in those areas.

4. Burning materials add carbon monoxide, carbon dioxide, particulate matter, oxides of nitrogen, oxides of sulfur, fluorides, and various other materials to the atmosphere. Ultraviolet radiations from the sun cause reactions that produce ozone and peroxyacetyl nitrate (PAN). These materials to varying degrees are detrimental to plant growth and human health and bring about billions of dollars worth of damage.

Review Topics and Questions

1. Discuss the manner in which population numbers, technology, and pollution are related.
2. Explain why biodegradable litter and wastes of various sorts create less of a problem than nonbiodegradable material.
3. Discharge of sewage into streams and lakes is detrimental even if there are no pathogenic organisms present. Explain.
4. Discuss the ways in which industrial discharges may be detrimental to an ecosystem.
5. Fertilizers and pesticides of various sorts are used to improve crop yield. In what way may this be damaging to an ecosystem?
6. Distinguish between eutrophic and oligotrophic lakes.
7. What is "monotypic agriculture"? What are advantages and disadvantages of this type of agriculture?
8. Why is sulfur dioxide considered to be an air pollutant, and what is its source?
9. Ozone and peroxyacetyl nitrate (PAN) are serious air pollutants of urban areas and yet they really are not wastes discharged directly from a technological process. Explain why the presence of these two pollutants is correlated with urban areas.

Suggested Readings

BENARDE, M. A., *Our Precarious Habitat* (New York, W. W. Norton, 1970).

OSBORN, F., *Our Plundered Planet* (Boston, Little, Brown, 1948).

OSBORN, F., *The Limits of the Earth* (Boston, Little, Brown, 1953).

OWEN, O. S., *Natural Resource Conservation: An Ecological Approach* (New York, Macmillan, Inc., 1971).

TURK, A., J. TURK, and J. T. WITTES, *Ecology, Pollution, Environment* (Philadelphia, Saunders, 1972).

WAGNER, R. W., *Environment and Man* (New York, W. W. Norton, 1971).

The Beginning

Perhaps a chapter entitled "The Beginning" should be numbered *one* rather than *thirty-three*. However, any discussion as to the origin of life cannot be fruitful without an understanding of what one means by "life." We have accumulated many chapters of background material upon which we can now build a picture of life's beginning.

Various suggestions have been made by numbers of people who attempted to explain how life could have originated. No attempt will be made to include all hypotheses, but any theory must be conducive to careful scrutiny in the light of present concepts of physics and chemistry. We must also make the basic assumption that, if conditions are favorable for life, life will come into existence. Obviously no one knows how "life" began, but the concepts and suggestions to be discussed are at least plausible in light of present knowledge.

33.1 Autotroph Hypothesis

Since all organisms require food, one frequently assumes that the first organism must have been able to manufacture its own food supply. This would mean an entity similar to the photosynthetic or chemosynthetic organism of today—an organism independent of other living organisms for its organic compounds (food). From this beginning would gradually (in a few millions of years) evolve organisms that could exist on food so nicely manufactured for them. Such an autotroph, however, requires a vast array of enzymes for the many syntheses necessary for life. Simple inorganic materials would have to be converted through many steps to the complex moiety we now know as the protoplast. With the increased study of cellular processes, one became more and more aware that autotrophic organisms are basically more complex than heterotrophic ones; more than likely, the former arose later in the evolutionary scheme of things.

One of the factors that led biologists to this conclusion was the fossil record—the oldest fossils are of a simpler basic structure than the more recent ones. Admittedly, the majority of fossils are those that already demonstrate considerable complexity when compared with what must have been a first living thing. But the general principle of simple leading to more complex is basic to evolution. There is no reason to doubt that this occurred at the very beginning as well as later.

33.2 Heterotroph Hypothesis

A heterotrophic organism forms its own structural components from preformed organic material (food), minerals, water, and oxygen. The first and last of these essential factors were the most difficult to obtain, since at the earth's beginning no oxygen gas molecules and no organic compounds were available. One of the important facets of the suggestion that heterotrophs were the first living things is an explanation of food formation that must have predated "life."

The understanding of the relationship of one organism to another makes it clear that autotrophs are more complex physiologically or metabolically because of their ability to start with extremely simple substances, whereas the physiologically simpler heterotrophs start with substances already partially syntheseized into cellular components. For example, a green plant readily produces proteins when supplied with water, minerals, carbon dioxide, and light. This is not true of a human, who must obtain either proteins or amino acids in order to form proteins. Many fewer reactions are involved in the human production of proteins, because the green plant has already done part of the work for us (i.e., the synthesis of amino acids). Since each step in the various metabolic processes is mediated by an enzyme, there must be more enzymes in the green plant. Hence, it is a more complex organism than a nongreen one. A basic premise of the evolutionary concept, as just mentioned, is that simple organisms gave rise to complex organisms. Therefore, heterotrophs must have existed before autotrophs if the conceptual scheme of evolution is a valid one. With these thoughts as a starting point, any scheme concerning the origin of life must logically include the evolution of organic compounds first, then heterotrophs, and finally autotrophs, which really saved the day, as will be seen later.

33.3 Organic Beginnings and Life

Most geologists, chemists, and astronomers generally agree that the earth was at some early stage a molten mass of exceedingly high temperatures. At these temperatures, atoms combine, separate, and recombine in a variety of ways. As the hot mass gradually cooled over millions of years, many interactions must have occurred, eventually producing an atmosphere of methane (CH_4), ammonia (NH_3), water vapor (H_2O), and hydrogen (H_2). At very high temperatures, oxygen and nitrogen could not have existed as free gases; they would react with hydrogen forming water (H_2O) and ammonia (NH_3) respectively. Most scientists consider this to have been the situation of our primitive earth. With further cooling, water vapor condensed as rain. Volcanic action and violent land upheaval would aid the mixing and distribution of materials.

Such an atmospheric mixture, when exposed to ultraviolet rays and cosmic rays which carry large amounts of energy, and violent electrical discharges (lightning), could result in the formation of a number of organic compounds, including amino acids—the basic building blocks of proteins that are a main component of the protoplast. In this fashion a tremendously great variety of organic compounds could have been formed, resulting (on further cooling and condensation of water vapor) in the seas of the world becoming in effect a vast colloidal soup without life.

In 1953 Stanley Miller circulated a mixture of CH_4, NH_3, H_2O, and H_2 past an electrical discharge at high temperatures. After a week his originally colorless solution had turned red. On analysis, he found the solution to contain many organic compounds including amino acids. Sidney Fox later (1957) demonstrated that amino acids could join together. He heated a mixture of amino acids to the melting point. When the mass had cooled, he discovered that many of the amino acids had bonded together to form chains with characteristics similar to those of protein structure.

Further evidence for the possible formation of rather complex organic compounds from a mixture of methane, ammonia, hydrogen, and water resulted from the work of Melvin Calvin. He was able to show that gamma radiation, acting upon such a mixture, produced amino acids, sugars, and probably purines and pyrimidines. The familiar compounds ATP, NAD, and nucleic acids are made from the latter two materials in living cells.

The intense bursts of energy necessary to bring about reactions that form organic compounds could be in the form of electrical discharges, gamma rays, cosmic rays, or ultraviolet rays. Since oxygen and ozone were lacking in the atmosphere of the primitive earth (at high temperatures, all oxygen would be in a combined form and not as free gas), such radiations from the sun were not filtered out as they are to a great extent now. This condition, combined with the high temperatures, could certainly have brought about the suggested reactions.

The various colloidal particles thus formed could come together in many instances to form clusters, or **coacervates,** held together by the electrical forces between them. The surfaces of these coacervates would differ much as the surface layer of water differs from the inner regions, a phenomenon referred to as surface tension and probably important in membrane structure.

Among the untold billions of colloidal particles and coacervates may have been some that were capable of utilizing less complex organic molecules in making duplicates of themselves in the manner that nucleic acids (DNA) reproduce. Such reproducing coacervates had a better chance of surviving as a type than would others. The slow accumulation of many changes over millions of years finally could have resulted in what is called "a living thing," which may have been very similar to a virus. Chemically, a virus consists of nucleic acid surrounded by protein, in other words, a nucleoprotein possibly similar to the nucleic acids in chromosomes. The virus and the nucleic acids obtain their substrate from the living cells in which they are found; the coacervate could obtain its substrate from the environment, which abounded in organic molecules. Further changes probably resulted in an organism similar to the fermentative bacteria of the present.

This discussion is speculative but is based on reactions we know can occur and upon sound principles of chemistry and physics. It is certainly possible for life to have originated in this fashion. It is logical and, after all, amino acids and proteins have been shown to form in an environment considered to be like that which once existed on earth.

33.4 The Early Food Problem

The increasing numbers of heterotrophic organisms brought about a food problem, which had not existed before. In fact, as the food supply became increasingly scarce, survival began to depend more and more upon the ability of an organism to utilize a greater variety of foods or to synthesize its complex structure from simple molecules. Finally, after millions of years of changes, some organisms appeared capable of using the vast supply of inorganic compounds. These were probably similar to certain organisms living today: (1) the sulfur bacteria, (2) the nitrifying bacteria, and (3) the green plants. Sunlight is a source of much more energy than is available from chemical compounds, and thus the photosynthetic organisms became the dominant autotrophs of the world. The appearance of these organisms ended the danger of the exhaustion of the supply of organic substances needed by the organisms that could not get their energy directly from the sun. Photosynthetic organisms saved the day for the type of life that existed before that process developed; the former now supply all the food material, as well as all oxygen, for all the organisms in the world. Once oxygen became available, the much more efficient aerobic metabolism dominated the scene, as is indicated by the predominance of aerobic organisms in the world today. The increased energy made available by aerobic respiration resulted in the evolution of larger and more complex organisms than would have been possible with strictly fermentative (anaerobic) processes.

33.5 Can Life Originate Now?

A basic premise made early in discussing the origin of life was that when conditions are suitable for life, life will come into existence. The conditions are certainly suitable for life now, and yet life cannot possibly begin again. This apparent paradox comes about because the living organisms now present would consume new organic compounds before they could possible evolve further. The time for life to originate was when no life existed, and further spontaneous generation on this earth is not possible.

The fact that life can no longer originate on Earth does not mean that such a beginning is impossible anywhere else. The universe is vast, containing untold numbers of galaxies (like the Milky Way) with their myriads of stars and planets. The existence of other planets similar to the Earth revolving about their own stars or suns is quite likely. However, most conditions seem to be weighted against the existence of life on most planets. If the planet is too near its sun, it will be too hot for life to exist; if it is too far away, it will be too cold. If it is much larger than the Earth, gravitational pull will retain too much atmosphere and result in a poisonous situation, as is probably found on Jupiter and Saturn. If it is much smaller, all atmosphere will have been lost, as is true of the Moon and Mercury.[1] Some astronomers feel that only Venus, Earth, and Mars of our own planetary system could possibly support life, that Mars is a planet where life has almost become extinct (possibly some scanty vegetation is left), and that Venus, although without life now, may be the home of life in the future.

Recent evidence obtained by transmissions from the Mariner 2 spacecraft, which passed within 21,600 miles of Venus in 1962, indicates a temperature of 800°F (427°C) and an atmosphere from 5 to 100 times denser than that of the Earth. The Russian spacecraft that landed instruments on Venus in 1967 transmitted data indicating a temperature of 100° to 500°F and an atmosphere (of almost pure CO_2) 15 times as dense as that of the Earth. Venus is closer to the sun than is Earth (67 million miles versus 93 million), and this could account for some of the temperature difference. In any event the situation does not appear to be conducive to life as we know it. The 1976 landing of a space craft on Mars has provided much information but no proof of the existence of what we call "living organisms." Further spacecraft explorations will, one hopes, provide us with additional information.

33.6 Spontaneous Generation

That spontaneous generation does not occur was difficult to demonstrate at first, and many years of controversy passed before Pasteur's famous experiment proved the matter conclusively. The appearance of maggots in putrefying meat was considered sufficient evidence that living creatures were generated spontaneously from nonliving materials until Francesco Redi(1668), an Italian biologist, covered meat with gauze and prevented maggots from forming. This did not settle the controversy because the covered meat did become putrid, and the natural conclusion was that even if maggots did not develop from meat, bacteria did. Of course, Redi's experiments could have quashed the idea of spontaneous generation if von Leeuwenhoek's microscopes had not enabled biologists to see the extremely small bacteria that developed so profusely as the meat decayed. The arguments pro and con waxed hot and heavy for some 200 years more until Pasteur boiled various media in a flask that had a long swan-shaped, or S-shaped neck. The flask was open to the air so that any "vegetative force" could still enter, and yet the curved neck prevented the relatively heavy bacteria and fungus spores from entering into the medium. No living organisms developed in such flasks after prolonged boiling and slow cooling. If the flasks were broken open or if dust were allowed to enter, microbes readily developed from those carried in the air. No doubt remained that, although spontaneous generation undoubtedly occurred mil-

---

[1] For a discussion of atmospheres, velocities of escape, and velocities of gas molecules, see the book by H. Spencer Jones or other astronomy texts.

lions of years ago when life originated, all organisms now develop from preexisting organisms and all cells develop from preexisting cells. This is one of the basic concepts of biology.

However, this still leaves open the question of where living organisms came from originally.

Summary

1. The autotroph hypothesis concerning the origin of life suggests that those plants that could manufacture their own food must have been the first to evolve. Since all other organisms now depend upon the autotrophs, this seemed to be a logical hypothesis. The fact became evident, however, that these organisms are basically more complex than are the heterotrophs.

2. The heterotroph hypothesis suggests that organic molecules evolved first and that the heterotrophs, which eventually arose, could then utilize these compounds. In this view, the food for the organisms was available before the organism existed. The more complex autotrophs then evolved from the preexisting heterotrophs.

3. The origin of autotrophs allowed the heterotrophs to continue living by providing food and then oxygen.

4. Earth's early environment was conducive to the formation of organic molecules from simple inorganic materials. Spontaneous generation did occur at one time but is no longer possible.

Review Topics and Questions

1. Why is the autotroph hypothesis of the origin of life not tenable?
2. How could a heterotrophic organism exist without autotrophs? Is this condition possible now? Explain what would happen if all autotrophic organisms died now.
3. What conditions are necessary before life could originate?
4. Explain why Pasteur had to cool his swan-neck flasks slowly rather than rapidly.
5. Explain why the evolution of an autotrophic type of metabolism was essential for the survival of life.

Suggested Readings

ADLER, I., *How Life Began* (New York, New American Library, 1959).
DOLE, S. H., *Habitable Planets for Man* (New York, Blaisdell, 1964).

HUANG, SU-SHU, "Life Outside the Solar System," *Scientific American* **202,** 55–63 (April 1960).

JONES, H. S., *Life on Other Worlds* (New York, New American Library, 1949).

KEOSIAN, J., *The Origin of Life* (New York, Reinhold, 1964).

OPARIN, A. I., *Life: Its Nature, Origin and Development* (London, Oliver and Boyd, 1960).

WALD, G., "The Origin of Life," *Scientific American* **191,** 44–53 (August 1954).

Glossary

See **Index** for terms not included here.

Abscisic acid: growth regulator that inhibits growth and promotes abscission and dormancy.

Abscission: dropping of leaves, fruits, or other plant parts, usually by the separation of cells in a zone at the base of a petiole or other structure.

Abscission layer: a zone of cells at the base of a petiole or other structure whose cells separate.

Absorption: the movement of substances into a cell or other structure from the outside.

Accumulation: the absorption of substances against a concentration gradient; requires respiratory energy.

Achene: a dry, indehiscent, one-seeded fruit formed from a single carpel; seed coat not fused with ovary wall.

Active absorption: absorption dependent upon respiratory energy and probably carrier molecules within the membrane; frequently results in accumulation.

Adaption: changes in an organism that result in its being better suited to an environment.

Adenosine diphosphate (ADP): a compound to which inorganic phosphate may be added to form a terminal high-energy phosphate bond; energy is used in the formation of such bonds; the resultant compound is ATP.

Adenosine triphosphate (ATP): a compound containing high-energy phosphate groups that may be removed with the liberation of usable cellular energy. ADP is formed when such energy is made available.

Adsorption: the concentration of molecules or ions on the surfaces of solid particles caused by forces of attraction between them.

Adventitious: describes those structures arising in an unusual place or abnormal position.

Aerobic respiration: respiration using gaseous oxygen as the final hydrogen acceptor.

Afterripening: metabolic processes that must occur in some seeds before germination will occur.

Agar: a gelatinous substance extracted from certain red algae (Rhodophyta); used as a solidifying agent in the preparation of nutrient media for the growth of microorganisms, and for a variety of other purposes.

Aggregate fruit: a fruit formed from a group of ovaries produced in a single flower.

Allele: one of the alternate forms of a gene at a particular locus (location) on homologous chromosomes.

Alternate leaves: a leaf arrangement with a single leaf at each node.

Alternation of phases: the development of spore-producing structures followed by the development of gamete-producing structures in the life cycle of a plant; the cells of the former contain twice as many chromosomes as do those the latter.

Amino acids: organic acids containing an amino group, NH_2; they are the units from which protein molecules are constructed.

Ammonification: the formation of ammonia during the decomposition of nitrogen-containing organic compounds by microorganisms.

Amylase: an enzyme that hydrolyzes (digests) starch to maltose.

Amyloplast: a plastid in which starch is formed; contains no pigment.

Anabolism: the constructive phases of metabolism; syntheses.

Anaerobic respiration: See **Fermentation.**

Anaphase: that stage of nuclear division during which the chromatids of each chromosome separate and move to opposite poles, or a similar separation of homologous chromosomes (as in meiosis).

Androecium: the aggregate of stamens in the flower.

Annual: a plant in which the entire life cycle is completed in a single growing season.

Annual ring: the amount of secondary xylem formed during one year's growth.

Annulus: a ring of specialized cells in the wall of a fern sporangium.

Anther: the pollen-bearing portion of a stamen.

Antheridium: a structure in which male gametes or male nuclei are produced; a male gametangium.

Anthocyanins: water-soluble pigments found in the cell sap, usually reds or blues.

Antibiotic: a substance produced by living organisms (usually bacteria and true fungi) that destroys or inhibits the growth of other organisms.

Antipodal cells: the cells, usually three, located at the end opposite the micropyle of the megagametophyte of flowering plants; functionless and vestigial.

Apical dominance: the influence of terminal buds in suppressing the growth of lateral buds.

Apical meristem: the meristematic tissue at the tip of a root or stem.

Apothecium: in Ascomycetes, a cup-shaped structure containing asci with ascospores; an ascocarp.

Archegonium: a multicellular structure in which an egg is produced; a female gametangium.

Ascocarp: in Ascomycetes, a structure in which asci are produced.

Ascogonium: the female gametangium of Ascomycetes.

Ascospore: a meiospore produced in an ascus.

Ascus: in Ascomycetes, a saclike structure in which ascospores are produced after nuclear fusion and meiosis.

Asexual reproduction: reproduction that does not involve the fusion of sex cells (gametes).

Assimilation: the conversion of nutrients into the protoplast.

ATP: see **Adenosine triphosphate.**

Autonomic growth movement: a spontaneous and self-controlled growth movement.

Autotrophic: capable of manufacturing its own food, such as green plants.

Auxin: a growth regulator that promotes cell elongation among other effects.

Axil: the upper angle between a petiole and the stem to which it is attached.

Axillary bud: a bud borne in the axil of a leaf.

Bacillus: a rod-shaped bacterium.

Bacteriophage: a virus that attacks bacteria.

Bark: the tissues of a woody stem outside of the vascular cambium.

Basidiocarp: in Basidiomycetes, a structure in which basidia are produced.

Basidiospore: a meiospore borne on a basidium.

Basidium: in Basidiomycetes, a cellular structure on which spores are borne after nuclear fusion and meiosis.

Berry: a simple fleshy fruit.

Biennial: a plant that grows vegetatively the first year, produces flowers and seed during its second year, and then dies.

Binomial: the generic name and specific name of an organism taken together as its official name.

Blade: the expanded portion of a leaf.

Bract: a modified or reduced leaflike structure.

Bud: a much compressed embryonic shoot.

Bud scale: one of the modified protective leaves surrounding the shoot apex and embryonic leaves of a bud.

Budding: (1) grafting in which the scion is a single bud; (2) vegetative reproduction in yeast.

Bulb: a short, underground stem with fleshy leaves.

Bundle sheath: a layer of cells surrounding the vascular strands of a leaf.

Callus: undifferentiated tissue; parenchymatous wound tissue.

Calorie: the amount of energy in the form of heat required to raise the temperature of 1 gm of water 1°C; it is useful as a measure of energy.

Calyptra: in mosses, a caplike structure covering the apex of the capsule; it consists of the upper portion of the enlarged archegonium.

Calyx: collective term for the sepals of a flower; outermost whorl of flower parts.

Cambium: see **Cork cambium** and **Vascular cambium.**

Capillarity: the movement of water through tiny passages as a result of surface forces.

Capillary soil water: water held in the soil against the force of gravity; the water is retained in spaces among and within the soil particles; the chief source of water absorbed by roots.

Capsule: (1) the sporangium of a bryophyte; (2) a simple, dry, dehiscent fruit formed from a compound pistil.

Carbohydrate: organic compound composed of carbon, hydrogen, and oxygen with the ratio of hydrogen to oxygen being 2:1.

Carnivorous: meat-eating.

Carotenes: yellow to orange or red carotenoid pigments found in plastids; one is a precursor of vitamin A.

Carpel: a floral organ bearing and enclosing ovules; a pistil may be composed of one or more carpels.

Catabolism: the destructive phases of metabolism in which organic compounds are broken down into simpler substances, as in respiration and digestion.

Catalyst: a substance that speeds the rate of a chemical reaction without being used up in the reaction.

Cell: the structural unit of plants and animals; generally in plants this consists of the protoplast surrounded by a cell wall.

Cell sap: the solution in the vacuole of a cell.

Cellulase: an enzyme that hydrolyzes (digests) cellulose.

Cellulose: an insoluble complex carbohydrate formed from glucose; it is the chief component of the cell wall in most plants.

Centromere: see **Kinetochore.**

Chemosynthesis: the synthesis of organic matter utilizing energy made available by the oxidation of inorganic compounds, such as ammonia, nitrites, sulfur, and so forth.

Chlorophylls: green pigments usually located in plastids; they absorb light and are involved in photosynthesis.

Chloroplast: an organelle that contains chlorophyll; found in cytoplasm.

Chromatid: the term applied to each member of a duplicated chromosome before they separate at anaphase.

Chromoplast: an organelle in the cytoplasm that contains pigments other than chlorophyll; the color is often yellow or red because of the carotenes and xanthophylls.

Chromosome: structural bodies made up of DNA and protein; found in the nucleus; the site of hereditary determiners or genes.

Cilium: short hairlike protoplasmic projection that functions as an organ of motility.

Class: in classification, a group of related orders.

Cleistothecium: a closed spherical structure composed of hyphae and containing one or more asci; an ascocarp.

Climax: the terminal community of a succession, which is maintained with little change provided the environment does not change significantly.

Coccus: a spherical bacterium.

Coenocytic: multinucleate; lacking cross walls.

Cohesion theory: a suggested explanation for the rise of water in plants; it is based upon the forces of attraction between water molecules.

Coleoptile: a sheathlike structure covering the epicotyl of grass seedlings.

644
Glossary

Collenchyma: a supporting and strengthening tissue in which the cells have irregular primary wall thickenings.

Colloid: a dispersion of particles ranging from 0.1 μm to 0.001 μm; the particles remain suspended and are usually large molecules or molecular aggregates.

Colony: a group of similar organisms living in close association, usually microorganisms; there is no division of labor.

Community: an assemblage of organisms living together.

Companion cell: a narrow, slightly elongated cell adjoining a sieve tube element; both arising from the same mother cell.

Compensation point: the light intensity or temperature at which the rates of photosynthesis and respiration are equal.

Complete flower: a flower containing sepals, petals, stamens, and at least one pistil.

Compound leaf: a leaf in which the blade is subdivided into two or more parts.

Conceptacle: in certain brown algae, a cavity in which gametes are eventually formed.

Cone: a close aggregation of spirally arranged sporophylls on which sporangia are produced.

Conidiophore: a hypha that produces conidia.

Conidium: an asexual fungus spore produced at the tip of a conidiophore.

Conjugation: (1) a type of isogamy in which one gamete moves toward another through a tube connecting the two cells; (2) transfer of genetic material in bacteria.

Convergent evolution: the independent development of similar structures in groups of organisms that are only very distantly related.

Cork: a protective tissue in which the cell walls are suberized; formed by a cork cambium; also called phellem.

Cork cambium: a meristematic tissue that produces cork cells; also called phellogen.

Corm: a bulky, short, vertical, underground stem that stores food.

Corolla: the petals of a flower taken collectively.

Cortex: the region of a root or stem between the epidermis and the vascular tissue; composed primarily of parenchyma-type cells.

Cotyledon: the leaf of an embryo; also called seed leaf.

Crossing-over: the interchange of corresponding gene segments between chromatids of homologous chromosomes during meiosis.

Cross-pollination: the transfer of pollen from the anther of one plant to the stigma of another, or from the male cones of one to the female cones of the other.

Cuticle: a fatty layer secreted by epidermal cells on their outer walls; composed of cutin.

Cutin: a fatty substance, rather impermeable to water.

Cytokinesis: cytoplasmic division and the formation of a new wall.

Cytokinin: a growth regulator that promotes cell division among other effects.

Cytoplasm: that portion of the protoplast exclusive of the nucleus and plastids.

Deciduous: (1) broad-leaved plants that drop their leaves at the end of each growing season; (2) parts that drop at the end of the growing season.

Denitrification: the conversion of nitrates into nitrogen gas by certain soil bacteria.

Deoxyribonucleic acid (DNA): hereditary material; self-replicating and determines RNA synthesis; in eukaryotes combined with protein to form chromosomes; in prokaryotes not combined with protein.

Dicotyledonous plant (dicot): a flowering plant characterized by two cotyledons or seed leaves in the embryo.

Dictyosome: small flattened cytoplasmic organelle having a secretory function.

Differentially permeable membrane: a membrane that allows some substances to pass through more readily than others; some substances may not be capable of penetrating at all.

Diffusion: the movement of molecules or ions from one location to another as a result of the kinetic energy of the molecules or ions.

Digestion: the enzymatic conversion (hydrolysis) of insoluble or complex substances into soluble or simpler substances.

Dihybrid cross: a cross (mating) between organisms differing in two pairs of genes.

Dioecious: having male and female reproductive organs on separate plants.

Diploid: having two sets of chromosomes, one set from one gamete and the second set from the other gamete of the two that fuse; the situation (2N) characteristic of the sporophyte phase in plants.

DNA: see **Deoxyribonucleic acid.**

Dominant character (factor): (1) a gene that masks or hides the presence of its allele; (2) that characteristic which excludes the appearance of its contrasting character in a hybrid (the latter is the recessive character).

Dormancy: a period of greatly reduced physiological activity in various plant parts, especially seeds and buds, even though the environment may be suitable for growth.

Double fertilization: refers to the situation in flowering plants where one sperm fuses with the egg and the second sperm fuses with the two polar nuclei.

Drupe: a fruit in which the exocarp is a thin skin, the mesocarp is the fleshy pulp, and the endocarp is hard (the pit).

Ecology: the study of interrelationships between organisms and their environment.

Egg: female gamete or sex cell.

Elaioplast: a plastid in which fats are formed; contains no pigment.

Embryo: the rudimentary sporophytic plant developed from a zygote within an archegonium or an ovule.

Endergonic reaction: one that requires an input of energy in order to proceed.

Endocarp: the inner layer of the pericarp or fruit wall (ovary wall).

Endodermis: a single layer of cells forming the innermost portion of the cortex; it is most obvious in the roots.

Endoplasmic reticulum (ER): the interconnected network of double membranes that constitute a part of the submicroscopic structure of the protoplast.

Endosperm: in flowering plants, a nutritive tissue surrounding the embryo; usually triploid.

Energy: the ability to do work.

Energy-rich phosphate bond: a bond joining certain phosphate groups to an organic molecule; energy is made available when the bond is broken; see **Adenosine diphosphate** and **Adenosine triphosphate.**

Enzyme: a proteinaceous catalyst produced by living cells.

Epicotyl: that part of the embryo above the point of attachment of the cotyledons; upon growth it forms the shoot of the plant.

Epidermis: the surface layer of cells before cork is formed.

Epigynous flower: a flower in which the ovary is embedded and the floral parts seem to arise from the top of the ovary.

Epiphyte: a plant that grows attached to some larger plant, usually on treee trunks or branches, without deriving nourishment from its host.

Erosion: the removal of soil by natural agents, especially by water and wind.

Ethylene: growth regulator that promotes fruit ripening and inhibits growth; simple hydrocarbon gas ($CH_2 = CH_2$).

Etiolation: the condition of a plant grown in the absence of light; the plant typically lacks chlorophyll, has an abnormally elongated stem, and poorly developed leaves.

Eukaryote: organism whose cells possess a membrane-bound nucleus.

Evolution: the relationships of groups of organisms; descent with modification.

Exergonic reaction: one in which there is a decrease in free energy; it yields energy.

Exocarp: the outer layer of the pericarp or fruit wall (ovary wall).

Eyespot: a small, pigmented, light-sensitive structure present in certain algae.

F_1: the first generation of progeny following a cross or mating; later generations are known as the F_2, F_3, and so forth.

Family: a taxonomic grouping of closely related genera.

Fat: organic compound consisting of carbon, hydrogen, and oxygen, with the last in much smaller proportionate amounts than in carbohydrates; if liquid at room temperature, they are called oils.

Fermentation: a complex series of cellular oxidation-reduction reactions in which energy is made available to the cell in the absence of oxygen; the incomplete oxidation of substrate with various organic materials as end products; also termed anaerobic respiration.

Fertilization: the fusion of two gametes resulting in a diploid zygote.

Fiber: greatly elongated, slender, tapering, thick-walled cell that serves to support and strengthen a structure.

Field capacity: the amount of water held in a soil against the forces of gravity.

Filament: (1) the stalk of a stamen; (2) a threadlike row of cells.

Fission: asexual reproduction of a unicellular organism in which the cell divides to form two organisms.

Flagellum: a rather long hairlike protoplasmic projection that functions as an organ of motility.

Flower: the sexual reproductive structure of Anthophyta; consists of sepals, petals, stamens, and pistil if all the parts are present.

Flower bud: a bud that develops into a flower.

Food: an organic substance that furnishes cellular energy (in respiration or fermentation) or that is transformed into the protoplast or cell secretions.

Food chain or **web:** a group of plants and animals interrelated by their food dependencies.

Fossil: any evidence of a former living thing.

Free energy: useful or available energy; the capacity to do work.

Frond: the leaf of a fern.

Fruit: a mature ovary or group of ovaries, usually containing seed, together with any adjacent parts adhering to them.

Fucoxanthin: the brownish pigment of certain algae.

Gametangium: a structure in which gametes are produced.

Gamete: a sex cell; gametes fuse in sexual reproduction, forming a zygote.

Gametophyte: the haploid, gamete-producing phase of the plant life cycle.

Gene: one of the determiners of hereditary characteristics, located on chromosomes; a portion of a DNA molecule.

Genera: plural of **Genus.**

Genetics: the study of heredity and variation.

Genotype: the genetic composition of an organism.

Genus: a taxonomic grouping of closely related species; the official name of an organism consists of its genus and species categories.

Geotropism: a growth response induced by gravity.

Germination: the resumption of growth by a seed, spore, bud, or other reproductive structure.

Gibberellin: a growth regulator that promotes cell enlargement.

Gills: in certain Basidiomycetes, the platelike structures beneath the cap of the basidiocarp and on which basidia are borne.

Girdling: the removal of a ring of tissue external to the cambium.

Glucose: a simple sugar, $C_6H_{12}O_6$; also termed dextrose or grape sugar.

Glycolysis: that part of the respiratory process during which glucose is changed to pyruvic acid with the production of a small amount of available energy and without involving free oxygen.

Grafting: the joining of two plant parts so that their tissues unite; the joining of scion to stock.

Grain: a dry, one-seeded fruit in which the seed coat is fused with the ovary wall.

Granum: one of the disclike bodies within the chloroplast; the chlorophyll and carotenoid pigments are located in the grana.

Gravitational water: water that moves downward in response to gravity.

Ground meristem: the immature cells of the stem or root apex, derived from the apical meristem, which will develop into the pith and cortex.

Growth regulator: organic compounds, other than nutrients, that influence growth and development in plants.

Growth ring: see **Annual ring.**

Guard cell: one of the two epidermal cells bounding a stoma.

Guttation: the exudation of liquid water from plants.

Gynoecium: the aggregate of pistils in a flower.

Habitat: the natural environment of an organism; the area in which it is usually found.

Haploid: having one set (or complement) of chromosomes; the situation (N) characteristic of the gametophyte phase in plants.

Head: a dense cluster of flowers crowded on a common receptacle.

Herb: a nonwoody plant, usually with a succulent annual stem.

Herbivorous: plant-eating.

Heredity: the transmission of characteristics from parent to progeny.

Heterogamous: producing two kinds of gametes.

Heterosis: see **Hybrid vigor.**

Heterosporous: producing two types of spores.

Heterothallic: a condition in which an individual gametophyte produces only one kind of gamete, usually in reference to fungi and algae; two different types of individuals are necessary for sexual reproduction.

Heterotrophic: organisms that cannot manufacture their own food but must obtain it from some external source.

Heterozygous: when the members of a gene pair are not alike; having two different alleles at the same locus on homologous chromosomes.

Hilum: a scar on the seed where the seed stalk had been attached.

Holdfast: the basal portion of an algal thallus that anchors it to a solid object.

Homologous chromosomes: chromosomes that associate in pairs during meiosis.

Homosporous: producing only one kind of spore.

Homothallic: a condition in which an individual gametophyte produces two kinds of gametes, usually in reference to fungi and algae.

Homozygous: when the members of a gene pair are alike; having identical alleles at the same locus on homologous chromosomes.

Hormone: an organic substance produced in one part of the plant which influences reactions in other parts.

Humus: a complex colloidal mixture of partially decomposed organic matter in the soil.

Hybrid: (1) the progeny of a cross between parents differing in one or more genes; (2) the progeny resulting from a cross between two species.

Hybrid vigor: the increased vitality of progeny that frequently results when inbred lines are crossed.

Hydathode: a pore, usually in leaves at the end of a vein, out of which liquid water may exude slowly.

Hydrolysis: the conversion of a compound into simpler compounds involving the uptake of water; digestion is a hydrolytic reaction.

Hydrophyte: a plant that grows in water or in very wet soils.

Hydroponics: the growth of plants in solutions containing the essential mineral elements.

Hygroscopic water: a thin film of water adhering to soil particles; not available to plants because of the great forces of attraction holding such water to the particles.

Hypha: the filament of a fungus.

Hypocotyl: that part of the embryo between the radicle and the point of attachment of the cotyledons.

Hypogynous flower: a flower in which the floral parts are attached below the ovary.

Imbibition: the uptake of water and the resultant swelling of a solid as a result of capillarity and the adsorption of water to the internal surfaces of the solid.

Imperfect flower: a flower that has stamens or pistils but not both.

Inbreeding: the breeding of closely related organisms; brings about the homozygous condition.

Incomplete dominance: the condition that results when the hybrid is intermediate in characteristics as compared with the homozygous parents.

Incomplete flower: a flower that lacks one or more of the floral parts.

Independent assortment: the random segregation of nonlinked genes to the meiospores during meiosis.

Indusium: a covering over the sorus (sporangial cluster) in ferns.

Inferior ovary: an ovary situated below the point of attachment of the other floral parts.

Inflorescence: a cluster of flowers.

Integument: the outer layer or layers of the ovule; develops into the seed coat.

Internode: the region of the stem between two successive nodes.

Interphase: the stage between mitotic divisions.

Ion: electrically charged atoms or groups of atoms formed by the dissociation (ionization) of a molecule.

Isogametes: sex cells (gametes) that are alike.

Isogamy: sexual reproduction in which there is a fusion of isogametes.

Isotope: one of several forms of an element all having the same number of protons and electrons in their atoms, and the same properties, but having different numbers of neutrons in their atoms; some are radioactive.

Karyogamy: the fusion of two nuclei.

Karyolymph: the more liquid portion of the nucleus.

Kinetic energy: the energy that results from movement of a body; motion energy.

Kinetochore: that portion of the chromosome to which a spindle fiber is attached.

Krebs cycle: a complex, cyclic series of oxidation-reduction reactions during which pyruvic acid is oxidized to carbon dioxide as it loses hydrogens; the latter go through terminal oxidations yielding water; much energy results from these oxidations.

Lamella: cellular membranes, usually in reference to those in chloroplasts; see also **Middle lamella.**

Lateral bud: a bud in the axil of a leaf (an axillary bud).

Leaching: the loss of soluble constituents, such as minerals, from the soil by percolating water.

Leaf: a lateral outgrowth of a stem; usually flattened and expanded, consisting of a blade, petiole, and frequently stipules; usually photosynthetic.

Leaf bud: a bud that develops into a stem with leaves.

Leaf gap: an area in the vascular region of a stem where parenchyma instead of vascular tissue differentiates; located immediately above a leaf trace.

Leaf primordium: a small lateral protuberance that develops from the apical meristem of a bud, and will expand to become a leaf.

Leaf scar: a scar left on a stem when the leaf falls.

Leaf trace: a vascular bundle connecting the vascular system of a leaf with that of the stem.

Leaflet: one of the parts constituting the blade of a compound leaf.

Lenticel: a pore in corky tissues of the stem and root.

Leucoplast: a nonpigmented plastid in which starch is usually formed.

Lignin: a complex organic compound frequently associated with the cellulose of secondary cell walls.

Linkage: the tendency for two or more genes to be inherited together because they are on the same chromosome.

Loam: a type of soil whose excellent garden characteristics are the result of a mixture of sand, silt, clay, and organic matter.

"Long-day" plant: a plant that flowers only when the photoperiod exceeds the critical amount.

Macronutrient: a mineral element required in relatively large amounts.

Megagametophyte: the female gametophyte.

Megaphyllous: having large leaves.

Megasporangium: the structure (sporangium) in which megaspores are produced; in flowering plants, frequently called the nucellus.

Megaspore: a haploid spore that develops into a female gametophyte.

Megaspore mother cell: See **Megasporocyte.**

Megasporocyte: a diploid cell that produces four haploid megaspores by meiosis.

Megasporophyll: a modified leaf that bears one or more megasporangia.

Meiocyte: a diploid cell that undergoes meiosis, giving rise to haploid cells or meiospores.

Meiosis: the two divisions during which homologous chromosomes separate and the resultant cells have one half the number of chromosomes that the original cell contained; there is a reduction from the diploid to the haploid number.

Meiospore: a haploid cell that results from meiosis.

Meristem: a tissue composed of cells that are capable of further divisions.

Mesocarp: the middle layer of the pericarp or fruit wall (ovary wall).

Mesophyll: the parenchyma tissue between the epidermal layers of a leaf; the cells usually contain chloroplasts.

Mesophyte: plant that grows in soils that contain moderate (or intermediate) amounts of moisture.

Metabolism: the sum total of the chemical processes that occur in a cell or an organism.

Metaphase: the stage of mitosis or meiosis when the chromosomes are at the central region of the spindle.

Microbody: any small, spherical, cytoplasmic organelle, bounded by a single membrane; apparently contains enzymes.

Microgametophyte: the male gametophyte.

Micronutrient: a mineral element required in very small amounts; sometimes called minor elements or trace elements.

Microphyllous: having small leaves.

Micropyle: the small opening in the integuments of an ovule.

Microsporangium: the structure (sporangium) in which microspores are produced.

Microspore: a haploid spore (meiospore) that develops into a male gametophyte.

Microspore mother cell: See **Microsporocyte.**

Microsporocyte: a diploid cell that produces four haploid microspores by meiosis.

Microsporophyll: a modified leaf that bears one or more microsporangia.

Microtubule: extremely small, elongated, slender, tubelike cytoplasmic organelle; may function in conduction of materials.

Middle lamella: a wall layer, primarily of pectic substances, which is common to two adjoining cells and serves to cement them together; it is located between the primary wall layers.

Mitochondrion: extremely small double-membrane-bound granular or rod-shaped cytoplasmic organelle, which is the site of the enzymes responsible for the Krebs cycle and terminal oxidation portions of respiration.

Mitosis: nuclear division in which the resultant nuclei have the same number and kind of chromosomes as the original nucleus.

Mixed bud: a bud that develops into both flowers and a stem with leaves.

Monocotyledonous plant (monocot): a flowering plant with one cotyledon or seed leaf in the embryo.

Monoecious: having male and female sex structures borne on the same plant but not in the same flower.

Monohybrid cross: a cross between organisms differing in a single gene (or character).

Multiple alleles: more than two kinds of alleles for a given locus (location) on a chromosome.

Multiple genes: two or more sets of alleles produce more or less equal and cumulative effects on the same characteristic, resulting in a quantitative gradation.

Mutation: a sudden unpredictable and inheritable change in a gene.

Mycelium: the mass of hyphae that forms the body of a fungus.

Mycorrhiza: the intimate association of a fungus with plant roots.

Naked bud: a bud without bud scales.

Nasties: growth movements independent of the direction of the stimulus.

Natural selection: occurs when agents other than man determine which individuals will survive.

Nectary: a glandular structure that secretes a sugary liquid.

Net venation: a type of vein arrangement in leaves in which the profusely branching veins form a network of vascular tissue.

Nicotinamide adenine dinucleotide (NAD): a coenzyme functioning as a hydrogen carrier.

Nicotinamide adenine dinucleotide phosphate (NADP): the phosphate of NAD, having a similar function.

Nitrification: the conversion of ammonia or ammonium compounds to nitrites and then to nitrates by the activity of certain soil bacteria.

Nitrogen fixation: the conversion of nitrogen gas into organic nitrogenous compounds by certain organisms.

Node: the region of the stem where one or more leaves are attached.

Nodule: enlargements on the roots of certain plants within which are found masses of nitrogen-fixing bacteria.

Nucellus: See **Megasporangium.**

Nucleic acid: the structural units of DNA and RNA.

Nucleolus: the small, spherical, deeply staining body located in the nucleus; probably center of ribosome formation.

Nucleoprotein: a substance formed by the combination of protein and nucleic acid.

Nucleus: the more or less spherical protoplasmic structure that contains the chromosomes in eukaryotes; it also governs the activities of the cell.

Oögamous: producing a large, nonmotile egg and a small sperm.

Oögonium: a one-celled structure in which one or more eggs are produced.

Operculum: in mosses, the lid or cover of the capsule (sporangium).

Order: a taxonomic group including closely related families.

Organelle: a formed body found in a cell and bound by a discrete membrane.

Organic compound: a compound containing carbon.

Osmosis: the net movement of water through a differentially permeable membrane from a region of high water potential to a region of lower water potential.

Osmotic concentration: the total concentration of solutes.

Osmotic pressure: the maximum pressure that may develop in a solution separated from pure water by a rigid membrane permeable to water only.

Ovary: the enlarged, basal portion of a pistil in which ovules are found; it develops into a fruit.

Ovule: in seed plants, a megasporangium surrounded by one or more integuments; it develops into a seed.

Oxidation: a loss of electrons; in biology, an energy-yielding process usually resulting from a loss of hydrogens or the addition of oxygen to a compound.

Palisade mesophyll: a leaf tissue composed of slightly elongated parenchymatous cells containing chloroplasts; located just beneath the leaf epidermis.

Palmate venation: a type of net venation in which the main veins originate at the base of the leaf blade and radiate outward.

Palmately compound leaf: a leaf in which the leaflets are attached at a common point at the top of the petiole.

Parallel evolution: similar evolution in different groups of organisms.

Parallel venation: a type of venation in which the main veins of a leaf are parallel or nearly so.

Parasite: a heterotrophic organism that obtains food from the living tissues of another organism.

Paratonic growth movement: a growth movement induced by external stimuli.

Parenchyma: a tissue composed of relatively unspecialized cells that are thin-walled and basically isodiametric; it is typically a storage tissue that retains meristematic capabilities.

Parthenocarpy: the development of fruit without fertilization.

Parthenogenesis: the development of an embryo from an egg without fertilization.

Pathogen: a disease-producing organism.

Pedicel: the stalk of an individual flower in a cluster.

Peduncle: the main stalk of a flower cluster or the stalk of an individual flower.

Penicillin: an antibiotic produced by certain species of *Penicillium,* an Ascomycete.

Perennial: a plant that lives for more than two years; it usually flowers annually after a period of vegetative growth.

Perfect flower: one that has both stamens and pistil.

Perianth: collective term for the calyx (sepals) and corolla (petals).

Pericarp: the wall of a fruit; developed from the ovary wall.

Pericycle: a parenchyma tissue between the vascular tissue and the endodermis; the branch roots develop from this tissue.

Periderm: collective term for cork, cork cambium, and phelloderm.

Perigynous flower: a flower in which the floral parts are fused and surrounding the ovary but not fused with it.

Peristome: a ring of toothlike structures surrounding the opening of a moss capsule (or sporangium).

Perithecium: a flask-shaped structure (ascocarp) with an opening and containing asci; it is composed of hyphae.

Permanent wilting point (PWP): the percentage of water remaining in the soil when a plant wilts and will not recover unless water is added to the soil.

Petal: one of the units of the corolla of a flower; frequently showy and conspicuous.

Petiole: the stalk of a leaf.

Phellum: see **Cork.**

Phelloderm: a secondary parenchyma tissue formed from the cork cambium on the inner side toward the vascular tissue.

Phellogen: see **Cork cambium.**

Phenotype: the external appearance of an organism.

Phloem: one of the vascular tissues whose main function is the conduction of organic materials (such as food).

Phloem ray: that portion of the vascular ray that extends through the secondary phloem.

Photomorphogenesis: the growth, development, and differentiation of a plant as influenced by light, independent of photosynthesis.

Photoperiodism: the influence of light exposures of different durations on plant growth and development.

Photosynthesis: the manufacture of carbohydrate, mainly sugar, from carbon dioxide and water in the presence of chlorophyll utilizing light energy.

Phototropism: a growth movement in response to the influence of light.

Pinnate venation: a type of net venation in which the secondary veins arise from each side of a single main vein.

Pinnately compound leaf: a leaf in which the leaflets are attached at intervals along the sides of a common axis (or rachis).

Pistil: the central structure of the flower, composed of one or more carpels and enclosing one or more ovules; typically consisting of stigma, style, and ovary.

Pit: a thin area in the cell wall.

Pith: the parenchyma tissue in the center of a dicotyledonous stem.

Placenta: that portion of the ovary wall to which the ovules are attached.

Plankton: free-floating aquatic organisms; mostly microscopic.

Plasma membrane: the outermost layer of cytoplasm; it is differentially permeable.

Plasmalemma: see **Plasma membrane.**

Plasmodesmata: minute cytoplasmic strands that extend from cell to cell through pores in the cell walls.

Plasmogamy: the fusion of protoplasts.

Plasmolysis: shrinkage of protoplast away from the cell wall as a result of water loss.

Plastid: specialized body in the cytoplasm, frequently involved in food manufacture and storage; bounded by a double membrane.

Polar nuclei: in flowering plants, the two central nuclei of the female gametophyte; they unite with a male nucleus (sperm) to form an endosperm nucleus.

Pollen grain: in seed plants, the young male gametophyte (microgametophyte) or germinated microspore.

Pollen tube: a tubular outgrowth of a pollen grain that carries male nuclei (sperm) into the ovule.

Pollination: in flowering plants, the transfer of pollen from anther to stigma; in conifers, the transfer of pollen from a microsporangium of a male cone to an ovule of a female cone.

Polyploid: having more than two sets of chromosomes.

Polyribosome or **Polysome:** a cluster or group of ribosomes.

Primary tissue: tissue that develops from an apical meristem.

Primordium: a rudimentary structure; the beginning of a structure.

Procambium: a strand of immature cells, derived from the apical meristem, which differentiates into the primary vascular tissue; also called provascular tissue.

Prokaryote: a primitive organism lacking a membrane-bound nucleus.

Prophase: the beginning stage of mitosis or meiosis in which the chromosomes become distinct and in which their doubled condition is visible.

Protein: complex, nitrogenous, organic compounds synthesized from amino acids and composed of carbon, hydrogen, oxygen, and nitrogen, and frequently sulfur.

Prothallus: gametophyte of ferns and similar plants.

Protoderm: the outermost layer of immature cells, derived from the apical meristem which develops into the epidermis.

Protonema: in mosses and similar plants, the filamentous growth forming the early stage of the gametophyte.

Protoplast: the living portion of the cell.

Pyrenoid: a proteinaceous body on the chloroplast of certain algae; it is a center for starch formation.

Rachis: the extension of the petiole that bears leaflets.

Radial section: a longitudinal section cut along a radius.

Radicle: the lower part of an embryo axis; it develops into the primary root.

Radioactive: elements that emit one or more types of penetrating radiations are termed radioactive elements.

Ray flower: in composites (sunflower fam-

ily), the marginal flower of a head; it has a strap-shaped corolla.

Receptacle: (1) the apex of a flower stalk from which the floral parts arise; (2) the swollen tips of certain brown algae within which sporangia are borne.

Recessive character: (1) a gene whose presence is masked by the presence of its dominant allele; (2) that characteristic whose appearance is excluded by the presence of its contrasting character in a hybrid.

Reduction division: see **Meiosis.**

Respiration: a series of complex oxidation-reduction reactions whereby living cells obtain energy through the breakdown of organic material and in which some of the intermediate materials can by utilized for various syntheses.

Rhizoid: threadlike cellular appendages in some of the lower plant forms that serve as absorbing and anchoring structures.

Rhizome: a horizontal, underground stem; frequently a storage organ.

Ribonucleic acid (RNA): nucleic acid found in the nucleolus and cytoplasm; one type carries genetic information to the ribosomes for protein synthesis.

Ribosome: submicroscopic granules located in the cytoplasm; they are rich in RNA and are the main site for the synthesis of proteins.

RNA: see **Ribonucleic acid.**

Root cap: a thimblelike, protective mass of cells over the root apex; it develops from the apical meristem of the root.

Root hair: long, tubular outgrowth from an epidermal cell in the lower portion of the region of maturation; it tremendously increases the absorbing surface of the cell.

Saprophyte: a heterotrophic organism that obtains its food from nonliving organic matter.

Sclerenchyma: strengthening tissue composed of thick-walled cells.

Secondary tissue: tissue that develops from a cambium.

Seed: the characteristic reproductive structure of seed plants; formed from the ovule and containing the embryo.

Seed coat: the outer layer of the seed which develops from the integuments of the ovule; also called testa.

Segregation: the separation of alleles during meiosis.

Self-pollination: the transfer of pollen from an anther to a stigma on the same plant.

Sepal: one of the units of the calyx of a flower, the outermost whorl of flower parts; frequently green and leaflike.

Sessile: lacking a stalk.

Sexual reproduction: involves the fusion of gametes, which is followed eventually by meiosis.

Shoot: collective term for a stem and its leaves.

"Short-day" plant: a plant that flowers only when the photoperiod is shorter than the critical amount; the continuous dark period must exceed a critical amount.

Sieve tube: in the phloem, a series of sieve tube elements arranged end to end; functions in conducting organic materials.

Sieve tube element: an elongated phloem cell having perforated end walls (sieve plates).

Silt: mineral particles of the soil, which vary in diameter from 0.02 to 0.002 mm.

Simple leaf: a leaf in which the blade is not divided.

Solute: a substance dissolved in a solvent.

Solvent: a liquid in which other substances are dissolved.

Sorus: in ferns, a cluster of sporangia.

Species: (1) the smallest taxonomic grouping of organisms; (2) a particular kind of organism.

Sperm: male gamete or sex cell.

Spirillum: a curved bacterium.

Spongy mesophyll: a leaf tissue composed of loosely packed, irregular parenchymatous cells containing chloroplasts.

Sporangiophore: the stalk that bears a sporangium.

Sporangium: a structure in which spores are produced.

Spore: an asexual reproductive structure, usually unicellular; in bacteria it is a resistant structure but not a reproductive body.

Spore mother cell: see **Sporocyte.**

Sporocyte: a diploid (2N) cell, which produces four haploid (N) meiospores as a result of meiosis; also called spore mother cell.

Sporophyll: a modified leaf that bears sporangia.

Sporophyte: the diploid, spore-producing phase of the plant life cycle.

Stamen: the pollen-producing structure of the flower, consisting of an anther and usually a filament.

Staminate flower: a flower having stamens but no pistil.

Starch: a complex, insoluble carbohydrate built up from many molecules of glucose (sugar), and which is a common storage product in plants.

Stigma: the part of the pistil, usually the apex, that receives pollen.

Stipe: a supporting stalk, as in the brown algae (Phaeophyta) or the Basidiomycetes.

Stipule: an appendage, usually green and leaflike, located near the base of a petiole.

Stolon: (1) a slender, horizontal stem that often develops new plants at its nodes; (2) horizontal, surface hypha of certain fungi.

Stoma: the pore or opening between two guard cells in the epidermis.

Strobilus: a cone-shaped aggregation of sporophylls.

Style: that part of the pistil between the stigma and ovary, usually elongated.

Suberin: a fatty substance deposited in certain cell walls, particularly cork cells.

Succession: an orderly sequence of one plant community replacing another until the climax is obtained.

Superior ovary: an ovary situated above the point of attachment of the other floral parts.

Symbiosis: an intimate living together of dissimilar organisms benefiting each.

Synapsis: the pairing of homologous chromosomes during meiosis.

Synergid: in flowering plants, one of the small cells lying near the egg in the megagametophyte.

Syngamy: fusion of gametes.

Tangential section: a longitudinal section cut at right angles to a radius.

Taxon: each group, regardless of rank, in any system of classification.

Teleology: the assigning of a purpose to natural processes.

Telophase: the final stage of mitosis or meiosis during which the chromosomes become organized into two new nuclei; the new cell wall usually begins forming at this time also.

Tendril: a slender, coiling, supporting structure; usually a modified leaf or stem.

Testa: see **Seed coat.**

Thallus: an undifferentiated plant body without true roots, stems, or leaves.

Thigmotropism: a growth movement in response to contact with a solid object.

Timberline: the highest elevational limit of tree growth.

Tissue: a group of cells, generally of similar structure and origin, that perform a common function.

Trace element: see **Micronutrient.**

Tracheid: a conducting and strengthening cell of the xylem; it is elongated, tapering, thick-walled, pitted, and contains no protoplast.

Translocation: the movement of organic material through the phloem.

Transpiration: the loss of water in vapor form from a plant.

Transverse section: a section cut at right angles to the long axis; a cross section.

Triploid: having three sets of chromosomes.

Tropism: a growth response resulting from a unidirectional external stimulus.

Tube cell: the large cell of the pollen grain that forms the pollen tube.

Tuber: an enlarged portion of a rhizome (underground stem).

Turgid: the plump, swollen, or distended condition of a cell resulting from the osmotic uptake of water.

Turgor movement: movements resulting from changes in the turgidity of cells rather than from growth.

Turgor pressure: the actual pressure that develops in a cell as a result of the osmotic uptake of water; this pressure is exerted against the cell wall.

Unicellular: an organism consisting of one cell.

Vacuolar membrane: the layer of cytoplasm that borders a vacuole.

Vacuole: a vesicle (or cavity) within the cytoplasm containing a solution of various substances.

Vascular bundle: a strand of xylem and phloem, the conducting tissues.

Vascular cambium: a meristematic tissue that produces secondary xylem and secondary phloem.

Vascular ray: ribbonlike sheet of parenchymatous cells extending radially (horizontally) through the secondary xylem and secondary phloem; produced by the vascular cambium.

Vascular tissue: conducting tissue; the xylem and phloem.

Vegetative: concerned with growth, development, and maintenance, rather than with sexual reproduction.

Vegetative reproduction: reproduction by a primarily vegetative part of the plant; not involving fusion of gametes.

Vein: a vascular bundle; usually used in reference to a leaf.

Venation: the arrangement of the vascular bundles in a leaf.

Vessel: a tubelike structure in the xylem composed of a vertical series of cells whose end walls are gone; it conducts water and minerals.

Vessel element: one of the cells of a vessel; somewhat elongated, pitted, thick-walled cell in which the end walls are perforated.

Virus: a submicroscopic pathogen that can reproduce only in living host cells; it consists of protein and nucleic acid, a nucleoprotein.

Vitamin: an organic substance that is essential for normal growth and development but is not a food or a source of energy; green plants synthesize their own vitamins.

Wall pressure: the pressure exerted by a cell wall upon the cell contents; it is equal and opposite to the turgor pressure.

Water table: the upper limit of the standing water in a completely saturated soil.

Whorl: a circle of three or more parts.

Wilt: a limp or flaccid condition resulting from a deficiency of water and a low turgor pressure within the cells of a structure.

Wood: the xylem.

Xanthophylls: yellow to orange carotenoid pigments associated with chlorophyll in chloroplasts; also found in some chromoplasts.

Xerophyte: a plant that grows in soils with a scanty water supply (i.e., in an arid habitat).

Xylem: that component of the vascular tissue that functions primarily to conduct water and minerals.

Xylem ray: that portion of the vascular ray that extends through the secondary xylem.

Zoosporangium: a structure (sporangium) in which zoospores are produced.

Zoospore: a motile spore; in algae and fungi.

Zygote: a diploid cell resulting from the fusion of two gametes; a fertilized egg.

Appendix:
Units of Measurement

The metric system is used in most of the world and for scientific work. A refinement of the traditional metric system, the Systeme International d'Unités (SI), was adopted by the Eleventh General Conference on Weights and Measures and endorsed by the International Organization for Standardization in 1960. The fundamental units are the meter and the kilogram, which have been defined specifically, and all other units are defined in relation to or derived from these standard quantities. It is a comprehensive and logical system in which all the derived units are multiples of ten. The prefixes in Table A-1 may be used to indicate fractions or multiples of the SI units.

Some of the units in the old centimeter-gram-second (cgs) system have been in use for such a long time that conversion to the International System is taking place somewhat slowly. Also, in the United States conversion to the metric system is proceeding at an extremely slow pace. It is still not unusual to find that many college students are not overly familiar with the metric system. To aid in the transition, Table A-2 presents a few of the more common measurements and their equivalents.

Table A-1 Prefixes for SI units

Fraction	Prefix	Symbol	Multiple	Prefix	Symbol
10^{-1} (0.1)	deci	d	10	deka	da
10^{-2} (0.01)	centi	c	10^2 (100)	hecto	h
10^{-3} (0.001)	milli	m	10^3 (1,000)	kilo	k
10^{-6} (0.000001)	micro	μ	10^6 (1,000,000)	mega	M
10^{-9} (0.000000001)	nano	n	10^9 (1,000,000,000)	giga	G

Table A-2 Some common measurement units

Quantity	Symbol	Value	Equivalent
Area:			
hectare	ha	10,000 m² (100 m × 100 m)	2.471 acres
acre	a	43,560 sq. ft.	0.4047 ha
sq. kilometer	km²	10^6 m² (1,000 m × 1,000 m)	100 ha = 0.386 sq. mile
Length:			
kilometer	km	1,000 m	0.6214 mile
meter	m	(=100 cm = 1,000 mm)	39.37 inches
decimeter	dm	0.1 m (=10 cm = 100 mm)	3.937 inches
centimeter	cm	0.01 m (=10 mm)	0.3937 inch
inch	in.		2.54 cm
millimeter	mm	10^{-3}m (=0.1 cm)	
micrometer	μm	10^{-6} m (=10^{-3} mm)	(micron, μ)
nanometer	nm	10^{-9} m (=10^{-3} μm)	(millimicron, mμ)
angstrom	Å	10^{-10} m (=0.1 nm = 10^{-4} μm)	
Mass:			
metric ton		1,000 kg	2,205 lbs (=1.1 U.S. ton)
kilogram	kg	1,000 g	2.2046 lbs
gram	g	(=1,000 mg = 10^6 μg)	0.0353 oz
pound	lb		453.6 g
ounce	oz		28.35 g
milligram	mg	0.001 g (=1,000 μg)	
microgram	μg	10^{-6} g (10^{-3} mg)	
Volume (solids):			
cubic meter	m³		1,308 cu. yards = 35.31 cu. ft.
cu. centimeter	cm³	10^{-6} m³	
cu. millimeter	mm³	10^{-9} m³	
Volume (liquids):			
liter	l	(=1,000 ml = 10^6 μl)	1.06 quarts
milliliter	ml	10^{-3} l (=10^3 μl)	
microliter	μl	10^{-6} l (=10^{-3} ml)	

Table A-3 Temperature

Centigrade = °C. Melting point of ice = 0 degrees at one atmosphere pressure
Boiling point of water = 100 degrees at one atmosphere pressure
Fahrenheit = °F Melting point of ice = 32 degrees at one atmosphere pressure
Boiling point of water = 212 degrees at one atmosphere pressure
To convert centigrade to Fahrenheit degrees: (°C × $\frac{9}{5}$) + 32 = °F
To convert Fahrenheit to centigrade degrees: (°F − 32) × $\frac{5}{9}$ = °C

Index

Page numbers in **boldface** refer to graphs, illustrations, or tables.